Blue versus Purple

1946

U.S. GOVERNMENT
OFFICIAL EDITION NOTICE

AUTHENTICATED
U.S. GOVERNMENT
INFORMATION
GPO

Use of ISBN Prefix

This is the Official U.S. Government edition of this publication and is herein identified to certify its authenticity. ISBN 978-1-935352-31-0 (e-book ISBN 978-1-935352-32-7) is for this U.S. Government Publishing Office Official Edition only. The Superintendent of Documents of the U.S. Government Publishing Office requests that any reprinted edition clearly be labeled as a copy of the authentic work with a new ISBN.

Legal Status and Use of Seals and Logos

For Sale by the Superintendent of Documents, U.S. Government Publishing Office
Internet: bookstore.gpo.gov Phone: toll free (866) 512-1800; DC area (202) 512-1800
Fax: (202) 512-2104 Mail: Stop IDCC, Washington, DC 20402-00001

ISBN: 978-1-935352-31-0 e-book ISBN: 978-1-935352-32-7

NAVAL WAR COLLEGE HISTORICAL MONOGRAPH SERIES NO. 24

The historical monographs in this series are book-length studies of the history of naval warfare, edited historical documents, conference proceedings, and bibliographies that are based wholly or in part on source materials in the Historical Collection of the Naval War College.

The editors of the Naval War College Press express their special gratitude to all the members of the Naval War College Foundation, whose generous financial support for research projects, conferences, and printing has made possible the publication of this historical monograph.

To my godparents, Mac and Rose Carp,
for demonstrating the value of education to me

Blue versus Purple: The U.S. Naval War College, the Soviet Union, and the New Enemy in the Pacific

1946

Hal M. Friedman

NAVAL WAR COLLEGE PRESS
NEWPORT, RHODE ISLAND
2017

The contents of this volume represent the views of the author. His opinions are not necessarily endorsed by the Naval War College or by any other agency, organization, or command of the U.S. government.

Printed in the United States of America

Historical Monograph Series

Library of Congress Cataloging-in-Publication Data

Names: Friedman, Hal M., 1965- author.
Title: Blue versus purple : the U.S. Naval War College, the Soviet Union, and the new enemy in the Pacific / Hal M. Friedman.
Other titles: U.S. Naval War College, the Soviet Union, and the new enemy in the Pacific
Description: Newport, Rhode Island : Naval War College Press, [2016] | Series: Naval War College historical monograph series ; no. 24 | Includes bibliographical references and index.
Identifiers: LCCN 2017007795 | ISBN 9781935352310
Subjects: LCSH: Naval War College (U.S.)—History—20th century. | War games, Naval—United States—History—20th century. | United States. Navy—Maneuvers—History—20th century. | Naval maneuvers—Pacific Ocean. | Naval strategy—Study and teaching—United States—History—20th century. | Naval education—United States—History—20th century.
Classification: LCC V420 .F748 2016 | DDC 359.4/8097309044—dc23
LC record available at https://lccn.loc.gov/2017007795

NAVAL WAR COLLEGE PRESS
Code 32
Naval War College
686 Cushing Road
Newport, R.I. 02841-1207

TABLE OF CONTENTS

LIST OF FIGURES AND MAPS

Figures

Maps

FOREWORD

The study of the history of naval warfare is an integral part of the Naval War College's educational programs. The importance of the discipline was firmly established with the foundation of the College in 1884 by the initial contributions of both the College's founder, Stephen B. Luce, and his successor as its President, Capt. Alfred Thayer Mahan. Historical research and analysis has continued as a recognized element of the academic life of the institution for over 125 years. Nowhere has the history of warfare at sea been more thoroughly investigated and analyzed for the professional purposes of the U.S. Navy than at the Naval War College. Nowhere is there a more logical requirement for a corpus of relevant source materials or for an academic research department devoted to new research in naval history.

On 1 January 2003, the College's Maritime History Department was established as part of the Center for Naval Warfare Studies to carry out this function. Predating this, a program for the publication of books and source materials on the history of naval warfare was formally established by the College in 1975, with the series of books known as the Naval War College Historical Monographs. To encourage and make more widely known the College's extensive collections for historical research, including its archives, historical manuscripts, and associated materials in the Naval War College Museum's collection, the series has been restricted to publications of book-length works that deal with the history of naval warfare and that are based wholly or in part on the source materials in the Naval War College library's Naval Historical Collection and in the Naval War College Museum. As the series has developed over the past forty years, these works have taken a variety of forms, including bibliographies and conference proceedings; many of them have been edited historical documents from the College's rich historical collections. This book series is now managed by the Maritime History Department in collaboration with the Naval War College Press and the head of the Naval War College's Naval Historical Collection.

Consistent with the earlier books that have appeared in this series since 1975, this study of naval war gaming at the Naval War College in 1946 is based primarily on materials in the Naval War College archives and historical collection. With this volume, Dr. Hal Friedman, of Henry Ford College in Dearborn, Michigan, directly complements his two earlier studies in this same series, *Digesting History: The U.S. Naval War College, the Lessons of World War Two, and Future Naval Warfare, 1945–1947,* published in 2010, and *Blue versus Orange: The U.S. Naval War College, Japan, and the Old Enemy in the Pacific, 1945–1946,* published in 2013. The first volume looked at the broad range of the educational activities at the Naval War College under its twenty-sixth President, Adm. Raymond Spruance. The second began a two-part study focused on the specific activity of war gaming, which the College had been developing since then-lieutenant William McCarty Little delivered the first lecture in Newport on the subject in 1886. By 1946 the Naval War College had, over a period of sixty years, developed the art of war gaming into one of its signature teaching and research tools for thinking about naval warfare. This, the third volume in the trilogy, *Blue versus Purple,* turns the emphasis to the process by which the Naval War College's gaming efforts slowly refocused from the war against Japan that had just ended to the new and emerging threat that the Soviet Union was beginning to pose in the opening days of the Cold War.

Today, after a span of another six decades, the Naval War College has developed that art of war gaming even further. In many ways, the war gaming conducted today, with the assistance of computers and in a very modern building, may seem completely different from and foreign to what is described in this book. Yet Friedman's close analysis of the procedures and practice reveals that the fundamental purpose was much the same—an attempt to understand human reactions and decision making. Up to now, few naval historians have been able to penetrate the process and explain clearly the practice of war gaming in this period. The Naval War College Museum and the Naval War College library have collections of both paper records and three-dimensional objects associated with war gaming from earlier periods. While experts in war gaming clearly understand these processes, explanations that are accessible to a broader audience interested in naval history have been lacking, particularly for the immediate post–World War II period.

With this volume, Dr. Friedman has made another significant contribution to both the history of the Naval War College and the history of naval war gaming. This volume, like his previous volumes, evokes for the reader the laborious and detailed process of war gaming in this era and helps us to understand why an automated approach with computers became attractive in the 1950s. Above all, it is particularly interesting to see the character of naval thinking in 1946, when the U.S. Navy was making the transition from total war to peace but was realizing that another conflict with a different enemy might be possible. In this, Dr. Friedman makes an important contribution by showing how the experience of naval operations in World War II continued to have a useful place in the curriculum of the Naval War College by illustrating some enduring aspects in naval warfare, while at the same time College faculty and students were beginning to imagine what sort of different challenges might be faced when fighting a major war in the context of the newly emerging world of atomic weapons.

JOHN B. HATTENDORF, D.PHIL.
Ernest J. King Professor of Maritime History
Chairman, Maritime History Department

PREFACE

This book is about the war-gaming activities of the Naval War College (NWC) in the late summer and fall of 1946 in Newport, Rhode Island. It is the third book in a trilogy on the immediate postwar Naval War College. My earlier works focused, first, very generally on how the College viewed future naval warfare in the 1945–47 period, then on the war games in the 1945–46 academic year. More particularly, my first look at the Naval War College was a general exploration of the College's reaction to the end of World War II and the beginning of the Cold War. Here I again direct my attention to the Pacific, as I did in the second book, because of its importance to the Navy for the half-century previous to the present. In effect, I want to look at how the end of World War II and the beginning of the Cold War impacted the Naval War College in terms of changing its focus from Japan to the Soviet Union as the primary enemy in the Pacific basin. This exploration may also prove especially insightful given that the 1940s was precedent setting in terms of the origins of the Navy's maritime strategy.[1]

More exactly, the first book set the context for the later works by looking at Admiral Raymond Spruance's public addresses, instructor and guest lectures, and student theses to determine what kind of future naval world these officers perceived and how they felt the United States would have to conduct its peacetime and wartime naval operations in that postwar world. The second book looked at the war games in the fall of 1945 and the winter of 1946, games in which Japan, or "Orange," was still the primary enemy. This present book explores how and why the Soviet Union, or "Purple," became the hypothetical enemy in the 1946–47 academic year.[2]

These studies are particularly significant given that the period falls between the end of the war and the early formulation of the Containment Doctrine against the Soviet Union. The importance of the events between the fall of 1945 and the beginning of containment as a coherent American foreign policy in the fall of 1947 is why I have chosen the book's chronological parameters. This transitional period is especially valuable as a window through which to explore institutions like the

Naval War College that were shifting from a hot war to a cold one.[3]

Adm. Raymond Spruance, President of the Naval War College, 1946–48

This immediate postwar transition entailed the reconstitution of the College on a full-time basis after its reduced wartime status, first under the presidency of Vice Adm. William Pye and then, from March 1946 onward, under Admiral Spruance. Spruance had been charged by Fleet Adm. Chester Nimitz, Chief of Naval Operations (CNO), with the strategic reformulation of American naval doctrine for the atomic and Cold War contexts of the postwar period. Some of these reforms began with Vice Admiral Pye before the war even ended. Pye, for instance, called for expanding the College so it was capable of teaching a tenfold increase in officers by way of a three-tiered educational structure consisting of a Command and Staff Course, War College Course, and Advanced Course. Pye presided until the end of his tenure over the six-month courses that had been the order of the day since before the war, but he began preparations for returning to a full course. By the time the war ended, the Naval War College had also started to provide joint-service education for officers from the other services, as well as personnel from the State Department. Most important for purposes of this study, however, is that because of the sudden end of the war, the 1945–46 curriculum could not be changed in time for the convening of classes to reflect the complete defeat of Japan or the rise of the Soviet Union as the nation's probable next enemy. That change had to await the 1946–47 academic year.[4]

Naval War College, 1946

This new focus on the Soviet Union in the College's maneuvers

Vice Adm. William Pye, President of the Naval War College, 1942–46

reflected late wartime Pacific Fleet practices and current fleet doctrine. In general, the maneuvers in the new academic year reflected strategic scenarios from the late Pacific War when Orange was the enemy, but they employed new scenarios if Purple was the opponent. Many of the scenarios involving Blue (the United States) against Purple entailed Blue strikes on or invasions of Kamchatka, and scenarios of Purple attacking Blue involved Purple strikes or invasions of the Aleutians. The operational and tactical aspects of the exercises and maneuvers still heavily reflected a combination of both interwar and wartime naval doctrine, once again equally emphasizing surface warfare and naval aviation, and submarines (SSs) again fulfilling primarily scouting and fleet strike functions. Another carryover was the pattern wherein the few aircraft carriers available to each side would be neutralized early in the maneuver, and both sides lacked the possibility of reinforcement by land-based air forces or submarines. Thus, many of the scenarios were again constructed such that the maneuvers became surface battles between Blue and Purple. While this may seem anachronistic, many of the scenarios in fact reflected the situation that had actually existed at the end of the battle of the Coral Sea in May 1942 and especially in the Solomon Islands by November 1942, when the United States and Japan had been obliged to rely heavily on surface forces in the absence of their exhausted carrier groups (CARGRUs). Another purpose of this book, therefore, will be to look at these various situations and explore why the Naval War College staff—that is, the military faculty—might have thought them pertinent to officer education.[5]

Fleet Adm. Chester Nimitz, Chief of Naval Operations, 1945–47

Though not as a major part of the analysis, the book also looks at some continuities in American naval doctrine across the interwar, wartime, Cold War, and post–Cold War periods. There is an interesting similarity in how the fleet was supposed to fight in these different periods. While the battleship remained the primary element of fleet doctrine up to the Japanese strike on Pearl Harbor, by then it was already heavily dependent on submarines for scouting and initial strikes and even more so on carrier air forces for scouting, gunfire spotting, strikes against enemy capital ships, and naval air cover vis-à-vis enemy carrier planes. While in the Pacific War the aircraft carrier became the obvious primary weapon, carrier groups still needed to be protected at night by battleships and other heavy surface forces, just as those surface forces needed protection provided by the carrier groups during the day. Carriers obviously remained at the center of American naval power during the Cold War and after—but not in navies lacking carriers. Soviet naval doctrine, for instance, stressed initial strikes against American carrier battle groups by, first, torpedo-firing, later missile-firing, submarines, followed up by strikes from long-range, land-based naval aviation, and then surface battle groups. Some of the 1946 scenarios at the Naval War College already reflected this pattern of doctrine. Finally, there are also potential insights in this war-college history for the U.S. Navy today and in the future. Given recent advances in Chinese antiship missile technology, it might not be as easy for American surface battle groups, carrier or otherwise, to venture into western Pacific waters during a conflict with the People's Republic of China unless missile-firing submarines and even land-based stealth bombers pave the way, much as submarines and carriers paved the way for battleships in late interwar scenarios.[6]

I have been selective in analyzing the exercises and Operations Problems. This book, like its predecessors, is focused on the Pacific, because I wanted to study the maneuvers that were conducted as the Naval War College curriculum transitioned from Orange to Purple as the primary hypothetical enemy. Orange had been the primary Pacific enemy for the Naval War College as far back as 1906, and war gaming had focused the most heavily on Orange since 1911, so one purpose of the book is to narrate how that transition happened. I have limited the exercises and Operations Problems to those conducted between July and November 1946. Actually, exercises and Operations Problems were held all through the academic year; however, one played after November 1946 focused again on Orange, in a scenario that repeated the Iwo Jima operation. The remaining exercises and Operations Problem during the 1946–47 academic year involved Purple, but in the Atlantic. Accordingly, my analysis stops in November 1946.

Not all the exercises were identically organized. A typical Operations Problem contained what the Naval War College called "Statements of the Problem" and "Staff Solutions" for both Blue and Purple, a record of the actual board game called the "History of the Maneuver," and even in a few cases "Maneuver Critiques." Most, however, lacked some of these elements. The chapters that follow, recording the exercises and Operations Problems in detail, are uneven in organization, therefore, because of the nature of the surviving records.[7] I have devoted one entire chapter and elements of a second to describing in detail what the Naval War College called "Maneuver Rules." These were the procedures by which the exercises and Operations Problems were conducted, including such matters as the computation of damage to ships and planes and the impact of weather. While my previous studies have outlined the maneuvers and the Rules in basic form, this book offers more detail.[8]

This book is a very straightforward, chronological narrative. I have not attempted to analyze deeply or "deconstruct" what the historical actors said but instead have taken great pains to record what they did at the time in a readable narrative. I think the value of this type of account is to allow readers to see what these war games represented at this time so that they can make their own judgments about the ideas enunciated in these years. Moreover, I have found myself a bit reluctant to judge these officers harshly where their ideas about future naval warfare turned out to be wrong or even wrongheaded. This was a very fluid period, and there was no quick formula to produce the "answers." Establishing the strategic, operational, and tactical constructs for the security of the Republic against a threat about which they knew almost nothing, and using potentially revolutionary new weapons about which they also knew almost nothing, was indeed an awesome challenge. Providing those answers at the conclusion of a major war as a new type of limited war was looming, and in a context of limited resources, must have been an even more daunting task.

In addition, a narrative account is in order since this work represents fairly new territory, historiographically speaking. Seminal works exist about the Naval War College in the interwar period and its impact on the Pacific War;[9] however, there have not been many studies on the post-1945 period. The works completed tend to be organizational histories and studies of the College.[10] Those that are strategic in nature focus on the College's role in a limited chronological window, either very early in the Cold War or much later in that conflict.[11] In addition, articles were written on American naval strategy during the Cold War under Naval War College auspices, as well as lectures by guests and instructors and theses by students.[12] So far, however, no monographs cover the Naval War College's contribution to strategic policy throughout the entire Cold War. My work does not fill this niche either, but I hope that by focusing on the immediate postwar period it complements the existing studies of the College both in the prewar years and during the Cold War as well as gives future historians a path by which to begin to forge a history of the College's contribution to U.S. strategy during the entire Cold War.

Throughout, unless otherwise indicated, figures, maps, and photographs are courtesy of the Naval Historical Collection, Naval War College, Newport, Rhode Island.

NOTES

1 Such historians as Michael Palmer, Richard Hegmann, and John Hattendorf have illustrated that the Maritime Strategy was not created by Secretary of the Navy John Lehman in the early 1980s but originated in the late 1940s and was developed throughout the Cold War. The strategy included a relegation of the Pacific as a subordinate theater after the Atlantic, the Mediterranean, and the Persian Gulf. The Pacific was nevertheless seen as an important theater since American naval forces were to keep Soviet forces in East Asia occupied so they could not reinforce Soviet forces in Europe and the Middle East; see Michael Palmer, *Origins of the Maritime Strategy: The Development of American Naval Strategy, 1945–1955* (Annapolis, Md.: Naval Institute Press, 1990). The strategy continued to develop in the 1950s and 1960s; see Richard Hegmann, "Reconsidering the Evolution of the US Maritime Strategy, 1955–1965," *Journal of Strategic Studies* 14 (September 1991), pp. 299–336. By the late 1970s, the Office of the Chief of Naval Operations's (OPNAV's) Washington, D.C.–based Strategic Concepts Branch, as well as OPNAV's Strategic Studies Group and the Naval War College's Center for Naval Warfare Studies—the latter two groups located in Newport—were resurrecting the northern and western Pacific as areas of American naval power projection that would be critical in preventing the movement of Soviet reinforcements from the Soviet Union's Far Eastern Maritime Provinces to Europe in case of a war with the North Atlantic Treaty Organization (NATO); see John Hattendorf, ed., *U.S. Naval Strategy in the 1970s: Selected Documents,* Newport Paper 30 (Newport, R.I.: Naval War College Press, 2007), and Hattendorf, *The Evolution of the U.S. Navy's Maritime Strategy, 1977–1986,* Newport Paper 19 (Newport, R.I.: Naval War College Press, 2004).

2 See Hal M. Friedman, *Digesting History: The U.S. Naval War College, the Lessons of World War Two, and Future Naval Warfare, 1945–1947* (Newport, R.I.: Naval War College Press, 2010), and *Blue versus Orange: The U.S. Naval War College, Japan, and the Old Enemy in the Pacific, 1945–1946* (Newport, R.I.: Naval War College Press, 2013). For the College's color coding of countries, see Michael Vlahos, *The Blue Sword: The Naval War College and the American Mission, 1919–1941* (Newport, R.I.: Naval War College Press, 1980), pp. 97–130.

3 For the concept of hot war to cold, see Jeffrey Barlow, *From Hot War to Cold: The U.S. Navy and National Security Affairs, 1945–1955* (Stanford, Calif.: Stanford Univ. Press, 2009).

4 John B. Hattendorf, B. Mitchell Simpson III, and John R. Wadleigh, *Sailors and Scholars: The Centennial History of the U.S. Naval War College* (Newport, R.I.: Naval War College Press, 1984), pp. 175–77. See also Nepier Smith, "Historical Analysis of the Organizational Success of the Naval War College during the Twenty-Five Years following the Second World War," 1974, pp. 11–67, Naval War College Advanced Research Project, Naval Historical Collection, Naval War College, Newport, R.I.; and Friedman, *Blue versus Orange,* chap. 1.

5 For detailed accounts of interwar American naval doctrine, see Joel Davidson, *The Unsinkable Fleet: The Politics of U.S. Navy Expansion in World War II* (Annapolis, Md.: Naval Institute Press, 1996), pp. 11–12, 14–16, 19–21, 23–24, 32, 34, 60, 96–97; Thomas Wildenberg, *Destined for Glory: Dive Bombing, Midway, and the Evolution of Carrier Airpower* (Annapolis, Md.: Naval Institute Press, 1998), pp. 48–64, 83–98, 126–28, 141, 155, 157–60, 163–64, 170–71; William McBride, *Technological Change and the United States Navy, 1865–1945* (Baltimore: Johns Hopkins Univ. Press, 2000), pp. 111–212; Thomas Hone and Trent Hone, *Battle Line: The United States Navy, 1919–1939* (Annapolis, Md.: Naval Institute Press, 2006), pp. 110–25; Craig Felker, *Testing American Sea Power: U.S. Navy Strategic Exercises, 1923–1940* (College Station: Texas A&M Univ. Press, 2007), pp. 61–75; John Kuehn, *Agents of Innovation: The General Board and the Design of the Fleet That Defeated the Japanese Navy* (Annapolis, Md.: Naval Institute Press, 2008), pp. 87–92, 173–75, 205; Joel Holwitt, *"Execute against Japan": The U.S. Decision to Conduct Unrestricted Submarine Warfare* (College Station: Texas A&M Univ. Press, 2009), pp. 22, 62–83; Albert Nofi, *To Train the Fleet for War: The U.S. Navy Fleet Problems, 1923–1940* (Newport, R.I.: Naval War College Press, 2010), pp. 25, 51–56, 60, 62, 64–66, 74, 76–77, 80, 85–86, 93–94, 100–102, 104–105, 110–17, 121–25, 129–36, 139–46, 151, 155–59, 169, 197–203, 207–16, 219–27, 229–37, 253–63, 287–88; and Timothy Wolters, *Information at Sea: Shipboard Command and Control in the U.S. Navy, from Mobile Bay to Okinawa* (Baltimore: Johns Hopkins Univ. Press, 2013), pp. 106–90. For

an excellent account showing that wartime practices were really a combination of interwar doctrine and wartime lessons, see Trent Hone, "U.S. Navy Surface Battle Doctrine and Victory in the Pacific," *Naval War College Review* [hereafter *NWCR*] 62, no. 1 (Winter 2009), pp. 67–105. Edwin Hoyt, *Blue Skies and Blood: The Battle of the Coral Sea* (New York: Paul S. Eriksson, 1975), pp. 126–27, 140, 165–66, is an older account that shows how American and Japanese on-scene commanders briefly sought a surface engagement after their carrier airpower had been nearly mutually neutralized. For a recent account of the battles of the Coral Sea, the eastern Solomons, and the Santa Cruz Islands, in which American and Japanese carrier forces essentially pummeled each other into near impotence, see Dallas Isom, *Midway Inquest: Why the Japanese Lost the Battle of Midway* (Bloomington: Indiana Univ. Press, 2007), pp. 79–91, 246–61. For a very similar account of this near-mutual neutralization of carrier airpower during the battles of the eastern Solomons and Santa Cruz and the first naval battle of Guadalcanal and of the subsequent heavy reliance on surface forces, see James Hornfischer, *Neptune's Inferno: The U.S. Navy at Guadalcanal* (New York: Bantam Books, 2011), pp. 114–18, 216–36, 250, 322, 324–25, 328, 335–36, 338–40, 367–68. For similar themes during the entire campaign for the Solomons, see John Prados, *Islands of Destiny: The Solomons Campaign and the Eclipse of the Rising Sun* (New York: Penguin Books, 2012), pp. 153, 158, 167–68, 260–349. For the same continued focus on surface forces throughout the war on the part of the Office of Naval Intelligence, see Douglas Ford, *The Elusive Enemy: U.S. Naval Intelligence and the Imperial Japanese Fleet* (Annapolis, Md.: Naval Institute Press, 2011), pp. 91, 121–22, 125–26. For the 1945–46 Blue and Orange war games, see Friedman, *Blue versus Orange*, passim.

6 For late interwar American naval doctrine, see Kuehn, *Agents of Innovation*, pp. 162–79. For the wartime period, see Hone, "U.S. Navy Surface Battle Doctrine," pp. 67–105. For Soviet naval doctrine, see Milan Vego, *Soviet Naval Tactics* (Annapolis, Md.: Naval Institute Press, 1992), pp. 169–85, 254–67. For speculation about a U.S.-Chinese naval confrontation in the western Pacific, see James Kraska, "How the United States Lost the Naval War of 2015," *Orbis* (Winter 2010), pp. 35–45; Vitaliy Pradun, "From Bottle Rockets to Lightning Bolts: China's Missile Revolution and PLA Strategy against U.S. Military Intervention," *NWCR* 64, no. 2 (Spring 2011), pp. 7–38; Jonathan Solomon, "Maritime Deception and Concealment: Concepts for Defeating Wide-Area Oceanic Surveillance-Reconnaissance-Strike Networks," *NWCR* 66, no. 4 (Autumn 2013), pp. 87–116; John Hanley, "Creating the 1980s Maritime Strategy and Implications for Today," *NWCR* 67, no. 2 (Spring 2014), pp. 11–29; and Jan van Tol, Mark Gunzinger, Andrew Krepinevich, and Jim Thomas, *AirSea Battle: A Point-of-Departure Operational Concept* (Washington, D.C.: Center for Strategic and Budgetary Assessments, 2010).

7 Many of the records from this period were disposed of in the mid-1960s in a highly misguided attempt to "clean up" the archives; e-mails with Dr. Evelyn Cherpak, curator of the Naval War College's Naval Historical Collection, 9 June 2010, 18 June 2010, 22 June 2010, and 19 July 2010.

8 Ronald Spector, *Professors of War: The Naval War College and the Development of the Naval Profession* (Newport, R.I.: Naval War College Press, 1977), pp. 71–87; Vlahos, *Blue Sword*, pp. 131–56.

9 Edward S. Miller, *War Plan Orange: The U.S. Strategy to Defeat Japan, 1897–1945* (Annapolis, Md.: Naval Institute Press, 1991); Vlahos, *Blue Sword*. For a more recent work that looks briefly at the interwar period as a prelude to the Pacific War, see Douglas Smith, *Carrier Battles: Command Decision in Harm's Way* (Annapolis, Md.: Naval Institute Press, 2006).

10 J. S. Hurlburt, "War Gaming at the Naval War College, 1969–1989," *NWCR* 42, no. 3 (Summer 1989), pp. 46–51; Hattendorf et al., *Sailors and Scholars;* Smith, "Historical Analysis," cited in Hattendorf et al., *Sailors and Scholars*, p. 248; Philip Crowl, "Education versus Training at the Naval War College: 1884–1972," *NWCR* 26, no. 3 (November–December 1973), pp. 2–10; James Barber, "The School of Naval Warfare," *NWCR* 21, no. 8 (April 1969), pp. 89–96.

11 For early in the Cold War, see Robert Fisher, "The U.S. Navy's Search for a Strategy, 1945–1947," *NWCR* 48, no. 3 (Summer 1995), pp. 73–86. For later, Hattendorf, *U.S. Naval Strategy in the 1970s* and *Evolution of the Navy's Maritime Strategy;* Robert H. Gile, *Global War Game: Second Series, 1984–1988*, Newport Paper 20 (Newport, R.I.: Naval War College Press, 2004); Bud Hay and Bob Gile, *Global War Game: The First Five Years*, Newport Paper 4 (Newport, R.I.: Naval War College Press, 1993); David Rosenberg, "Being 'Red': The Challenge of Taking the Soviet Side in War Games at the Naval War College," *NWCR* 41, no. 1 (Winter 1988), pp. 81–93; Robert Wood, "The Conceptual Framework for Strategic Development at the Naval War College," *NWCR* 40, no. 2 (Spring 1987), pp. 4–16; James Barber, "Mahan and Naval Strategy in the Nuclear Age: A Lecture Delivered at the Naval War College," *NWCR* 24, no. 7 (March 1972), pp. 78–88; and Donald White, "Admiral Richard L. Conolly: A Perspective on His Notions of Strategy," *NWCR* 24, no. 3 (November 1971), pp. 73–79. For an example of this type of literature from the post–Cold War period, see Kenneth Watman, "Global 2000," *NWCR* 54, no. 2 (Spring 2001), pp. 75–88.

12 These articles numbered in the dozens during the Cold War, but a sample from primarily the 1950s includes Robert Carney, "Role of the Navy in a Future War," *NWCR* 6, no. 10 (June 1954), pp. 1–12; George Phelan, "Sea Power and Strategies for the Control of the Seas," *NWCR* 6, no. 10 (June 1954), pp. 15–36; James Field, "Seapower and Military Strategy Today," *NWCR* 8, no. 8 (April 1956), pp. 21–39, and "Origins of Maritime Strategy and the Development of Sea Power," *NWCR* 7, no. 7 (March 1955), pp. 1–23, and "The Influence of Sea Power on Modern Strategy," *NWCR* 10, no. 1 (September 1957), pp. 31–52; J. F. McInteer, "The Significance of Seapower to the United States," *NWCR* 12, no. 1 (September 1959), pp. 1–32; Joseph Bredestege, "Limited Nuclear War at Sea," *NWCR* 19, no. 6 (February 1967), pp. 4–31; and Harry James, "An Analysis of Limited Maritime War," *NWCR* 19, no. 6 (February 1967), pp. 33–74.

ACKNOWLEDGMENTS

I have incurred many debts in the process of producing this book, as I have with my previous books. None of the individuals or organizations mentioned here is in any way responsible for the opinions asserted in this work; responsibility for interpretations or errors is mine alone. However, each of the people cited assisted me in a very significant way and deserves thanks for that assistance.

Like my earlier works, this book would not have been possible without my wife Lisa. I cannot imagine what things would be like without her. She has been not only my soul mate in terms of our life together but the most active supporter of career aspirations that any spouse could ask for. This support has been given while she has pursued her own career and has taken the central role in our son's activities. I simply do not know what I would have done without her during these over twenty-five years of personal and professional life.

My son, Jeffrey, is now an adult. My hope is that this work will add further to his appreciation of studying the past. More exactly, I hope it will further demonstrate to him what his grandparents' world was like. I also hope that this book gives him some idea of what I did during all of those hours in front of the computer and what contributions I am trying to make to my profession. Finally, I hope that he personally will never have to experience war or its aftermath firsthand, and I hope that his country's imperial adventures are kept to a minimum in his lifetime.

My parents, Irving and Elaine Friedman, continue to be my personal connection with the past. It has been almost two decades since my father's death and over ten years since my mother's. It is strange for my siblings and me to be without parents. Studying 1940s America brings them a bit back to life for me, if only momentarily and in certain dimensions. Their lessons to me about the Great Depression and the war taught me more about history—both writing it and teaching it—than they probably knew. In all kinds of ways, my parents' absence from my life will be particularly felt as I write additional monographs in the years to come.

I continue also to gain great support from my in-laws, Ronald and Carolyn Sampsell, who are fortunately both still with us and hold special places in the creation of my thesis, my dissertation, and my books. They provided Lisa and me with various kinds of support during graduate school. Since then, they have been the best kind of inspiration to our son; I am so glad that he has been able to grow up in close physical proximity to at least one set of grandparents, as I did. They have provided me with many historical lessons as well, since they were part of the last sizable generation in the United States to grow up on farms. My in-laws also continue to be advisers to me concerning my career, as they were throughout the 1990s.

I continue greatly to benefit from my interaction with my siblings about these matters. My sister Karen and my brother-in-law Doug have been examples to me by virtue of their hard work in their careers and their attention to professionalism and detail. My sister Nancy was particularly important in instilling in me the idea of attending college and obtaining a professional position in whatever field I chose. I especially thank my sister Margaret and my brother-in-law Willard for encouragement early in my professional life and for career advice even as they transitioned to new careers of their own. I've also been helped immensely by my sister Phyllis and brother-in-law Mark through their support, which included a day on the nuclear-powered aircraft carrier *Ronald Reagan,* thanks to their friend, then the ship's executive officer, Capt. (now Vice Adm.) Herman Shelanski. Days spent like that, and walks around the battleship *Wisconsin* in downtown Norfolk, Virginia, truly inspired me to start writing about the Naval War College in the postwar period and the College's ideas about future wars. I also want to thank my brother Alan for introducing me to the late Professor John Bowditch, formerly chair of the History Department at the University of Michigan, and for always listening when I talked about how difficult it is to publish professionally while teaching full-time at a community college.

Two individuals who should have been thanked for helping me with my first book but were not through my own oversight were my godparents—my great-aunt Rose and great-uncle Mac Carp. They have, like my parents, represented particular examples to me about pursuing a career as a historian. They—both now, unfortunately, deceased—provided me financial support through my undergraduate years and moral support afterward. Uncle Mac immigrated to Canada from imperial Russia in 1913; with his thick accent, his ability to converse in several languages, and his constant interest in historical and political affairs, he always reminded me of the typical—or perhaps stereotypical—European intellectual living in the United States. He was, in fact, the intellectual of our family. Similarly, Aunt Rose, raised in a Jewish farming colony in Bad Axe, Michigan, then educated as a school librarian, was the teacher of the family. She was also a feminist in her own time; as one cousin said, she was the "Katherine Hepburn" of the family. Standing six feet tall, deciding to attend college and then graduate school in the 1920s and 1930s, establishing her career, wearing slacks, smoking in public, and marrying late, Aunt Rose was a unique example of living history in and of herself. I greatly miss our conversations about the past but I'm glad she was alive during my early years, and I'm glad I'm able to dedicate this book to the two of them.

I must also thank many colleagues at the former Henry Ford Community College (HFCC), now Henry Ford College (HFC). Of these, now-retired President Andrew Mazzara; his successors, Interim President Sally Barnett, President Gail Mee, and President Stan Jensen; now-retired Vice President / Dean of Academic Education Edward Chielens; his successors Interim Vice-President / Dean of Academic Education Lynne Hensel, Vice-President of Academic Affairs for Arts and Sciences Reginald Gerlica, and Vice-President for Academic Affairs Tracy Pierner; Social Science (subsequently the Humanities & Social Science Division and then the Social Science, Arts & Fitness Division) Associate Deans Henry Bowers and Kim Schopmeyer (the latter Interim Vice-President for Arts and Sciences for much of this writing); Humanities & Social Science Interim Associate Dean Anthony Perry and Social Science, Arts & Fitness Associate Deans Randall Knight and Cynthia Stiller; and John McDonald, President of the Henry Ford Community College Federation of Teachers (American Federation of Teachers Local 1650) stand out. John McDonald is deserving of a special thanks, since he has been a champion of faculty contractual access to professional development opportunities, particularly at community colleges, where time for research and scholarship is at a premium. He negotiated the provision in our contract for faculty sabbaticals, and Presidents Mazzara's and Mee's support in approving the sabbaticals for the fall semester of 2003 and the winter semester of 2011, respectively, are greatly appreciated.

The Executive Board and the rank and file of Local 1650 have also gone far in creating numerous sources of funding for instructors to pursue research for professional development, especially by presenting papers at conferences. Vice-President/Controller Marjorie Swan, now-retired Vice-President / Dean of Student Services Michael Meade, and the current Vice-President for Financial Services, John Satkowski, have additionally approved several of my requests for such funding, in which ideas for this book were presented and refined. I have also received helpful hints and ideas from our college library staff, both active and retired members, especially Kathy Cunningham, Pat Doline, Dan Harrison, Bruce Koldys, Barb Lukasiewicz, Terrence Potvin, Nancy Widman, and Bill Williamson. A special thanks goes to Terry Potvin for major assistance with interlibrary loan requests.

Current colleagues Maggie Anderson, Tom Anderson, Rodney Barnhart, Shatha Baydoun, John Burks, Mario Di Ponio, Kimberly Lark (now at Schoolcraft College), Keith Jacobson, Cris Lewandowski, Lynda Litogot, Gary Mitchell, Sam Plaza, Pamela Sayre, Bill Secrest, and Ken Shepherd, along with former colleagues Hani Bawardi,

Kathryn Beard, Saeed Khan, Devissi Muhammad, Richard Marquis, and Reginald Witherspoon, retired colleagues Virginia Caruso, Bill Hackett, Michael Johns, Wendy Osthaus, Dennis Ringle, Morris Taber, Bill Wilkinson, Bill Williamson, and Sue Zimmerman, and deceased colleagues Raye Howe, Armen Ovhanesian, Berny Rogers, Vince Solomon, Robert Spiro, Art Thomas, and Bill Watson, have gone even farther in fostering this type of professional environment. Whether they realize it or not, they created and sustained in the History Department at Henry Ford Community College and the History Section of Henry Ford College an environment that was and continues to be the most conducive to unencumbered exchange about pedagogy, research, writing, and historiography that I have ever encountered in the field. I value our comradeship more than I can express in words, and I hope it continues for a very long time. I have particularly to thank Pamela Sayre for taking on much of the day-to-day departmental administration that is so typical in small colleges. Her willingness to deal with these tasks freed much of the time I needed to write my books. I have also especially valued the mentoring provided to me by Mario Di Ponio and Robert Spiro, as well as their advice to keep teaching, service, scholarship, family, and life "in perspective."

To the list of HFCC-HFC historians must be added the names of three retired and two deceased HFCC political scientists, Brian Coyer, the late Umesh Pant, the late Tom Payette, John Smith, and Barbara Suhay. In particular, my conversations with Brian about the literature on bureaucratic rivalry were quite useful, and John lectured me—and still does—about the importance of publishing as a community-college scholar. In addition, I have to acknowledge several current and retired division colleagues with whom I have had general conversations about political and international affairs that have added to this monograph, especially Nabeel Abraham, Henry Bowers, Alison Buchanan, Pete Cravens, Kalvin Harvell, Tarek Joseph, Dan Kearney, Elaine Louisell, Kathy Orlich, Greg Osowski, Tony Perry, Eric Rader, Kim Schopmeyer, Meena Sharma, Brian Smith, and Cynthia Stiller. Special thanks are due to Tony Perry and Kim Schopmeyer, as well as Randall Knight; Linda Prisza, Social Science, Arts & Fitness Division Academic Affairs Assistant; and Thomas Berryman, Social Science, Arts & Fitness Division Assistant to the Associate Dean, for helping me with computer problems that developed during the preparation of the manuscript. Along the lines of technical assistance, Patricia Sekulidis, HFC Graphics Specialist, was of great help with document scanning and did duty well above and beyond by preparing the individual shots of staff and students from the group photos that were supplied by the Naval War College Museum.

Numerous people, currently or formerly at the Naval Historical Center (now the Naval History and Heritage Command) in Washington, D.C., helped me get a start on this project in the 1980s and 1990s and continued on to help me finish it in the early twenty-first century. These individuals include Dr. Dean Allard and Bernard

Cavalcante, both retired directors of the Naval Historical Center, and Kathy Roar, former Operational Archives Branch Assistant Chief. More particularly, with this project as well as my previous ones, I must thank Kathy Lloyd, now-retired Chief, Operational Archives Branch, for her flexibility in providing me with research facilities. Included also are Dr. Regina Akers, John Hodges, Ken Johnson (now at the National Archives), Allen Knechtmann (now at the Navy Department Library), Tim Pettit, and Mike Walker (now retired), all of whom have made my research excursions to the Operational Archives so successful. In addition, this last bit of research would not have been very successful if it were not for now-retired Senior Historian of the Navy Ed Marolda, former Contemporary History Branch Chief Gary Weir, Contemporary Branch historians Jeff Barlow (now retired) and John Sherwood, former Contemporary Branch historian Randy Papadopoulos (now Secretariat Historian, Office of the Under Secretary of the Navy), and Dr. Robert Cressman, Editor, *Dictionary of American Naval Fighting Ships,* in the Command's History and Archives Division.

The staff at the U.S. Naval War College, in Newport, Rhode Island, were indispensable to the creation of this book. Thanks go first to philanthropist Edward Miller for generously establishing the Edward S. Miller Research Fellowship in Naval History. That award allowed me to travel to Newport in October 2003 and obtain primary materials. Thanks must also go to Rear Adm. Joseph Strasser, USN (Ret.), former executive director and today a trustee of the Naval War College Foundation, and especially to Dr. John Hattendorf—at the time, Ernest J. King Professor of Maritime History, chair of the Maritime History Department, and director of the Naval War College Museum—for coordination of the award and especially for encouragement in my research efforts. Professor Hattendorf was particularly helpful in the logistics of administrative support, as was his assistant in 2003, Patricia Cormier, who did more photocopying than I had a right to ask of her. Professor Hattendorf was also helpful in making comments that strengthened my previous manuscripts with the Naval War College Press and in getting those manuscripts funded and published. In addition, faculty members associated with the Strategy and Policy Department and Fleet Seminar Program—Doug Smith, Stan Carpenter, and Nick Sarantakes—allowed me to air ideas to them at critical times. A special thanks goes to Stan for helping me define succinctly the "reverse course" maneuver for readers.

Putting in equally wonderful efforts when I was in Newport were Evelyn Cherpak, Curator, Naval War College's Naval Historical Collection (now retired), and Teresa Clements, archival technician, both of whom pulled numerous records for me as I raced the research clock. At subsequent meetings and conferences, as well as through numerous e-mails, Dr. Cherpak has been extraordinarily supportive of my work, especially in answering numerous requests for additional records and questions about the College in the 1940s. The staff at the Naval War College Press have similarly made this endeavor a very fruitful and easy professional project. Thanks go especially to Dr. Carnes Lord, director of the Press; Pelham Boyer, managing editor, now retired; his successor, Robert Ayer; Nicole Gresham and Kate Acosta, associate editors; Phyllis Winkler, former book review editor; Robert Cembrola, Exhibits Manager of the Naval War College Museum; John Kennedy, Director of Museum Education and Public Outreach; Art Lamoureux, visual information specialist; Jeremiah Lenihan, Lead Specialist, Desktop Publishing Division; Robin Lima, Interlibrary Loans, Naval War College Library; and Lori Almeida, Naval War College Press administrative assistant.

At Michigan State University and the University of Michigan, a number of people again need to be mentioned. To Don Lammers, Professor Emeritus of History at Michigan State, thanks are again due for being such a fine professional example, not only as a dissertation director but as a teacher, scholar, and colleague as I embarked on these additional monographs. To John Shy, Professor Emeritus of History at the University of Michigan—I still have the urge to thank you for helping me get my career off the ground, even though you don't want to be thanked. In the end, suffice it to say that you continue to inspire me with your example as a scholar and that I am still paying you back by "paying forward."

Other individuals were also key to this study being completed. Chris Hamel taught me what being a historian is all about. Not only did he instruct me in how and why historians analyze and write the way they do, but his help in classifying my early work into strategic and bureaucratic dimensions was the starting point for my postdissertation monographs. Our long conversations about these matters were a basic building block on which this work flourished, and his assistance during graduate school was vital to the beginning of my career. I continue to admire him as well for having put up with the often ridiculous issues of day-to-day academic administration while staying focused on the Big Picture and finding the classroom to be a worthwhile charge.

Along these same lines, several colleagues have provided career advice and assistance at numerous points. Michael Palmer of East Carolina University, Gary Weir, now chief historian of the National Geospatial-Intelligence Agency, and Carol Reardon of Penn State University have been informal mentors to me in the last twenty-five years, either through books they wrote that helped matters gel for me or for pointers they offered in numerous conversations over the phone, on e-mail, or in person at professional conferences regarding teaching, publishing, and job searches. In addition, I have to thank Jonathan Winkler of Wright State University for numerous conversations about the strategic and national security history of the United States and, especially, the logic of the Naval War College exercises and maneuvers in the 1940s. I also have James Levy of Hofstra University to thank, for his feedback and amplification about research on U.S. Navy surface warfare in the early Pacific War; John Lundstrom, for conversations about the 1942 Pacific War battles and their impact on immediate postwar NWC exercises; and Trent Hone, Director of Software Engineering at Trimble Mobile Resource Management. Trent's articles on U.S. Navy tactical surface doctrine in the interwar and Pacific War periods, as well as our e-mail conversations, put everything into perspective for this book. Most especially, I need to thank (belatedly) Marc Gallicchio, Chair, History Department at Villanova University. Marc not only wrote a letter of recommendation for my application for the Miller Fellowship in 2003 but was courteous enough to look the application package over before I submitted it and suggest revisions. In addition, he has always been ready with advice about my place in the profession whenever I've asked him in the last twenty-plus years. I was also assisted by Dr. David Hogan of the U.S. Army Center of Military History and Mike Coombes of the UK Navy List Research in finding information on certain members of the 1946–47 Naval War College classes.

My professors at Eastern Michigan University (EMU), my undergraduate alma mater, are perhaps most central to this study, since they first introduced me to serious historical study and taught me that college teachers can also be first-rate scholars. I must thank the late Richard Abbott for first discussing the idea of the United States in the 1940s Pacific with me during the 1980s and for continuing to discuss this project with me as I pursued it in graduate school and as he scuba dived in Micronesia. I still find it ironic and more than a bit intimidating that I taught his course in U.S. military history on an occasional basis during and after his terminal illness, and I am trying to follow in his footsteps as much as possible. He continues to be greatly missed.

I must also thank Don Briggs, George Cassar, Robert Grady, Mike Homel, Jim Johnson, Roger King, Karen Linderman, Jim Magee, Walter Moss, Joe Ohren, James Pfister, Leonis Sabaliunas, Les Shearer, Janice Terry, Jiu Upshur, Reinhard Wittke, and the late Richard Goff for instilling in me the professionalism and high standards that I was able to hone in graduate school and am now able to impart to students at Henry Ford College. The History and Political Science Departments at Eastern Michigan in the 1980s were the most encouraging environments I can think of in which a first-generation college student from a working-class background could get started in the profession.

Similarly, I have to thank several individuals from the current EMU History and Philosophy Department who have been instrumental in my graduate education, my career as an instructor, and my development as a historian. These include—but are not limited to—Marsha Ackerman, Kathy Chamberlein, Mark Higbee, Jim Holoka, Richard Nation, Gersham Nelson, Philip Schmitz, and Matt Schumann. Especially helpful in these matters has been Michael Homel, whose History Reader's

Group has been a wonderful outlet of professional activity for me and a magnificent release from the rigors of teaching.

Most especially for my EMU association, I have to thank Rob Citino, now Senior Historian, National World War II Museum. Rob has been a professional model for me to emulate. In thirty years of teaching, always with heavy classroom and administrative loads, Rob has managed to publish nine monographs, is working on additional ones, and has become a recognized specialist in his area, all while maintaining an extraordinary reputation for excellence in the classroom. His professional example continues to awe me, as it does many in our profession.

I would also like to acknowledge the assistance that resulted from my membership in the Michigan War Studies Group at the University of Michigan's Department of History. Since I joined that informal weekly group in January 1991, it has been a forum for communicating my ideas, my hopes, and my fears about professional life in general. Special thanks go to past and present members (some already named in other connections) Saiful Islam Abdul-Ahad, Keith Arbor, Wil Blythe, Bill Boardman, Denver Brunsman, George Cassar, Rob Citino, Tom Collier, Ken Estes, David Fitzpatrick, Paul Forage, Jim Hill, Jim and Chris Holoka, Bob Jefferson, Doron Lamm, Sheldon Levy, Gerald Lindermann, Jonathan Marwil, Dennis Ringle, Mike Riordon, Pamela Sayre, Matt Schumann, Stanley Shapiro, Alan Sherzer, the late Jack Sherzer, Ken Slepyan, and Bruce Zellers, for constantly providing me with new perspectives on military history and strategic thought.

The Metro Detroit Historians Collegium has been another outlet for my professional activities. Consisting of historians from departments at small two-year and four-year colleges in the Detroit area, the collegium has become a local forum for history instructors at institutions that have minimal resources for professional development. The collegium was kind enough to hear a presentation on my first book and continues to be significant to my professional development. Within the collegium, I especially have to thank Jayne Morris-Crowther, formerly of Madonna and Oakland Universities; Tom Klug, from Marygrove College; Tony Baracco, Shawn Dry, Valerie Emanoil, Ed Gallagher, Tim Koerner, Marilynn Kokoszka, Nancy Shockley, and Mike Vollbach, from Oakland Community College; Steve Berg, Kimberly Lark, Evan Garrett, Michael Johns, and Michael Swope, from Schoolcraft College; Roy Finkenbine, from the University of Detroit Mercy; and Duane Ashley, from Wayne County Community College. In addition, I wish to thank my colleagues from HFCC who helped start the collegium—John Burks, Virginia Caruso, Mario Di Ponio, Michael Johns, Richard Marquis, Devissi Muhammad, Wendy Osthaus, Pamela Sayre, Bill Secrest, Ken Shepherd, and Bob Spiro.

ACRONYMS

AA	antiaircraft
AEW	airborne early warning
AGC	amphibious force flagship
AKA	attack cargo ship
APA	attack transport
AS	submarine tender
AV/AVP	seaplane tender
BATDIV	battleship division
BB	battleship
CA	heavy cruiser
CAP	combat air patrol
CARDIV	carrier division
CARGRU	carrier group
CB	battle cruiser
CINCPOA	Commander in Chief, Pacific Ocean Areas
CL	light cruiser
CLAA	antiaircraft cruiser
CNO	Chief of Naval Operations
COM CVG	Commander, Carrier Air Group
COMAIR	Commander, Air Forces
COMAIR3FLT	Commander, Air Forces, 3rd Fleet [or any numbered fleet]
COMBATRON	Commander, Battleship Squadron
COMCARDIV	Commander, Carrier Division
COMCRUDIV	Commander, Cruiser Division
COMDESRON	Commander, Destroyer Squadron
COMFAIRWING	Commander, Fleet Air Wing
CRUDIV	cruiser division
CTF	Commander, Task Force
CTG	Commander, Task Group
CTU	Commander, Task Unit
CV	fleet carrier

CVB	battle carrier
CVE	escort carrier
CVL	light carrier
DD	destroyer
DESDIV	destroyer division
DESFLOT	destroyer flotilla
DESRON	destroyer squadron
DM	light minelayer
FAIRWING	Fleet Air Wing
FIDO	fog intensive disposal equipment
GP	general-purpose
LORAN	long-range navigation
MESA	Maximum Engine Speed Allowed
NATO	North Atlantic Treaty Organization
NHC	Naval Historical Collection, Naval War College, Newport, R.I.
NWC	Naval War College
NWCR	*Naval War College Review*
OINC	officer in charge
ONI	Office of Naval Intelligence
OPNAV	Office of the Chief of Naval Operations
OTC	Officer in Tactical Command
RG	Record Group
SOPA	Senior Officer Present Afloat
SS	submarine
SUBDIV	submarine division
SUBRON	submarine squadron
TCAP	Top Combat Air Patrol
TF	task force
TG	task group
TU	task unit
USA	U.S. Army
USAAF	U.S. Army Air Forces
USMC	U.S. Marine Corps
USN	U.S. Navy
USNR	U.S. Naval Reserve
VA	Attack [squadron]
VA(Q)	carrier tactical electronics warfare squadron
VB	dive-bomber
VF	Fighter [squadron]
VHF	very-high-frequency [radio]
VP	Patrol [squadron]

NAVAL WAR COLLEGE
STAFF 1946-47

I *Context at the College*

The Naval War College Curriculum in Perspective, 1946–1947

Most of this study explores in detail the processes, procedures, and pedagogy of the war games, or "Operations Problems," played at the Naval War College during the 1946–47 academic year. However, to give some sense of the larger curriculum, this chapter looks at selected academic materials that were offered to the junior and senior classes of June 1947. In particular, it looks at the lecture, discussion, and reading activities related to studies meant to help the student officers prepare for the actual Board and Chart Maneuvers that they were to practice in their exercises and Operations Problems.

The Theses

In late June 1946, Rear Adm. Allan Smith, chief of staff to the President of the Naval War College, issued a directive about the thesis projects the students of both classes were to complete during the academic year. The first thesis was due by 15 January 1947 and was to address "Relations between Russia and the United States, and Their Influence on U.S. Foreign Policy." The second was due by 1 May 1947 and was to be on "The Influence of the Atomic Bomb on Future Naval Warfare." A suggested length for each thesis was nine thousand words; six thousand was the minimum; an original and carbon copy were to be submitted by the students, and the original would be returned before the end of the course. Further details could be found in "Annex A, Hints on Thesis Writing." Classified material had to be cleared through the head of the Naval War College's Department of Intelligence, Capt. Richard Hartung.[1]

Annex B of Smith's instruction was a suggested bibliography of unclassified readings on these subjects. Students were expected to augment them with their own research and draw from "appropriate" presentations and lectures in the College curriculum. While theses were to demonstrate that students had prepared themselves in their subjects and were familiar with the opinions of "prevailing" authorities, their research was to be extensive and "profound" enough to develop original thoughts and opinions: "Most to be desired is a comprehension on the part of the student of the fundamental principles of the subject under study and a discussion in his thesis of his views on their significance in national and naval affairs."

Rear Adm. Allan Smith, USN

Capt. Richard Hartung, USN

On the first subject, no summary of relations with Russia was needed; any historical introduction was to be brief, not more than 20 percent of the total length of the thesis. "Chief attention" was to be given to the factors that had controlled such relations in the past, the "validity" of such factors in the present, and the opinion of the student about their application for the future. As to the second subject, "This is an unlimited field and the student has considerable latitude of treatment. Particular attention should be focused on the effects of this new weapon on our strategical and tactical concepts."[2]

Annex A's "hints" began by stating that the purpose of student theses was to elicit opinions on "matters of importance" to naval affairs as well as to cause the students to work with research materials and exercise their minds in collating, evaluating, and condensing them in support of definite conclusions. "The accent is placed upon research and original thought by the student." The resources were to be the Naval War College's archives and library, as well as a "careful reading" of newspapers and magazines. The College's librarians would be able to supply material beyond that listed in Annex B; students could obtain classified material from the archives. The archivist, in fact, had a list of classified material for "convenient" reference by the students. Students did not have to read books from cover to cover but should skim, using the tables of contents, indices, and first and final chapters, so as to allow fairly wide reading on given subjects. "Rapidly dissecting" books in this manner, students were to allow themselves to be led from one source to another and thus complete a general survey of the field in which they were interested.[3]

After browsing, students were to begin formulating a methodological approach to their project. They were especially to begin creating an outline, in terms of an "Introduction," "Scope of Thesis," "Development of a Theme," and "Conclusions." The actual subdivision would depend on the student's knowledge, but outlines in general would save much energy and time in creating a "unified" work. The outline was, in fact, the "framework" for the finished thesis; after creating it, the student was to begin "serious reading," taking notes on index cards, summarizing or quoting information directly, and adding the student's thoughts. These "thesis cards" were to emphasize the student's original thoughts, be dated or otherwise organized, and include "nearly everything" needed to write the thesis. In effect, the students' thesis cards, outlines, and tables of contents were to establish what they were going to say in their theses as well as at what points their ideas would be expounded.[4]

The next step was to recheck the outline for unity, logical framework, coherent connections, and proportionality of each section to its importance. Students were to cut out "ruthlessly" sections that were not germane and arrange their materials as was best for their reader. The students were not merely to narrate a story or create a work of "pure exposition" but to contribute an exposition and argument. As to writing itself, the advice was simply to start at the beginning. Some theses, however, were best written as separate sections to be fitted together later, but generally, "consecutive" thought would probably be clearer if the paper was written from the beginning through to the end.[5]

Admiral Smith detailed format, paper size, pagination, margins, binding, handling of quotations, citations, submission, and copies. He reiterated that the paper represented an endeavor to outline "a method of approach and attack in writing a

ANNEX "B"

BIBLIOGRAPHY

First Subject: Relations between Russia and the United States, and their Influence on U.S. Policy.

- - - -	America and the Future, Fortune, Jan. 1945
Army Service Forces	Geographical Foundations of National Powers - Section I. 18 April 1944
Chamberlin	The Russian Enigma. 1943
Cressey	The Basis of Soviet Strength. 1945
Dallin	Russia and Post War Europe. 1943
Dallin	Soviet Russia's Foreign Policy. 1942
Dallin	The Real Soviet Russia. 1944
Davies	Mission to Moscow. 1941
Dean	Russia at War. 1942
Dennis	The Foreign Policies of Soviet Russia. 1924
Dulles	The Road to Teheran. 1944
Duranty	U.S.S.R. 1944
Earle	Makers of Modern Strategy
Encyclopaedias	Americana, Volume 24 Britannica, Volume 19
Fischer	The Soviets in World Affairs. 1930
Fox	The Super-Powers. 1944
Hard	Eight Things to do about the Soviet Union. Reader's Digest, September 1945.
Harris Lectures	The Soviet Union and World Problems. 1935
Hopper	Narkomindel and Comintern. Foreign Affairs, July 1941.
Johnston	My talk with Joseph Stalin. Reader's Digest, October 1944.
Kerr	The Russian Army. 1944
Lauterbach	These are the Russians. 1945

Lippmann	U.S. Foreign Policy. 1943
Lippmann	U.S. War Aims. 1944
NavPers-16022	Readings on "The Foundations of National Power." Booklet No. 8 - The Soviet Union as a World Power.
Pares	Russia. 1943
Pares	Russia and the Peace. 1945
Schuman	American Policy Toward Russia Since 1917. 1928
Schuman	Soviet Politics. 1941
Simonds & Emeny	The Great Powers in World Politics. 1939 ed. pp. 321-350
Snow	The Pattern of Soviet Power. 1945
Sorokin	Russia and the United States. 1944
Spykman	The Geography of the Peace. 1944
Standley	Stalin and World Unity. Colliers, p. 17, June 30, 1945
Stevens	Russia is no Riddle. 1945
Strong	The Soviets Expected It. 1941
- - - -	Tripartite Conference at Moscow, 19 Oct. 1943 International Conciliation, December 1943
Vernadsky	A History of Russia. 1944
Vernadsky	Trends in Soviet Foreign Policy. Yale Review, June, 1944.
Welles	The Time for Decision. pp. 306-336. 1944
White	Report on the Russians Reader's Digest, December 1944
Willkie	One World. 1943 pp. 50-87

Fig. 1
Annex B: Bibliography

ANNEX "B"

BIBLIOGRAPHY

Second Subject: The Influence of the Atomic Bomb on Future Naval Warfare.

Cooke & Markus	Electronics Dictionary. 1945
Dietz	Atomic Energy in the Coming Era. 1945
Masters & Way	One World or None. 1946
Pendray	The Coming Age of Rocket Power. 1945
Smyth	Atomic Energy for Military Purposes. 1945
Zim	Rockets and Jets. 1945
Acheson	Safeguarding atomic energy against misuse. U.S. News, May 3, 1946.
Allen	Use of the atomic bomb. Infantry Journal, January 1946.
Brodie	American security and the atomic bomb. Yale Review, March 1946.
Condon, E.V.	Atom ABC's. All Hands, April 1945.
DeVore	What the atomic bomb really did. Collier's, March 2, 1946.
Finletter	Timetable for world government. Atlantic, March 1946.
Fortune	Fortune survey; use of atomic bomb. Fortune, December 1945.
Huie	How the next war will be fought. American Mercury, April 1946.
International Conciliation	Atomic energy and American policy. Internation Conciliation, December 1945.
Jones	Can atomic power be controlled. Harper, May 1946.
Nimitz	Atomic age navy. Collier's, May 11, 1946.
Pepperburg	Operation Crossroads. Seapower, March 1946.
Science News Letter	Atomic bomb Science News Letter, August 18, 1945.
Spaatz	Air power in the atomic age. Collier's, December 8, 1945.
State Dept.	A report on the international control of atomic energy.
Stowell, E.C.	Laws of war and the atomic bomb. American Journal International Law, Oct. 1945.
Walker, S.H. ed.	The first one hundred days of the atomic age. 1945. Woodrow Wilson Foundation.
War Department Military Int. Division	Preparedness in the atomic age. Intelligence Bulletin, March 1946.
Wintringham	Last Weapon Virginia Quarterly Review, April 1946.
- - - -	Aviation Yearbook. 1946.
- - - -	"Cast". Naval Aviation News, April 1946.
Hershey	Three new military planes give forecast of future developments. Aviation News, January 28, 1946.
Joint Board on Scientific Information DuPolicy	Electronics Warfare. A report on radar countermeasures. 1945.
Durand	Jet motor's future. Aviation News, December 10, 1945.
- - - -	First all-jet fighters announced by Navy; FD-1 Phantom. Aviation News, January 14, 1946.
Streid	Development of the I-40 jet propulsion gas turbine. Aviation, January 1946.
- - - -	Pilotless aircraft. Naval Aviation News, January 1946.
- - - -	"Pushbutton" world. A preview of the future Navy. All Hands, May 1946.
Roger	Push-button warfare. Flying, February 1946.
Garrison, Earle	Rockets and their capabilities. Military Review, June 1945.
- - - -	The German long range rocket. Journal of the Royal United Service Institute, February 1945.
Healy, Roy	Aeronautical supremacy demands jet and rocket research. Part II. Aviation, August 1945.
Martin, P.W.	The future of the rocket in naval warfare. U.S. Naval Institute, August 1945.
War Department Mil. Int. Div.	Guided Missiles - the weapon of the future. Intelligence Bulletin, April 1946.
De Seversky	Navies are finished. American Mercury, February 1946.
- - - -	Future fighting forces. Naval Institute, May 1946.
Harris, C.T.	Old weapons and new. Army Ordnance, January-February 1945.
LeMay, Curtis	The AAF's war of research. Flying, June 1946.
Naval Institute	Professional Notes.
Schwartz	New aerial weapons. Army Ordnance, March-April 1946.
Steinhardt	The role of operations research in the Navy. Naval Institute, May 1946.
Taffrail	The Navy's future. The Navy, April 1946.
Thursfield	Science and sea warfare. National Review, November 1945.

Fig. 1, continued

thesis." Experience indicated that the greatest difficulty was simply getting started; once reading and preparation of notes were under way, the project should pick up momentum. Finally, additional material was to be found in such publications and periodicals as *State Department Bulletins, Vital Speeches, Foreign Policy Reports, Soviet Russia Today, Foreign Affairs, Time, Newsweek,* the *Saturday Evening Post, Collier's,* the *New York Times,* and *Periodical Indexes.*[6]

The Staff Presentations

A major part of the 1946–47 Naval War College curriculum comprised Staff Presentations, or lectures, delivered to the students between early July 1946 and early

Cdr. Albert Pelling, USN

Capt. Fred Dickey, USN

Cdr. Lauren Johnson, USN

Capt. Donald Evans, USN

May 1947. In total, there were forty, and they fell into several categories. First, there were lectures on straightforward naval topics. For instance, on 2 July 1946, Cdr. Albert Pelling, an instructor in the Department of Tactics, lectured on "Employment of Ship's Weapons," addressing doctrine, the use of guns in naval surface actions, night actions, fire effect, and even the "Atomic Gun Ship" of the future.

About two weeks later, Capt. Fred Dickey, an instructor in the Department of Intelligence, gave a lecture on "Fast Carrier Task Forces," in which the centrality of aircraft carriers to naval power was clearly recognized. Captain Dickey introduced the students to carrier doctrine and procedures but also pointed out the continuing roles of heavy surface ships.

Three days later, Cdr. Lauren Johnson, of the Department of Tactics, introduced students to "Shore and Tender Based Aircraft, Their Organization and Employment." Commander Johnson's lecture focused on the important role played by naval patrol planes and naval patrol bombers in strike, search and patrol, antisubmarine warfare, air-sea rescue, minelaying, and photographic reconnaissance. He spoke against the background of calls by some in Washington, D.C., for the dissolution of Navy shore-based aviation.[7]

On 3 January 1947, Capt. Donald Evans, an instructor for the junior class, delivered a long lecture on "Organization and Functioning of Naval Staffs," which introduced the students to basic functions and organizations of naval staffs, pointed out differences from Army staff organization, and gave examples of naval staff organization and operation from both the wartime Atlantic and Pacific theaters.

On 15 March, Commander Pelling lectured on "Naval Gunfire Support Plans," addressing not only such issues as target analysis, logistics, schedules of fire, and communications but also how important naval gunfire support had been during the war and what its future might be. On 20 February, Cdr. Earl Hydeman, a member of the senior class of June 1947, spoke on "Submarine Operations in the Sea of Japan," looking at the strategic significance of these operations, the technology that had been developed to carry them out, and the operation plan for what the Navy had termed the "Japan Sea Patrol Pack."

Cdr. Earl Hydeman, USN

Capt. Henry Moran, USN

Cdr. James Reed, USN

Lt. Col. Donald Dunlap, USAAF

On 29 April, Capt. Henry Moran, an instructor in the Department of Intelligence, gave an unusual lecture—"The Antarctic Region" and the Navy's recent expedition to the area, the Antarctic Developments Project of 1947, known as Operation HIGHJUMP.

The final lecture on what might be considered fairly traditional naval topics was given on 1 May by Cdr. James Reed, an instructor for the junior class, on "The Anti-submarine Warfare Situation Today." Commander Reed concentrated on the antisubmarine problems created for the U.S. Navy by the German Type XXI U-boat.[8]

There were also a small number of lectures on what today would be termed "joint military operations" and at the time "defense unification." On 18 July 1946, Lt. Col. Donald Dunlap, U.S. Army Air Forces (USAAF), of the Department of Intelligence, delivered a lecture on "Strategic Bombing." Given the controversial aspects of this subject, Colonel Dunlap gave significant credit to the other services but also wanted students to understand how important his own service's aerial offensives had been in both the European and Pacific theaters to eventual Allied victory.

On 26 November, Lieutenant Colonel Dunlap gave another lecture, this time on "Air Force Trends in the United States, Great Britain and the USSR," laying out and comparing the orders of battle, material development, aircraft production, and future trends of the air forces of the three nations. On 9 April 1947, Capt. Richard Bates, head of the Department of Research and Analysis, delivered "Estimate of the Situation: Part I," in which students were reintroduced to the Naval War College's pamphlet *Sound Military Decision,* but with a new joint-service interpretation from the Armed Forces Staff College. On 10 May 1947, Colonel Dunlap spoke again, on "Army Air Force Operations." The students viewed a film, *Air Force Highlights in World War II,* after which Dunlap and his colleagues Col. Lewis Lyle, a student in the senior class of June 1947, and Lt. Col. Richard Leary, a student in the junior class of June 1947, both of the Army Air Forces, devoted most of the presentation to tactical operations in Europe and to the Eighth Air Force's operations against Schweinfurt, Germany, of 17 August 1943.[9]

Capt. (previously Commodore) Richard Bates, USN

Col. Lewis Lyle, USAAF

Lt. Col. Richard Leary, USAAF

Col. Feodor Schmidt, USA

Out of the forty Staff Presentations that academic year, eleven were historical studies of various military operations. For instance, on 30 August 1946, Col. Feodor Schmidt, U.S. Army, of the Department of Intelligence, delivered the first of a series of lectures entitled "Strategic Employment of Military Forces: World War II." Part I focused on the campaigns between 1939 and 1941 in Poland, Finland, Norway, Western Europe, and the Balkans.

On 11 September 1946, Cdr. Dale Mayberry, an instructor for the junior class, lectured on "The Search for and Sinking of the *Bismarck*," providing a brief review of the naval operations leading up to this episode, German naval strategy, a narrative of the ship's sinking, and lessons learned about radio intelligence, aerial search, weather, and British naval influence on German strategy.

On 25 October, Colonel Schmidt delivered his second lecture, this time on the major wartime diplomatic conferences in Washington, London, Casablanca, Quebec, Cairo, Yalta, and Potsdam, with stress on the global situations and overall strategic picture. Three days later, Capt. John Sweeney, the head of the Department of Tactics, spoke on "The Battle of the River Plate," recounting the British destruction of the German heavy cruiser ("pocket battleship") *Graf Spee,* especially the role of South America in both British and German naval strategies, the tactical aspects of the engagement, and the principles demonstrated by the British naval victory.

The next day, Capt. Cornell Sullivan, head of the War Gaming Section in the Department of Tactics, presented "The Battle of the North Cape," the engagement that had seen the destruction of the German battle cruiser *Scharnhorst.* After looking at the German cruiser's operations against British sea lines of communication, Captain Sullivan provided a detailed account of the action, which proved to be the last major German surface threat to the Allied Arctic supply line.[10]

These history-based "lessons learned" lectures continued on 13 November with Colonel Schmidt's "Strategical Employment of Military Forces, World War I." Most of this lecture was focused on the major early ground campaigns of the western front and German failure there, but Colonel Schmidt provided information as well on the major early eastern front campaigns, such as Tannenberg, and later

Cdr. Dale Mayberry, USN

Capt. John Sweeney, USN

Capt. Cornell Sullivan, USN

Capt. Paul Crosley, USN

campaigns of the Great War. Richard Bates, then a commodore and head of the Department of Research and Analysis, spoke on 23 November on "The Battle of the Coral Sea," discussing the preliminaries, including the Doolittle Raid on Japan, the Coral Sea area, Japanese plans and dispositions for the region (to the extent that they were yet understood in the United States), and Allied command organization and positions prior to the battle. On 16 December the students received a lecture from Capt. Paul Crosley, an instructor in the Department of Strategy, on "Jutland Strategy." Captain Crosley started with the global situation in 1916, British and German naval strategies, events at sea and on land leading up to the classic naval engagement, and the "sortie" itself. Colonel Schmidt delivered two more lectures on the strategic employment of military forces in World War II. The first of these, on 31 January 1947, treated the Russo-German war between the German invasion in June 1941 and the Russian summer and winter offensives of 1943–44.

Colonel Schmidt also devoted part of his lecture to the war in North Africa up to the German defeat in 1943, including Allied strategic plans and concepts for the theater. His final lecture in this series took place on 19 April and looked at the Allied invasions of Sicily, Italy, and Western Europe, as well as the Western Allied and Soviet culminating offensives in 1944–45. The last of the historically based lectures was by now-captain Bates, on the battle of the Coral Sea, recounting the strategic and tactical aspects of the battle itself, including plans, uses of technologies, and "battle lessons."[11]

The next significant section was on administration, logistics, and supply, a total of eight lectures. The first of these was delivered on 11 October 1946 by Capt. James Bierer of the Navy Supply Corps, an instructor in the Department of Logistics; entitled "The General Supply System," it was a general outline of the organization and functions of, and future trends in, Navy supply, accounting, and finance. A day later, Col. Woodrow Vaughan, U.S. Army, and Col. Robert Wood, U.S. Army Air Forces, both students in the senior class of June 1947, introduced naval officers especially to the organizations, functions, and terminology of their respective services' supply systems.

Capt. James Bierer, USN

Col. Woodrow Vaughan, USA

Col. Robert Wood, USAAF

Cdr. Constantin Mathas, USN

Three days later, Cdr. Constantin Mathas of the Department of Logistics lectured on "Theater Logistic Planning Agencies in the Pacific Ocean Area." He gave a historical background and, especially, joint aspects worked out from 1943 to 1945. He also looked at the organization of the J-4, or Supply, division on the staff of the Commander in Chief, Pacific Ocean Areas (CINCPOA), the distinctions between the types of commands, and the organization and duties of Commander, Service Force, Pacific Fleet.[12]

On 17 October, Captain Bierer spoke on "Ships and Advanced Base Supply," significant in terms of how the Pacific War had developed. After looking at these two types of supply, he focused on the exercise of control over them by Commander, Western Sea Frontier. Cdr. Charles Daily, Civil Engineer Corps, an instructor in the Department of Logistics, gave a lecture on 19 October on "Logistic Functions Performed by the Bureau of Yards and Docks," specifically about the functions and departments of the bureau and especially its vast wartime expansion.

On the same day, Commander Mathas took up "Personnel Procurement and Training," not only wartime matters, but also, especially, how the Navy was going to handle the postwar period in terms of procurement and training for the officer and enlisted categories. On 21 October, Captain Crosley lectured on a topic absolutely vital to morale, "Mail as a Logistic Problem in Operation Plans," focusing on coordination with other government agencies and with operational activities and movements. Finally, on 8 May 1947, Captain Bierer delivered a talk on "Logistic Support of Joint Overseas Operation," especially the responsibilities, organization, and planning procedures at the theater-commander level.[13]

Four lectures were devoted late in the academic year to various aspects of amphibious operations. On 20 March 1947, Lt. Cdr. James Curran of the U.S. Naval Reserve (USNR), an instructor in the Department of Intelligence, presented a lecture on "Landing Plans" that looked at the planning process at the attack-force level, the numerous stages of planning and execution, and the criticality of amphibious operations to Allied victory.

Cdr. Charles Daily, USN

Lt. Cdr. James Curran, USNR

Col. Robert Bare, USMC

Cdr. (later Captain and Rear Admiral) Henry Eccles, USN (left); other man unidentified

A week later, Commander Curran, in "Landing Ships and Landing Craft," employed the U.S. invasion of Saipan in June 1944 as a detailed case study of the planning process, the numerous types of assault and landing ships and craft developed during the war, and the mammoth logistical effort that these operations entailed. Two days later, Col. Robert Bare, U.S. Marine Corps, of the Department of Intelligence, delivered "Amphibious Operations in the Central Pacific, Part I," on the general situation in the Central Pacific in the spring of 1943, the planning and execution phases for the operations against the Gilberts in the fall of 1943, and conclusions about naval gunfire support, artillery support, amphibian tractors, command ships, and underwater demolition teams. On 5 April, Colonel Bare delivered Part II of his lecture, applying the same topics to the operations against the Marshalls in early 1944.[14]

There were also three lectures on base development during the war and its future implications. On 15 October, Captain Bierer gave a lecture on "Advance Bases and Their Problems." After outlining the organization of base development within the Navy and aspects of military government, Captain Bierer took the students through the assault and landing phases of amphibious operations and the role of bases in both the initial posthostilities period and more advanced development phases. He also looked at the problem of disestablishing bases during and after the war. The next day, Commander Mathas spoke on "Base Development Planning," especially how little prewar experience the Navy had possessed, early wartime experiences in the South Pacific, and then the massive efforts by CINCPOA in the latter stages of the war. Commander Mathas concluded with the argument that base planning needed the same attention to detail as operational planning. Capt. Henry Eccles, head of the Department of Logistics, gave a lecture on 1 May on "The Development of Over-Seas Bases." Captain Eccles, the architect of the logistical aspects of the Naval War College's early postwar curriculum, spoke on the vital importance of the Navy's bases for its prosecution of the war, command relationships and organization, planning, transportation, allocation of real estate, construction,

Lt. Col. Michael Sampas, USMC

Lt. Cdr. Valenti Holzapfel, USN

Cdr. Richard Antrim, USN

sanitation, personnel, floating versus shore-based facilities, the future, and even the "psychological" aspects of base development, which he called "esprit de corps."[15]

Finally, there came a miscellaneous group of three lectures. On 16 July 1946, Lt. Col. Michael Sampas, U.S. Marine Corps, an instructor in the Department of Intelligence, presented on photographic reconnaissance—support of amphibious operations, British wartime methods of aerial photography, types of cameras used by American combat aircraft, and especially training.

The next day, Lt. Cdr. Valenti Holzapfel, a junior-class instructor, gave a lecture on "Rockets," including a history of British development of the weapon, the types of air and surface rockets in use in 1946, and trends for future use in shore bombardment, ground assault, and aerial attack.

Finally, on 20 February 1947, Cdr. Richard Antrim, a student in the senior class of June 1947, presented "Observed Effect of International Law on the Japanese," a lecture that looked at the role, and sometimes ineffectiveness, of international law during wartime, giving especially examples of Japanese violations of international law in the treatment of prisoners and wounded.[16]

Spruance and the Welcome Aboard

In mid-July 1946, Admiral Spruance welcomed the students of the senior and junior classes of June 1947 to the Naval War College. In his remarks he informed them that an atomic bomb had been exploded the day before at Bikini Atoll, in the Marshall Islands, in a "test of great proportions which should do much to further the science of naval warfare." Spruance argued that while the present was being called the "Atomic Era," that was simply a new name for something that had been going on for centuries—the "gradual increase" in the effectiveness of weapons of war. Over the centuries the basic principles of war had not changed, only their applicability. One of the main purposes of the Naval War College was, Spruance pointed out, to offer students the opportunity to master these principles and apply them so that the services could prepare to defend the country and in wartime defeat its enemies through "proper planning, clear thinking, sound decisions and skillful application of the means available."

Spruance informed the students that the Naval War College was a "school of ideas" and that an important "device" to that end, developed through the years, was the Estimate of the Situation. The more ideas that were expounded, the more useful the Naval War College became. Exposition, however, was better than argument: "State your case and then let the other person state his but if you cannot get to this level then argument is better than no ideas at all." Naval War College staff and student officers were all "students of war" in a setting of "freedom of expression."[17]

The eleven-month course, the first since 1940–41, he explained to the students, was meant to "gradually unroll" their minds, through reading, seeing, and hearing a panorama of both successful and unsuccessful warfare. The basis for their development and, he thought, their greatest opportunity was the series of "problems" that had been devised for the year. Much effort had been put into the problems so that the students could extract much from them. Students would be required to produce solutions and one of the solutions would be selected for play in the war-game phases of the problems. The student officers would make constant estimates, reach sound decisions, and then implement them. They would need to be fully aware of the capabilities and limitations of weapons systems to employ them adequately. The problems would exercise their abilities in ways not otherwise possible short of war. Accordingly, they were not to "spare" themselves; the more time spent on the solutions, the better. He warned them that they would generally be dissatisfied with their solutions but assured them that they would see improvement as the problems went by.[18] As a background for the year's course, Spruance told them, they would hear a series of lectures and participate in a series of studies on international affairs. "Well known" lecturers in such fields as journalism, science, diplomacy, education, and public information would appear. Each lecture would end with a question period that Spruance thought would be "quite" challenging and interesting. He also introduced them to the two thesis topics for the year and generally reiterated for them the "wonderful" opportunity they had to improve themselves for the good of their services. Except for the first week in July, they would be given Wednesday afternoons off during July, August, and September so they and their families could enjoy Newport; Spruance was certain they would enjoy the town, the weather, and the people. In fact, he was equally certain that their families would have "improved" themselves by association with the local community.[19]

The Naval Communications Course

The Naval Communications section of the course occurred between 8 and 17 August 1946. On the first day of the course, students drew publications from the archives, went through an orientation, studied the topic "General Communications," and participated in seminars on that subject. Study and seminar discussion continued each day of the course, on most days interspersed with lectures on topics such as "Employment of Naval Aircraft vs. Warships," "Radio Communications,"

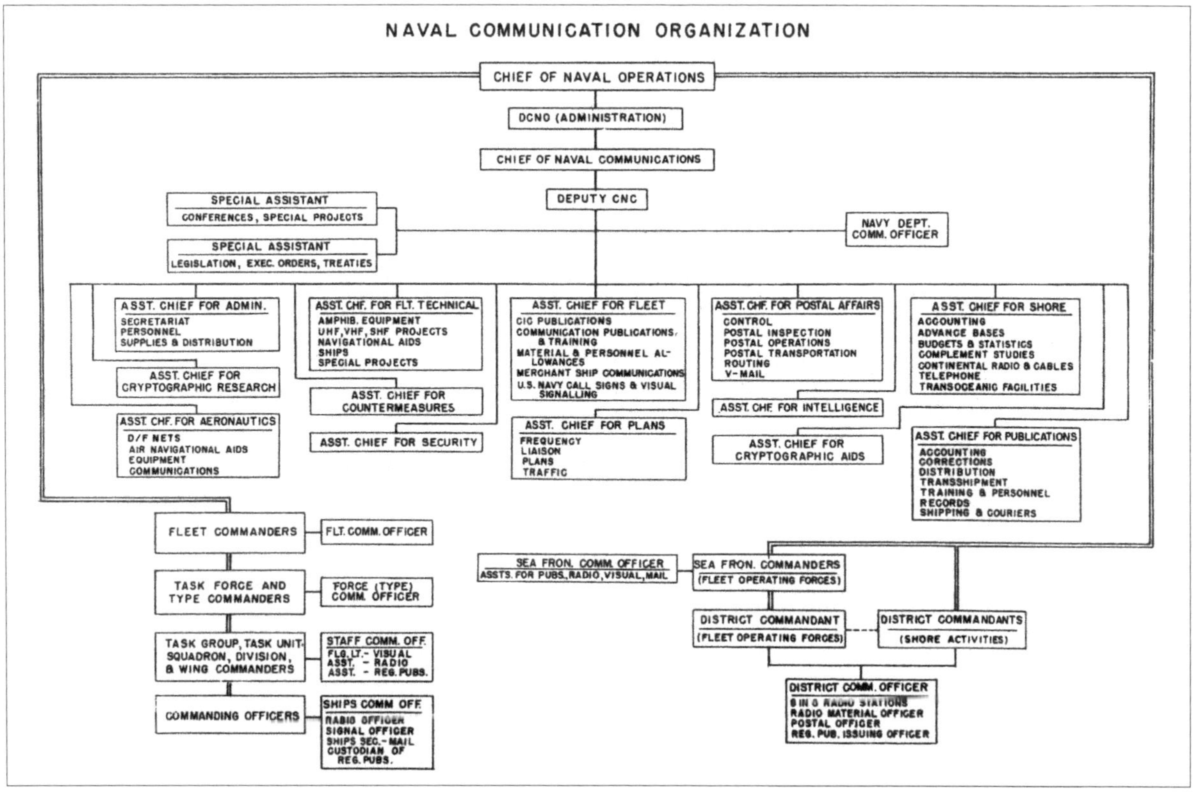

Fig. 2
Naval Communication Organization

"Communication Security," "Communication Intelligence," "Amphibious Communications," and "Communications and Electronics in the Exercise of Command." Organization charts introduced students to the role of naval communications in the larger Navy. Lecture, study, and seminar activity were also devoted to "Voice Procedure Recordings," "Radio Telephone Procedure Drill," and "Message Drafting Exercises." The section ended with a seminar, the Operations Problem Communication Planning Exercise, and a final postproblem seminar.[20]

Numerous references were listed for the students to use, including *Navy Regulations,* the *U.S. Navy Visual Call Sign Book,* the *U.S. Navy Voice Call Sign Book,* the *U.S. Navy Mail Service Manual,* the *U.S. Navy Radio Call Sign Book, Notes on Communication Security, Communications Instructions, Wartime Instructions for Merchant Ships* (both visual and radio), and the *International Code of Signals* (again, both visual and radio). Most of the students were issued one copy of each publication. Seminars were conducted in the Pringle Auditorium, the exercises and planning seminars in the Maneuver Room of Pringle Hall. Except when it joined the senior class in the auditorium, the junior class would conduct its activities in the Old Maneuver Room in Luce Hall.[21]

The "General Objective" was to familiarize student officers with the capabilities, limitations, and techniques of naval communication as they affected the exercise of

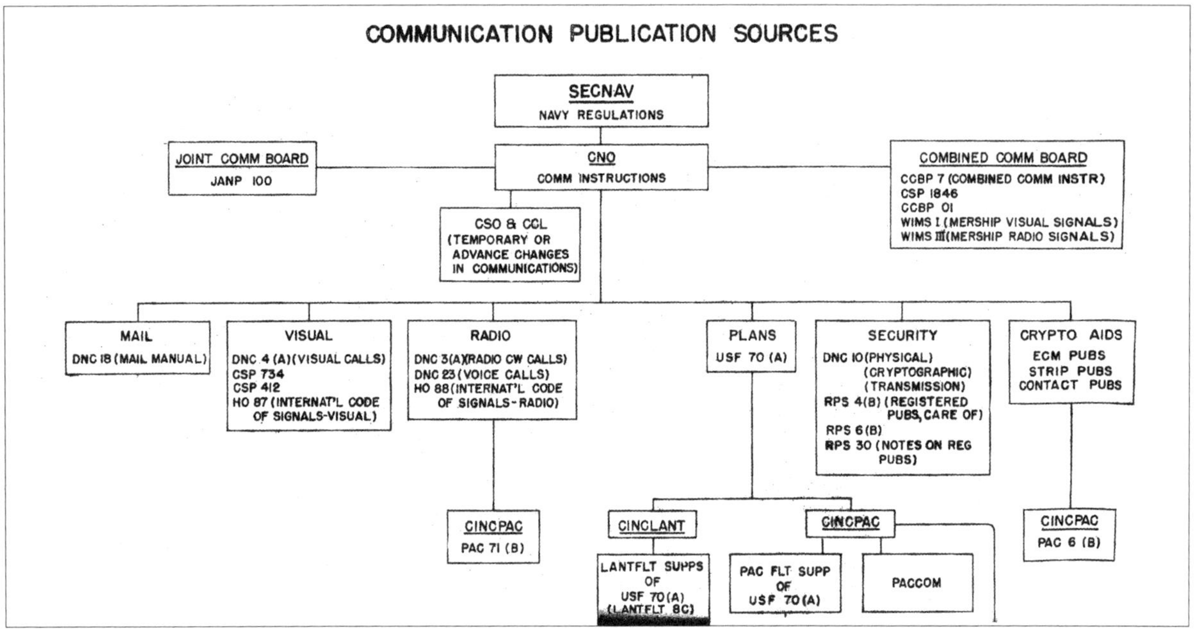

Fig. 3
Communication Publication Sources

command in naval, joint, and combined operations. "Special Objectives" included appreciation of the vital role of communications in the exercise of command and of the coordination required between operational and communication planning and training; of the dependence of command on communications in the conduct and supervision of operations; and of the potentials of communications security and communications intelligence. "Special Objectives" also included, in this case, understanding of the complexities of communication problems, of the problems involved in message drafting and in radiotelephone procedure and control, and of the principles of communication security.[22]

A schedule of events indicates how intensive these courses were. On the first day, for instance, students received staff lectures on "Some Communications Failures Vital to Command during World War II" as an introduction to communications publications. Study of these references was followed late the same morning by seminars on "Items of Communication Knowledge Essential to Flag Officers and Staff Officers" and "Items of Communication Knowledge Essential to Commanding Officers." These seminars involved student officers in discussion and were followed by additional reading and seminars on the duties of shipboard communication officers and staff communication officers. The second day of the course involved more readings, a lecture on the "Employment of Naval Aircraft vs. Warships," and seminars on "Naval Aircraft and Submarine Force Communications." In addition, Army student officers, not familiar with Navy topics like visual signaling, assembled in the Old Maneuver Room for a training film. The third day focused on communications related to convoys and merchant ships, as well as communications of "importance" to infantry division and airborne division commanders. The day concluded with reading material on "Radio Material and Wave Propagation," as well as seminars on "Some Aspects of Radio Wave Propagation Associated with Command Functions" and "Electronic Logistics." The fourth day entailed additional readings on fleet and shore radio organization, seminars on shore radio organization and fleet radio communications, and readings on radio procedures. The students then heard recordings of radiotelephone procedures in the auditorium and then participated in a "Radio Telephone Procedure Drill."[23]

The fifth day of the course saw readings and then seminar discussions on radio countermeasures, communications security, and physical security, as well as a lecture on "Communications and Electronics in the Exercise of Command." The students received lectures on cryptographic aids and communications intelligence, as well as readings on cryptographic and transmission security,

Assignment of Subjects for Talks

by Student Officers

NAVAL COMMUNICATIONS

(8 August - 16 August, 1946)

Note: 1) A talk may supplement material previously studied, or it may be the main source of information on the subject for the class. Examine the "Detailed Schedule and Assignments" and determine the relationship of your subject to the material read and to the subjects of other talks. Govern yourself accordingly.

2) In general, a talk should be based on personal experience or observation, supplemented by reading of pertinent reference material, which may or may not have been included in the class reading assignments.

3) Talks should be delivered within the time allowance indicated, in order that there may be some time available for subsequent discussion.

Senior Class

	Subject	Time	Assigned to
1.	Items of Communication Knowledge Essential (a) to Flag Officers and (b) to Staff Officers	15 m	Capt. Ricketts
2.	Items of Communication Knowledge Essential to Commanding Officers.	15 m	Capt. Madeira
3.	Duties of the Ship's Communication Officer.	15 m	Comdr. Hay
4.	Duties of a Staff Communication Officer.	15 m	Comdr. Hay
5.	Organization of the Navy Mail System.	10 m	Capt. Taylor
6.	Visual Publications and Visual Communications.	15 m	Comdr. Matter
7.	Naval Aircraft Communications.	15 m	Comdr.Roudebush
8.	Submarine Force Communications.	15 m	Comdr. Stevens
9.	Convoy and Merchant Ship Communications.	15 m	Capt. Heming
10.	Importance of Communications to an Infantry Division Commander.	15 m	Col. Johnson
11.	Importance of Communications to an Airborne Infantry Division Commander.	10 m	Col. Bidwell
12.	Some Aspects of Radio Wave Propagation Associated with Command Functions.	30 m	Capt. Goulett
13.	Electronics Logistics.	20 m	Capt.Goulett
14.	Permanent Shore Radio Organization.	20 m	Capt.Stryker
15.	Communications During Movement to Objective.	20 m	Capt.Sweeney
16.	Pre Dog Day Naval Gunfire Communications.	20 m	Capt. Price
17.	Naval Attack Force Communications.	20 m	Capt. Higgins
18.	Landing Force Communications.	20 m	Lt.Col. Shell
19.	Naval Gunfire Support Communications.	20 m	Capt. Price
20.	Radar and Radar-Countermeasures.	20 m	Capt. Smith

Fig. 4
Senior class student talks, Naval Communications Course, August 1946

Junior Class

	Subject	Time	Assigned to
1.	Communication Publications and Their Relationship.	15 m	Comdr. Foley
2.	Items of Communication Knowledge Essential to Commanding Officers.	15 m	Comdr. Dalton
3.	Duties of the Ship's Communication Officer.	15 m	Comdr. Nibbs
4.	Duties of a Staff Communication Officer.	15 m	Comdr. Church
5.	Visual Publications and Visual Communications.	15 m	Comdr. Rogers
6.	Naval Aircraft Communications.	15 m	Comdr. Owers
7.	Submarine Force Communications.	15 m	Lt.Com.Taeusch
8.	Convoy and Merchant Ship Communications.	15 m	Comdr.Loveland
9.	Importance of Communications to an Infantry Division Commander	15 m	Lt.Col.Downey
10.	Pre Dog Day Naval Gunfire Communications.	20 m	Lt.Cdr.Bonner
11.	Pre Dog Day Air Support Communications.	20 m	Comdr. Richardson
12.	Joint Expeditionary Force Communications.	20 m	Comdr. Kirkpatrick
13.	Naval Attack Force Communications.	20 m	Comdr. Kirkpatrick
14.	Landing Force Communications.	20 m	Col. Edson
15.	Naval Gunfire Support Communications.	20 m	Lt.Cdr.Bonner
16.	Air Support Communications.	20 m	Comdr. Richardson
17.	Fighter Direction at Assault Area.	20 m	Comdr.Baldridge
18.	Radar and Radar-Countermeasures.	20 m	Lt.Cdr. Howell

Fig. 5
Junior class student talks, Naval Communications Course, August 1946

Rear Adm. (in photo, Captain) Thomas Inglis, USN (right); other man unidentified

Capt. (in photo, Vice Admiral) Carl Espe, USN

while the day after that saw a message-drafting exercise and additional readings and lectures on communications security. The second-to-last day of the course was focused on amphibious communications, specifically readings and seminars on "Planning and the Movement to an Objective," "Naval Gunfire," "Air Support," "Joint Expeditionary Force," "Naval Attack Force," "Landing Force," "Fighter Direction," and "Radar and Radar-Countermeasure Communications." The final day finished with staff lectures on the "Mechanics of NWC Communications," "Operation Problem Communication Plans," an exercise in "Operation Problem Communication Plans," and a seminar discussion.[24]

Guidance entitled "Assignments of Subjects for Talks by Student Officers" suggested that student talks might supplement material previously studied or even serve as the main sources of new information. A student talk was to be based on personal experience or observation, supplemented by the reading of "pertinent" reference material, which might but need not be course readings. Talks were to leave time for subsequent discussion. Finally, a short staff questionnaire on the benefits of the Naval Communications Course was given to solicit recommendations for changes or improvements.[25]

The Naval Intelligence Course

Late in August 1946, Admiral Smith prepared instructions for a course in Naval Intelligence Studies for the junior and senior classes of June 1947. These instructions again illustrate how the College's curriculum operated in conjunction with the Operations Problems. Phase I of the course was oriented toward both classes; only the senior class would participate in Phase II. Phase I would entail lectures on naval intelligence by representatives of the Office of Naval Intelligence (ONI); study of the publication *Naval Intelligence, February 1946,* with seminars led by ONI officers; study of intelligence annexes from World War II operations plans; and preparation and presentation of intelligence estimates and plans by student officers. The overall course studied the application of procedures and practices in intelligence during the war by some of the more successful American naval commanders. It also served as an exercise in the practical use of the operational functions of command and in the solution of "typical" military problems.[26]

Phase I took place between 4 and 9 September. The first day saw the drawing of publications, an orientation from Captain Hartung (also the officer in charge [OINC] of the Command and Staff class of June 1947), and a lecture from Rear Adm. Thomas Inglis, Chief of Naval Intelligence, on "Intelligence: Organization and Mission."

The next day, students received a lecture on "Strategic and Operational Intelligence" from Capt. Carl Espe, head of the Operational Intelligence Branch of the Office of the Chief of Naval Operations (OPNAV); Captain Espe led a seminar on the topic later in the day. The 6th followed the same pattern, with the study of chapters in *Naval Intelligence, February 1946* on naval air intelligence and amphibious intelligence, then lectures on those subjects by, respectively, Cdr. John Lamade, assistant to the OINC of OPNAV's Air Intelligence Branch, and Capt. John Mosher, U.S. Naval Reserve, an instructor in intelligence at the National War College. Later in the day, Captain Mosher and Commander Lamade led seminar discussions on their lecture topics. The fourth day of Phase I brought a staff lecture on "Current Intelligence," along with the study of further chapters; the last day of Phase I saw additional readings, a lecture on "Current Intelligence" by Capt. Wallace Wharton, a reservist, and a seminar discussion led by Captain Wharton.[27]

Phase II was conducted for the senior class between 10 and 13 September, with the first day devoted to an orientation and a study of operations plans, specifically their intelligence annexes. The next day, Phase II groups prepared for a later seminar by reviewing the intelligence annexes of a number of operations plans and heard a staff presentation on the "Search for and Sinking of the *Bismark* [*sic*]." Day three was devoted to additional study and preparation for the next day's seminar, which would involve individual as well as group presentations. The project divided the senior class into twelve groups of five to six students each, each group noting the "good" and "bad" points of its assigned plan. Groups 1 and 2 took Operation FORAGER, the invasion of the Marianas, while Groups 3 and 4 handled Operation STALEMATE II, the assaults on Palau and Morotai. Groups 5 and 6 had Operation INDUCTION, the invasion of Lingayen in the Philippines, Groups 7 and 8 Operation DETACHMENT, the seizure of Iwo Jima. Operation

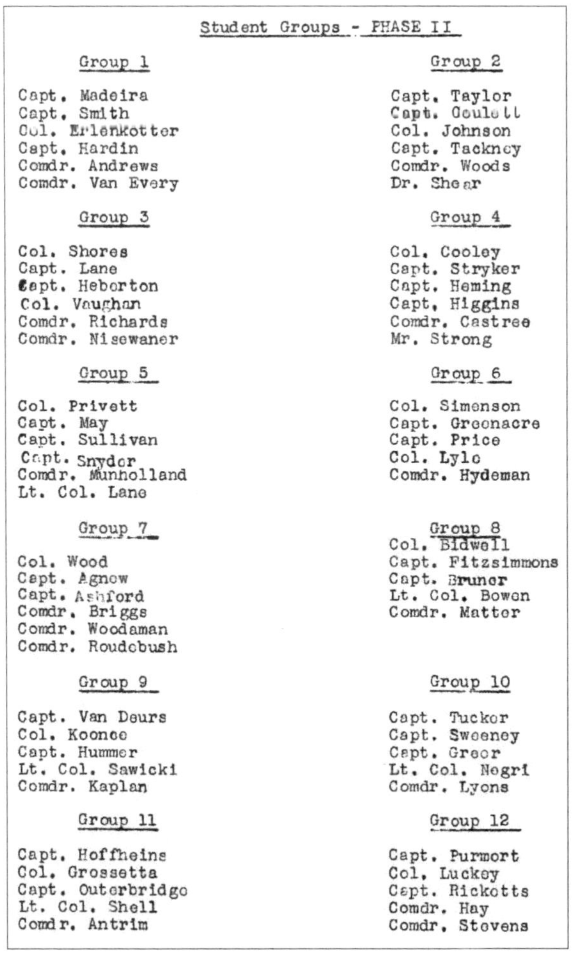

Student Groups - PHASE II

Group 1

Capt. Madeira
Capt. Smith
Col. Erlenkotter
Capt. Hardin
Comdr. Andrews
Comdr. Van Every

Group 2

Capt. Taylor
Capt. Goulett
Col. Johnson
Capt. Tackney
Comdr. Woods
Dr. Shear

Group 3

Col. Shores
Capt. Lane
Capt. Heberton
Col. Vaughan
Comdr. Richards
Comdr. Nisewaner

Group 4

Col. Cooley
Capt. Stryker
Capt. Heming
Capt. Higgins
Comdr. Castree
Mr. Strong

Group 5

Col. Privett
Capt. May
Capt. Sullivan
Capt. Snyder
Comdr. Munholland
Lt. Col. Lane

Group 6

Col. Simenson
Capt. Greenacre
Capt. Price
Col. Lyle
Comdr. Hydeman

Group 7

Col. Wood
Capt. Agnew
Capt. Ashford
Comdr. Briggs
Comdr. Woodaman
Comdr. Roudebush

Group 8

Col. Bidwell
Capt. Fitzsimmons
Capt. Bruner
Lt. Col. Bowen
Comdr. Matter

Group 9

Capt. Van Deurs
Col. Koonce
Capt. Hummer
Lt. Col. Sawicki
Comdr. Kaplan

Group 10

Capt. Tucker
Capt. Sweeney
Capt. Greer
Lt. Col. Negri
Comdr. Lyons

Group 11

Capt. Hoffheins
Col. Grossetta
Capt. Outerbridge
Lt. Col. Shell
Comdr. Antrim

Group 12

Capt. Purmort
Col. Luckey
Capt. Ricketts
Comdr. Hay
Comdr. Stevens

Fig. 6
Phase II student groups

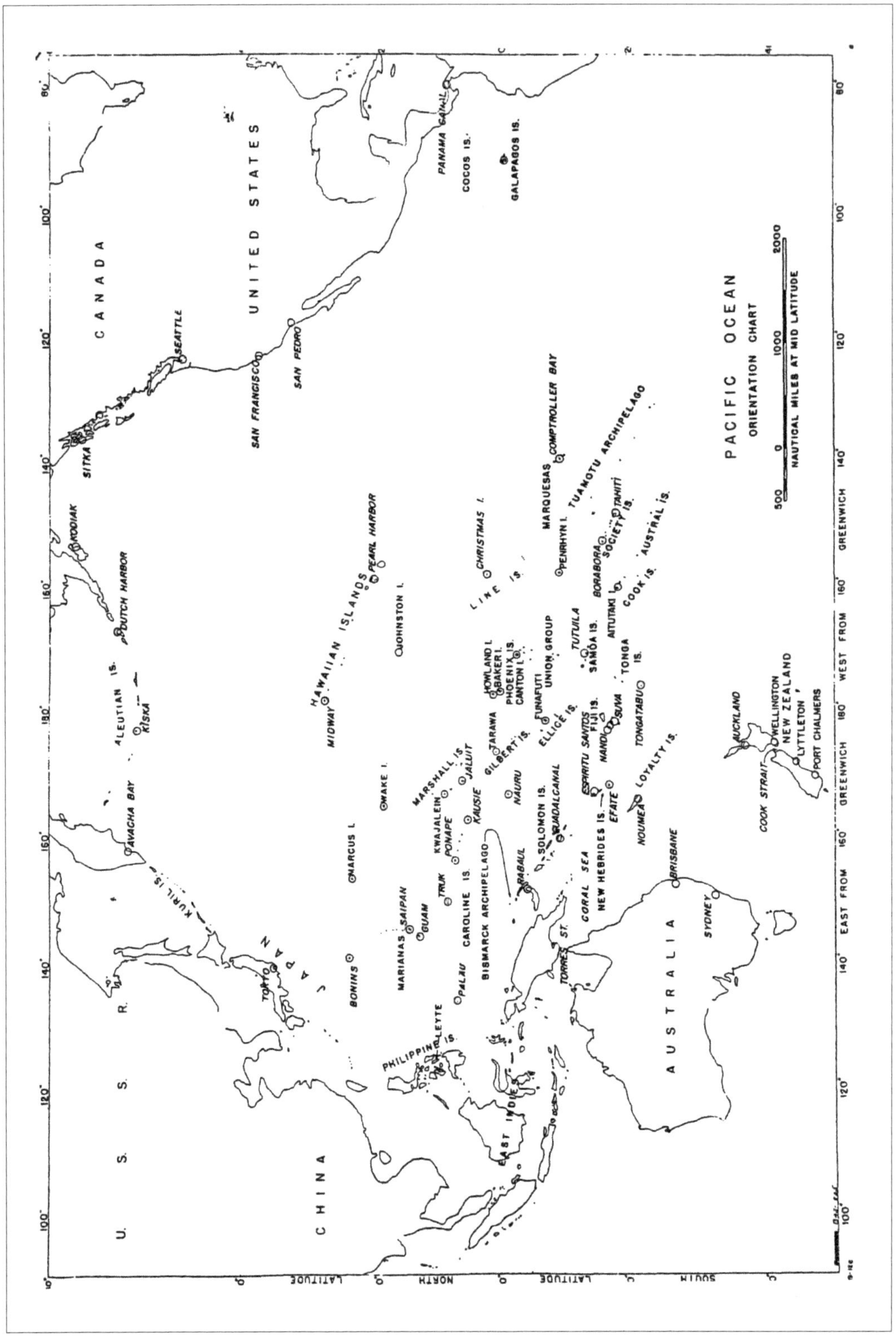

Map 1
The Pacific basin

Iceberg, the invasion of Okinawa, was done by Groups 9 and 10. Finally, the intelligence annex for the projected (but of course never conducted) invasion of the home islands of Japan, Operation Olympic, was reviewed by Groups 11 and 12.[28]

The students, acting as if they were the intelligence officers on the staffs of the theater commanders involved, analyzed joint staff studies, fleet intelligence bulletins, and studies of terrain, theater, and engineering aspects, etc., related to the operations and their requirements; studied the intelligence material available; and considered the intelligence that had been available at the time to the Joint Intelligence Center, Pacific Ocean Areas. Student officers then prepared intelligence estimates for the operation and a plan for obtaining the needed operational information. Students subsequently compared their work with the actual intelligence annexes and prepared brief comments on the strengths and weaknesses of their own work, again by the criteria of *Naval Intelligence, February 1946.* One student from each group would be selected to present his estimate, plan, and comment for class discussion.[29]

The Theses Again

In early January 1947, Admiral Smith issued to the staff a procedural reminder about the approaching deadline for the theses on "Relations between Russia and the United States, and Their Influence on U.S. Foreign Policy." The originals of the theses were to be distributed to assigned reviewing officers for comment (but without actual comments being written on the original theses) and then finally approval by Admiral Smith.[30] Three weeks later, Admiral Smith issued another memo to the Naval War College staff about the next step in the process. By this time, the Topic 1 theses had been read once; the reviewing officers were now "desired" to give them a second reading to ensure "correct evaluation" and to prepare for a critique.[31]

By early April, the staff and students were preparing for the Critique on Topic 1. Its purpose was to assist the staff and students in synthesizing their ideas on Russia into a concept that would be useful in their service relations, their duties, their civilian interactions, and the creation of an understanding of Russia that the United States had not enjoyed concerning Japan in World War II (understandings that today, at least, College representatives might be asked to brief to policy makers in Washington, D.C.). The "General Plan" was to lay a foundation during the first half of the period allotted for the Critique in historical, geographical, "racial" (Smith's word), military, political, and economic contexts through Staff Presentations and guest lecturers. Student presentations would then take place to achieve a consensus of student thought, one that represented both conventional and unusual approaches. In effect, the goal was to create a "Russian Estimate," or a military "Estimate of the Situation," concerning American relations with the Soviet Union.[32]

In the second reading, the reviewing officers were not to assign marks but merely to comment briefly on merit, especially the quality of the development of ideas

\- 1 -

Encl. (A) to N.W.C.
Memo of 10 Jan. 1947

THESES READING DETAIL

COMMODORE DEES

Capt. D. L. Madeira, USN
Capt. E. A. Taylor, USN

CAPTAIN JOHNSON

Col. Von. R. Shores, USA
Col. G. P. Privett, USA

CAPTAIN EVANS

Col. S. E. Manzo, USA
Col. H. J. Hawthorne, USA

CAPTAIN MORAN

Capt. W.B. Goulett, USN
Capt. J. M. Lane, USN
Capt. J. W. Stryker, USN
Capt. E. F. May, USN

CAPTAIN DOLE

Capt. J. D. Sweeney, USN
Col. A.V. Grossetta, USA
Col. R. B. Luckey, USMC
Col. R. Erlenkotter, USA

CAPTAIN SULLIVAN

Capt. H. M. Heming, USN
Capt. E. J. Sullivan, USN

CAPTAIN CROSLEY

Capt. J. D. Greer, USN
Capt. W.W. Outerbridge, USN
Capt. C.V. Ricketts, USN
Capt. J.T. Hardin, USN

CAPTAIN BIERER

Lt. Col. W. W. Vaughan, USA
Comdr. T. V. Briggs, RN
Lt. Col. R. C. Bowen, USMC
Lt. Col. S. Sawicki, USA

COLONEL SCHMIDT

Col. A. D. Cooley, USMC
Capt. G. Van Deurs, USN
Capt. T. T. Tucker, USN

CAPTAIN HARTUNG

Col. E. G. Simenson, USA
Col. R. L. Wood, USA

COLONOL BARE

Col. B. W. Bidwell, USA
Capt. W. L. Hoffheins, USN
Capt. G. L. Purmort, USN
Capt. R. S. Smith, USN

CAPTAIN MONTGOMERY

Capt. A. J. Greenacre, USN
Capt. D. M. Agnew, USN
Capt. J. P. Fitzsimmons, USN
Col. A. C. Koonce, USMC

CAPTAIN SWEENEY

Col. W. P. Johnson, USA
Capt. C. M. Heberton, USN

CAPTAIN McKILLIP

Capt. W. H. Price, USN
Capt. F. Bruner, USN
Capt. W. H. Ashford, USN
Capt. H.R. Hummer, USN

CAPTAIN McILHENNEY *

Capt. S.N. Tackney, USN
Capt. P.W. Snyder, USN
Capt. R.D. Higgins, USN
Col. L.E. Lyle, USA

LT. COL. SAMPAS

Lt. Col. P. J. Negri, USMC
Lt. Col. G.R.E. Shell, USMC
Comdr. R. R. Hay, USN
Comdr. R. S. Andrews, USN

\- 2 -

THESES READING DETAIL (Cont'd.)

Encl. (A) to N.W.C.
Memo. of 10 Jan. 1947

COMMANDER ACKER

Comdr. W. P. Woods, USN
Comdr. W. L. Richards, USN
Comdr. J. F. Castree, USN
Comdr. J. Munholland, USN

COMMANDER MAYBERRY

Col. H. C. Edson, USA
Col. R. J. Mason, USA
Lt. Col. L. Wallace, USA
Lt. Col. M. H. Thompson, USA

COMMANDER DAILY

Comdr. C. B. Stevens, USN
Comdr. S. A. Van Every, USN
Comdr. T. A. Nisewaner, USN
Lt. Col. J. J. Lane, USA

COMMANDER THOMSON

Comdr. W. W. Bridewell, USN
Comdr. J. E. Owers, USN
Comdr. R. C. Morton, USN
Comdr. C. A. O'Connell, USN

COMMANDER JOHNSON

Comdr. C. D. Simonsen, USN
Comdr. N. C. Gillette, USN
Comdr. W. S. Kirkpatrick, USN

COMMANDER HASKELL

Comdr. R. W. Rynd, USN
Comdr. E. F. Baldridge, USN
Comdr. H. J. P. Foley, USN
Comdr. G. F. Dalton, USN

LT. COMDR. CURRAN

Comdr. R. P. Webber, USN
Lt. Cdr. E. P. Bonner, USN
Lt. Cdr. J. P. Aymond, USN
Lt. Cdr. J. D. Ramage, USN

LIEUT. COL. DUNLAP

Comdr. E. T. Hydeman, USN
Comdr. R. J. Woodaman, USN
Comdr. A. R. Matter, USN

COMMANDER PELLING

Comdr. A. D. Kaplan, USN
Comdr. C. M. Lyons, USN
Comdr. R. N. Antrim, USN

COMMANDER STALEY

Comdr. J. Roudebush, USN
Mr. R. C. Strong
Comdr. G. S. Coleman, USN

COMMANDER KIRKPATRICK

Comdr. K. Loveland, USN
Comdr. A. M. Nibbs, USN
Comdr. G. H. Miller, USN

COMMANDER MATHAS

Lt. Col. S. W. Downey, USA
Comdr. D. C. Richardson, USN
Lt. Col. R.E. Leary, USA
Comdr. C. B. Lundgren, USN

COMMANDER REED

Comdr. H. Marvin-Smith, USN
Comdr. W. K. Rogers, USN
Comdr. A. T. Church, USN

LT. COMDR. HOLZAPFEL

Lt. Col. F. L. Street, USA
Comdr. T. S. White, USN
Lt. Cdr. W. Y. Howell, USN
Lt. Cdr. F. L. Taeusch, USN

*Note: If Capt. McIlhenny is not available, the Head, of the Intelligence Department will reassign his theses within the Intelligence Department.

Fig. 7
Theses Reading Detail, Topic 1, first reading

(all in language suitable for transcription as the chief of staff's comment to the student). Staff officers' comments were also to recommend theses for presentation in the Critique in early May, especially those showing that the student "has employed reasonable assumptions and that the development has been logical and sound."[33]

On the same day, Captain Hartung sent Capt. Gail Morgan, Naval War College Secretary, a memo warning of the need for precautions when students had civilians outside the College (or perhaps their wives) type their theses. Hartung reminded Morgan that theses could use classified material, but if they did they had to be classified and cleared through the head of the Department of Intelligence. Hartung wanted student officers cautioned about fully observing security regulations in hiring uncleared civilian typists.[34]

Enc.(B) to N.W.C. Memo.
of 10 January, 1947

Comments on Thesis

Writer ______________________ Senior Class / Junior Class

Subject: Relations between Russia and the United States, and their Influence on U.S. Foreign Policy.

Reader ______________________

____

Signature of Reader

Fig. 8
Comments on Thesis, Topic 1, first reading

Encl. (A) to N.W.C. Memo.
of 31 Jan. 1947

THESES READING DETAIL

CAPTAIN DEES

Col. A.D. Cooley, USMC
Capt. G. Van Deurs, USN

CAPTAIN JOHNSON

Capt. T.T. Tucker, USN
Col. E. G. Simenson, USA
Col. R.L. Wood, USA

CAPTAIN EVANS

Col. E.O. Edson, USA
Col. R.J. Mason, USA

CAPTAIN MORAN

Col. B.W. Bidwell, USA
Capt. W.L. Hoffheins, USN
Capt. G.L. Purmort, USN

CAPTAIN DOLE

Capt. R.S. Smith, USN
Capt. A.J. Greenacre, USN
Capt. D.M. Agnew, USN

CAPTAIN SULLIVAN

Capt. J.P. Fitzsimmons, USN
Col. A.C. Koonce, USMC
Col. W.P. Johnson, USA

CAPTAIN CROSLEY

Capt. C.M. Heberton, USN
Capt. W.H. Price, USN
Capt. F. Bruner, USN

CAPTAIN BIERER

Capt. W.H. Ashford, USN
Capt. H.R. Hummer, USN
Capt. S.N. Tackney, USN

COLONEL SCHMIDT

Capt. D.L. Madeira, USN
Capt. E.A. Taylor, USN
Col. Von R. Shores, USA

CAPTAIN HARTUNG

Col. G.P. Privett, USA
Capt. W.B. Goulett, USN

COLONEL BARE

Capt. J.M. Lane, USN
Capt. J.W. Stryker, USN
Capt. E.F. May, USN

CAPTAIN MONTGOMERY

Capt. J.D. Sweeney, USN
Col. A.V. Grossetta, USA
Col. R.B. Luckey, USMC

CAPTAIN SWEENEY

Col. R. Erlenkotter, USA
Capt. H.M. Heming, USN
Capt. E.J. Sullivan, USN

CAPTAIN McKILLIP

Capt. J.D. Greer, USN
Capt.W.W.Outerbridge,USN
Capt. C.V. Ricketts, USN

CAPTAIN McILHENNEY

Capt. J.T. Hardin, USN
Capt. P.W. Snyder, USN
Capt. R.D. Higgins, USN

CAPTAIN HOWE

Col. L.E. Lyle, USA
Lt.Col. W.W. Vaughan, USA
Comdr. T.V. Briggs, RN

LIEUT. COLONEL SAMPAS

Lt.Col. R.O. Bowen, USMC
Lt. Col. S. Sawicki, USA
Comdr. W.P. Woods, USN

LIEUT. COLONEL DUNLAP

Comdr. R.S. Andrews, USN
Comdr. W.L. Richards, USN
Comdr. J.F. Castree, USN

COMMANDER PELLING

Comdr. C.B. Stevens, USN
Comdr. S.A. Van Every, USN
Comdr. T.A. Nisewaner, USN
Comdr. Munholland, USN

COMMANDER STALEY

Comdr. C.M. Lyons, USN
Comdr. R.N. Antrim, USN
Lt.Col. L. Wallace, USA
Lt.Col. M.H. Thompson, USA

COMMANDER KIRKPATRICK

Lt.Col. J.J. Lane, USA
Comdr. W.W. Bridewell, USN
Comdr. J.E. Owers, USN
Comdr. R.C. Morton, USN

COMMANDER MATHAS

Comdr. C.D. Simonsen, USN
Comdr. N.C. Gillette, USN
Comdr. W.S. Kirkpatrick, USN
Comdr. H. Marvin-Smith, USN

COMMANDER HASKELL

Comdr. W.K. Rogers, USN
Comdr. A.T. Church, USN
Lt.Col. F.L. Street, USA
Comdr. T.S. White, USN

LIEUT. COMDR. CURRAN

Comdr. H.J.P. Foley, USN
Comdr. G.F. Dalton, USN
Lt. Cdr. W.Y. Howell, USN
Lt. Cdr. F. L. Taeusch, USN

COMMANDER ACKER

Lt.Col. P.J. Negri, USMC
Lt.Col. G.R.E.Shell, USMC
Comdr. R.R. Hay, USN

COMMANDER MAYBERRY

Col. S.E. Manzo, USA
Col. H.J. Hawthorne, USA
Comdr. K. Loveland, USN

COMMANDER DAILY

Comdr. E.T. Hydeman, USN
Comdr. R.J. Woodaman, USN
Comdr. A.R. Matter, USN
Comdr. A.D. Kaplan, USN

COMMANDER THOMSON

Comdr. J. Roudebush, USN
Mr. R.C. Strong
Comdr. G.S. Coleman, USN
Comdr. A.M. Nibbs, USN

COMMANDER JOHNSON

Comdr. G.H. Miller, USN
Comdr. C.A. O'Connell, USN
Lt.Col. S.W. Downey, USA
Comdr. D.C. Richardson, USN

COMMANDER REED

Lt.Col. R.E. Leary, USA
Comdr. O.B. Lundgren, USN
Comdr. R.W. Rynd, USN
Comdr. E.F. Baldridge, USN

LIEUT. COMDR. HOLZAPFEL

Comdr. R.P. Webber, USN
Lt. Cdr. E.P. Bonner, USN
Lt. Cdr. J.P. Aymond, USN
Lt. Cdr. J.D. Ramage, USN

Fig. 9
Theses Reading Detail, Topic 1, second reading

The same format for a thesis critique was used about a month later for Topic 2. Here, the focus was on ideas about the capabilities and limitations of the atomic bomb so that strategic and tactical concepts could be developed for the employment of the weapon, as well as insight into its possible influence on future naval warfare. In this case, the Staff Presentations would be on the results of Operation CROSSROADS (the 1946 Bikini tests) and the probable impact of the atomic bomb on fleet disposition and movement. The presentations from eight "outstanding" student theses presented a "cross-section" of student thought by officers from various branches and arms.[35]

Enc.(B) to N.W.C. Memo.
of 31 January, 1947

Comments on Thesis

Writer ______________________ Senior Class / Junior Class

Subject: Relations between Russia and the United States, and their Influence on U.S.Foreign Policy.

Reader ______________________

____

____

____

____

____

____

____
Signature of Reader

Fig. 10
Comments on Thesis, Topic 1, second reading

Spruance and the Full Circle of Graduation

In early June 1947, Admiral Spruance gave a short address to the graduating classes before introducing Fleet Admiral Nimitz, the CNO, as the keynote speaker. Spruance recalled the

Capt. Gail Morgan, USN

work the students had done in the past year, especially the problems the staff had devised to test them, the lectures to which they had listened, and the theses they had written. All of these activities had been geared to bringing the students up to date with conditions as they "actually" existed in the postwar world and to set them thinking into the future. No one could foresee the future for the United States in the area of international affairs, he argued, but he was certain that global conditions would be "decidedly" unsettled for the present. He hoped that the United Nations would be able to solve issues of disagreement between nations in satisfactory and peaceful ways; he was certain that if war came, it would not be of American making. He was also certain that the United States would do everything "honorable" to avoid war but again, if it came, it would be up to officers such as those in the graduating classes to win that war with minimal losses to the United States.[36]

He further reminded them that modern wars were won through the united efforts of all the branches of the armed forces, "each contributing its share toward

Encl. (A) to N.W.C.
Memo of 22 April 1947

ATOMIC BOMB

THESES READING DETAIL

FIRST READING

COLONEL SCHMIDT

Capt. W.B. Goulett, USN
Capt. J.M. Lane, USN
Capt. J.W. Stryker, USN

CAPTAIN EVANS

Comdr. K. Lovcland, USN
Comdr. G.H. Miller, USN
Comdr. R.P. Webber, USN

CAPTAIN ECCLES

Col. G.P. Privett, USA
Capt. E.F. May, USN
Col. R. Erlenkotter, USA

CAPTAIN MONTGOMERY

Capt. J.D. Greer, USN
Capt. W.W. Outerbridge, USN
Capt. C.V. Ricketts, USN

CAPTAIN SWEENEY

Capt. J.D. Sweeney, USN
Col. A.V. Grossetta, USA
Col. R.B. Luckey, USMC

CAPTAIN McKILLIP

Capt. J.P. Fitzsimmons, USN
Col. W.P. Johnson, USA
Capt. J.T. Hardin, USN

CAPTAIN CROSLEY

Capt. P.W. Snyder, USN
Col. A.C. Koonce, USMC
Capt. R.D. Higgins, USN

CAPTAIN BIERER

Col. B.W. Bidwell, USA
Lt.Col. P.J. Negri, USMC
Lt.Col. G.R.E. Shell, USMC

CAPTAIN HARTUNG

Col. A.D. Cooley, USMC
Capt. G. Van Deurs, USN

COLONEL BARE

Capt. T.T. Tucker, USN
Col. E.G. Simenson, USA
Col. R.L. Wood, USA

CAPTAIN MORAN

Capt. D.L. Madeira, USN
Capt. E.A. Taylor, USN
Col. Von R. Shores, USA

CAPTAIN DOLE

Capt. G.L. Burmort, USN
Capt. H.M. Heming, USN
Capt. E.J. Sullivan, USN

CAPTAIN SULLIVAN

Capt. R.S. Smith, USN
Capt. A.J. Greenacre, USN
Capt. D.H. Agnew, USN

CAPTAIN HOWE

Capt. C.M. Heberton, USN
Capt. W.H. Price, USN
Capt. F. Bruner, USN

CAPTAIN McILHENNEY

Capt. W.H. Ashford, USN
Capt. H.R. Hummer, USN
Capt. S.N. Tackney, USN

LT. COL. SAMPAS

Col. L.E. Lyle, USA
Lt.Col. W.W. Vaughan, USA
Comdr. T.V. Briggs, RN

Enc. (A) to N.W.C.
Memo of 22 Apr.1947

LT.COL. DUNLAP

Comdr. C.B. Stevens, USN
Comdr. S.A. Van Every, USN
Comdr. T.A. Nisewaner, USN

COMMANDER PELLING

Capt. W.L. Hoffheins, USN
Lt.Col. R.O. Bowen, USMC
Lt.Col. S. Sawicki, USA
Comdr. W.L. Richards, USN

COMMANDER STALEY

Comdr. W.F. Woods, USN
Comdr. E.T. Hydeman, USN
Comdr. R.J. Woodaman, USN
Comdr. J. Munholland, USN

COMMANDER THOMSON

Comdr. C.M. Lyons, USN
Comdr. R.N. Antrim, USN
Comdr. A.R. Matter, USN
Comdr. A.D. Kaplan, USN

COMMANDER JOHNSON

Lt.Col. M.H. Thompson, USA
Lt.Col. L. Wallace, USA
Comdr. W.W. Bridewell, USN
Comdr. J.E. Owers, USN

COMMANDER HASKELL

Lt.Col. R.E. Leary, USA
Comdr. O.B. Lundgren, USN
Comdr. W.S. Kirkpatrick, USN
Comdr. H. Marvin-Smith, USN

LIEUT. COMDR. CURRAN

Comdr. W.K. Rogers, USN
Comdr. A.T. Church, USN
Lt.Col. F.L. Street, USA
Comdr. T.S. White, USN

COMMANDER MAYBERRY

Comdr. A.M. Nibbs
Lt.Col. J.J. Lane, USA
Comdr. C.A. O'Connell, USN
Lt.Col. S.W. Downey, USA

COMMANDER DAILY

Comdr. R.R. Hay, USN
Comdr. R.S. Andrews, USN
Comdr. J.F. Castree, USN

COMMANDER KIRKPATRICK

Col. S.E. Manzo, USA
Col. H.J. Hawthorne, USA
Col. H.O. Edson, USA
Col. R.J. Mason, USA

COMMANDER MATHAS

Comdr. J. Roudebush, USN
Mr. R.C. Strong
Comdr. G.S. Coleman, USN
Lt.Cdr. E.P. Bonner, USN

COMMANDER REED

Comdr. R.C. Morton, USN
Comdr. D.C. Richardson, USN
Comdr. C.D. Simonsen, USN
Comdr. N.C. Gillette, USN

LT. COMDR. HOLZAPFEL

Comdr. R.W. Rynd, USN
Comdr. E.F. Baldridge, USN
Comdr. H.J.P. Foley, USN
Comdr. G.F. Dalton, USN

LIEUT. COMDR. BLACK

Lt.Cdr. W.Y. Howell, USN
Lt.Cdr. J.P. Aymond, USN
Lt.Cdr. J.D. Ramage, USN
Lt.Cdr. F.L. Taeusch, USN

Fig. 11
Atomic Bomb Theses Reading Detail, First Reading

Encl. (B) to N.W.C.
Memo of 22 Apr 1947

ATOMIC BOMB
THESES READING DETAIL
SECOND READING

COLONEL SCHMIDT

Col. G.P. Privett, USA
Col. E.G. Simenson, USA
Col. R.L. Wood, USA

CAPTAIN EVANS

Comdr. G.S. Coleman, USN
Comdr. A.M. Nibbs, USN
Comdr. C.A. O'Connell, USN

CAPTAIN ECCLES

Capt. T.T. Tucker, USN
Capt. J.M. Lane, USN
Capt. J.W. Stryker, USN

CAPTAIN MONTGOMERY

Col. B.W. Bidwell, USA
Capt. W.L. Hoffheins, USN
Capt. G.L. Purmort, USN

CAPTAIN SWEENEY

Capt. C.V. Ricketts, USN
Capt. A.J. Greenacre, USN
Capt. D.M. Agnew, USN

CAPTAIN McKILLIP

Col. A.C. Koonce, USMC
Capt. H.M. Heming, USN
Capt. E.J. Sullivan, USN

CAPTAIN CROSLEY

Capt. H.R. Hummer, USN
Col. L.E. Lyle, USA
Comdr. T.V. Briggs, RN

CAPTAIN BIERER

Capt. J.T. Hardin, USN
Capt. P.W. Snyder, USN
Capt. R.D. Higgins, USN

CAPTAIN HARTUNG

Capt. D.L. Madeira, USN
Capt. E.A. Taylor, USN

COLONEL BARE

Capt. W.B. Goulett, USN
Capt. G. Van Deurs, USN
Col. V.R. Shores, USA

CAPTAIN MORAN

Col. A.D. Cooley, USMC
Capt. R.S. Smith, USN
Col. R. Erlenkotter, USA

CAPTAIN DOLE

Capt. E.F. May, USN
Capt. J.D. Greer, USN
Capt. W.W. Outerbridge, USN

CAPTAIN SULLIVAN

Capt. J.D. Sweeney, USN
Col. A.V. Grossetta, USA
Col. R.B. Luckey, USMC

CAPTAIN HOWE

Capt. J.P. Fitzsimmons, USN
Col. W.P. Johnson, USA
Capt. W.H. Ashford, USN

CAPTAIN McILHENNY

Capt. C.M. Heberton, USN
Capt. W.H. Price, USN
Capt. F. Bruner, USN

LT. COL. SAMPAS

Capt. S.N. Tackney, USN
Comdr. C.B. Stevens, USN
Comdr. S.A. Van Every, USN

Encl. (B) to N.W.C.
Memo of 22 April 1947

LT. COL. DUNLAP

Lt. Col. W.W. Vaughan, USA
Lt. Col. P.J. Negri, USMC
Lt. Col. G.R.E. Shell, USMC

COMMANDER PELLING

Comdr. W.P. Woods, USN
Comdr. R.R. Hay, USN
Comdr. R.S. Andrews, USN

COMMANDER STALEY

Comdr. J.F. Castree, USN
Comdr. T.A. Misewaner, USN
Comdr. A.R. Matter, USN
Comdr. A.D. Kaplan, USN

COMMANDER THOMSON

Comdr. R.J. Woodaman, USN
Comdr. J. Munholland, USN
Col. S.E. Manzo, USA
Col. H.J. Hawthorne, USA

COMMANDER JOHNSON

Col. R.J. Mason, USA
Lt. Col. J.J. Lane, USA
Comdr. J. Roudebush, USN
Mr. R.C. Strong

COMMANDER HASKELL

Lt. Cdr. J.D. Ramage, USN
Lt. Cdr. F.L. Taeusch, USN
Comdr. D.C. Richardson, USN
Lt. Cdr. W.Y. Howell, USN

LT. COMDR. CURRAN

Comdr. R.W. Rynd, USN
Comdr. E.F. Baldridge, USN
Comdr. C.D. Simonsen, USN
Comdr. N.C. Gillette, USN

COMMANDER MAYBERRY

Comdr. G.H. Miller, USN
Comdr. R.P. Webber, USN
Comdr. W.W. Bridewell, USN
Lt. Cdr. J.P. Aymond, USN

COMMANDER DAILY

Lt.Col. R.O. Bowen, USMC
Lt. Col. S. Sawicki, USA
Comdr. W.L. Richards, USN
Comdr. E.T. Hydeman, USN

COMMANDER KIRKPATRICK

Comdr. R.C. Morton, USN
Lt.Col. S.W. Downey, USA
Lt.Col. R.E. Leary, USA
Comdr. O.B. Lundgren, USN

COMMANDER MATHAS

Comdr. C.M. Lyons, USN
Comdr. R.F. Antrim, USN
Comdr. J.E. Owers, USN
Col. H.O. Edson, USA

COMMANDER REED

Lt. Col. M.H. Thompson, USA
Lt. Col. L. Wallace, USA
Comdr. W.S. Kirkpatrick, USN
Lt. Cdr. E.P. Bonner, USN

LT. COMDR. HOLZAPFEL

Comdr. K. Loveland, USN
Comdr. H. Marvin-Smith, USN
Comdr. W.K. Rogers, USN
Comdr. A.T. Church, USN

LT. COMDR. BLACK

Lt. Col. F.L. Street, USA
Comdr. T.S. White, USN
Comdr. H.J.P. Foley, USN
Comdr. G.F. Dalton, USN

Fig. 12
Atomic Bomb Theses Reading Detail, Second Reading

the proper employment of the weapons and resources placed by the country at its disposal for the prosecution of the war." He hoped, therefore, that the work the students—officers of the Army, USAAF, Navy, and Marine Corps—had been doing together for the past year had given them better insight into the many problems that they would have to solve jointly in the future. Noting that the Naval War College for the first time had had a Foreign Service officer as a student, Spruance hoped that everyone had been reminded of the need always to keep in mind the close relationship that had to exist in both peace and war between foreign and military policy.[37]

Enc.(C) to N.W.C. Memo.
of 22 April, 1947.

Comments on Thesis

Writer______________________ Senior Class / Junior Class

Subject: The Influence of the Atomic Bomb on Future Naval Warfare.

Reader ______________________

____

____

____

____

____

____

____
Signature of Reader

Fig. 13
Comments on Thesis, Topic 2

NOTES

1 Allan Smith, "Directive for Theses," 25 June 1946, p. 1, folder 2591, box 136, Record Group [hereafter RG] 4, Naval Historical Collection, Naval War College, Newport, R.I. [hereafter NHC]. The fact that there were two thesis topics corrects this author's earlier work on the immediate postwar Naval War College. I was under the mistaken impression that there was only one thesis topic for the year, on atomic weapons and naval warfare. See Hal M. Friedman, *Digesting History: The U.S. Naval War College, the Lessons of World War Two, and Future Naval Warfare, 1945–1947* (Newport, R.I.: Naval War College Press, 2010), p. 220.

2 Smith, "Directive for Theses," pp. 1–2.

3 Smith, "Directive for Theses," annex A, "Hints on Thesis Writing," 25 June 1946, pp. 1–2, folder 2591, box 136, RG 4, NHC.

4 Ibid., pp. 2–3.

5 Ibid., pp. 3–4.

6 Ibid., pp. 4–5.

7 "Outline of Staff Presentations," 2 July 1946–10 May 1947, folder 2615, box 142, RG 4, NHC.

8 Ibid.

9 Ibid. For an examination of "Sound Military Decision," see Frank Snyder, introduction to *Sound Military Decision: U.S. Naval War College* (Annapolis, Md.: Naval Institute Press, 1992).

10 "Outline of Staff Presentations."

11 Ibid. Per Navy personnel policy at the time, officers like Bates who held the temporary, wartime grade of commodore were reduced in peacetime to their permanent grade, captain—though in Bates's case that demotion did not take place until 1947. See Rear Adm. Richard W. Bates Biographical Sketch, Officer Biographical Files, Navy Operational Archives, Naval History and Heritage Command, Washington Navy Yard, Washington, D.C.

12 "Outline of Staff Presentations."

13 Ibid.

14 Ibid.

15 Ibid.

16 Ibid.

17 Raymond Spruance, "Opening," 13 July 1946, p. 1, folder 6, box 1, RG 16, Addresses, NHC.

18 Ibid.

19 Ibid., pp. 1–2.

20 "Naval Communications: Schedule and Assignments, Junior and Senior Classes of June 1947," 31 July 1946, pp. 1–2, folder 2584, box 135, RG 4, NHC.

21 Ibid., pp. 5–6. The Pringle Auditorium is still in use today; the Pringle Maneuver Room, devoted since those years to various purposes, is now used by the Naval Staff College (the International Program's junior course). The "Old Maneuver Room" in Luce Hall was presumably one of two large gaming spaces, originally two stories high, on the second floor. Both have since disappeared in successive renovations and are now subdivided into offices for students of the College's senior international course, as well as classrooms.

22 Ibid., p. 7.

23 Ibid., pp. 7–11.

24 Ibid., pp. 11–13.

25 Ibid., pp. 14–16; "Communications Course Comments and Recommendations," 8 August 1946, pp. 1–2, folder 2584, box 135, RG 4, NHC.

26 "Instructions for Naval Intelligence Studies, Senior and Junior Classes of June 1947," 29 August 1946, p. 1, folder 2585, box 135, RG 4, NHC.

27 Ibid., p. 2.

28 Ibid., pp. 3–5.

29 Ibid., pp. 5–12.

30 Allan Smith, "Memorandum for Officers of the Naval War College Staff," 10 January 1947, p. 1, folder 2591-A, box 136, RG 4, NHC.

31 Allan Smith, "Memorandum for Officers of the Naval War College Staff," 31 January 1947, p. 1, folder 2591-B, box 136, RG 4, NHC.

32 Allan Smith, "Critique of Student Theses on Relations between Russia and the United States, and Their Influence on U.S. Foreign Policy," 4 April 1947, p. 1, folder 2591-B, box 136, RG 4, NHC.

33 Allan Smith, "Memorandum for Officers of the Naval War College Staff," 22 April 1947, p. 1, folder 2591-C, box 136, RG 4, NHC.

34 Hartung to Morgan, "Memorandum for Secretary," 22 April 1947, p. 1, folder 2591-C, box 136, RG 4, NHC.

35 Allan Smith, "Critique of Student Theses on the Influence of the Atomic Bomb on Future Naval Warfare," 12 May 1947, p. 1, folder 2591-B, box 136, RG 4, NHC.

36 Raymond Spruance, "Remarks of Admiral R. A. Spruance, U.S.N. President, Naval War College at Graduation Exercises," 3 June 1947, p. 1, folder 8, box 3, RG 16, NHC.

37 Ibid., pp. 1–2.

SIMS HALL
U.S. NAVAL WAR COLLEGE

II *The Maneuver Rules, June 1946*

In late June 1946, Rules for Chart and Board Maneuvers were issued for the incoming 1946–47 classes. The Rules paint a very detailed portrait of how Naval War College staff and students conducted war-game simulations in the first full academic year after the end of World War II. This chapter describes in greater detail than previous works on the Naval War College precisely how war simulations were conducted and demonstrates how war games of this type were simulated before the postwar revolution in electronics.[1]

The Rules began with a foreword characterizing Chart and Board Maneuvers as serving to "create successive changes in military situations under study." These changes were to simulate the "realities" of war, at least to the extent that those realities were brought about by the actions of the opposing commanders and their subordinates in the games. The foreword pointed out, however, that the Maneuver Rules "merely" prescribed methods and "conventions" for conducting the maneuvers, that the Rules were based on "well founded" opinions, and that they represented a "conservative" average of "true" conditions while "simple" enough to prevent prolonged delays in application. "When accepted, interpreted and employed in this spirit they fulfill their purpose."[2]

"Chart Maneuvers," the foreword explained, covered "vast areas" and were "strategical" in nature. "Board Maneuvers," in contrast, covered more-restricted areas and were used for more-tactical exercises. However, in either type of maneuver, student officers who functioned as Student Commanders saw only those ships and aircraft that were within their effective visibility or radar range. All maneuvers were divided into sequentially numbered "moves" representing successive intervals of time. Generally, moves for Board Maneuvers were of three minutes' (game time) duration or multiples of three, while Chart Maneuvers involved moves embracing the actions of several hours or even days. There were many points of similarity in the conduct of Chart and Board Maneuvers but also differences requiring different procedures, materials, and staffs. (Page i.)

Student Commanders were to be "scrupulous" in avoiding transmission of information during a Chart or Board Maneuver except by methods that would be available at sea under game circumstances, since "unauthorized conversations between student commanders tend to destroy the value of a maneuver, and to relieve a subordinate of making decisions on his own responsibility and initiative." Once all of the lessons to be derived from a maneuver had been "clearly demonstrated," in the opinion of the Director of the Maneuver, it was to be terminated. The instruction emphasized that while it was customary to call the maneuvers "games," they were in "no sense" contests to see who won or who lost. (Pages i–ii.)

Section A of the Rules: Conduct of Chart and Board Maneuvers

Preliminary to a maneuver, a specific Problem depicting a military situation was drawn up for the student officers to solve. Solutions for the Chart or Board Maneuver were then selected, the solutions were made available to the Student Commanders, and other student officers were assigned as their Subordinate Commanders. Each Problem was to include a Statement of the General Causes that created the game Situation, such as the broad progress of operations to date, instructions or orders that had been received by higher authority, the composition and locations of own forces, the state of defense of one's own bases and outposts, information on enemy forces (including size, general location, and bases), the "attitude" of neutral powers, logistics support available, and meteorological, hydrographic, and topographic data. Each Problem also included any Special Situations or conditions that would have a bearing on the Problem, as well as an explanation of what was expected of the student officers. (Page a-1.)

For a Chart Maneuver, one end of the Maneuver Room was screened off to form a "Master Plot," the headquarters of the Maneuver Staff. Student Commanders were not to visit this enclosure during a maneuver. The Maneuver Staff itself kept the Master Plot of the maneuver. Each Student Commander handling a force kept a Strategic Plotting Chart in his assigned room. This plot tracked the movements of all elements of his force as well as those of other friendly and enemy forces with which the Student Commander was concerned. Forms to be employed by the Student Commanders, found outside the Maneuver Room, included Aircraft Flight, Fuel Account Message, and Record of Move Forms, as well as Fuel Work Sheets. Documents for use by the Maneuver Staff included Aircraft Casualty and Information Forms, Air Damage and Umpire Communication Records, Move Umpire Check-Off Sheets, Radar Contact, and Umpire Damage Reports. Members of the Maneuver Staff were appointed from among the student officers by the Naval War College staff. The Director of the Maneuver was in charge of the maneuver's conduct and responsible for the coordination of the Maneuver Staff; the Director was the arbiter in all cases, and his decisions were final. He also conducted the Post-Maneuver Critique. (Pages a-1 to a-2.)

In addition, there were two Assistant Directors who supervised the Master Plot and took charge of successive moves in alternation. In general, the Assistant

Directors announced the lengths of the moves, the weather conditions, and other General Conditions of exercises on bulletin boards. They announced when forms and "flimsies" (tracings on tissue paper) were to be turned in, announced Maneuver Time during moves, determined contacts, transmitted "appropriate" information to Student Commanders, consulted with Damage Umpires regarding damage to be assigned to various forces, supervised the History of the Maneuver, and initiated discussion during the Critique. Liaison Officers were also assigned, as intermediaries between the Director and the Student Commanders, on the one hand, and an Air Umpire, on the other, the latter having one or more assistants. The Air Umpire had a number of duties, including the general oversight of the operation of aircraft and the maintenance of records of aircraft available to each side. The Air Umpire also received Aircraft Flight Forms, assisted the Director with respect to aircraft operations and contacts, and determined the results of aerial attacks, the latter subject to change by the Director. The Air Umpire further informed the Director, the Plotting Officer, the Damage Umpire, and the Recorder of the damage done by these air attacks. Finally, the Air Umpire maintained a separate plot of aircraft operations. (Pages a-2 to a-3.)

The Move Umpire received flimsies of ship and aircraft movements as well as the Record of Previous Moves. In addition, he transmitted these flimsies to the Master Plot as well as the Records of Moves to the Recorder of the History of the Maneuver and the concerned Assistant Director. The Communication Umpire and his assistants supervised the application of the communication rules: they received and examined messages from the Student Commanders, determined from the Student Commanders' Communication Plans which addressees could be reached on which circuits and which required relays, and recorded each message on the Communication Record Form. The Communication Umpire and Communication Assistants computed the time of delivery of the message, had Marine Corps orderlies prepare copies of the messages and deliver them to originators and addressees via pneumatic tubes when Maneuver Time reached the proper time of delivery, and kept "dummy" clocks to show the current Maneuver Time. The Recorder, under the supervision of the Director and Assistant Directors, recorded the History of the Maneuver. Moreover, when the maneuver started, the Director published "necessary" information concerning the first move and established Maneuver Time for the beginning and length of the move. Having noted the announced length of a move and the weather conditions, each Student Commander updated his Strategic Plotting Chart, including the intended movements of his vessels and aircraft for the entire period of the move. (Pages a-3 to a-4.)

A number of conventions were followed in plotting. The position of the Main Body was indicated by a dot within a small square, other detachments and single ships by dots within small circles. A wavy line over a circle indicated a submerged submarine. Tracks of vessels proceeding on the surface were drawn with full lines, tracks of submerged submarines and aircraft in flight with dotted lines. All tracks were labeled to show the ship, aircraft, or force; course and speed; and the direction of unit movement, with an arrowhead for ships and a *T* for aircraft. Each point at which a change of speed or course was made was indicated and labeled with the time of such change; each point where a submarine left or returned to the surface or an aircraft landed or took off had to be shown.

Tracings of data on the Strategic Plotting Charts were made on tissue paper with soft black-lead pencils. All flimsies indicated proposed tracks as completed at the end of the current move and had two reference points, usually latitudes and longitudes. A legend in the upper right-hand corner in colored pencil (blue indicating Blue forces and red for opposing forces) marked the move number and Maneuver Time, the composition of the force and its task designator, the name of the Student Commander (game role and actual name), and the room number. Each flimsy contained all information with respect to composition and movements of forces by the Master Plot to determine contacts and results. If radar or antisubmarine listening devices were being used during the exercise, that was stated, identifying ships and arcs of coverage. (Pages a-4 to a-5.)

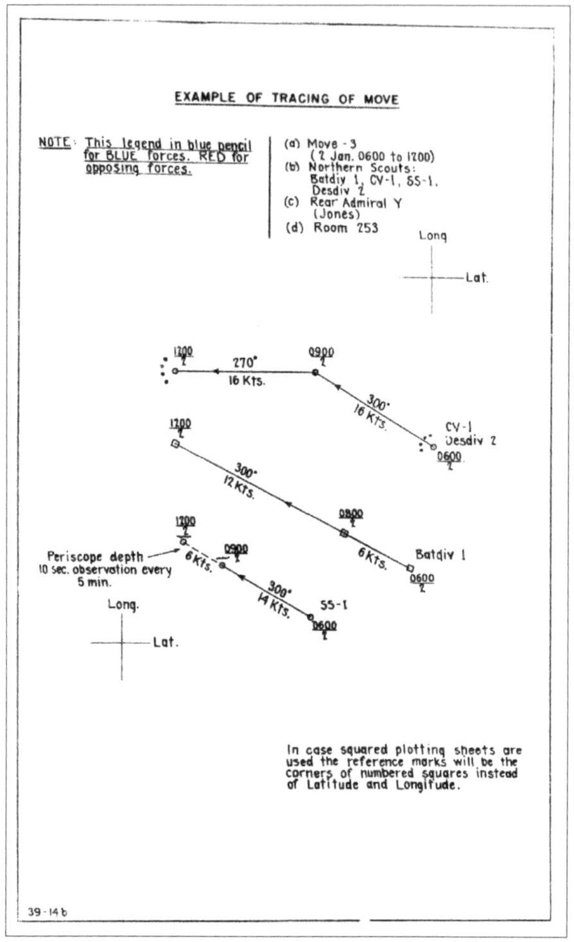

Fig. 14
Example of tracing a move

Zigzags of ships had a specific symbol; actual speed, speed of advance on the base course, and the zigzag plan in use were noted. In addition, when vessels were accompanied by air patrols, the flimsy noted the composition, altitude, station, and purpose of the patrol, and the presence of the patrol was indicated on the track. The presence of an antisubmarine screen was indicated on the flimsy by dots or a broken circle around the circle that indicated the initial position, the number of dots representing the number of screening vessels. (Page a-5.)

Student Commanders were at liberty to change their intended movements at any time, for "good reason." In such case, they submitted a "Modified Move," warning the Director as soon as possible so that Maneuver Time would not be advanced before the change notification was made. The flimsy numbered the modification of a move, such as "Move 1, Mod. 2" and supplied the revised tracks of the forces affected, from the time of modification to the end of the current move, crossing out the original. Flimsies for all moves were to be submitted to the Master Plot promptly when called for. Records of Moves for the just-completed moves were submitted at the same time. (Pages a-5 to a-6.)

At the beginning of each move, Aircraft Flight Forms and flimsies were to be submitted for all flights planned for that move. If flights were not planned but became necessary at some point during the move, the Air Umpire was notified and the Aircraft Flight Form and flimsy were prepared and submitted as soon as possible. The same procedure was to be used if a flight was to be modified during a move. In order to expedite the computation of times required for aircraft to complete various legs of their flights, the lengths of legs were given to the nearest ten miles and the times of flight to the nearest five minutes. When flights were repeated—provided the weather did not change—a single drawing of the airplane tracks sufficed; the ship flimsy showed the points and times of the beginnings of each repeated flight. The airplane flimsy was marked "Repeated Flight," with a table showing the individual planes of each flight, start times, course changes, and completion times. If the flight legs were "very short," as in the case of antisubmarine patrols, a figure was to be drawn around the ship or formation at the position of the first launch, showing the outline of the patrol, the number of planes, and the hours during which it was to be flown. Flimsies for submarines showed whether the submarines were surfaced or submerged and, if the latter, depth, exact times of employing the periscope—if employed—and the periods of listening for radio traffic, with or without a vertical antenna exposed. (Pages a-6 to a-7.)

A Student Commander could ask the Master Plot to hold Maneuver Time to permit him to send a message. During Chart Maneuvers, all flimsies, Records of Moves, messages, and other papers were sent to the Master Plot by pneumatic tube. The Plotter, who was a civilian draftsman or technical assistant, transferred to the Master Plot the movements shown on all flimsies, entering course, speed, and times. The Director and Assistant Director then examined the Master Plot to determine whether any interactions had occurred. If none had occurred and all communications had been handled, the Assistant Director advanced Maneuver Time to the end of the move and called for the next move. If contact had been made or an important message was yet to be delivered, the Director or Assistant Director advanced the maneuver clock to or beyond the time of the contact or of delivery of the message and announced the new Maneuver Time over a loudspeaker. Information concerning interactions and action subsequently taken was conveyed to and from the Student Commanders on "appropriate" forms in duplicate. Student Commanders then decided on their actions, wrote synopses of their decisions on the forms, and returned one copy to the Master Plot. Student Commanders could change planned movements of their forces and submit Modified Moves. The Director might authorize the exchange of information by Liaison Officers, message,

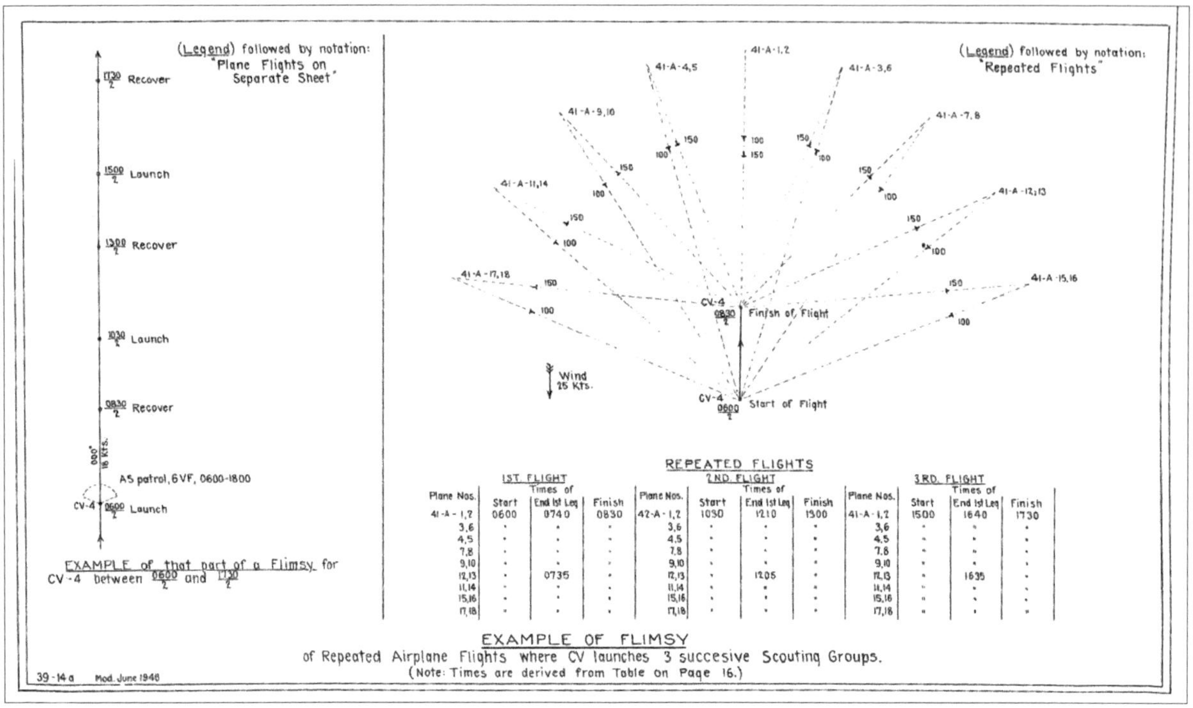

Fig. 15
Example of a flimsy

or phone, but the results of the interactions were decided by the Director before the maneuver resumed. When the end of a move was approaching, the Director might broadcast the date for the next move in advance so that Student Commanders might complete their plotting and thereby expedite the submission of flimsies when the next move was called for. (Pages a-7 to a-8.)

Various records were also to be kept by the Student Commanders: Orders, Plans, and Special Instructions; track charts; a file of messages sent; a file of messages received; a file of Records of Moves; fuel accounts; a file of Aircraft Flight Forms; a location for all carrier planes; and copies of diagrams showing cruising and other dispositions.

As soon as possible after the completion of the last move, all student officers and Maneuver Staff assembled for a Critique. This began with comments on student officer solutions, followed by a brief History of the Maneuver in which the Master Plots for each move were shown on slides. The Director or an Assistant Director discussed "interesting" features and situations, the merits and defects of decisions and movements by each side, and strategic and tactical points demonstrated. During the discussion, student officers and the Maneuver Staff were free to make such comments as they desired; it was assumed that a "healthy difference" of professional opinion would arise. The Director and the Maneuver Staff were to give their personal views and call for those of the students. The discussion, however, was not to stray from the point at issue or become "undignified." (Pages a-8 to a-9.)

Type Designation	Model Size	Distinctive Features
BB	Large	1 Mast, 1 Stack
CC, CB	do	1 Mast, 1 Stack White Stripe
CV, (all types)	do	Flat Flight Deck
CA	Intermediate	1 Mast, 2 Stacks
CL	do	1 Mast, 4 Stacks
DD, DE, DM	Small	1 Mast, 2 Stacks
SS		As indicated by the Director
Auxiliaires	Intermediate	1 Stack

Fig. 16
Model ship type designations and features

For a Board Maneuver, the equipment required was a Maneuver Board, model ships, "turning cards," a plotting table with drawing instruments and protractors, screens, and tactical plotting sheets. The Maneuver Board represented a portion of the earth's surface, either sea or land. The board was divided into large squares and each large square subdivided into a hundred small squares. Each small square represented a thousand yards on a side. The tracks of the forces moved were shown by white chalk, those of torpedoes in the colors (blue or red) of the ship models that had "fired" them. Model ships were shown in several sizes and shapes to indicate types, as seen in figure 16. Tracks of single ships were dotted, those of two or more ships in column in solid lines. Positions at the end of the moves were indicated by numerals corresponding to the number of the move.

Model ships had metal strips with which could be assembled formations of a division, squadron, or fleet of any size, at intervals of three hundred, six hundred, seven hundred, or a thousand yards. Additionally, turning cards constructed to the scale of the Maneuver Board were used to plot the track of a ship during a move. The straight edge of the card represented the distance that a ship would move on a straight course during three minutes at a speed in knots as indicated; a circle on the turning card represented the turning circle of the ship, assuming a tactical diameter of one thousand yards. The card allowed for the retardation of speed due to the turn; straight lines tangent to the turning circle represented the new paths of a ship making changes of course in multiples of fifteen degrees. The numbers along the tangent sides represented the position the ship would reach in three minutes at the average engine speed in knots during the move. In use, the card's straight edge was laid in the direction of the actual course at the beginning of the move, with the zero point at its position at that time. If a turn had begun in a previous move and had now to be completed, the card was laid from the beginning of the turn and the position of the ship fixed by carrying the card along from its position in the turn at the end of the previous move. (Pages a-9 to a-11.)

All target data, such as target angle, course, and speed, were determined from the plots kept by the Student Commanders. Tactical Plotting Sheets and Range and Target Angle Indicators were issued by the Drafting Room for this purpose. Student Commanders were to keep clear of view of their enemy's model ships; in fact, they were not to be permitted to see portions of the Maneuver Board beyond the effective visibility from their ships. Screens were employed, or Student Commanders were placed out of the sight of the board. When a Student Commander so separated moved his force on a Plotting Sheet, his moves were handled in the same way as those of a Chart Maneuver, except that the Director could have them transferred to the Maneuver Board, the Master Plot, or both. Student Commanders thus separated from the board were given information in the same way as in a Chart Maneuver. (Page a-11.)

Forms to be used by Student Commanders in Board Maneuvers, found on shelves outside the Maneuver Room, included those for Aircraft Flight, Message, Mine Laying, Move and Gunfire, and Torpedo Fire. Documents used by the Maneuver Staff included Air Damage and Umpire Communication Records; Aircraft Casualty, Information, Radar Contact, Ship Data, Submarine Information, and Target Indicator Quick Decision Maneuver Forms; Damage Computer Work, Fire Distribution, Ammunition Expenditure Tally, and Move Umpire Check-Off Sheets; Fire Effect Tables for 1946; and Force Damage Summaries.

The Maneuver Staff for a Board Maneuver was appointed from the Naval War College staff or student officers. The Director was charged with the conduct of the maneuver and the Critique. He decided the length of the moves, weather and visibility conditions, and such matters as damage from torpedo or mine hits. Liaison Officers aided in the conduct of the maneuver, while the Communication Umpire, with two assistants, supervised the application of the Communication Rules. The Chief Damage Computer was to supervise gunfire damage computing as well as the recording and reporting of all damage and ammunition expenditure. The Chief Damage Computer was also to assist the Director in the application of the Gunfire Section of the Maneuver Rules on the Maneuver Board. (Pages a-12 to a-13.)

The Air Umpire saw that the moves of aircraft, aircraft carriers, or other vessels flying planes on or off were plotted on the Master Plot. He also received the Aircraft Flight Form from the Aircraft Commanders and recorded combat and operational losses. In addition, the Air Umpire decided results of aerial engagements and resolved aerial interactions. He additionally decided the damage inflicted by aerial bombs and torpedoes and informed the Chief Damage Computer about such damage. All final decisions about damage, however, were up to the Director. Similarly, the Plotter, again a civilian draftsman or technical assistant, was responsible for plotting the moves of all forces on the Master Plot; informing the Director and Submarine and Air Umpires of all contacts between forces on the Maneuver Board; and plotting the tracks of torpedoes fired during the maneuvers. He also plotted the locations of minefields as well as supervised the preparation of blueprints and slides for use in the Critique. The Historian was to keep the History of the Maneuver, obtaining from the Assistant Directors and the various umpires such notes and comments as they desired to be included. In order to facilitate the preparation of the Critique, the Historian divided the maneuver into periods that corresponded to moves. There were also Move Umpires, each of whom received all forms submitted by Student Commanders of his side, scrutinized them "carefully" for compliance with Maneuver Rules, and required corrections when necessary before submitting them to the Maneuver Staff. Each Move Umpire reported to the Director when all forms for his side had been submitted and when all moves for his side had been made on the Maneuver Board. (Pages a-13 to a-14.)

When a move was announced, each Student Commander made out and submitted to the appropriate umpire a Move and Gunfire Form. This form was to show exactly what movement his vessels would make during the impending move and what gunfire, if any, they were to deliver. The Student Commander was also to place on the Maneuver Board arrows pointing to his targets in time to enable the enemy commander to know that he was under fire. Also submitted were any messages from his ships during the move, as well as any Mine Laying, Torpedo Fire, or other forms that covered actions to be taken during the move. Forms thus submitted were to be complete and "definite," not contingent on what other forces might do. A Move and Gunfire Form once submitted could not be altered without the approval of an Assistant Director. Even if altered, forces moved for at least one minute in accord with the form originally handed in. When Student Commanders had turned in their Move and Gunfire Forms and the forms had been "properly" verified, Move Umpires informed the Assistant Directors, the Director, and the Communication Umpire that the Student Commanders in question were ready to make the move. After receiving the "ready" reports, the Director broadcast, "Make Move No. ——," whereupon students moved their units strictly in accordance with their Move and Gunfire Sheets, and the Maneuver Time was advanced to the end of the move. Should a new move be ordered before the damage of the last one had been determined, the Director would decide when to apply the effects. (Pages a-14 to a-15.)

Aircraft Moves, however, were to be called for separately and were usually half an hour or more in length. For each Aircraft Move, all flights had to be specified on the Aircraft Flight Form. As soon as damage inflicted by aircraft was determined, if the total was 10 percent or more, each commander was to be notified of the effect on vessels under his immediate command. A commander could ascertain damage to ships of his own side but not under his personal command only by signal or message to or from the commander of the damaged ship. The state of damage to ships of one side could be made known to commanders of the other only if the Director so desired. In addition, all damage during a three-minute move was to become effective at the end of the move in which it had been received. After all of this was resolved, the Critique of the Board Maneuver followed, according to the Rules for the Critique of a Chart Maneuver. (Page a-15.)

Section B: General Rules

Ships and aircraft in Problems were designated by letter and number symbols. Data for ships, aircraft, and weapons were given in current Naval War College publications or in the Statement of the Problem. This data included the size of vessels—sizes of combatants expressed in tons standard displacement, and sizes of merchant vessels expressed in gross tons. There were five classes of ships. Large Ships had displacements of eighteen thousand tons or more, while Intermediate Ships were between eight thousand and 17,999 tons. Small Ships had displacements or tonnage of between three thousand and 7,999. Destroyers (DDs) were placed in a 200-to-2,999-ton category. A fifth class, Submarines, was not associated with a displacement or tonnage; all surface craft of 199 tons or less were classed with Submarines, due to the similarity of their silhouettes. The unit of distance used in Operations Problems was the nautical mile, two thousand yards, and the unit of range for gun or torpedo fire was one thousand yards. The unit of speed was the knot; the speed of ships was adjusted to the nearest half-knot, while the speed of aircraft was specified to the nearest five knots. Wind force was expressed as 0 to 12 on the Beaufort scale; the velocity of wind in knots was taken to be five times the given figure. The state of the open sea in winds of from Force 0 to Force 3 were considered Smooth, in Forces 4–6 Moderate, in Forces 7–9 Rough, and Forces 10–12 Heavy. The direction of the sea was to be considered the same as that in which the wind was blowing. Students were warned that navigational positions given to them by the Director would rarely be exact and would in fact be deliberately in error when the Director deemed warranted. (Pages b-1 to b-2.)

Damage to ships was classified as Above Water Damage, that caused by hits from gunfire, rockets, or bombs; Under Water Damage, owing to torpedoes, mines, rockets, bombs, ramming, or grounding; and Total Damage, the sum of the two. Relatedly, the prospective "life" of a ship was expressed as a whole number or a fraction of fourteen-inch penetrative hits it might receive before sinking. The remaining life of a ship at any instant was a percentage, to the nearest 10 percent. Life of the ship lost due to Above Water Damage and Under Water Damage was recorded separately as percentages of total life. During Chart Maneuvers, the Director could arbitrarily assign damage in combat—to ships as a percentage of life lost, and to aircraft as a number of planes destroyed. His decisions, in general, would be guided by the Rules set forth in the following chapter. (Page b-2.)

NOTES

1 For previous works, see James Barber, "The School of Naval Warfare," *NWCR* 21, no. 8 (April 1969), pp. 89–96, and "Mahan and Naval Strategy in the Nuclear Age: A Lecture Delivered at the Naval War College," *NWCR* 24, no. 7 (March 1972), pp. 78–88; Donald White, "Admiral Richard L. Conolly: A Perspective on His Notions of Strategy," *NWCR* 24, no. 3 (November 1971), pp. 73–79; Philip Crowl, "Education versus Training at the Naval War College: 1884–1972," *NWCR* 26, no. 3 (November–December 1973), pp. 2–10; Nepier Smith, "Historical Analysis of the Organizational Success of the Naval War College during the Twenty-Five Years following the Second World War," 1974, pp. 11–67, Naval War College Advanced Research Project, NHC; Ronald Spector, *Professors of War: The Naval War College and the Development of the Naval Profession* (Newport, R.I.: Naval War College Press, 1977); Michael Vlahos, *The Blue Sword: The Naval War College and the American Mission, 1919–1941* (Newport, R.I.: Naval War College Press, 1980); John B. Hattendorf, B. Mitchell Simpson III, and John R. Wadleigh, *Sailors and Scholars: The Centennial History of the U.S. Naval War College* (Newport, R.I.: Naval War College Press, 1984), pp. 175–77; Robert Wood, "The Conceptual Framework for Strategic Development at the Naval War College," *NWCR* 40, no. 2 (Spring 1987), pp. 4–16; David Rosenberg, "Being 'Red': The Challenge of Taking the Soviet Side in War Games at the Naval War College," *NWCR* 41, no. 1 (Winter 1988), pp. 81–93; J. S. Hurlburt, "War Gaming at the Naval War College, 1969–1989," *NWCR* 42, no. 3 (Summer 1989), pp. 46–51; Edward S. Miller, *War Plan Orange: The U.S. Strategy to Defeat Japan, 1897–1945* (Annapolis, Md.: Naval Institute Press, 1991); Bud Hay and Bob Gile, *Global War Game: The First Five Years,* Newport Paper 4 (Newport, R.I.: Naval War College Press, 1993); Kenneth Watman, "Global 2000," *NWCR* 54, no. 2 (Spring 2001), pp. 75–88; John B. Hattendorf, *The Evolution of the U.S. Navy's Maritime Strategy, 1977–1986,* Newport Paper 19 (Newport, R.I.: Naval War College Press, 2004); Robert H. Gile, *Global War Game: Second Series, 1984–1988,* Newport Paper 20 (Newport, R.I.: Naval War College Press, 2004); Douglas Smith, *Carrier Battles: Command Decision in Harm's Way* (Annapolis, Md.: Naval Institute Press, 2006); and Hattendorf, ed., *U.S. Naval Strategy in the 1970s: Selected Documents,* Newport Paper 30 (Newport, R.I.: Naval War College Press, 2007). See also Hal M. Friedman, *Digesting History: The U.S. Naval War College, the Lessons of World War Two, and Future Naval Warfare, 1945–1947* (Newport, R.I.: Naval War College Press, 2010) and *Blue versus Orange: The U.S. Naval War College, Japan, and the Old Enemy in the Pacific, 1945–1946* (Newport, R.I.: Naval War College Press, 2013).

2 "Maneuver Rules," 24 June 1946, p. i., folder 2567, box 131, RG 4, NHC; subsequent in-text page references in this chapter are to this source. See also Friedman, *Blue versus Orange,* chap. 1.

III *The Maneuver Rules Continued, June 1946*

The distributed Maneuver Rules specified "ideal" underway conditions for passages in smooth seas. Ideal conditions meant no damage, no current, a clean bottom, no paravanes streamed, and nothing being towed. A condition of not having to keep station in formation or maintain what was called "reserve steam" also was defined.

Section C of the Maneuver Rules: Speed and Logistics

Engine speed was to correspond to the engine revolutions being made under ideal underway conditions; engine speed was considered the same as speed through the water and, in ideal conditions, over the ground. Ground Speed was Engine Speed corrected for deviations from ideal underway conditions and was to be the speed plotted on the Maneuver Board or Charts. Reserve Speed was the difference between the highest Engine Speed immediately available and the engine speed being made. Maximum Engine Speed Allowed (MESA) figures and losses due to sea conditions are shown in figure 17.

The Director was in charge, however, of the application of other rules, such as the maximum speed allowed for a vessel under tow. Vessels in formation, except the Guide, had to have a minimum Reserve Speed for station keeping depending on the Formation Engine Speed: for ten knots or less, one knot of Reserve Speed; for eleven to twenty-two knots, two knots; for twenty-three to thirty knots, three knots; and for thirty-one knots or more, four knots. If Formation Engine Speed was so increased as to reduce a ship's Reserve Speed below that required, the vessel was to break formation. Moreover, the Formation Engine Speed plus the required Reserve Speed had to be within the MESA range. With a paravane streamed, a vessel could not make more than twenty-five knots without losing the paravane itself; vessels streaming or recovering paravanes were restricted to twelve knots engine speed for six minutes. A submarine under way was classified as either in its surface condition or submerged. While in a surface condition, it was always to be ready for a quick dive. Changing between surface and submerged was considered to take only one minute, and during that minute the submarine was capable of making a visual or radar contact.*

For large increases in speed in Chart Maneuvers, forty-five minutes was allowed to raise steam in one or more boilers. A steam vessel of destroyer size or less in port was to have enough steam available for half maximum speed (not to exceed fifteen knots) and be able to get under way thirty minutes after commencing preparations. Steam vessels of intermediate and small size in port were to have enough steam for up to ten knots and be able to get under way in forty-five and thirty minutes, respectively. A vessel with an internal combustion engine, in contrast, was able to get under way within ten minutes but could make only Engine Speeds allowed by the MESA table. Once under way, steam vessels could increase speed as soon as additional boiler power was available. When additional boiler power became available, engine speed was considered to be increased immediately (rather, that is, than work up gradually) to the level intended. (Pages c-3 to c-4.)

Speed changes took place at uniform rates during a three-minute move; the mean speed during the move was to be used in plotting. Under Water Damage reduced a ship's designed speed in proportion, in 10 percent increments. The reduction was in addition to any owing to Above Water Damage; loss of speed was to be expressed in a percentage of original designed speed, converted to the nearest half-knot. All losses of speed were effective at the end of the move in which the damage was received. Above Water Damage up to 29 percent resulted in no loss of speed, nor would 30–49 percent Above Water Damage in a battleship (BB), heavy cruiser (CA), battle cruiser (CB), or battle carrier (CVB). The same percentage of damage in other types of ships resulted in a 10 percent loss of speed. Above Water Damage of 50–69 percent in battleships, heavy cruisers, battle cruisers, and battle carriers cost 20 percent of speed, 30 percent in other types of ships. A 50 percent loss of speed was suffered by all types receiving 70–79 percent Above Water Damage. A ship that received a Total Damage of 80 percent or more would be unable to make more than five knots. (Page c-5.)

6215-5715
24/June/46cl

- c-2 -

SECTION C
RESTRICTED

Classification		Angle between sea and course. at beginning of a move: 0°-30°	31°-80°	81°-120°	121°-180°
Moderate Sea					
Large	MESA	(M)*	(M)*	(M)*	(M)*
	Loss	(2)	(1)	(0)	(0)
Intermediate	MESA	(M)	(M)	(M)	(M)
	Loss	(2)	(1)	(0)	(0)
Small	MESA	(31)	(M)	(M)	(M)
	Loss	(3)	(2)	(0)	(0)
Destroyer	MESA	(29)	(32)	(34)	(M)
	Loss	(4)	(3)	(1)	(0)
Submarine	MESA	(M-2)	(M-1)	(M)	(M)
	Loss	(3)	(2)	(1)	(0)
Rough Sea					
Large	MESA	(20)	(26)	(M)	(M)
	Loss	(4)	(3)	(1)	(0)
Intermediate	MESA	(18)	(23)	(28)	(M)
	Loss	(4)	(3)	(2)	(1)
Small	MESA	(18)	(20)	(25)	(30)
	Loss	(5)	(4)	(3)	(1)
Destroyer	MESA	(17)	(19)	(23)	(27)
	Loss	(7)	(5)	(3)	(2)
Submarine	MESA	(M-3)	(M-2)	(M-2)	(M-2)
	Loss	(4)	(3)	(2)	(2)
Heavy Sea					
Large	MESA	(10)	(12)	(15)	(18)
	Loss	(10)	(5)	(3)	(2)
Intermediate	MESA	(10)	(10)	(12)	(16)
	Loss	(10)	(5)	(3)	(2)
Small	MESA	(10)	(10)	(12)	(16)
	Loss	(10)	(8)	(5)	(3)
Destroyer	MESA	(10)	(10)	(12)	(15)
	Loss	(10)	(10)	(6)	(3)
Submarine	MESA	(10)	(10)	(10)	(12)
	Loss	(9)	(9)	(8)	(7)

Fig. 17
MESA table

* "Maneuver Rules," 24 June 1946, pp. c-1 to c-3, folder 2567, box 131, RG 4, NHC. Subsequent pages references in this chapter are to this source.

CHANGE OF SPEED - BOARD MANEUVERS

Maximum designed Speed	19 or less	20-29	30 or more
Large	3	4	5
Intermediate	4	5	6
Small	5	6	8
Destroyer	6	8	10
Submarine	6	8	

Engine Speed being made	19 or less	20-29	30 or more
Large	3	4	5
Intermediate	4	5	6
Small	5	6	8
Destroyer	6	8	10
Submarine	6	8	

Fig. 18
Change of Speed: Board Maneuvers

Designed Max. Speed of Ship	Engine Speed Being Made			
	Designed Maximum	1 kt. below designed maximum	2 kts. below designed maximum	3 kts. below designed maximum
12-19	24 hrs.	X*	X*	X*
20-25	6 "	12 hrs.	X*	X*
26-29	6 "	12 "	24 hrs.	X*
30 plus	6 "	12 "	16 hrs.	24 hrs. **

*Can be maintained indefinitely or as long as fuel lasts.
**Same as * for 4 kts. below designed maximum.

Fig. 19
Engine Speed Being Made

The number of hours that a steam vessel could maintain various engine speeds at or near its designed maximum was as shown in figure 19. When those limits were reached, a steam vessel was to slow for six hours to a speed it could maintain indefinitely. It could then return to high speed. The Director could impose engineering casualties at any time on ships running at or near maximum speeds. His decision would depend on the type of propulsion plant, the duration of high speed, battle damage incurred, the state of the sea, and presumed fatigue of personnel. Engineering casualties could involve anything from intermittent smoking to complete breakdown. The students were reminded that engineering and other casualties imposed could be "excessive" if current fleet doctrine concerning the upkeep of material and the health of personnel was disregarded, especially in the case of submarines. (Pages c-5 to c-6.)

For purposes of computing fuel expenditure, Port Condition meant that main engines were not in use, with steam maintained for in-port use only. Stand-By Condition meant that the main engines were ready for use, Underway Condition that main engines were actually in use. Fuel expenditures for Port and Underway Conditions were calculated from data in the fleet tactical publication *Fuel Expenditure Tables of Fleets*. Stand-By Expenditures were double those for Port Conditions; under way, fuel expenditure increased 10 percent while raising or keeping steam for five or more knots above the engine speed actually being made. For vessels in close formation, fuel expenditure was to be increased by 3 percent to allow for station keeping, while vessels with paravanes streamed increased fuel expenditure by 5 percent. As regarded refueling, a fuel oil tanker, gasoline tanker, and aviation gasoline barge could deliver aviation gas to a carrier or tender at the rate of thirty thousand gallons an hour. These ships could deliver and receive fuel simultaneously. Submarines making ten knots or more while charging batteries would expend fuel as if they were under way at maximum surface speed. If a submarine stopped or was making less than ten knots while charging batteries, the fuel expended would be half that for maximum speed. (Pages c-6 to c-9.)

Rule C-21. OILING AT SEA

Oiler (Delivering Ship) Type	Max. Rate of Delivery (Barrels per hr.)	Receiving Ship Type	Max. Number at one time that can receive oil.	Max. Rate each vessel can receive (Barrels per hr.)	Sea	Speed while oiling	Time in hrs. to come alongside connect & disconnect.
AO, BB & CB	4200-5600	DD	2(Broadside)	1800	S	12	1/2
		"	"	1800	M	10	1/2
		"	1(Astern)	1400	R	8	3/4
		"	-	-	H	oiling	impossible
CVB, CV CVL, CVE CA&CL	2100-2800	DD	1(Broadside)	1400	S	12	1/2
		"	"	"	M	10	1/2
		"	1(Astern)	1100	R	8	3/4
		"	-	-	H	oiling	impossible
AO	4200-5600	BB CB CA& CL	2(or 1 and one of any other type) (Broadside)	2100	S	12	1/2
		"	"	2100	M	10	1/2
		"	1(Astern)	1400	R	6	3/4
		"	-	-	H	oiling	impossible
		CVB CV CVL & CVE	1 and 1 of any other type (Broadside)	2100	S	12	1/2
		"	"	2100	M	10	1/2
		"	1(Astern)	1400	R	6	3/4
		"	-	-	H	oiling	impossible

Fig. 20
Oiling at Sea

Rule C-22. OILING IN PORT

Oiler (Delivering Ship) Type	Max. Delivery (Barrels per hr.)	Receiving Ship Type	Max. Number at one time that can receive oil.	Max. rate each vessel can receive (Barrels per hr.)	Time in hrs. to come alongside connect up and disconnect
AO, BB & CB	4200-5600	DD	2	2100	1/4
CVB, CV CVL, CVE, CA, CL	2100-2800	DD	2	1400	1/4
AO	4200-5600	BB, CB CA, CL	2(or 1 and 1 of any other type)	2500	1/2
		CVB, CV CVL & CVE	1 and 1 of any other type	2500 (DD2100)	1/2 (DD 1/4)
AO, AS	1400-1800 (Diesel)	SS	2	280	1/4

Fig. 21
Oiling in Port

CHARGING AND DISCHARGING OF SUBMARINE STORAGE BATTERIES

(a) Submarine storage battery capacity is expressed in percentage of a full charge.

(b) The capacity of a submarine storage battery may be depleted at the following arbitrary rates:

Submarine speed when using the battery	Percent of depletion of total battery for each minute at designated speed.
over 8 knots	1.6%
8 "	1.3%
7 "	.9%
6 "	.6%
5 "	.5%
4 "	.4%
3 "	.2%
2 "	.07%

(c) While charging batteries a submarine can not make more than 15 knots.

(d) Submarine storage batteries may be charged at the following arbitrary rates:

From a capacity of 1% to 72% in one hour.
From a capacity of 73% to 88% in two hours.
From a capacity of 89% to 100% in three hours.

Fig. 22
Charging and Discharging of Submarine Storage Batteries

Section D: Contact Rules

For purposes of adjudicating "contacts"—detections and interactions—"meteorological visibility" was defined as the maximum distance at which a large object could be seen without artificial illumination. Visibility, which depended on light from "heavenly" bodies and the density of the atmosphere, was tabulated in the Statement of the Operations Problem, along with sunrise, sunset, duration of twilight, times for moonrises and moonsets, and phases of the moon. During a maneuver, the Director might announce changes in meteorological visibility, perhaps as one mile or less in daytime in fog, rain, mist, or falling snow or on a moonless night in slight haze. Visibility was considered to be three miles during the day in light haze or rain and at night if there was a clear atmosphere and moonlight, and the object being observed was away from the moon. Five miles visibility during the day was possible in haze, three miles at night. Ten, fifteen, and twenty miles were listed as possible during the day given average-to-high barometric pressure and broken or clear sky; high barometer, clear sky, and dry air during the day would allow for meteorological visibility of thirty, forty, or fifty miles. "Low visibility" meant nine miles or less of meteorological visibility while "normal visibility" was from ten to twenty miles and "high visibility" was twenty-one or more. Phases of twilight would be indicated by increased or decreased visibility as the occasion required. "Effective visibility" was the range at which an object had a reasonable chance of being seen. Effective visibility was dependent on meteorological visibility, the nature of the object, the background of the object, the position of the sun or moon, and the position of the observer. It was to be measured in nautical miles or yards, whichever happened to be more convenient. (Pages d-1 to d-2.)

All ships on the surface could sight other ships as unrecognized types at distances up to meteorological visibility at that time. However, sightings could not occur beyond certain limits of effective visibility. For instance, Large Ships could see other Large Ships at twenty-five miles, while Large and Intermediate-sized Ships could see each other at twenty-two miles. Large Ships could see smaller ones, and vice versa, at twenty miles, the same distance at which Intermediate and Small Ships could see each other. Intermediate Ships could see each other at twenty-two miles, and Small Ships could see each other at twenty miles. Submarine periscopes could see but not identify Large Ships at fifteen miles, Intermediate Ships at thirteen miles, and Small Ships at ten miles; recognition for any type was possible only within 80 percent of meteorological visibility or the limits noted above for effective visibility, whichever was less. Effective visibility of unrecognized ship types from aircraft was according to the table in figure 24; effective visibility from aircraft was reduced one mile for every additional thousand feet of altitude. (Pages d-2 to d-3.)

6215-5715
24/June/46gq

- d-3 -

SECTION D
RESTRICTED

(e) the effective visibility for sighting unrecognized objects from aircraft is given in the following table:

Target	Altitude in feet	Meteorological Visibility in Nautical Miles								
		1	3	5	10	15	20	30	40	50
Stationary Any Ship	500	0.2	0.7	1.5	3.0	4.0	4.5	5.0	5.5	5.5
	1000		0.7	1.5	3.0	4.5	5.0	6.0	6.5	7.0
	2000			1.5	3.5	5.0	6.0	7.5	8.0	8.5
	3000				3.5	5.5	6.5	8.0	9.0	11.0
	5000					5.5	7.0	9.0	10.5	11.5
	7000						7.0	10.0	12.0	13.0
	10000(1)							10.0	13.5	15.0
Moving Small Ships SS or DD	500	0.8	3.0	4.0	7	8	9	10	11	12
	1000	0.8	3.5	5.0	8	10	11	13	14	15
	2000		3.5	5.5	9	11	13	16	17	18
	3000			6.0	10	12	14	17	19	20
	5000				11	14	16	19	21	23
	7000					16	18	21	23	25
	10000(1)						20	23	25	27
Moving Intermediate Ships	500	0.8	3.5	5.0	11	14	16	19	21	23
	1000	0.8	3.5	6.0	12	15	18	23	25	27
	2000		3.5	6.5	13	17	21	26	29	32
	3000			6.5	13	18	22	28	32	35
	5000				13	20	24	31	35	39
	7000					20	25	32	37	42
	10000(1)						26	33	39	43
Moving Large Ships	500	0.8	3.5	6.0	12	15	18	23	25	27
	1000	0.8	3.5	6.0	13	17	21	26	29	32
	2000		3.5	6.5	14	19	23	29	34	35
	3000			6.5	14	20	24	32	37	40
	5000				14	21	26	34	40	45
	7000					21	27	36	42	47
	10000(1)						28	37	44	50
Moving Convoy Or Task Force	Same as for the largest ship present, plus the radius of the disposition.									

(1) Reduce effective visibility one mile for every additional 1000 feet altitude.

Fig. 23
Meteorological Visibility in Nautical Miles

Making smoke was also governed by rules. Oil-burning ships were considered to exhaust gas at all speeds and intermittent smoke at speeds within two knots of maximum speed, while coal-burning ships made light smoke at all speeds. Smoke was visible by ships and aircraft forty miles away if heavy, thirty miles away if intermediate or light, and nine miles away if from a submarine snorkel. Submarines using periscopes would be able to see smoke in these three categories from fifteen, ten, and three miles, respectively. Aircraft could be seen in the air

Class Seen	Number in Group	Col. 1 Same Altitude or higher than observer			Col. 2 Lower Than observer		
		High vis.	Norm. vis.	Low vis.	High vis.	Norm vis.	Low vis.
VP	1	6	5	1.7	4	3	1
"	2-6	8	7	2.3	5	4	1.3
"	7 plus	9	8	2.6	6	5	1.7
VF, VO, VA	1	5	4	1.3	3	2	.7
" " "	2-6	7	6	2	4	3	1.0
" " "	7 plus	8	7	2.3	5	4	1.3

Fig. 24
Aerial visibility distances and altitudes

from other aircraft at distances given in figure 24, provided the difference of altitude between the two aerial observers was not over two thousand feet. The distance of sighting was to be decreased half a mile for every additional thousand feet of difference in altitude. Airplanes could be seen during daylight by observers on surface vessels, through submarine periscopes, or by observers on land one mile farther than the distances given in figure 24. Airships, on the other hand, could be seen by any observer at the same distance at which a large vessel could be seen. (Pages d-3 to d-4.)

Mines floating on the surface were considered visible from vessels on the surface up to five hundred yards in smooth sea and daylight. Under other conditions, mines could not be seen at sufficient distance by ships proceeding at fifteen knots or more to avoid them. Mines floating on the surface or below the surface and antennae floating on the surface could be seen by observers in aircraft within a radius of five hundred yards, provided that the altitude of the aircraft was not more than a thousand feet and the observer was concentrating on sighting such objects at that time. Minelaying by surface ships or aircraft was considered detectable at the same ranges at which ships or aircraft could be recognized; mine explosions were visible and audible at ranges up to ten miles. A submarine could be seen by any observer within six thousand yards, unless the submarine remained submerged at periscope depth or greater; did not expose the periscope when making more than three knots; did not expose the periscope within a thousand yards of surface ships or aircraft, or did not expose the periscope longer than specified in figure 25. The exposed periscope of a submerged submarine, however, could not be observed at all more than six thousand yards away; if less than one minute intervened between two periscope exposures, the duration of the two exposures plus the intervening time was to be regarded as one continuous exposure. The snorkel of a submarine could be sighted as if it were a periscope; false sightings of periscopes or snorkels by surface ships or aircraft could be introduced into the exercise. (Pages d-4 to d-6.)

Sea Smooth, no white caps Wind 0-3		Sea Choppy with white caps Wind 4 and above	
Distance (yds)	Time (sec.)	Distance (yds)	Time (sec.)
1000-2000	10 sec.	1000-2000	30 sec.
2000-4000	20 sec.	2000-4000	60 sec.
4000-6000	30 sec.	4000-6000	90 sec.

Fig. 25
Visibility distances of submarine periscopes

The commander of a submerged submarine was to enter on his flimsy the Maneuver Time at which his periscope was raised, whether he observed on a specified bearing or took an "all around look," and the duration of periscope exposure. A look on a specified bearing was to take ten seconds; ten seconds was also the time allotted to shifting from an observation of one ship or unit to another. As an example, a thirty-second exposure permitted the submarine commander to look at each of two bearings. A ten-second "all around look," however, would give bearings only of vessels within 1,200 yards, distinguish whether other vessels were nearer or farther than eight thousand yards, and identify a sighted vessel as Large, Intermediate, Small, or Very Small. For a look of ten seconds or longer on a designated bearing, information would be given by the Director: within a seven-thousand-yard actual range, the type of vessel, whether friendly or enemy, exact compass bearing, and target angle—the observer's relative bearing as seen from the target (see a more technical definition below)—on the center ship of a group. If the actual target angle was close to zero degrees or 180 degrees, the target angle was to be given to the nearest five degrees. As the actual target angle approached ninety or 270 degrees, it could be to the nearest fifteen degrees. In general, target angles were considered more accurate the greater the range, up to seven thousand yards. Ranges to targets for submarines between seven and five thousand yards were given within a thousand yards of the actual, between 5,000 and 2,500 yards within five hundred, from 2,500 to 1,500 yards within a hundred, and 1,500 yards or less as accurately as could be measured on a plot. (Pages d-6 to d-7.)

Meteorological visibility	distance seen
5	30
3	15
1 (or less)	1 (or less)

Fig. 26
Searchlight, star shell, and parachute flare meteorological visibility

The speed of a target was to be normally given as determined by the Student Commander from his plot. If supplied, the accuracy of the information was governed by the range and the number of observations taken. For instance, for a target seven to twelve thousand yards distant, it was considered that the Student Commander should be able to ascertain the approximate type and, if a large vessel, whether friendly or enemy. At that range, it should also be possible to obtain an exact compass bearing; target angles, as noted above, for ranges within seven thousand yards; and ranges within one or two thousand yards of the actual figure, inaccuracy varying with the range. Speed, however, was not obtainable at this range. Above twelve thousand yards actual range, it would be possible to obtain only "very meager" information about the composition of the force, an exact compass bearing, and a range figure within two to four thousand yards of the actual. Target angles, though, could not be obtained. Within the context of visibility, a submerged submarine could be identified one minute after firing a colored smoke bomb but smoke bombs could not be seen at a distance greater than four miles. (Page d-7.)

Gun	Maximum Range of Bursts	Diameter of Illumination in yards.	Duration of Illumination in minutes.	Minimum Burst Per Min.
6".1 6" 5".9	17,000	1,000	1	4
5".5	15,500	800	1	5
5".1 5"	14,500	800	.8	6
4".7 4".1 4" 3".9	8,000	600	.7	7
3".5 3" 2".9	6,500	500	.4	8
Flares	------	2,000	3.0	2

Fig. 27
Star shell standards for illumination of a target

Effective visibility at night could be increased by the use of searchlights, the amount of searchlight needed depending on the existing meteorological visibility. Effective visibility at night, however, was not to exceed nine thousand yards for Large or Intermediate Ships, 7,500 yards for small vessels, six thousand yards for Destroyers, and three thousand yards for Submarines. Searchlights, star shells, and parachute flares were visible at night in terms of a number of factors (figure 26). Star shell firing necessary to maintain illumination of targets making fifteen knots was as in figure 27; if the target was moving faster, a proportionately higher rate of fire was necessary. The rate of fire, the horizontal range of burst, and the bearing limits of any search spread were stated by the Student Commander on the Move and Gunfire Form. (Pages d-7 to d-8.)

If properly illuminated by star shells or parachute flares, a target was visible for both recognition and gunfire at the ranges given in figure 28. Star shells or flares could, by accident or design, completely obscure vision beyond them. The deliberate use of star shells or flares for this purpose, however, had to be in accordance with a definite plan submitted to the Director in advance.

Similarly, there were various ranges for audibility. Ranges of audibility for gunfire, for instance, were twenty miles for twelve-inch guns or higher, fifteen miles for guns eight-inch to eleven-inch, ten miles for six- to seven-and-a-half-inch, seven miles for four- to five-and-a-half-inch, and five miles for three-inch guns or smaller. These ranges varied according to wind, one's own gunfire, and one's own speed. If continued noise could be heard, true bearing could be estimated to the nearest multiple of forty-five degrees. Student Commanders furnished the Director with sketches or descriptions of their Underwater Sound Plans; the Director would announce the "assured sonar echo-range of the day," that which would be obtained at sea by a bathythermograph. (Page d-9.)

Object seen	Bursts 00 to 2500 yds. beyond target Meteorological Visibility		Bursts 2500 to 4000 yds. beyond target Meteorological Visibility	
	(5 miles)	(3 miles)	(5 miles)	(3 miles)
Large or Int. Ship	15000 yds.	6000 yds.	9000 yds.	4000 yds.
Small Ships	12000 "	5000 "	8000 "	4000 "
Destroyers	9000 "	4000 "	6000 "	3000 "
Submarine	5000 "	2000 "	3000 "	1000 "

Fig. 28
Ranges for star shell and parachute flare target illumination

Echo-ranging performance could also vary from the assured sonar echo range of the day, the variance to be decided by the Director. When ranges were obtainable, bearings could be measured within plus or minus one degree. Turbulent masses of water, such as those caused by depth charge explosions or propellers at high speed, could, in the line of the sound beam, interrupt echo ranging from five to twenty minutes. The speed of one's own ship would also reduce sonar range; eighteen to twenty knots would reduce the echo range by 25 percent, twenty-one to twenty-two knots by 50 percent reduction, and twenty-three or twenty-four knots by 75 percent. Ships making over twenty-four knots would not make any sound contacts. Passively, bearings within plus or minus five degrees and general information about a type and speed could be obtained by listening. The greatest range attainable for underwater sound listening was ten thousand yards; the Director could reduce that range as he deemed appropriate for conditions at the time. (Pages d-9 to d-10.)

Radar was assumed to be installed in all ships, all ships and aircraft were considered equipped with Identification Friend or Foe, and all shore stations possessed radar, as set forth in the Statement of the Problem. A nomogram (figure 30) was used to obtain theoretical maximum radar range under "perfect" operating conditions; radar ranges given to Student Commanders never exceeded them. The theoretical maximums were reduced by the Director, the amount of reduction varying

	Surface Search	Air Search	Fighter Director	Main Battery Fire Control	Double Purpose Fire Control
BB CB	2	2	0	2	4
CVB CV	1	2	1	0	2
CVL	1	1	1	0	0
CVE	1	1	1	0	1
CA CL	2	1	0	2	2
DE DD	1	1	0*	0	1
SS	1	1	0	0	0
AUX	1	1	0	0	0

* Carried by Division Flagships only.

Fig. 29
Shipboard radars by type and number

directly with the theoretical maximum range and inversely with the number of aircraft in a group to be detected. In addition, there were theoretical maximum radar search ranges for detection between ships. (Pages d-10 to d-11.)

Maximum ranges of main-battery fire-control radars were equal to those of surface search radars. Maximum ranges of double-purpose fire-control radars were thirty-eight thousand yards for a large target or over seven planes, thirty-five thousand yards for an intermediate target or three to seven planes, and twenty-four thousand yards for a small target or one to two planes. The average ranges, in miles, that might be expected from airborne search radar were as shown in figure 31. (Pages d-11 to d-12.)

Object of Search	Searching Ship			
	Large	Intermediate	Small and DD	SS
TASK FORCE	65,000	60,000	42,000	24,000
LARGE SHIP	50,000	45,000	38,000	20,000
INTERMEDIATE SHIP	43,000	38,000	35,000	18,000
SMALL SHIP AND DD	35,000	30,000	24,000	13,000
SS	9,000	8,000	7,000	6,000

Fig. 30
Theoretical maximum radar search ranges in yards

The ranges given in tables could also be reduced by the Director, by amounts varying directly with the ranges given in the tables and inversely with the size of the target. The approximate bearing of ships or aircraft operating radar could also be given to the Student Commander of any unit carrying radar-direction-finding equipment, again at the discretion of the Director. Bearings could be obtained at ranges up to one and a half times the theoretical maximum range of the operating radar; the Director would decide on the effectiveness of radar interference and deception. During a maneuver, the Director would supply data obtained by radar. It was not considered practical, however, to give continuous information, as would be the case in actual service. However, "pertinent and timely" information, such as marked changes in course and disposition, that would be disclosed by the use of radar would be given to the Student Commander concerned. All ships suffering 30 percent or more Above Water Damage were considered to have lost all their radar equipment. (Pages d-12 to d-13.)

Object of Search	Range from Searching Aircraft.				
	VF (N)	VO VA	VA (N)	VP	Air Early Warning
Well defined shore line	65 mi.	80 mi.	80 mi.	100 mi.	200 mi.
Group of ships (4 or more)	50 mi.	50 mi.	50 mi.	75 mi.	200 mi.
Large ship	40 mi.	40 mi.	40 mi.	55 mi.	200 mi.
Intermediate ship	35 mi.	35 mi.	35 mi.	40 mi.	200 mi.
Small ship	30 mi.	30 mi.	30 mi.	35 mi.	175 mi.
Submarine, surfaced	15 mi.	15 mi.	15 mi.	20 mi.	150 mi.
Conning Tower	5 mi.	7 mi.	7 mi.	10 mi.	65 mi.
Aircraft, single	5 mi.	7 mi.	7 mi.	10 mi.	65 mi.
Aircraft, group (6 or more)	8 mi.	10 mi.	10 mi.	15 mi.	120 mi.
Aircraft, (tail warning)	800 yds. (tail warning)		800 yd.	3000 yd.	3000 yd.

Fig. 31
Average ranges of airborne search radar

The Maneuver Rules indicated that smoke screens could be laid by any oil-burning vessel, by any vessel or aircraft carrying special smoke apparatus, or by aircraft carrying smoke bombs. For a smoke screen laid by ships to be dense enough to prevent vision by the enemy, it had to be laid by at least three vessels not over five hundred yards apart. For the smoke to lie on the surface, atmospheric conditions had to be favorable and the relative wind had to be fifteen knots or more. The

Director had to be informed at least three minutes beforehand (in Maneuver Time) that a smoke screen was being started or stopped. Aircraft smoke bombs or smoke-bomb clusters were carried by aircraft in lieu of hundred-pound bombs; smoke bombs were assumed to produce smoke for nine minutes but only when dropped on water. To obscure vision, at least six bombs had to be laid and not over two hundred feet apart. Large smoke tanks could be carried—in lieu of 1,600-pound bombs—by naval patrol planes, naval patrol bombers, or similar shore-based aircraft; one large smoke tank was able to produce a screen 2,400 yards long. Small smoke tanks could be carried in lieu of five-hundred-pound bombs by patrol planes, naval patrol bombers, carrier-based aircraft, or similar shore-based types and could produce screens 1,800 yards long. A smoke screen, however, was not considered to obscure a vessel during the three-minute move in which the screen was begun. Such laying of smoke, however, could be employed to interfere with gunfire. (Pages d-13 to d-14.)

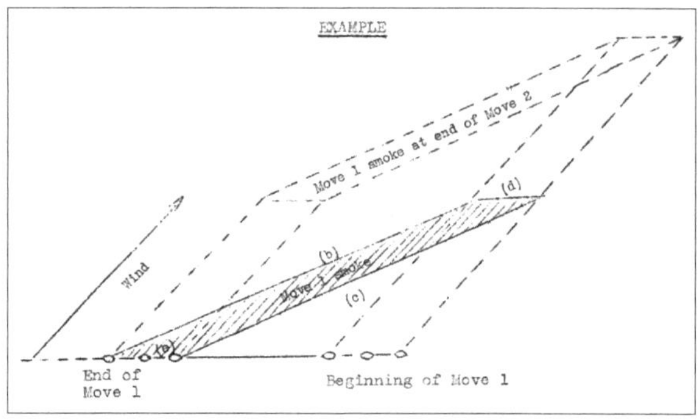

Fig. 32
Smoke screen area: example 1

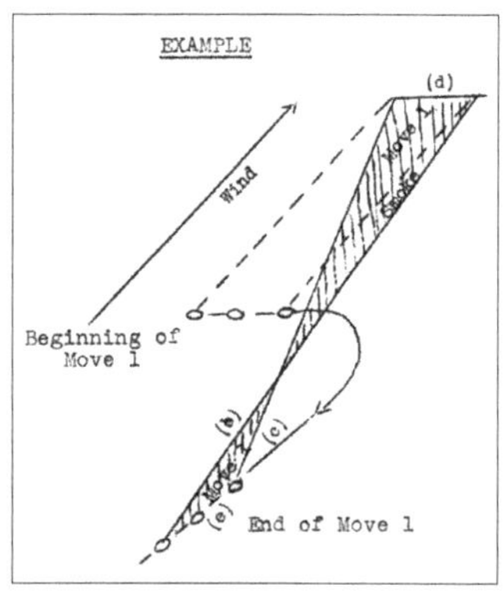

Fig. 33
Smoke screen area: example 2

The smoke emitted during a three-minute move was carried directly to leeward at the speed of the wind for three subsequent three-minute moves, during which it would obscure vision. By the end of the fourth three-minute move, including that in which it was first emitted, the smoke would have entirely dissipated. The shape of a smoke area at the end of a three-minute move during which smoke was begun was determined from the position of the leader at the beginning of the move. A line was drawn in the direction in which the true wind was blowing and equal in length to the distance a particle of smoke would travel during the move. That point was joined with the position of the leader at the end of the move, defining the forward edge of the smoke area—called "line (b)" (see figures 32, 33). In the same way, students determined a line for the rear vessel, the after edge of the smoke area, or "line (c)." Connecting the outer ends of lines (b) and (c) with a line parallel to the shape of the formation at the beginning of the move would result in "line (d)." The line connecting the inner ends of (b) and (c) through the formation as it existed at the end of the move would be "(e)." The smoke area was bounded by lines (b), (c), (d), and (e). This area was usually in the form of a parallelogram, but when the smoking unit made a marked change of course during the move, the smoke area frequently appeared as two triangles with apexes touching. Moreover, in plotting smoke screens, it was "convenient" to lay out the smoke for one move and then translate this block of smoke bodily in each successive move by the distance the smoke would be moved by the true wind. It was found that blocks of smoke laid out in second and subsequent moves would connect with the first block and with each other, forming a continuous smoke screen. Each block of smoke was labeled with the number of the move in which it was laid. (Pages d-14 to d-15.)

Section E: Communication Rules

Communication systems were to be simulated in Chart and Board Maneuvers in various ways. For instance, radiotelephone, telephone, and loud-hailers were simulated by calls on specified telephone equipment during periods announced by the Director. In addition, visual, keyed radio and radar, wire and cable, radioteletype, message drops, and underwater sound were simulated by Message Forms via the Master Plot. Finally, mail, messenger, and other systems were simulated in writing, also via the Master Plot. Communication equipment in ships and shore stations and channels available to them were shown in confidential supplements to the Maneuver Rules. Communications of both opposing forces were conducted in accordance with U.S. Navy communication instructions, as well as the Communication Plans of the respective Student Commanders. Communication Plans had to include systems the Student Commander desired to use. They outlined radio channels and frequencies as well as conditions of radio and radar silence. Additionally, there were plans for radio interception, countermeasures, and schedules for submarines and other types. There were also visual chains of responsibility (that is, for relaying), procedures for special equipment, voice codes, voice radio calls, and authentications. (Page e-1.)

Radio transmissions of certain powers could be intercepted and enemy Student Commanders given various information about them. For example, they might be allowed to listen in on voice transmissions or be given a copy of keyed messages transmitted in plain language. Finally, the enemy commander might be informed about the frequency, time, length, procedure, or general form of an encrypted transmission. The effective range of any communication channel was to be decided by the Director, as the situation warranted. The interval between the instant the drafting of a message commenced and its delivery to the addressee took into account loss of time for administration, operations, and "abnormal" conditions. Administrative and operational delays imposed were based on the number of plain-language words; messages sent in plain language or automatically encrypted and decrypted were considered to be drafted at the rate of five words per minute, to the nearest minute, a rate that included all "normal" administrative delays. Messages were transmitted or received at the rate of ten words per minute, to the nearest minute, a figure reflecting all "normal" operational delays. The Director might also introduce "appropriate" additional delays for "abnormal" conditions: circuit congestion, disregard or sudden change of a Communication Plan; encryption or decryption that was not automatic, the need to wait for a specified schedule of transmission, damage to the transmitting or receiving ship or station, and relays. (Page e-2.)

Adding all delays to the normal Maneuver Time of drafting would give the Maneuver Time of delivery to any particular addressee. The minimum time to deliver any message, other than by voice, was three minutes. For communications purposes, Student Commanders were considered to be present in all units of their command to which no other student officers had been assigned to game roles as subordinates. The exchange of messages was to be simulated in accordance with the "spirit" of the Communication Rules even if the originator and the addressee were the same student officer. The Maneuver Time at which the drafting of a message commenced was the "Time of Origin"; the file number of the message was converted to Greenwich Civil Time, plus the day of the month for Chart and Board Maneuver messages. (Page e-3.)

Section F: Gunfire Rules

Radar Director Control could be used to the extreme range of the guns under control. Optical Director Control could be used to the limits of effective visibility, and pointer fire could be used by ships that had "Top Spots" (platforms high in the upperworks from which the fall of shot could be observed and aim corrections, or "spots," given) but no directors (which tracked target movements, processed them with analog computers, and aimed the guns). Only Local Control (each gun aimed by its own crew) was available for vessels that had neither directors nor Top Spots (in effect, only submarines). Fire Effect was the damage done to a ship by a single target, expressed in percentage of target life lost in three minutes. Fire Effect was to vary with the weapon, the target, the range and target angle, and the method of spotting. These factors were found in the pamphlet *Fire Effect Tables*. Range and target angle were measured at the beginning of each three-minute move; moves of greater duration were divided into three-minute increments. Ranges themselves were to be measured to the nearest thousand yards. Target angle was the angle between the line of fire and the keel of the target ship, and measured to the nearest forty-five degrees. Line of fire was the bearing of the point of aim from the center of the firing ship. If within thirty degrees of the bow or stern, only bow or stern guns could bear on the target. Otherwise, all guns were considered to bear. A division of destroyers in close formation was considered to be a single firing unit or a single target; the center of the division was the point of aim, and all ranges and bearings were to be measured from or to that. The course of the division was assumed to be that of the Guide. (Page f-1.)

Values in the *Fire Effect Tables* were computed for "normal" conditions of gunfire. Values were increased or decreased as conditions differed from "normal" by multipliers, *M*, more or less than unity, in accordance with certain rules. The Fire Effect of a ship or unit on a target equaled the Fire Effect of one of its guns, times the number of guns firing at a given target, times the multiplier from the *Fire Effect Tables*. The number of ships' guns firing at the target was the number that the commander wished to fire, without regard to any damage that may have been received. Above Water Damage would reduce *M* one-tenth for every 10 percent of ship life lost. There were also ranges for Radar Spot and Top Spot. These could be reduced by the Director, but in general a twelve-inch, fourteen-inch, or sixteen-inch gun had a Radar-Spot range of thirty-five thousand yards and a Top-Spot range of twenty-six thousand. Eight-inch, six-inch, and five-inch guns had Radar-Spot and Top-Spot ranges equal to their maximums. A Plane Spot could be used when definite assignment of a spotting plane had been made and the plane was in the air. Top Spot could be used in all ships except submarines; when more than four ships were concentrating fire on a target, all types of spotting were considered effective. When neither Radar, Top, nor Plane Spots were considered effective, *M* was reduced by five-tenths. During the move in which fire was opened, *M* was also reduced five-tenths; when fire was shifted to a new target, *M* was reduced three-tenths for that move, unless the new target was within 1,500 yards of the old one. (Page f-2.)

These reductions continued in other contexts. For example, guns firing in Local Control had their *M*s reduced by one-tenth; *M* was increased by one-tenth for ships not under fire. Multipliers of ships undergoing air attack were reduced by an amount, in tenths, decided by the Director at the time. Moreover, the shock effect of a torpedo, mine, or bomb hit reduced *M* five-tenths for the three-minute move during which the hit was received. When a target ship undertook evasive maneuvers, the multipliers of the ships firing at the moving target and of the target ship itself were reduced by five-tenths if the Director decided that the evasive maneuver was effective. Similarly, when a target or a firing ship changed course and speed, or both, during a move to a degree that affected the gunfire of both, the Director decided the amount of reduction for the *M*s. Pitch, roll, yaw, and spray could also collectively reduce the *M* of a ship—one-tenth in a moderate sea, two-tenths in a rough sea, and four-tenths in a heavy sea. (Page f-3.)

Visibility was taken into account but impacted *M* only when Radar Director Control or Radar Spot were not employed. In these cases, *M* was to be reduced one-tenth if gun or funnel gases drifted across the line of fire. When the target was obscured by a smoke screen, *M* was reduced by a number of tenths proportional to the time the target was obscured—unless indirect fire was used, in which case *M* was reduced by three-tenths. For two hours after sunrise and before sunset, glare reduced *M* by two-tenths when the sky was clear and the target bore within fifteen degrees of the sun. At night, *M* was reduced—by one-tenth when Radar Director Control was used and target illumination was effective, by two-tenths with Radar Director Control but no effective illumination. Without Radar Director Control at night but effective illumination, a three-tenths reduction was imposed for ranges up to five thousand yards, four-tenths over five thousand. (Pages f-3 to f-4.)

Vessels other than the target received "enfilade" damage if within a thousand yards of the target and a hundred yards of the line of fire. The amount of damage received by the enfiladed vessel was half that inflicted on the target. The total damage done to a division of destroyers under enfilade fire was twice the normal damage to the target. In addition, the fire of a ship was masked if it attempted to fire within a hundred yards over or past a ship that was a thousand yards or less from it. *M* in this case was reduced by a number of tenths proportional to the period masked in a three-minute move. At night at close range, or during a day or night melee, the *M* of firing ships might be further reduced as the Director chose.

The Fire Effect of naval guns against shore targets was determined by the Director in each case; the Fire Effect of a single shore-battery gun against a ship was that given in the Radar Spot column of the *Fire Effect Tables* for the gun type nearest to that fired and the type of ship fired on. A summary of the losses to equipment, firepower, and speed due to damage was printed on Ship Damage Reports during a Board Maneuver. Finally, at the discretion of the Director, a "Chance Factor" could be imposed on all the above to determine damage. A restricted publication, "Memorandum on the Application of the Chance Factor to the Scoring of Gunfire Damage on the Maneuver Board," was used for this purpose. (Pages f-4 to f-5.)

Section G: Torpedo-Fire Rules

The types of torpedoes and their characteristics were given in confidential publications issued separately. Rules included the inability to train submerged tubes in azimuth. Tubes mounted on a surface ship's deck near the side could be brought to bear within thirty degrees of the fore-and-aft line, while centerline tubes could be aimed within forty-five degrees. Every torpedo, except those carried by aircraft, could be fired at an angle up to ninety degrees from the direction of the tube. A vessel with deck tubes and spare torpedoes could reload in smooth seas in fifteen minutes, thirty minutes in

TABLE OF REDUCTIONS AND ADDITIONS TO MULTIPLIER "M."

Symbol	CAUSE	Rule No.	Reductions or Additions	
a	ABOVE WATER DAMAGE	F-5		For each 10 percent life lost reduction to "M" is -.1.
b	SPOTTING NOT EFFECTIVE	F-6	-.5	Applies when Radar, Top or Plane Spot is not effective. Also when more than 4 ships are concentrating on a target. Does not apply if R, T or P. Spot is lost through Damage.
c	OPEN FIRE	F-7	-.5	Applies for move during which fire is opened.
d	SHIFT FIRE	F-8	-.3	Does not apply if new target is within 1500 yards of old target.
e	LOCAL CONTROL	F-9	-.1	Does not apply if ship uses LOCAL CONTROL because of loss of Directors through Damage.
f	NOT UNDER FIRE	F-10	+.1	
g	AIR ATTACK	F-11		Reduction As decided by Director.
h	SHOCK EFFECT	F-12	-.5	Reduction for move caused by BOMB, MINE or TORPEDO HIT
i	EVASIVE MANEUVERS	F-13	-.5	Effectiveness of evasive maneuvers to be decided by Director.
j	CHANGE OF COURSE and SPEED	F-14		Reduction as decided by Director.
k	SEA CONDITION	F-15		Moderate Sea -.1; Rough Sea -.2; Heavy Sea -.4
l	* GUN or FUNNEL GASES	F-16(a)	-.1	Applies if gun or funnel gases drift across line of fire.
m	* TARGET HID by SMOKE SCREEN	F-16(b)		Reduction proportional to time except if Indirect Fire is used, then reduction is -.3
n	* SUNGLARE	F-16(c)	-.2	Two hours after sunrise and two hours before sunset. Target within 15 degrees of SUN.
o	* TARGET SILHOUETTE	F-16(d)	+.2	At twilight, when target bears with 30 degrees of SUN. Other conditions, as ruled by DIRECTOR.
p	NIGHT ACTION	F-17		(a.) Radar Control and with Effective Illumination. -.1; (b) Radar Control but without Effective Illumination. -.2; (c) Without Radar Control but with Effective Illumination: Ranges less than 5,000 yards. -.3; Ranges over 5,000 yards. -.4
q	ENFILADE	F-18		Vessels within 1,000 yards of target and 100 yards of line of fire. Damage is one half that of target ship - Des Div receives twice normal damage.
r	MASKED FIRE	F-19		Applies when ship attempts to fire over, or within 100 yards of, another ship which is 1,000 yards or less away. Reduction is proportional to time fire is masked.
s	MELEE	F-20		Night or Day – close range action. Reduction decided by Director.
	EFFECTIVE † SPOTTING RANGES	F-6		Caliber Gun / Radar Spot / Top Spot: 16"-14"-12" 35,000 yds. 26,000 yds.; 8"-6"-5" Max. Range Max. Range

* No Fire Control Radar available

† Radar Director Control is effective to maximum range of the gun.

Fig. 34
Table of Reductions and Additions to Multiplier M

moderate seas, in rough seas not at all. A submarine could fire all its loaded tubes in a three-minute move; in any of a submarine's torpedo rooms two tubes could be reloaded in six minutes and four tubes in nine. Before a torpedo could be fired from a submerged submarine, however, the periscope had to be exposed for at least ten seconds, unless the torpedo was being fired by sound. Torpedo fire itself was to be indicated by filling out a Torpedo Fire Form and handing it to the Torpedo Umpire. (Page g-1.)

Single vessels could fire on their own initiative during a three-minute move, handing in the Torpedo Fire Form after the move had been made but before the next was called for. Otherwise, the form had to be handed in before the firing move was made on the board and could not specify torpedo fire earlier than the end of the first minute of movement on the Board subsequent to submission of the form. Such fire was to be considered complete at the Maneuver Time designated on the form—provided it conformed to the above restriction—unless the Torpedo Umpire was informed to the contrary immediately after this time was reached. If fire was held, it could be completed at any subsequent Maneuver Time by so informing the Torpedo Umpire. If firing settings were to be modified, a modified form was to be handed in. Data for torpedo fire were obtained by Student Commanders from their tactical plots; the Director, however, could furnish bearings and ranges that would be available under actual conditions. (Pages g-1 to g-2.)

A vessel receiving 50 percent or more Above Water Damage lost half of its deck torpedo tubes on each side. A vessel receiving 70 percent or more Above Water Damage lost all of its deck tubes. The tracks of a torpedo salvo fired from a division in formation were plotted from the center of the division, except that at ranges below five thousand yards the division commander might elect to have the torpedoes plotted from individual firing ships. Of the total number of torpedoes fired from a vessel or dropped from aircraft not themselves under Effective Fire, 90 percent—to the nearest whole number—were considered to run hot, straight, and normal. If the firing ship or dropping aircraft was under Effective Fire, only 70 percent to the nearest whole number ran hot, straight, and normal. In any torpedo salvo, one torpedo was plotted on the right limit of the spread, one on the left limit, and the remainder equally spaced between them. When the launching or the track of a torpedo could be seen from the target, the Director was to inform the Student Commander concerned, who was then permitted to take action, modifying a previous move if necessary. (Page g-2.)

From the plots of the tracks and positions of the torpedoes and of such ships as might be in "torpedo water" and whose drafts exposed them to danger, the Director would determine the number and distribution of hits. He based his decision on the type of torpedo, the type of target ship, the shape of the formation, and maneuvers taken to avoid. The effects of successive torpedo hits within fifteen minutes of preceding hits were to be increased—by one-third for the first additional hit and two-thirds for any succeeding ones. A torpedo hit might cause a number of types of temporary or permanent casualties, as judged by the Director: one or all engines out of commission, one or all propellers destroyed, the rudder jammed, or a list preventing gunfire or the launching or recovery of planes. The Director could also, as in the case of gunfire, employ the Chance Factor. (Page g-3.)

Section H: Mine and Depth-Charge Rules

Mines were classified according to their method of use, either moored, drifting, or ground. They were also classified according to their method of firing: contact, influence (acoustic or magnetic), antenna, or controlled. Moored mines could be either contact, influence, antenna, or controlled. In addition, moored mines could be laid by surface ships or submarines in any depth of water up to 250 fathoms and could be set to watch at any desired depth. A Student Commander who directed ships or planes to carry mines could select the type of mine for each ship or plane but had to inform the Director before the beginning of the maneuver. This rule also applied to dummy mines. In a current, the submergence depth of moored mines was to be increased 5 percent of the length of the mine's anchor cable times the strength of the current in knots. With a current faster than four knots or in rough seas, moored mines might drag, in which case they became inoperative. Whether all or part of a moored mine dragged and whether or not it was inoperative was determined by the Director. For drifting mines, a suspended antenna was thirty feet long unless otherwise stated. Successive actuations of influence mines were limited to ten feet; the degree of protection against this type of mine was determined by the Director. Moored antenna mines, unless otherwise stated by the student, had upper antenna wires of ten feet shorter than the intended low water depth of the mine, not to exceed seventy feet. The downward antenna section of the anchor line was a hundred feet long. (Pages h-1 to h-2.)

Mines could become inoperative in various ways. Moored mines could go adrift, and drifting mines themselves became inoperative one hour after being laid unless otherwise stated. Influence mines became inoperative after an interval stated by the Student Commander on the Mine Laying Form. Five percent of contact mines, 10 percent of antenna mines, and 15 percent of influence mines would be initially defective; 10 percent of both antenna and influence mines would experience premature explosions; and 2, 5, and 10 percent of contact, antenna, and influence mines, respectively, would experience subsequent defects, per month. Mines laid by surface ships had to be laid at intervals of three seconds or more, but they could be laid at any speed up to thirty knots. Submarines could lay mines through their torpedo tubes either surfaced or submerged, but at intervals not less than one minute between mines carried at the same end of the boat and in different tubes. Submarines were not allowed to lay mines from forward tubes at speeds faster than four knots, and they had to have at least forty feet of water under their keels when laying mines. They could carry only two mines per tube; reloading of tubes required three minutes. (Page h-2.)

In a Chart or Board Maneuver, for each mine operation a Student Commander handling the minelaying ships or aircraft was to deliver to the Director a filled-out Mine Laying Form. The Student Commander would submit a plan of his anticipated minefield; this Mine Plan could be drawn on a maneuvering board or on a flimsy made from a plot on a large-scale chart, in either case showing the navigational reference point to which the minefield was fixed. In addition, during the time that mines were intended to be effective, the Student Commander was to show his minefield on his Strategic Plotting Chart and on all flimsies of ship movements. In a Board Maneuver, the Student Commander laying mines was to indicate those mines on the Move and Gunfire Form for the ships concerned. Paravanes were assumed to be carried by all ships of more than three thousand tons. For a ship proceeding on a straight course, paravanes were considered to afford "complete" protection against moored contact mines, unless the ship struck the mine head-on. Paravanes, however, gave no protection against any other type of mine. Contact of a paravane with a mine antenna would destroy the paravane, whatever damage it might itself do. (Pages h-2 to h-3.)

If a ship entered a mined area, the Director had its track plotted on the minefield diagram and determined the number of mines, if any, that had detonated. Whether a ship was damaged by mines depended on the ship's underwater structure (specifically, the draft of the ship or the depth of a submerged submarine). Damage could also be determined by the water levels at or through which the mines were set to watch and by the vertical length of the antenna wire, if the mine had any. Damage or absence thereof was affected also by whether or not the paravanes were streamed and whether the ship was on a straight or changing course. In the latter situation, the length of the ship and whether or not it had been degaussed (its magnetic field compensated for) could influence how much damage was done. A mine explosion did the same damage as a torpedo explosion. It was assumed, however, that a detonating mine could not damage more than one ship. A mine inflicted Under Water Damage equal to three fourteen-inch penetrative hits, and second and third hits in the same field each equated to four fourteen-inch penetrative hits, in the case of ground mines and at the discretion of the Director. Its permanent effect on the life and capabilities of the ship was added to damage from other causes.

Relatedly, a Student Commander initiating a depth-charge attack on a submarine had to inform the Director of the number of surface ships or aircraft in the attack group and of the general plan of attack itself. The results of a depth-charge attack against a submarine would be decided by the Director. (Pages h-3 to h-4.)

Section I: Aviation Rules

Replacement airplanes delivered to combatant units were in "combat condition" and with crews. Landings and takeoffs at night from either airfields or ships entailed using navigation lights that were visible for five miles on a clear, dark night. Land planes setting down on the water, seaplanes landing on land, and airplanes of any kind shot down by antiaircraft fire or in aerial combat were considered to have been totally destroyed. A Student Commander operating carrier-based planes was to be prepared at all times to inform the Director as to the location of his planes on the carrier and the readiness of both the airplanes and the flight deck. At the beginning of flight operations for each (game time) day of the maneuver, the Student Commander furnished the Director a memorandum indicating the types and numbers of bombs, rockets, and torpedoes loaded on each plane and thereafter any changes that were made. (Page i-1.)

Elevators on aircraft carriers could carry in one trip the number of airplanes indicated in figure 35. An elevator's round trip took one minute and any carrier could keep on its hangar deck half the aircraft of each type in its allowance. Half the allowance of each type could be "free launched" (i.e., without catapults) without moving airplanes between the hangar and flight decks. In an escort carrier (CVE), once half the allowance had been recovered, no more could land until planes had been struck below; in all other types of carriers, the total allowance of assigned operating aircraft could be landed without such delay. If the Student Commander chose to deckspot more than half the number of any one type of airplane, he wrote out for the Director the number of each type of airplane involved, the armament loading contemplated, and the amount of wind over the deck while launching. To free-launch planes from the flight deck, the apparent wind (as measured over the deck of a moving ship) was to be within ten degrees of the ship's course and the wind over the deck not less than thirty knots for battle carriers, fleet carriers (CVs), or light carriers (CVLs) and not less than twenty knots for CVEs. To land planes, the apparent wind had to be within five degrees of the ship's course and the wind over the deck not less than twenty-seven knots for battle, fleet, and light carriers, or twenty knots for escort carriers. (Pages i-1 to i-2.)

	CVB			CV			CVL		CVE	
Type Air Plane	Fwd Elev	Deck Edge Elev	Aft Elev	Fwd Elev	Deck Edge Elev	Aft Elev	Fwd Elev	Aft Elev	Fwd Elev	Aft Elev
VF	2	2	2	2	1	2	2	2	1	1
VA	2	1	2	1	1	1	1	1	1	1

Fig. 35
Aircraft carrier elevator capabilities

When the carrier's course was more than twenty degrees off the true wind, five minutes were required to turn into the wind and begin landing or launching aircraft. No change of course, however, was necessary to launch planes from flight-deck catapults, provided the apparent wind was forward of the beam. The number of planes that could be launched from a carrier while it was landing airplanes was to be decided by the Director in each case. During daylight and in smooth or moderate seas, airplanes could be free-launched at intervals of fifteen seconds and landed

From Condition	To Condition X	XI	XII
XIII	30 minutes	20 minutes	10 minutes
XII	20 minutes	10 minutes	*********
XI	10 minutes	************	**********

Fig. 36
Flight deck readiness condition changes

every thirty seconds. At night and during daylight in rough seas, these intervals were doubled. In addition, land planes could be launched from flight-deck catapults at the rate of one plane per catapult per minute. Unscheduled launchings could start in accordance with a series of aircraft and flight deck conditions of readiness: in Condition Ten, aircraft could be launched as soon as wind conditions were satisfactory; in Condition Eleven, launches could take place in ten minutes; Condition Twelve meant twenty minutes, and Condition Thirteen meant thirty minutes. To change from one readiness condition to another required meeting the conditions noted in figure 36. A carrier could no longer launch or land planes if it had received 50 percent Above Water Damage or its after flight deck had been destroyed. (Pages i-2 to i-3.)

Wind Force	Method Employed BAKER.	CHARLIE.	DOG.
1	1 each side every 5 min.	1 on lee side every 5 min.	1 each side every 5 min.
2-3	1 each side every 5 min.	1 on lee side every 5 min.	Cannot be employed
4-5	1 on lee side every 10 min.	1 on lee side every 10 min.	Cannot be employed
6	Director will decide.	Director will decide.	Cannot be employed

Fig. 37
Battleship/cruiser/seaplane recovery conditions

Seaplanes could be launched by catapult from cruisers, other surface ships, and submarines if there was at least five knots of apparent wind down the catapult. A catapult could launch a seaplane every five minutes. However, it would take a submarine five minutes after surfacing to float a seaplane off its deck and another five to recover it. A ship stopped to hoist a seaplane out of the water could do so with each crane every five minutes if the wind was not above three knots. A ship could hoist a seaplane out of the water on the lee side every five minutes no matter the wind condition; a battleship or cruiser could recover seaplanes according to figure 37. A battleship or cruiser could not launch or recover aircraft after receiving 50 percent Above Water Damage. (Pages i-3 to i-4.)

Type and Location of airplane	VF,VA,VO	VP(ML) VP(MS)	VP(HL) VP(HS)	Very Heavy Bombers
Landplanes at an air-field	15 min. plus 1 min.per plane	15 min. plus 2 min. per plane	15 min. plus 4 min. per. plane	15 min. plus 6 min. per plane
Seaplanes on a park-ing apron	15 min. plus 5 min.per plane.	30 min. plus 10 min. per plane.	30 min. plus 15 min. per plane.	
Seaplanes at a moor-ing	15 min. plus 6 min.for each 4 planes	30 min. plus 10 min. per plane.	30 min. plus 15 min. per plane.	
Seaplanes at an AV	15 min. for each 4 planes min.time 15 min.	30 min. plus 15 min. per plane.	30 min. plus 20 min. per plane.	
Seaplanes at an AVP	15 min. per plane	30 min. plus 15 min. per plane.	30 min. plus 20 min. per plane.	

Fig. 38
Refueling times for shore- and tender-based aircraft

The time required for refueling shore- and tender-based airplanes at a standard land-plane/seaplane base or larger air base is indicated in figure 38. When it came to rearming these planes, machine-gun ammunition, torpedoes, bombs, mines, and eleven-inch and five-inch rockets could be replaced in the same time required for refueling. Groups of land planes could be rearmed with the first hundred bombs in the time required for refueling, but each additional hundred bombs or any increment of a hundred required an additional forty minutes. Seaplanes on parking aprons required the same time to rearm as did land planes; each seaplane at a mooring could be rearmed with the equivalent of four five-hundred-pound bombs in the refueling time. Each additional sixteen five-hundred-pound bombs, their equivalent, or any increment thereof would require an additional forty minutes. Each seaplane could also be rearmed by a seaplane tender (AV/AVP) with four five-hundred-pound bombs during the time required for refueling, but each additional four five-hundred-pound bombs, their equivalent, or any increment thereof would take an additional twenty minutes for each plane. (Pages i-4 to i-5.)

Battleships, cruisers, and submarines required thirty minutes to refuel and rearm one or more seaplanes after hoisting them aboard. At night and in rough seas, this time was increased by 25 percent. Carrier planes could be landed, refueled, rearmed, and launched again within various times under various conditions outlined in figure 39. One hour was required on a carrier to rig and install smoke-filled bombs on an airplane and two hours were required to fill the smoke tanks themselves. (Pages i-5 to i-6.)

<u>Rule I-6</u>. (a) Carrier based airplanes may be landed, refueled, rearmed and launched within the following time limitations:

(1) Landing time	1/2 min. X No. of planes.
(2) Refueling and respotting	1 min. per plane. Minimum time 15 minutes.
(3) Machine gun ammunition may be replaced during respotting and refueling. Rearming with other weapons may proceed concurrently with these operations but will require the following time, starting with the landing of the last plane:	
One bomb per plane	1 min. per bomb
Additional Bombs per plane	1/2 min. for each additional bomb.
Mines	Same time as for bombs.
Torpedoes	5 min. per plane. Minimum time required 15 min.
Rockets 5"	15 sec. per rocket.
Rockets 11"	Same time as for bombs.

(b) One hour is required on a carrier to rig and install filled smoke tanks in an airplane. Two hours are required to refill smoke tanks.

Fig. 39
Carrier landing, service, and launching times

Airplane losses due to noncombat operational casualties could be assigned to any flight as the Director deemed appropriate. These losses, like those impacting other aspects of the maneuver, could become “excessive” if fleet doctrine concerning upkeep and personnel rest was disregarded. Airplanes that expended more than 80 percent of their fuel, for instance, would crash if the Director so determined. The rules also extended to speed and navigation. Time of flight was taken to the nearest five minutes, ground speed used in plotting aircraft flights to the nearest five knots. Ground speed, moreover, was to be obtained by subtracting the full wind velocity from the airspeed of the aircraft when the true wind was against the base course of the aircraft. Ground speed was obtained by adding the full wind velocity to the airspeed of the aircraft when the wind was with the base course of the aircraft; for plotting purposes, aircraft would be considered to make good their steered courses; lateral drift was disregarded. (Page i-6.)

When photographic missions were planned, Flight Forms indicated precise information desired. All aircraft had gun cameras that might be used for damage evaluation; handheld cameras were available for each observation plane and each patrol plane. Patrol planes were equipped with radarscope cameras. Quick prints could be made from a limited number of selected exposures from all of this equipment within one hour after return to the base or carrier. Three hours, however, were required to make prints other than quick prints. Information obtained from photographs was to be furnished to the Student Commander by the Director. (Pages i-6 to i-7.)

Airships could enter or leave hangars when the wind was Force 4 or less, or in Force 5 or more provided that the wind was within forty-five degrees of longitudinal axis of the hangar. The airship could moor to a mast in wind from any direction and could land in wind up to Force 4. Airships could be reserviced in two to six hours, depending on the duration of the previous flight. Finally, aircraft losses due to aerial combat and antiaircraft gunnery, along with percentage of damage done by bomb or rocket hits, were recorded in separate confidential pamphlets. (Page i-7.)

The rules detailed above illustrate the parameters in which staff and students were to conduct the war games. Bear these rules and parameters in mind as exercises and Operations Problems are recounted in the following chapters, since the Maneuver Rules set the context in which the students encountered the hypothetical enemies placed before them.*

* For an even more extensive examination of Naval War College war-game Maneuver Rules of the period, see Hal M. Friedman, *Blue versus Orange: The U.S. Naval War College, Japan, and the Old Enemy in the Pacific, 1945–1946* (Newport, R.I.: Naval War College Press, 2013), chaps. 2–5, 11.

IV Search Exercises, Chart Maneuvers, and Relative Movement Exercises, July 1946

In early July 1946, the junior and senior classes of June 1947 began preparations for a Search Exercise—including the demonstration of a Chart Maneuver—that was to take place the following month. As recounted by Rear Admiral Smith in an instruction dated the 2nd, the purpose of the exercise was to acquaint students with the principles, forms, and standard methods of naval search. In addition, the exercise was to provide instruction in the mechanics of a Chart Maneuver and familiarize student officers with the publications and materials used. These publications included the Naval War College Maneuver Rules (summarized in the previous chapters) and such manuals as *Aerial Search for Surface Targets, Fast Carrier Task Force Instructions, The Scouting Manual,* and the *U.S. Navy Combat Airplane Characteristics and Performance Data Sheets.* Materials included Strategic Plotting Charts 7 and 8, drawing instruments, tracing paper for flimsies, and such standard forms as the Aircraft Flight Form, Fuel Account and Fuel Work Sheets, Message Form, and the Record of Move. The exercise itself was to take place between 2 and 7 August 1946.

On 2 August, students would draw the Statement of Exercise, study the statement and the above publications, and assemble in the auditorium to commence the exercise. Move 1 and part of 2 would occupy 3–5 August; the remainder of Move 2 and all of Moves 3 and 4 would take place on 6 August. Students would turn in flimsies, Flight Forms, Fuel Account and Fuel Work Sheets, Message Forms, and Records of Moves to the Master Plot as in a regular maneuver. On completion, they would draw the Staff Solutions and assemble in the auditorium for a discussion. Students worked in assigned rooms, sending forms to the Master Plot by pneumatic tube or messenger and receiving special announcements or instructions over the general announcing system. Members of the staff would visit student officers' rooms during the exercise to answer individual questions; forms and paper for flimsies were found in racks outside the Maneuver Room in Pringle Hall, as well as in racks in the corridors of Pringle and Luce Halls.[1]

Students unfamiliar with chart work were given plotting instructions. The nautical mile was taken to be six thousand feet, two thousand yards, or one minute of latitude (measured at the mid-latitude of a track). Courses were to be plotted in degrees, measured clockwise from 000 degrees at the north, and wind was described by the direction from which it was blowing.[2]

Statement of the Exercise: The Blue Situation

In the exercise, Blue, or the United States, was in a tense relationship vis-à-vis a coalition of western Pacific powers collectively known as Black. War appeared imminent, and Blue assessed that if Black decided to open hostilities, a sneak attack on the Panama Canal might be made to destroy canal installations and deny its use to Blue. Intelligence reports indicated that Black might have fitted out four or five merchant ships of about eighteen knots maximum speed as heavily armed auxiliary cruisers, "well disguised" with "outwardly innocent" merchant-ship silhouettes. It was also reported that these ships carried six small but "highly efficient" floatplanes, capable of delivering attacks with weapons of a "new and highly destructive" type. These ships, last reported in western Pacific bases, had not been located for the last two weeks.

Black naval forces were "generally concentrated" in the western Pacific and Blue fleet dispositions were such that any major Black move to the east could be detected early. In view of the threat to the Panama Canal, Blue Naval Air Station Seymour Island, in the Galapagos Islands, had been activated. Fleet Air Wing (FAIRWING) 20 was to be established there; two patrol squadrons, VP-14 and VP-15, each consisting of fifteen PBM-5 Mariner naval patrol bombers, were currently present for duty. In addition, Blue Task Group (TG) 18.1 was off the coast of Baja California en route to the Gulf of Panama to provide additional protection for the canal. This task group consisted of Carrier Division (CARDIV) 3, with two fleet carriers under Rear Admiral BC; Cruiser Division (CRUDIV) 11, four light cruisers (CLs) under Rear Admiral BD; and Destroyer Squadron (DESRON) 19, eight destroyers under Captain BJ. The entire task group was commanded by Admiral BC as Commander, Task Group (CTG) 18.1, and was in a standard cruising disposition on course 125 degrees at a speed of fifteen knots. By 1800 on 15 July, Cruiser Division 11 and Destroyer Squadron 19 had just completed refueling from a tanker, and Task Group 18.1 was at latitude 19°25′ N, longitude 107°07′ W.[3]

Move 1 covered from 0500 on 15 July to 0600 on 16 July. The sea was smooth, meteorological visibility was fifteen miles, and the wind was Force 2, from 150. Cloud cover was 50 percent and broken, while sunrise was assumed to take place at 0600 and sunset at 1800 for all moves. Late in the afternoon of 14 July, Captain BK, Commander, FAIRWING (COMFAIRWING) 20 at Seymour Island, received instructions from his immediate superior to search for possible Black raiders approaching the Panama Canal. He was to assume that the maximum speed of the raiders was eighteen knots and establish a barrier patrol on the ninetieth meridian West from an initial point at latitude 00°00′, longitude 90°00′ W to a northern

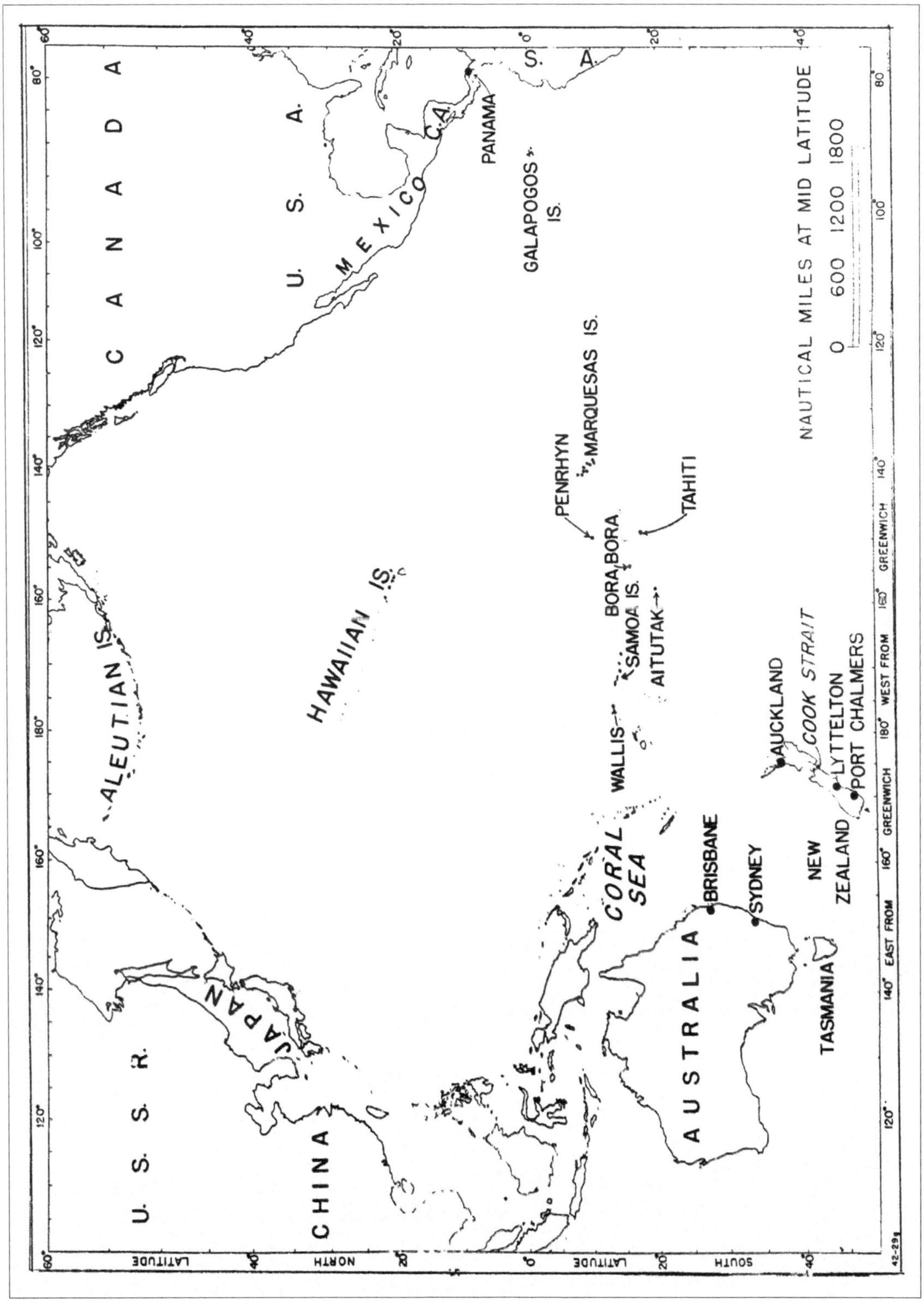

Map 2
Hawaii, Panama, and the South Pacific

limit of latitude 11°00′ N, in order to detect passage of ships from the west. This patrol was to be conducted by single-plane flights and a 60 percent probability of detection was assumed. This patrol was to commence at 0600 on 15 July. There was also to be a daily eight-plane parallel sweep, with the southernmost plane's track between Point A (5°00′ S, 90°00′ W) and Point B (07°00′ S, 83°00′ W), with a 62.5 percent probability of detection. Search planes were to pass through their initial points at 0800 daily. At 1445 on 15 July, Barrier Naval Patrol Plane 14-P-1 sighted a suspicious merchant ship similar in appearance to the suspected Black raiders. Sighting the plane, the ship turned west at a high speed of about eighteen knots. As this move ended, students filled out the Plots and Flight Forms for the barrier patrol and parallel-sweep searches, approximated the position of the Black raider when sighted, and plotted the track of TG 18.1 commencing at 1800 on 15 July. They used plus-six time, maintained the Fuel Account and Fuel Work Sheets for light cruiser CL-81 throughout the exercise and were mindful of the Fuel Expenditure Table for that class of ship.[4]

Fuel expenditure table for CL-81 class:

SPEED	10	11	12	13	14	15	16	17	18	19	20
BBLS/HR	15.9	17.4	18.9	20.8	22.8	25.1	27.8	30.8	34.1	38.0	42.3

SPEED	21	22	23	24	25	26	27	28	29	30	31
BBLS/HR	47.7	52.9	59.3	66.2	73.3	80.5	89.3	100.3	114.0	132.8	163.0

Fig. 40
Fuel expenditure tables for CL-81 class

In Move 2, from 0600 on 16 July to 0600 on 17 July, weather conditions remained the same, except for intermittent rain squalls to the northwest of the Galapagos. A special search was sent out from Seymour Island at 0600 to regain and develop the contact of 15 July. At about 1600, Naval Patrol Plane 14-P-7 sighted a merchant vessel disappearing to the westward into a heavy rain squall at 5°15′ N, 95°55′ W. At 2100, Rear Admiral BD, Commander, CRUDIV 11, ordered his ships to have boiler power available for thirty knots. He informed them that at 2200 on 16 July the division would form a scouting line and conduct a search for the suspicious vessel using the Flank Sector Method. This scouting-line search assumed that the vessel was at 5°20′ N, 97°45′ W at 2200 on 16 July on a westerly course and that it would not pass northward of a line connecting its 2200 position and Clipperton Island. This search was to commence at 1800 on 17 July, using assumed target speeds between fifteen and eighteen knots. Carrier Division 3 and DESRON 19 were to rendezvous with a tanker, refuel, and by 0600 on 18 July be in position at 10°00′ N, 106°00′ W to launch a dawn search. While the students were not to plot the movements of Carrier Division 3 and Destroyer Squadron 19 before 0600 on 18 July, they were to track Cruiser Division 11, estimate the track of the Black raider, and record Move 1 for all forces. They were not required to plot the search flights from Seymour Island.[5]

In Move 3, from 0600 on 17 July to 0600 on 18 July, the seas had become moderate; the cloud cover was 40 percent, high, and scattered; and the wind was Force 4 from 170. There were no new developments; students tracked the movements of Cruiser Division 11 and recorded Move 2 for Task Group 18.1. They used twenty-five knots as the speed for the scout on the western station; the scouts changed course every four hours to follow their search curves; they did not retire at night in reduced visibility. In Move 4, from 0600 18 July to 0600 on 19 July, the wind was Force 2 from 090, seas were smooth; clouds were 50 percent, high and broken; and Meteorological Visibility was twenty miles. At 0600, fleet carrier CV-12 launched a sector search covering the southwestern quadrant, twelve SB2C-5 Helldiver attack planes (VAs) in Scout Loading Condition. The radius of the search was four hundred miles, between 130 degrees and 270 degrees true from the 0600 position of the carrier. After launching these planes, Carrier Division 3 and Destroyer Squadron 19 steamed on course 250 degrees at fifteen knots until it was time to recover the planes. The movements of the carrier division during the launch and recovery could not be shown, due to the small scale of the plotting sheets, but students plotted the movements for each ship of Cruiser Division 11, Carrier Division 3 as a whole, and DESRON 19, as well the sector search for six of CV-12's planes.[6]

In Move 4 Modification One, at 0805 Attack Plane 12-VA-1 reported two merchant ships in company moving eastward at high speed at 6°30′ N, 109°50′ W. When the plane closed to investigate, the two ships reversed course to 260 degrees at about eighteen knots. At 0820, Attack Plane 12-VA-11 reported a single ship on a westerly course at about sixteen knots at 4°05′ N, 107°00′ W. Ten minutes later, Rear Admiral BC, Commander, Task Group 18.1, directed Rear Admiral BD to discontinue CRUDIV 11's search and send one cruiser to close, investigate, and trail the contact made by 12-VA-11. Admiral BD himself, with the other cruisers, was to conduct a search with the Rear Direct Method for the two ships reported by 12-VA-1; these ships were to form a scouting line by 0400 on 19 July, scout to a distance of thirty-eight miles, or twice their maximum radar range, and maintain a scouting speed of twenty-four knots once the line was formed. Since his cruisers were radar equipped, BD was to form a scouting line at the point of the enemy's assumed position at 0400. BD was also to assume that the enemy's lowest speed was sixteen knots and search accordingly on course 260. Admiral BD detailed light cruiser CL-83 to trail the contact made by Attack Plane 12-VA-11. The students tracked light cruisers CL-80, -81, and -82; turned in the Contact Report by Attack Planes 12-VA-1 and 12-VA-11 on Message Forms; and recorded Move 4 for Task Group 18.1.[7]

Staff Solution of Search Exercises

In late July, a Staff Solution to the Search Exercise was issued, along with numerous sheets recording the moves noted above. The probability of detection was 60 percent; one plane cycle took 12.9 hours, and 3.4 cycles per day were required to cover the barrier continuously, which required that a plane start the patrol every seven hours. Since the Barrier Patrol was to detect the passage of ships from the west, the first leg was laid out from 00°00′, 90°00′ W in a direction 8.6 degrees to the right of the ninetieth meridian and extending to the eleventh parallel of north latitude. The next leg of the patrol was laid off to the west 55.2 miles along that latitude, and the following leg 8.6 degrees to the east of south to the equator. From this point, the plane returned directly to its base.[8]

AIRCRAFT FLIGHT FORM

FORCE FAIRWING 20 WEATHER AREA –

STUDENT ROE MOVE NO. 1

ROOM NO. 400 MANEUVER TIMES OF MOVE 0500/15 - 0600/16 July

List of Task Groups: Base Ship or Station	Squadron	Plane Numbers	Time of Start of Flight	Time of End of Flight	Duration of Flight	Air Speed, Knots	Altitude, Feet	Fuel in Planes at Start of Flight or Move	Bombs, Torpedoes or Special Equipment	Orders to Aircraft Task Group Commanders and Radio Frequency.
NAS SEYMOUR ID	14	14-P-1	0545	1849	13.1 hrs.	120	2000	2989/150	NONE	- - - -
"	14	2	1245	(NEXT DAY) 0149	13.1 hrs.	120	2000	2989/150	NONE	- - - -
"	14	3	1945	(NEXT DAY) 0849	13.1 hrs.	120	2000	2989/150	NONE	- - - -
NAS SEYMOUR ID	15	15-P-1 2 3 4	0530 0600 0630 0653	1730 1705 1640 1620	12.0 hrs 11.1 " 10.2 " 9.5 "	120 " " "	2000 " " "	2989/150	NONE	- - - -
"	15	5 6 7 8	0708 0704 0658 0634	1610 1610 1635 1705	9.0 " 9.1 " 9.6 " 10.5 "	" " " "	" " " "	2989/150	NONE	- - - -
CV 12	1	VA-1 2 3	0600 " "	1144 1142 1140	5.7 hrs " "	130 " "	5000 " "	553/32	NONE	- - - -
		4 5 6	" " "	1141 1142 1144	" " "	" " "	" " "	" " "	" " "	- - - -
		7 8 9	" " "	1147 1149 1153	5.8 hrs " 5.9 hrs	" " "	" " "	" " "	" " "	- - - -
		10 11 12	" " "	1157 1152 1143	6.0 hrs 5.9 hrs 5.7 hrs	" " "	" " "	" " "	" " "	- - - -

46-42

Fig. 41
Aircraft Flight Form

For the Parallel Sweep, with the probability of detection still 62.5 percent, spacing was set at sixty miles. For the Flank Sector Method search, scouts headed directly for their starting points on the limiting enemy radius when the signal to form the Scouting Line was received. Cruiser Division 11 was to scout from left to right in the order CL-83, -81, -80, and -82. The sector search by carrier aircraft was to be carried out with twelve planes, each having a 7.5-degree sector to search. Plane 1 was to have the right-center sector, facing in the direction of the search, and Plane 7 the left-center sector; Planes 2, 3, 4, 5, and 6 were to have the sectors in that order from the right of Plane 1, while Planes 8, 9, 10, 11, and 12 were to have the sectors in that order from the left of Plane 7. Since the maximum chord of each search sector was greater than twice the visibility, each plane had to fly three legs. Planes 8, 9, 10, 11, and 12 would have to fly a fourth leg to cover all of their assigned sectors.[9]

Finally, for the Rear, Direct Method search, students were instructed to have the scouts head directly for their initial positions on the Scouting Line, which was a line normal to the assumed enemy course drawn through his assumed 0400 position. The order of the Scouting Line from left to right was to be light cruisers CL-81, -80, and -82, the distance between them thirty-eight miles, twice the maximum radar range of the ships.[10]

Exercise in Developing Task Force Dispositions

Also late in July, the junior and senior classes of June 1947 were put through an exercise in "Developing Task Force Dispositions." Admiral Smith again introduced

Message Form
NWC, May 1946

MANEUVER MESSAGE FORM

FROM: PILOT (TITLE) 12-VA-1 (SHIP) DOE (STUDENT COMMANDER) 500 (ROOM)

Check One: Action	Check One: Info	Addressees: Title	Addressees: Student Commander	Addressees: Room	Delays in Minutes due to: Time to Draft (RULE E-6)	Delays in Minutes due to: Time to Transmit (RULE E-7)	Delays in Minutes due to: #Other Causes (RULE E-8)	Maneuver #Time of Delivery (RULE E-9)
x		CTG 18.1	DOE	500	4 min	2 min	-	0811

Maneuver time drafting commenced 0805
Greenwich Civil Maneuver "Time of Origin" (plus date) 181405 (Rule E-12)

Two	freighters	latitude	zero	six	5
thirty	north	longitude	one	zero	10
nine	fifty	west	course	two	15
six	zero	speed	seventeen		20
					25
					30
					35
					40
					45
					50
					55
					60
					65
					70
					75
					80
					85
					90
					95
					100

Number of words to the nearest ten 20
System (Rule E-1) Keyed radio
Circuit (Rule E-2, 4) -
#UMPIRE'S File No. 2

To be filled in by the UMPIRE.
All other data to be filled by the student commander and verified by the UMPIRE.

DO NOT USE BACK OF THIS FORM FOR ADDITIONAL TEXT

Directions for preparing messages for duplication:– Place a sheet of "special duplicating carbon", face up, under message form. Use typewriter or hard pencil with celluloid sheet or other hard surface beneath carbon. Illegible writing or typing may cause further delays

16-43

Message Form
NWC, May 1946

MANEUVER MESSAGE FORM

FROM: PILOT (TITLE) 12-VA-11 (SHIP) DOE (STUDENT COMMANDER) 500 (ROOM)

Check One: Action	Check One: Info	Addressees: Title	Addressees: Student Commander	Addressees: Room	Delays in Minutes due to: Time to Draft (RULE E-6)	Delays in Minutes due to: Time to Transmit (RULE E-7)	Delays in Minutes due to: #Other Causes (RULE E-8)	Maneuver #Time of Delivery (RULE E-9)
x		CTG 18.1	DOE	500	4 min	2 min	-	0826

Maneuver time drafting commenced 0820
Greenwich Civil Maneuver "Time of Origin" (plus date) 181420 (Rule E-12)

One	freighter	latitude	zero	four	5
zero	five	north	longitude	one	10
zero	seven	zero	zero	west	15
course	two	seven	zero	speed	20
sixteen					25
					30
					35
					40
					45
					50
					55
					60
					65
					70
					75
					80
					85
					90
					95
					100

Number of words to the nearest ten 20
System (Rule E-1) Keyed radio
Circuit (Rule E-2, 4) -
UMPIRE'S File No. 1

To be filled in by the UMPIRE.
All other data to be filled by the student commander and verified by the UMPIRE.

DO NOT USE BACK OF THIS FORM FOR ADDITIONAL TEXT

Directions for preparing messages for duplication:– Place a sheet of "special duplicating carbon", face up, under message form. Use typewriter or hard pencil with celluloid sheet or other hard surface beneath carbon. Illegible writing or typing may cause further delays

-43

Fig. 42
Maneuver Message Form

FUEL WORK SHEET

Vessel(s) CL-81

Move No.	1	2	2	2
From (date and hour)	1800/15	0600/16	2100/16	2200/16
To (date and hour)	0600/16	2100/16	2200/16	0600/17
(1) HOURS	12	15	1	8
(2) ENGINE SPEED MADE	15	15	15	24.2
(3) BARRELS - PER HOUR, normal	25.1	25.1	25.1	67.6
Incrd. by add. steam (10%)	-	-	2.5	6.8
Station-keeping (3%)	0.8	0.8	0.8	-
Paravanes (5%)	-	-	-	-
(4) BARRELS - PER HOUR	25.9	25.9	28.4	74.4
(5) FUEL EXPENDITURE (1)x(4)	310.8	388.5	28.4	595.2
(6) PORT EXPENDITURE	-	-	-	-
(7) TOTAL FOR MOVE	310.8	388.5	28.4	595.2

Vessel(s) CL-81

Move No.	3	3	4	4
From (date and hour)	0600/17	1800/17	0600/18	0830/18
To (date and hour)	1800/17	0600/18	0830/18	0400/19
(1) HOURS	12	12	2.5	19.5
(2) ENGINE SPEED MADE	26.2	23.3	22.3	24.5
(3) BARRELS - PER HOUR, normal	82.3	61.4	54.8	69.8
Incrd. by add. steam (10%)	8.2	6.1	5.5	7.0
Station-keeping (3%)	-	-	-	-
Paravanes (5%)	-	-	-	-
(4) BARRELS - PER HOUR	90.5	67.5	60.3	76.8
(5) FUEL EXPENDITURE (1)x(4)	1086	810	150.8	1497.6
(6) PORT EXPENDITURE	-	-	-	-
(7) TOTAL FOR MOVE	1086	810	150.8	1497.6

Use one column for each period of a move during which conditions as to speed and fuel expenditure remain unchanged.

FUEL WORK SHEET

Vessel(s) CL-81

Move No.	4			
From (date and hour)	0400/19			
To (date and hour)	0600/19			
(1) HOURS	2			
(2) ENGINE SPEED MADE	24			
(3) BARRELS - PER HOUR, normal	66.2			
Incrd. by add. steam (10%)	6.6			
Station-keeping (3%)	-			
Paravanes (5%)	-			
(4) BARRELS - PER HOUR	72.8			
(5) FUEL EXPENDITURE (1)x(4)	145.6			
(6) PORT EXPENDITURE	-			
(7) TOTAL FOR MOVE	145.6			

Vessel(s)

Move No.				
From (date and hour)				
To (date and hour)				
(1) HOURS				
(2) ENGINE SPEED MADE				
(3) BARRELS - PER HOUR, normal				
Incrd. by add. steam (10%)				
Station-keeping (3%)				
Paravanes (5%)				
(4) BARRELS - PER HOUR				
(5) FUEL EXPENDITURE (1)x(4)				
(6) PORT EXPENDITURE				
(7) TOTAL FOR MOVE				

Use one column for each period of a move during which conditions as to speed and fuel expenditure remain unchanged.

Fig. 43
Fuel Work Sheet

Blue T.G. 18.1
(Task Group, as BLUE Main Body)

DOE
(Name of Officer, as Jones)

F U E L A C C O U N T F O R M O V E

Blue CruDiv 11
(Organization unit, as BLUE Batron One)

VESSEL	CL-81				
On hand, beginning of move	13850				
Received: Amount	-				
Place	-				
From: (Ship)	-				
Time: (Hours)	-				
Expended	310.8				
On hand, End of Move	13539.2				

Total during Maneuver up to End of this Move

Expended	310.8				
Received	-				

VESSEL					
On hand, beginning of move	13539.2				
Received: Amount	-				
Place	-				
From: (Ship)	-				
Time: (Hours)	-				
Expended	1012.1				
On hand, End of Move	12527.1				

Total during Maneuver up to End of this Move

Expended	1322.9				
Received	-				

VESSEL					
On hand, beginning of Move	12527.1				
Received: Amount	-				
Place	-				
From: (Ship)	-				
Time: (Hours)	-				
Expended	1896				
On hand, End of Move	10631.1				

Total during Maneuver up to End of this Move

Expended	3218.9				
Received	-				

Blue T.G. 18.1
(Task Group, as BLUE Main Body)

DOE
(Name of Officer, as Jones)

F U E L A C C O U N T F O R M O V E

Blue CruDiv 11
(Organization unit, as BLUE Batron One)

VESSEL					
On hand, beginning of move	10631.1				
Received: Amount	-				
Place	-				
From: (Ship)	-				
Time: (Hours)	-				
Expended	1794				
On hand, End of Move	8837.1				

Total during Maneuver up to End of this Move

Expended	5012.9				
Received					

VESSEL					
On hand, beginning of move					
Received: Amount					
Place					
From: (Ship)					
Time: (Hours)					
Expended					
On hand, End of Move					

Total during Maneuver up to End of this Move

Expended					
Received					

VESSEL					
On hand, beginning of Move					
Received: Amount					
Place					
From: (Ship)					
Time: (Hours)					
Expended					
On hand, End of Move					

Total during Maneuver up to End of this Move

Expended					
Received					

Fig. 44
Fuel Account for Move

the exercise, which took place between 26 and 31 July: a Statement of Exercise was issued on 26 July, which students were to study; an orientation in the auditorium of Pringle Hall was to take place on the same day; and a series of exercises was to take place over the subsequent days. Students also received lectures on the "Capabilities and Limitations of Fleet AA [Antiaircraft] Defense" and "Guided Missiles," and "Anti-Submarine Warfare" and viewed a training film, *Battle of Empress Augusta Bay*. On the last day of the evolution, senior and junior students assembled separately for discussions of Student Solutions, which were to be turned in after the discussions.[11]

One of the objectives of the exercise was to familiarize students with various publications, including *War Instructions, General Tactical Instructions, Current Tactical Orders and Doctrine, U.S. Fleet,* and the tactical publications for ship and aircraft types. The exercise was also to familiarize students with the employment of the various types of ships and aircraft, especially in mutual support. Students drew diagrams of task force (TF) dispositions on maneuvering board sheets, using a circle spacing of a thousand yards and a scale that showed the entire disposition on one sheet. Diagrams showed the position of each surface ship or each unit of small ships in close formation, as well as the position and schedule of air patrols or air task units (TUs), fighter-direction ships, assignment of radar guards (including radar pickets) and their sector assignments, and radar-intercept ships. It was assumed that battleships, cruisers, carriers, and fighter-direction ships were fitted with radar-intercept equipment.[12]

Exercise A, the Cruising Disposition, involved a balanced fleet organization of three fleet carriers and one light carrier, organized as Carrier Division 3, under Admiral BA. Rear Admiral BD commanded two battleships, Battleship Division (BATDIV) 9, while Rear Admiral BE commanded Cruiser Division 16, two battle cruisers. Cruiser Division 14, under Rear Admiral BF, was normally a unit of four light cruisers but in this case had only two ships. Rear Admiral BG commanded Cruiser Division 11, normally six light cruisers but in this case consisting of a single antiaircraft cruiser (CLAA). Destroyer Squadron 47, under Captain B-1, had nine destroyers, five of them under B-1 as Destroyer Division (DESDIV) 93 and the other four comprising DESDIV 94, under Captain B-2. DESRON 54, under Captain B-3, completed the organization; it had the five destroyers of DESDIV 107, under B-3, and another four under Captain B-4, Destroyer Division 108.[13]

Aircraft included, first, fifty-seven F6F Hellcat fighter planes, twenty-four Helldivers, and twenty TBM-3 Avenger attack planes, organized into two fighter and two attack squadrons on fleet carrier CV-33. Fleet carrier CV-21 carried twenty-five F8F Bearcat fighters, along with thirty-two Hellcat fighters, and twenty-four Helldivers and twenty Avengers, with a similar squadron organization. Fleet carrier CV-34, also with two fighter squadrons and two attack squadrons, carried forty-nine F4U-1 Corsair fighters, along with four Hellcat night fighters and four Hellcat photoreconnaissance planes (which the other fleet carriers also carried), as well as twenty Helldivers and twenty Avengers. Light carrier CVL-27 carried twenty-two Bearcat fighters, two Hellcat photographic planes, and nine Avengers in one fighter squadron and one attack squadron. In addition, Battleship Division 9 carried an observation detachment of four SC-1 Seahawk observation planes; Cruiser Divisions 16 and 14 each had similar detachments on board, and Cruiser Division 11 had a detachment of the same type of plane but only two of them.[14]

The situation for Exercise A included a formation that was within range of enemy aircraft and in an area where enemy submarines might be operating. Weather conditions included visibility variable out to ten miles, frequent rain squalls, a ceiling of five thousand feet, a temperature of 45 degrees Fahrenheit, and a wind from the southeast at Force 3. Sunrise was at 0745, sunset 1645. Enemy submarine torpedoes could travel at forty-five knots and had a range of five thousand yards. Students listed the assigned units in a tactical organization, showing the command relationship, and produced a diagram for a Day Cruising Disposition.[15]

In Exercise B, the Approach Disposition, Carrier Division 8 under Admiral BA—again three fleet carriers and one light carrier—was accompanied by Battleship Division 6 (three battleships under Rear Admiral BC) and Battleship Division 9 (now two battleships under Rear Admiral BD). Rear Admiral BE commanded Cruiser Division 16 of two battle cruisers, Rear Admiral BF commanded the four light cruisers of Cruiser Division 14, and Captain B-5 commanded Destroyer Squadron 25, five of its destroyers in Destroyer Division 49, under him, and four more in DESDIV 50, under Captain B-6. DESRON 47 consisted of five ships from DESDIV 93 under Captain B-1 and four under Captain B-2 in DESDIV 94; DESRON 54 under Captain B-3 similarly comprised five ships directly under B-3 in DESDIV 107 and four under Captain B-4 in DESDIV 108. Finally, eight destroyers of DESRON 60 were commanded by Captain B-7, four of them directly under B-7 as Destroyer Division 119 and the other four under Captain B-8 as DESDIV 120. Exercise B involved the same aircraft as Exercise A, plus a detachment of six observation planes for Battleship Division 6 and four more for Cruiser Division 14. The situation entailed a major enemy force fifty miles distant forming an Approach Disposition, evidence that it was accepting a surface engagement. At this point, fleet carriers CV-21 and CV-34 and light carrier CVL-27 had been sunk. The enemy force consisted of six battleships similar in characteristic to the *Indiana*- and *South Dakota*–class fast battleships, four heavy cruisers similar to the *Baltimore* class, four light cruisers similar to the *Cleveland* class, and twenty-four destroyers similar to the *Fletcher* class. Enemy air strength had been neutralized except for an estimated eight fighter planes and eight attack planes. The enemy force was on a course of 230 degrees at twenty knots and bore 050. The nearest enemy base was five hundred miles away to the northeast; neither force had land-based air support available, nor were additional ship-based aircraft available to either. Weather conditions, sunrise, and sunset were the same as in Exercise A. The students listed the assigned ships, showing command relationships, as in Exercise A, and produced a diagram for an Approach Disposition.[16]

Exercise C was a Battle Disposition. The fleet and aircraft organizations were the same as for the start of Exercise B, except that fleet carriers CV-21 and CV-34 and light carrier CVL-27 had been sunk and Blue had no carrier aircraft left. Weather conditions were the same. By this time, the enemy force was deploying for battle. Both forces had high-speed torpedoes available with ranges of twenty-six thousand yards at fifty knots and 40,600 yards at forty knots. Enemy air strength by this time had been neutralized, but Blue still had observation planes on its battleships and cruisers for spotting. Students drew a diagram of a Battle Disposition for the forces listed and a diagram for a "normal" action.[17]

Exercise D entailed another Battle Disposition. By now, the fleet consisted of six light cruisers under Rear Admiral BF, as Cruiser Division 14, and twelve destroyers

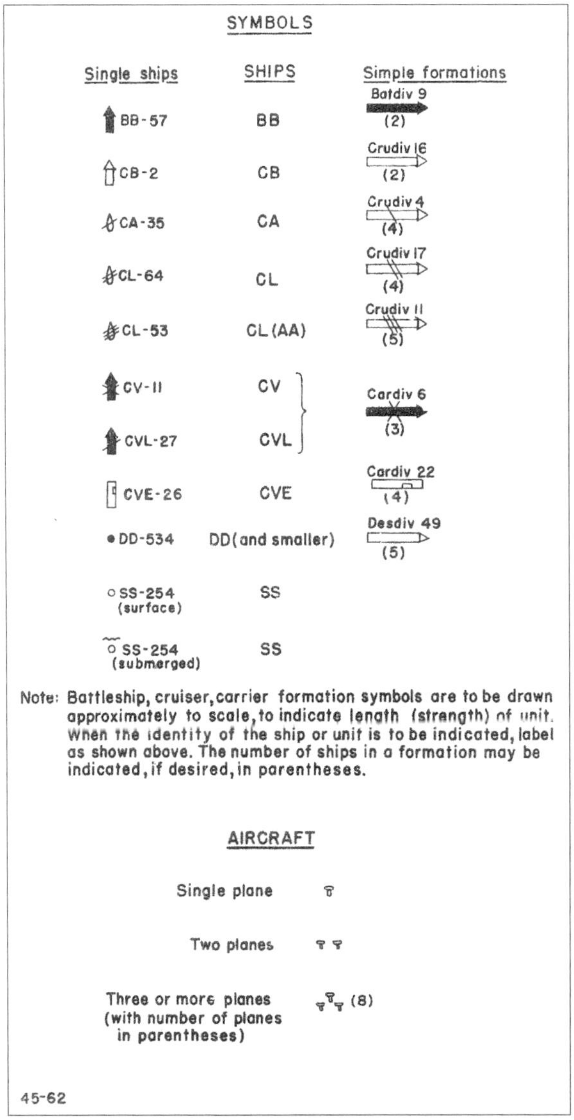

Fig. 45
Ship and formation symbols

of Destroyer Squadron 25, under Captain B-5 (five ships directly under him as DESDIV 49, four under Captain B-6 as DESDIV 50, and three under Captain B-7 as DESDIV 119). It was 2000; an enemy task group consisting of one fast battleship, four light cruisers, and eight destroyers had been sighted at sunset, bearing north a hundred miles from Blue's transport loading area. The enemy's course was 180 degrees, its speed twenty-five knots. A land-based radar search plane was tracking the enemy force and reporting continuously to Admiral BF. Admiral BF's mission was to protect the landing of Blue troops and supplies. There was almost continuous light rain, visibility of zero to two thousand yards, a ceiling of zero to five hundred feet, and a wind from the northeast with Forces 1 to 3. Sunrise was at 0700, and the sun had set at 1700. Students produced a diagram of a Blue formation for a surprise night light-force (i.e., light cruisers, destroyers, and destroyer escorts) torpedo attack on the enemy.[18]

Exercise E involved Blue Task Force 18, consisting of Task Groups 18.1, 18.2, and 18.3. Each task group was similar in number and type to the ships and aircraft listed in Exercise A, and they occupied Stations 1, 2, and 3, respectively, in the Task Force Disposition. The task force was approaching a group of enemy-held islands to conduct a daylong sustained strike operation to inflict maximum damage on shore installations and on any enemy aircraft present. Two of the islands had airstrips from which shore-based patrol planes might operate; enemy fighter, attack, and patrol aircraft were known to be based there. The number of these planes, however, had not been ascertained since photographic intelligence of the target area was "meager." Two nights before, one of Blue's night-search patrol planes had reported six ships in the harbor of the largest island and had received heavy antiaircraft fire from the ships and the shore installations. Air searches by Blue shore-based patrol planes during the last week had not located any enemy surface force near enough to threaten the proposed operation; these searches were being continued in "sufficient" scope to have detected any departure of enemy ships from the target area. A Blue submarine was also to operate on lifeguard duty (i.e., pilot rescue) within a designated area in the immediate vicinity of the islands.[19]

At the time of the first strike's launch, the general target area was to be 150 miles distant on a bearing of 015 true. The launching position for the first strike was to be reached by 0500. Task Group 18.3 had been assigned to strike one of the islands containing an airstrip with possible enemy aircraft present. The other two task groups had been assigned "suitable" targets, including ships in the harbor. Fast Carrier Task Force Special Picket Destroyers and Rescue Combat Air Patrol were to be provided by these two task groups. No Anti-Submarine Patrol was required but a TCAP—Top (high-altitude) Combat Air Patrol—was to be maintained throughout the day. The latter could be reduced in size as damage to the target area mounted. Times of launch and order and types of strikes by TG 18.3 corresponded with the general plan of the Task Force Commander. Weather called for twenty-mile visibility and cloud coverage of five-tenths, with top clouds at seven thousand feet and the base at five thousand. The wind was from the southeast at Force 3, sunrise was at 0745, and sunset was at 1645. Students produced a diagram of the Task Force Disposition, including pickets, aircraft radio homing beacon approach sectors for returning aircraft, and the paths of outgoing and incoming strikes. The diagram was additionally to show the Air Plan of the Day for Task Group 18.3 and to indicate the location, altitude, and composition of all Combat Air Patrols (CAPs) employed by TG 18.3, except for TCAP.[20]

NOTES

1 Allan Smith, "Search Exercise: Including a Demonstration of a Chart Maneuver," 2 July 1946, pp. 1–3, folder 2581, box 135, RG 4, NHC. Pringle and Luce Halls were and are connected at the second-floor level by an enclosed walkway, visible in the photograph on the back cover of Hal M. Friedman, *Digesting History: The U.S. Naval War College, the Lessons of World War Two, and Future Naval Warfare, 1945–1947* (Newport, R.I.: Naval War College Press, 2010).

2 Smith, "Search Exercise," p. 4.

3 Ibid., p. 5. At the Naval War College, Black typically meant Germany, but here it is obviously either a resurgent Japan or the Soviet Union, the latter often obliquely referred to in the United States in this period as "any other power." For country color codes, see Michael Vlahos, *Blue Sword: The Naval War College and the American Mission, 1919–1941* (Newport, R.I.: Naval War College Press, 1980), p. 163. For American perceptions of potential threats from Japan and the Soviet Union between 1945 and 1947, see Hal Friedman, *Creating an American Lake: United States Imperialism and Strategic Security in the Pacific Basin, 1945-1947* (Westport, Conn.: Greenwood, 2001), pp. 1–62.

4 Smith, "Search Exercise," p. 6.

5 Ibid., p. 7.

6 Ibid., p. 8.

7 Ibid., p. 9.

8 "Staff Solution of Search Exercises including a Demonstration of a Chart Maneuver," 29 July 1946, p. 1, folder 2581-A, box 135, RG 4, NHC.

9 Ibid., pp. 1–2.

10 Ibid., p. 2.

11 "Exercise in Developing Task Force Dispositions: Schedule of Employment," 26 July 1946, p. 1, folder 2583, box 135, RG 4, NHC.

12 Ibid., pp. 2–3.

13 Ibid., p. 3.

14 Ibid., pp. 3–4.

15 Ibid., p. 4.

16 Ibid., pp. 5–6.

17 Ibid., pp. 6–7.

18 Ibid., pp. 7–8.

19 Ibid., pp. 8–9.

20 Ibid., pp. 9–10.

V *Maneuver Rule Supplements, Naval Communications, and More Relative Movement, August 1946*

On 10 August 1946, copies of "Board Maneuver Rules: Supplement to Maneuver Rules" were issued to the students. The first section, "Conduct of the Maneuver," described the Supplement as offering a way to "speed up" the conduct of Tactical Board Maneuvers. The Maneuver Rules, however, were to remain the basis for the maneuver except as modified by the Director or as modified by the paragraphs that followed. Move Sheets (no longer "Forms") were to be filled out by a member of the Student Commander's staff according to the latter's wishes; the staff member, or "Move Plotter," was to be designated. Gunfire Sheets (also replacing "Forms") were to be filled out and damage computed by a member of the Maneuver Staff, the "Damage Computer."[1]

The primary purpose of the Move Sheet was to help the Move Plotter keep track of his commander's ships on the Maneuver Board. The Move Sheet also served as a record of a ship's movements. It was subject to review by the Move Umpire, who would also check to determine that ships were being maneuvered according to their capabilities. Upon conclusion of the maneuver, the completed Move Sheets were to be turned into the Historian for use in the History of the Maneuver. The Gunfire Sheet was intended as an aid to the Damage Computer in determining the gunfire damage inflicted by the ships of the command to which he was assigned.

Prior to the end of a move during which his ships were firing, the Damage Computer was to notify the Force Damage Recorder of the opposing forces—another member of the Maneuver Staff—of the amount of damage sustained by target vessels during the move. The Force Damage Recorder was to keep track of the damage inflicted on vessels of his force for each move and the total damage inflicted on each ship for the maneuver. The Gunfire Sheets were subject to review by the Chief Damage Computer, who was to satisfy himself that the damage was being correctly computed and recorded. Upon conclusion of the maneuver, all Gunfire Sheets were turned over to the appropriate Force Damage Computer, who prepared a summary of the damage inflicted on his force. The Historian was then to receive the sheets. (Pages 1–2.)

MOVE SHEET

UNIT BB 58
MOVE PLOTTER Capt. B.A. SMITH

MOVE NO.	MANEUVER TIME	SHIP	SPEED at START (due to ENGINES)	CHANGE of SPEED (due to ENGINES)	SPEED at END (due to ENGINES)	AVERAGE SPEED (due to ENGINES)	LOSS on account of SEA	Final AVERAGE SPEED during move	COURSE at START of move	CHANGE of COURSE	COURSE at END of move
1	2	3	4	5	6	7	8	9	10	11	12
1	00-03	BB 58	20----	----	----	----	----	----	000	----	----
2	03-06	" "	20	+5	25	22.5	0	22.5	000	R90	090
			Fig.	1 (Example	page	4)					

46-67a

Fig. 46
Move Sheet

GUNFIRE SHEET

UNIT BB 58
DAMAGE COMPUTOR Capt. E.E. JONES

MOVE NO.	MANEUVER TIME	FIRING SHIP	GUNS FIRING Number & Caliber	TARGET SHIP	RANGE	REDUCTION for A.W. DAMAGE	REDUCTION by DIRECTOR	Total REDUCTION	MULTIPLIER	NORMAL DAMAGE (N.G.D. Tables)	DAMAGE (N.G.D. Tables)	CHANCE DAMAGE	TOTAL DAMAGE
1	2	3	4	5	6	7	8	9	10	11	12	13	14
1	00-03	BB 58	9-16"	BB 1	21	0	0	0	1.	10	10	10	10
2	03-06	" "	" "	" "	20	.1	.2	.3	.7	20	14	20	30
				Fig.	2 (Example page 4)								

46-67b

Fig. 47
Gunfire Sheet

Special Speed and Spotting Rules held that there was to be no reduction in a ship's speed because of sea conditions except as specifically imposed by the Director. All combatant vessels except destroyers could accelerate five knots during a three-minute move; destroyers could accelerate ten knots. The rate of deceleration was equal to that of acceleration. Vessels steaming singly could make the maximum speed that was given for the fleet or in the Statement of the Problem; vessels steaming in formation kept a three-knot speed reserve for station keeping. Any reduction in speed because of damage sustained was specified in the Ship Damage Report. Only the tracks of Division Guides were plotted when ships were in line of bearing. The tracks of individual ships were to be plotted by Drafting Room personnel, as required by the Director, when "convenient," so as not to interfere with the progress of the maneuver. (Pages 2–3.)

To compute Gunfire Damage, the tables in the Supplement would hereafter be used instead of the *Fire Effect Tables.* Reductions to multiplier *M* were prescribed by the Director in each instance and were in general based on those listed in Section F of the Maneuver Rules. The reduction of a ship's Fire Effect as given in Rule F-5 was assumed by the ship's commander as soon as damage was ascertained. It was applied in damage computations for the move following that during which it was sustained; computations were not held up pending the posting of the latest damage reports. The Total Damage listed for the unit concerned at the time of damage computation governed; this also applied to the losses of ship's equipment and speed incident to the damage sustained as tabulated in the Ship Damage Report. (Page 3.)

An example given for Gunfire Damage Computation had Blue's fast battleship BB-58 opening fire on Purple battleship BB-1 during Move 1 with nine sixteen-inch guns at twenty-one thousand yards. No reductions are assigned by the Director, the speed of the ships is twenty knots, no changes of course or speed occur during the move, and the student is to compute the Total Damage at the end of Move 2. The total reduction to multiplier *M* for Move 1 is zero; the Normal Gunfire Damage inflicted during a three-minute move on a B-class target (a thirty-five-thousand-ton battleship) such as Purple BB-1 at twenty-one thousand yards is 10 percent. Since BB-1 is penetrated by sixteen-inch shells, the student enters the Chance Sinking Tables with 10 percent damage. Since the first unused letter is *N*, the ship is not sunk. Since Chance Damage might be more or less than 10 percent, the student is to enter the Chance Damage Tables with that argument. Here, the first unused number is a one, so the Chance Damage due to gunfire for Move 1 is 10 percent and is in addition to other Above Water Damage that might have been inflicted by bombs or rockets.

In this example, the Director assigns a reduction to Blue fast battleship BB-58's multiplier *M* of two-tenths, because of the ship's change of course and speed during

SHIP DAMAGE REPORT

UNIT..

STUDENT.. END OF MOVE NO. ....

SUMMARY OF LOSSES DUE TO DAMAGE

30% ABOVE WATER DAMAGE	50% ABOVE WATER DAMAGE	70% ABOVE WATER DAMAGE	80% TOTAL DAMAGE
FIRE CONTROL ALL FIRE CONTROL & SEARCH RADAR destroyed SPEED SHIPS other than BB,CC,CB,CVB Lose 10% SPEED COMMUNICATIONS Time to deliver outgoing messages doubled	FIRE CONTROL CAs & smaller ships Lose GUN DIRECTORS & TOP SPOT, must use LOCAL CONTROL. ALL SHIPS Lose PLANE SPOT, SEARCHLIGHTS, SECONDARY & A.A. BATTERIES SPEED BB,CC,CB,CVB Lose 20% of SPEED, Other Types Lose 30% of SPEED. COMMUNICATIONS Lose all but EMERGENCY & SHORT RANGE VHF RADIO. Cannot transmit by SEARCHLIGHT. TORPEDOES Lose ½ of DECK TORPEDO TUBES on each side. MINES MINES can no longer be laid. AIRCRAFT ALL SHIPS cannot launch or land AIRCRAFT.	FIRE CONTROL ALL SHIPS Lose GUN DIRECTORS & TOP SPOT, must use LOCAL CONTROL. SPEED ALL SHIPS Lose 50% SPEED. COMMUNICATIONS Cannot transmit by RADIO or FLAG-HOIST. TORPEDOES Lose all DECK TUBES & TORPEDOES. MINES MINELAYERS with 50% of MINES on board will blow up.	FIRE CONTROL ALL SHIPS can no longer fire GUNS. SPEED ALL SHIPS reduced to 5 kts. maximum. COMMUNICATIONS Can no longer transmit or receive messages by RADIO, VISUAL or UNDERWATER SOUND.

ABOVE WATER DAMAGE	FIRE EFFECT MULTIPLIER "M" is reduced one tenth for each 10% of ABOVE WATER DAMAGE									
	10%	20%	30%	40%	50%	60%	70%	80%	90%	100%

UNDER WATER DAMAGE	SPEED LOSS is proportional to UNDER WATER DAMAGE and additional to that resulting from ABOVE WATER DAMAGE									
	10%	20%	30%	40%	50%	60%	70%	80%	90%'	100%

TOTAL DAMAGE	10%	20%	30%	40%	50%	60%	70%	80%	90%	100%

46-7, REV. 8/46

Fig. 48
Ship Damage Report

TABLE NO. 2 - TYPE "B" TARGET

BB (35,000 Ton)

Normal Gunfire Damage - Percentage

Target Life Lost Per 3 Minutes

No. guns firing	9	12	9	*9	12	10
Caliber	16"	14"	12"	*8"	6"	5"
RANGE (1000 yards)						
5-6	180	170	80	10	25	20
7-8	135	130	65	7	10	8
9-10	100	95	45	5	8	8
11-12	70	70	30	4	6	6
13-14	50	50	20	3	3	4
15-16	35	35	9	2	2	2
17-18	30	20	7	2	2	2
19-20	20	10	6	1	1	
21-22	10	6	4	1	1	
23-24	10	6	3	1	1	
25-26	5	2	1	.5	1	
27-28	6	2	1	.5		
29-30	6	3	2	.3		
31-32	6	5	3	.1		
33-34	3	2	1			
35-36	3					
37-38	2					
39-40	2					
41-42	1					

NOTE: Use range to the nearest thousand yards. Do not interpolate.

*For CAs with Automatic Loading, multiply values in this column by 2.

No penetration in range band between heavy lines.

Fig. 49
Table 2: Type B Target

30 PERCENT LIFE DESTROYED

Use Each Entry in Order, Proceeding Across the Rows from Left to Right. Mark off Each Entry as it is Used.

CHANCE SINKING

S: Sunk N: Not Sunk

NSNNN	NNNNN	NSNNN	NSNNN	SNSNN	SNNNN	NNNNS
NNSNN	SNNSS	NNNSN	NNNNS	NNNSS	NNNNN	NNNNN
NSNSN	NNNNS	SNNSS	NNNNS	NNSNN	NNNSN	SNNNS
SNSNN	NSSSN	NSNNN	SNNNN	NNSSN	SNNSS	NNSNS
NNNNN	SSNNN	NNSNN	SNNNS	SSNNN	NNNNN	NNSSN
NNNNN	NSSNN	NNSNN	NNNNN	NNNNS	NNNNN	NNNSN
NNNNS	NNNSN	NNSNN	NNNNN	NNNNN	SNSSN	NSSNN
SSSSN						

CHANCE DAMAGE

02205	11311	20002	32122	14222	00112	13302
30412	20112	01113	41231	31202	12103	11123
12133	15400	31100	20201	43232	35112	10002
00010	44110	12040	10011	21221	12023	21100
01201	11230	01212	02230	01000	03234	11001
12100	31322	22021	12222	00044	23013	03201
11231	02111	11121	33211	20101	11111	03322
12110						

Fig. 50
30 Percent Life Destroyed Table

TABLE NO. 11 - AMMUNITION EXPENDITURE

Total Rounds Per 3 Minutes

No. guns firing	9	12	9	* 9	12	10
Caliber	16"	14"	12"	* 8"	6"	5"
RANGE (1000 yards)						
5-6	54	63	54	72	252	399
7-8	48	63	48	66	24	411
9-10	42	57	42	66	23	587
11-12	39	54	39	60	225	375
13-14	36	48	36	57	225	363
15-16	33	45	33	54	210	351
17-18	33	42	33	54	207	342
19-20	30	39	30	51	204	
21-22	30	39	30	51	201	
23-24	30	39	30	48	198	
25-26	27	36	27	45	198	
27-28	27	36	27			
29-30	27	36	27			
31-32	27	36	27			
33-34	24	33	24			
35-36	24					
37-38	24					
39-40	21					
41-42	21					

NOTE: Use range to the nearest thousand yards. Do not interpolate.

*For CAs with Automatic Loading, multiply values in this column by 2.

Fig. 51
Table 11: Ammunition Expenditure

Move 2. The reduction to *M* for damage sustained on Purple battleship BB-1 is minus 1, and the total reduction to *M* for Move 2 is 0.3. By Move 2, the range has closed to twenty thousand yards and the Normal Gunfire Damage at this range was 20 percent. The damage inflicted by BB-58 on its target is now 0.7 times 20 percent, or 14 percent. The Chance Sinking Tables help the student determine that BB-1 is not sunk by this damage but sustained Chance Damage of 20 percent and Total Damage during Moves 1 and 2 equal to 30 percent. (Page 4.)

The assessment of Torpedo Damage was, in many cases, made by the Director. When operations were rapid and complex—with damage accumulating on attacking units—arbitrary assessments that promptly allocated definite damage could be made, and the concerned ships notified of such damage. It was thought that this procedure, by affording immediate results, was frequently preferable to attempting to figure out damage more accurately by the use of tables. Percentage of Hits Tables, however, for various conditions of torpedo fire were furnished for guidance and use when "practicable."

TORPEDO FIRE -- PERCENTAGE OF HITS FROM DESTROYERS AND CRUISERS

Firing Range (1000 Yards)	Target Speed 0-10 Kts. Unit Spread Used		Target Speed 10-20 Kts. Unit Spread Used		Target Speed 20-30 Kts. Unit Spread Used	
	1½°	1°	1½°	1°	1½°	1°
3-4	8%	9%	7%	8%	6%	7%
4-5	7%	9%	6%	8%	5%	7%
5-6	7%	9%	6%	8%	5%	7%
6-7	7%	8%	6%	7%	5%	6%
7-8	7%	8%	6%	7%	5%	6%
8-9	6%	8%	5%	7%	4%	6%
9-10	6%	8%	5%	7%	4%	6%
10-11	6%	7%	5%	6%	4%	5%
11-12	6%	7%	5%	6%	4%	5%
12-13	6%	7%	5%	6%	4%	5%
13-14	6%	7%	5%	6%	4%	5%
14-15	5%	6%	4%	5%	3%	4%
15-16	5%	6%	4%	5%	3%	4%
16-17	5%	5%	4%	4%	3%	3%
17-18	5%	5%	4%	4%	3%	3%

For each 5000 yard increase in range above 18000 yards decrease values in above table by 1%.

MULTIPLY VALUES FROM ABOVE TABLE BY FOLLOWING FACTORS ACCORDING TO TARGET CLASS

(BB, 45,000 tons)	(BB, 35,000 tons)	(OBB)	(CVB)	(CB)	(CV)	(CA)	(OCA 10,000 tons)	(CL, CVL)	(CVE) and misc.	(CL AA)	(DD) and misc.
A	B	D	E1	E	E1	F1	G	H	J	K	L
1.3	1.0	0.9	1.4	1.1	1.2	1.0	0.8	.9	.8	.8	.5

Fig. 52
Table 12: Torpedo Fire—Percentage of Hits from Destroyers and Cruisers

TORPEDO FIRE: PERCENTAGE OF HITS

FIRING UNIT, SUBMARINES

Relative Bearing or Smoke Bomb	Firing Range (1000 yards)	Target Speed in knots 0-15	15-20	20-30
0° (360°) to 45° (315°)	1-2	65	55	45
	2-3	56	48	39
	3-4	50	42	34
	4-5	43	35	29
	5-6	34	29	24
	6-7	27	23	18
	7-8	20	16	13
	8-9	12	10	8
	9-10	5	4	3
45° (315°) to 90° (270°)	1-2	75	70	65
	2-3	69	62	58
	3-4	58	54	50
	4-5	50	46	42
	5-6	42	38	34
	6-7	33	30	27
	7-8	24	22	20
	8-9	16	14	13
	9-10	8	7	5
90° (270°) to 135° (225°)	1-2	70	65	60
	2-3	62	57	53
	3-4	54	50	46
	4-5	47	43	39
	5-6	39	36	32
	6-7	31	29	25
	7-8	24	22	19
	8-9	16	14	12
	9-10	8	7	5
135° (225°) to 180°	1-2	55	45	35
	2-3	48	39	30
	3-4	41	33	25
	4-5	34	27	20
	5-6	27	22	0
	6-7	20	16	0
	7-8	12	0	0
	8-9	5	0	0
	9-10	0	0	0

Fig. 53
Table 13: Torpedo Fire: Percentage of Hits—Firing Unit, Submarines

The tables shown in figures 52 and 53 outline the conditions that were considered in making damage decisions and how percentages given could vary to fit the conditions. The tables also took into account the number of torpedoes fired that ran "hot, straight, and normal." Concerning Table 12 of the Rules (figure 52), if the target group took effective evasive action, the student was to multiply by 0.5 (i.e., reduce the percentage of hits by half). Each ship in the target group was subject to the percentage of hits obtained from this table; the hits were distributed by the Director. Fractional hits were not to be assessed; instead, one vessel of the target group was given a percentage chance (equal to the percentage of a hit) of receiving one hit. Table 16 (figure 56, below) was used to determine whether or not a hit had been scored, by means of a single throw of two dice by the students. Table 13 (figure 53) applied to zigzagging targets; if a target was not zigzagging, a factor of 1.3 (i.e., increasing the percentage of hits) was applied. If the target ship maneuvered "radically" immediately after torpedoes were fired, the student was to multiply by 0.7 (i.e., in favor of the target); if the target ship was in a formation of three or more vessels, the student was to multiply by 1.2, the result not to exceed 100 percent. If a destroyer or destroyer escort was in position to start "depth bombing" one minute

DAMAGE INFLICTED BY EACH TORPEDO HIT

ALL CLASSES

Target Class	Percentage of Life of Target
A (BB, 45,000 ton)	12
B (BB, 35,000 ton)	15
D (OBB)	20
E (CB)	25
E1 (CVB)	5
F1 (CV)	20
F (CA)	30
G (OCA, 10,000 ton)	35
H (CL or CVL)	35
J (CVE, and misc. types)	50
K (CL AA)	50
L (DD and misc. types)	60

Fig. 54
Table 14: Damage Inflicted by Each Torpedo Hit, All Classes

MINE HIT ON SINGLE VESSEL IN A COLUMN

PASSING THROUGH A MINEFIELD LAID IN ROWS.

Mine Rows Passed through	One of Two Ships in Column (Total of Two Dice)	Three or More Ships in Column (Total of Two Dice)
1	5	3 or 4
2	2 or 7	6 or 7
3	6 or 7	2,4,5, 6 or 8

Fig. 55
Table 15: Mine Hit on Single Vessel in Column

or more before a submarine fired a smoke bomb, the student was to multiply by zero—that is, there would be no hits. (Pages 16–18.)

To obtain target damage inflicted per torpedo hit, the hits determined from the preceding tables were to be multiplied by the percentages contained in Table 14 (figure 54). This was Under Water Damage, and it affected the ship's speed immediately. (Page 19.)

Whenever mines were laid, a report was made immediately to the Director, giving the location, direction, and number of lines, the distance between lines, and the depth of and spacing between mines. Damage caused by a mine hit was to be half that caused by a torpedo. In the case of mines laid in rows encountered by a formation in column, a single throw of two dice would determine whether or not one of the ships struck a mine. In the case of mines laid in rows and a formation in column open order, line of bearing, or a compound formation, the number of hits depended on the type of formation, its course, and the direction of the minefield. In this case, the decision as to the number of mine hits and the ships hit belonged to the Director. Distribution of mines laid by aircraft was also taken into account. For "practical" purposes, these patterns were considered random, and the percentage chance of an individual ship actuating one of the mines when passing through the field was obtained from the formula

$$\text{Percentage Chance} = (M \times T)/W,$$

in which M is the number of mines laid, T is the ship's beam, and W is the width of the minefield. For example, if thirty mines were laid in a field six thousand feet wide, a vessel sixty feet in beam had a Percentage Chance of actuating one mine equal to $(30 \times 60)/6000 = 0.3$ and so would incur 30 percent damage. (Pages 20–21.)

The probability of an event was expressed in terms of percentage. Table 16 (figure 56) gives these percentages and corresponding numbers resulting from a single throw of two dice. Thus, if the Percentage Chance of a vessel striking a mine while traversing a minefield is 30 percent and a six or seven resulted from the throw of the dice, the ship is considered to have struck the mine. Increments of 5 percent were accounted for by adding three to the numbers corresponding to the 10 percent divisions. Therefore, if the Percentage Chance of a ship striking a mine is 35 percent and either a three, six, or seven results from the throw of the dice, the vessel is considered to have struck a mine. Tables 16 was also used in any other situation to determine whether or not an event took place when the chance of its occurring was known. (Page 21.)

Additionally, Target Ship Classes were designated (see figure 57). For this purpose, combatant and auxiliary ships of the U.S. fleet were divided into ten groups of ships according to tonnage, armament, and ability to resist damage. For "convenience," alphabetical designations were given to each class, from A to L but omitting I. (Appendix A, pages 1–2.)

The scoring procedure by which chance factors were applied to Gunfire or Torpedo Damage was the subject of a special memorandum. If the ship was receiving penetrative hits, the student entered the Chance Sinking Tables and took the first letter encountered. If that letter was *S*, the ship was counted as sunk and removed from the board; students thereafter took letters in order across the page, marking out each as it was used. If the ship was not being penetrated, the student was to use the Chance Damage Tables with the percentage of life lost

TABLE NO. 16

PERCENTAGE CHANCE VERSUS NUMBERS

RESULTING FROM A SINGLE ROLL OF TWO DICE

Percentage Chance	Number
5	3
10	5
20	2 or 7
30	6 or 7
40	2,4,5,6 or 11
50	2,4,5,6 or 8
60	2,4,5,6,7 or 10
70	2,4,5,6,7,8 or 12
80	2,4,5,6,7,8,9 or 12
90	2,4,5,6,7,8,9,10 or 11

Fig. 56
Table 16: Percentage Chance versus Numbers Resulting from a Roll of Dice

TARGET CLASS	TYPE	SHIP CLASS
A	BB (45,000 ton)	IOWA CLASS
B	BB (35,000 ton)	SOUTH DAKOTA CLASS NORTH CAROLINA CLASS
C	(Omitted)	
D	OBB	WEST VIRGINIA CLASS COLORADO CLASS TENNESSEE CLASS NEW MEXICO CLASS
E	CB	ALASKA CLASS
E 1	CVB	MIDWAY CLASS
F 1	CV	ESSEX CLASS
F	CA	BALTIMORE CLASS DES MOINES CLASS OREGON CITY CLASS
G	CA (10,000 tons)	NEW ORLEANS CLASS PORTLAND CLASS CHESTER CLASS WICHITA CLASS
H	CL and CVL	ALL CLASSES
J	CVE,AGC,AK,AKA,APA,XAP,ARG, AP,AKV,ARB,AO,AKN,AKS,AS,ARL, AR,AVS,AV,ARH,CM,AG,AF,AE,AD, CM, and other auxiliaries above 6000 tons.	
K	CL(AA)	SAN DIEGO CLASS OAKLAND CLASS
L	DD,DE,DM,APD,AVD,SS,PF,PG,AVO, AVP,LCS,LSD,LST,LSV,LSM,AKN,AN, ARB,ARV,ASR,ARL,AVD,DMS,AOG,AM, ACM, and other auxiliaries below 6000 tons.	

Fig. 57
Target Ship Classes

during the move (unless the damage was equal to or less than 5 percent), again taking the first unused entry, this time a number, as the Chance Damage for the move. This was added to the damage that had been accumulated in previous moves in order to produce the total. (Appendix B, page 1.)

As an example, a battleship of thirty-five thousand tons, a "type B" target, is under fire from nine sixteen-inch guns for three minutes at a range of twenty-two thousand yards. No reductions to multiplier *M* are effective. The student needs to work out whether the ship is sunk or, if not, what the Chance Damage is. Table 2 of the Normal Gunfire Damage Tables shows that the proportion of life lost is 10 percent. Sixteen-inch shells are penetrative at this range, so the student must enter the Chance Damage Table covering 10 percent of the ship's life lost. Since the first unused letter in the table is *N*, the ship is not sunk. The first unused number was 1; therefore Chance Damage is 10 percent Above Water Damage for the three-minute move. This was the reduction in firepower in tenths (in this case, one-tenth) for the move. Taking the numbers in order across the page, the student marks out each as it is used. Students also kept a cumulative record of Chance Damage for scoring the future Fire Effect of the ship in case of Above Water Damage. This method was also used for determining speed reduction in the case of Under Water Damage. (Appendix B, page 2.)

Torpedo or Mine Damage was considered Under Water Damage. If the ship could be penetrated by torpedoes (Target Classes F to L, i.e., heavy cruisers to destroyers) such that magazines or gasoline storage tanks could explode and cause the loss of the ship, students entered the Chance Sinking Table with the percentage of target life lost, derived from a table. If the first unused letter was *S*, the ship was sunk; if the first unused letter was *N*, the student entered the Chance Damage Table and took the first unused figure. This figure was the Under Water Damage caused by the torpedo or mine; it did not affect the ship's Fire Effect, as Above Water Damage did. In the example provided, a heavy cruiser is torpedoed, and the table indicates that the proportion of life lost is 30 percent; entering the Chance Sinking Table with 30 percent life lost, the student finds the result *N*, and in the Chance Damage Table, the student would find the first unused number is a 2. Therefore, Under Water Damage inflicted by the torpedo is 20 percent. (Appendix B, page 3.)

Another complication requiring special elaboration was the "Application of a Chance Factor to the Scoring of Gunfire Damage on the Maneuver Board." Percent of Life Destroyed was tabulated for 5, 10, 15, 20, 25, 30, 40, 50, 60, 70, 80, and 90 percent. When the Naval War College Fire Effect Diagrams were recomputed by the Princeton Statistical Research Group, it had become evident that in certain circumstances, 100 percent (or even more) of a ship's life could be destroyed in a single move, but that did not mean statistically (a ship's "life" being an average value) that the ship was certain to be sunk in the move. Actually, its probability of destruction was only 0.63. Therefore, alternative scoring sequences were provided in the event that a ship had 100 percent or more of its life destroyed in a single move. If the percentage of life destroyed in a move was between 95 and 125, a special 100 percent sequence was used; if between 125 and 150, a 150 percent sequence was applied; if the life destroyed was greater than 150, the ship was removed from the game board whether or not it had been subjected to penetrative hits.[2]

NOTES

1 "Board Maneuver Rules: Supplement to Maneuver Rules," 10 August 1946, title page and p. 1, folder 2577 H, box 133, RG 4, NHC. Subsequent page references in this chapter, until the next endnote, are to this source.

2 Ibid., app. B, "Explanation of 100% and 150% Tables," 21 December 1945.

VI *More Exercises and More Rules, August–September 1946*

The Naval War College curriculum included not only gaming, of course, but direct, real-world instruction and demonstrations—the numerous lectures already mentioned, live exercises and demonstrations in fighter direction (with actual ships and aircraft, at training facilities in the Newport area), communications drafting and encryption training, and the like. One such event, formally a communications exercise, is recounted here because its scenario and planning as well as its decision factors serve to introduce readers to the actual gaming that is the true focus of this book.

Communications Exercise: Operation Plan No. 3-44

This exercise entailed the communications-related requirements of a simulated operation plan and battle plan. Operation Plan No. 3-44 placed Vice Admiral BF, Commander, Task Force (CTF) 31, the Raiding Force, and Commander, Air Forces, 3rd Fleet (COMAIR3FLT), in charge of Task Group 31.1, the Striking Group, which consisted of four fast battleships organized as Battleship Division 5; two fleet carriers organized into Carrier Division 2; four heavy cruisers as Cruiser Division 6; eight light cruisers as Cruiser Divisions 9 and 11; twenty-six destroyers organized into Destroyer Squadrons 5, 12, and 14; and the nine destroyer leaders (large destroyers, often with extra space and communications for embarked commanders) of Destroyer Squadron 21. Task Group 31.2, the Air Support Group, was under the command of Captain B-15 and consisted of Fleet Air Wings 1 and 10, thirty-six scout planes and forty-nine naval patrol planes, respectively. The Air Support Group also included twenty-four dive-bombers of the Army's 3rd Bombardment Group. Finally, Admiral BF commanded Task Group 31.3, the Submarine Group, which consisted of twelve submarines organized into Submarine Squadron (SUBRON) 6, under Captain B-14.

In the General Situation, Blue was extending control of the sea to the west to and including the line of Kiska–Marcus–Truk–Rabaul. The 3rd Fleet was to capture Truk Atoll and use it as an operational base. Enemy forces consisted of an Orange naval force composed of several battleships, fifteen to twenty cruisers, and about thirty-five destroyers. These fleet units were at Truk; Blue knew that twenty to

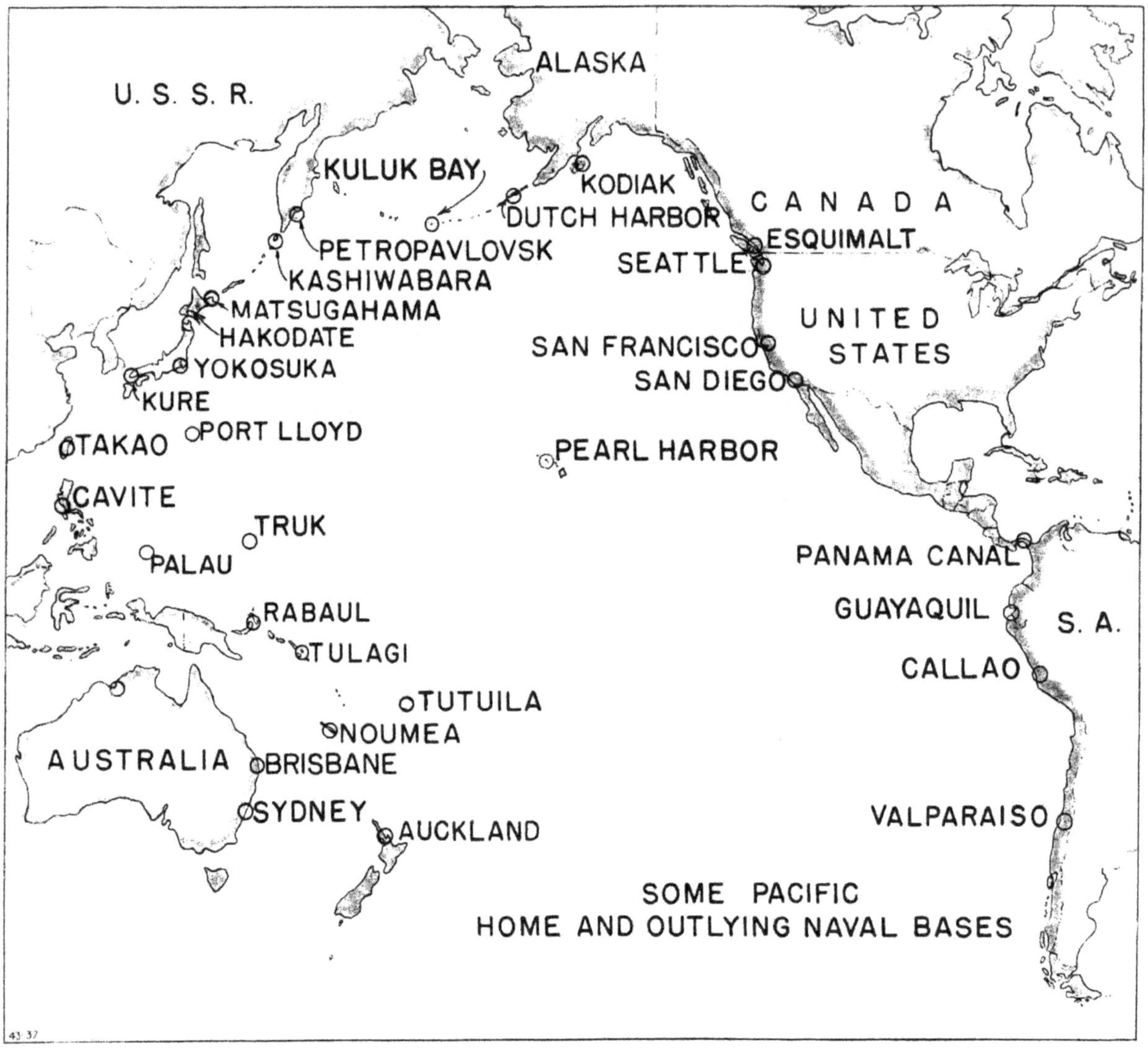

Map 3
Some Pacific Home and Outlying Naval Bases

twenty-five Orange submarines were also operating in the Truk area. Orange also had an undetermined number of seaplanes, land planes, auxiliaries, and defense craft based at Truk. Orange shore defenses were principally batteries of fourteen six-inch guns with maximum ranges of twenty-five thousand yards, distributed on islands of the atoll. There were also about twenty-four mobile three-inch antiaircraft guns and about a hundred mobile .50-caliber antiaircraft machine guns. Sixteen-inch guns on the islands of Moen and Tol were out of action. Orange cruisers and destroyers were usually equipped with mines for tactical use, which made it dangerous to follow them or pass through waters recently used by them. Their aircraft were also equipped with mines, of either drifting or anchored type.[1]

Orange aircraft strength on Truk was estimated at eighteen PBY Catalina naval patrol bombers, twenty Hellcat fighters, and ten Avenger and twelve Helldiver

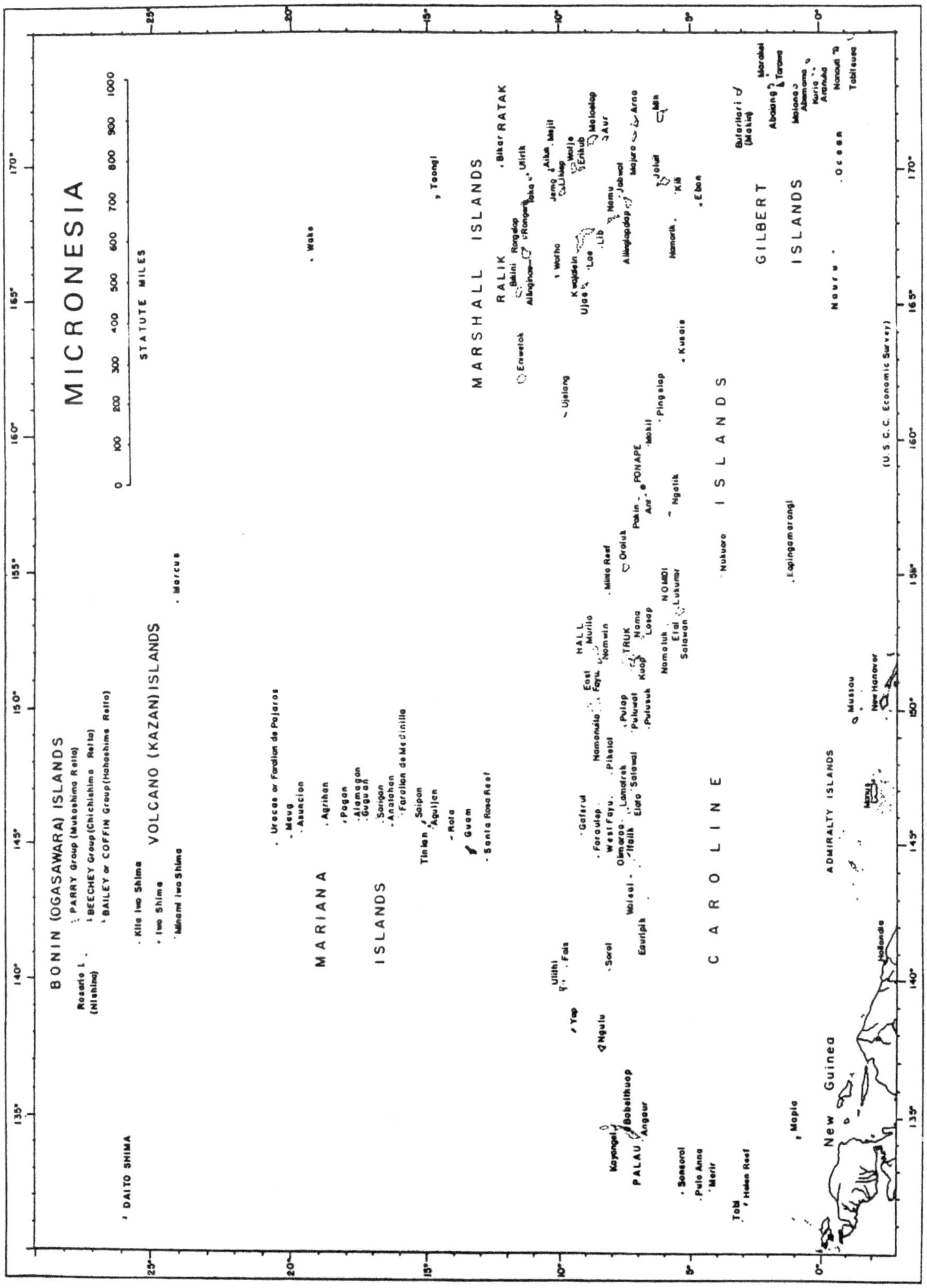

Map 4
Micronesia, including Truk Atoll

attack planes. The island of Ponape had no fighters or bombers, but it did have five PBM Mariner naval patrol bombers. Eniwetok had an additional four Mariners, along with thirteen B-17 Flying Fortress heavy bombers, and there were another thirty B-17s and eight Mariners elsewhere in the Marshalls. In the Marianas, Orange had about fourteen Catalinas but was not believed to have any fighters or bombers. Admiral BF ordered the Submarine Division (SUBDIV) to depart its patrol station at Truk at 1800 on 18 September and return to base. By the same time, the Blue Amphibious Force was to be in the vicinity of Rabaul, prepared to capture Truk after a successful action. Blue Task Force 51 was to capture Wotje and Jaluit, covered by Task Force 52 in the process. Task Groups 52.1 and 52.2 would raid Marcus Island and the Marshalls, respectively, in order to bring about maximum destruction of the air and shore installations there. Task Forces 51 and 52 would depart Pearl Harbor on 4 September. Admiral BF also assumed that all available Orange combatant units would defend Truk against the Blue threat, that Orange forces were in strength, approximately as reported, and that a major action would be fought in daylight on 9 September. He further assumed that visibility would be high, that smoke would lie on the surface, that aircraft could be flown, and that the present "excellent" weather would continue. He additionally assumed that Orange submarines would be present; that Orange would exert his maximum effort to carry out attrition attacks on Blue forces at night by air, surface, and submarine forces; and that Blue surface forces would be discovered by Orange prior to 9 September.[2]

Admiral BF's Striking Group was to destroy or drive off Orange sea forces and reduce Orange air forces by destroying enemy aircraft and reducing Orange air facilities at Truk by the time of the Blue air attack at 0600 on 9 September. Air strikes by the Striking Group and the Air Support Group on heavy ships (i.e., battleships, battle cruisers, and heavy cruisers) were also to be coordinated, and the Air Support Group was to assist in destroying or driving off Orange air and sea forces in the area as well. In particular, the latter unit was to locate and track Orange surface forces on the night of 8–9 September, conduct searches in accordance with the supplied Air Search Diagram, Annex B (see figure 58), destroy Orange aircraft, reduce Orange air facilities at Truk at 0330 and 0700 on 9 September, and reduce the fighting strength of Orange heavy ships by aerial torpedo attacks coordinated with air strikes from the Blue carrier units after 1000 on 9 September, as directed by Commander, Striking Group. The Air Support Group was also to maintain reconnaissance of Truk. In addition, this unit was to establish a land-plane base at Kavieng and seaplane bases at Rendezvous Blossom and Custard on the night of 8–9 September.[3]

The Submarine Group was to provide radio beacons for the Air Support Group at Rendezvous Apricot from 0245 to 0315, assist in destroying Orange sea forces in the Truk Area, and observe all entrances to the atoll commencing at 1800 on 8

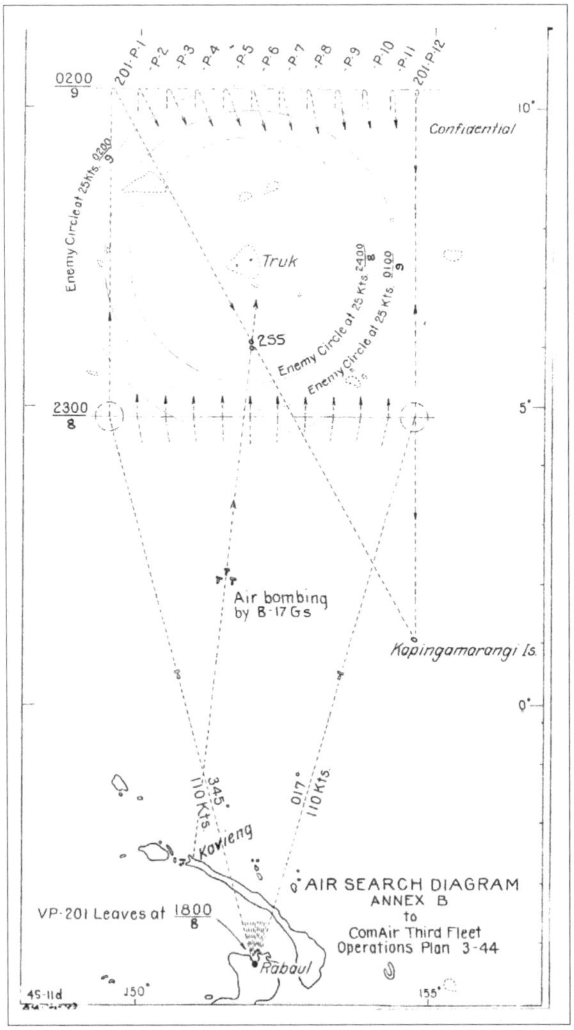

Fig. 58
Air Search Diagram, Annex B, COMAIR3FLT Operation Plan No. 3-44

September. The Submarine Group was also to attack Orange surface forces during their sortie from Truk, during which these forces would probably use the atoll's northeast pass. Further, the submarines were to trail enemy task forces during darkness and interpose themselves between the Orange task forces and Truk during the day action. Additionally, the Submarine Group was to keep clear of Blue surface forces, remain within fifteen miles of the Truk barrier reef during daylight on 9 September, and destroy damaged Orange fleet units returning to Truk. Battleships were the submarines' primary targets, but the latter were also to guard against enemy submarines and drifting mines. Because of where the Submarine Group was to be stationed during the main action, there was a total restriction against bombing submarines within fifteen miles of the Truk barrier reef during the daylight hours of 9 September.[4]

Operation Plan No. 3-44 was to become effective upon receipt, as were the 3rd Fleet Standard Logistic Plan Love and Raiding Force Special Fueling Plan, Annex Dog. In addition, Communication Plan, Annex Charlie and Rendezvous List, Annex Able were to be used. Admiral BF informed his subordinates that his flagship would be the fast battleship BB-57, with the Striking Group. (The complexities of the delivery of these plans and orders to a lengthy, multiechelon distribution list was one of the points of the communications exercise.)[5]

Battle Plan No. 4-44 gave a more detailed task organization. The four battleships of Battleship Division 5 were designated Task Unit 31.1.1, the "Battleline," and were under the command of Rear Admiral B-1. The four heavy cruisers of Cruiser Division 6 and the nine destroyer leaders of Destroyer Squadron 21 were designated Task Unit 31.1.2, the Center Force, and placed under the command of Rear Admiral B-3. Task Unit 31.1.3, the Right Flank Force, was under Captain B-20 and consisted of two light cruisers—less CL-55 and CL-56—from Cruiser Division 9 and the nine destroyers of Destroyer Squadron 5. Task Unit 31.1.4, the Left Flank Force under Rear Admiral B-4, consisted of CL-55 and CL-56 from Cruiser Division 9 and the nine destroyers from Destroyer Squadron 12. Task Unit 31.1.15, the Carrier Group, under Rear Admiral B-2, consisted of the two fleet carriers organized into Carrier Division 2, the four antiaircraft cruisers of Cruiser Division 11, and the eight destroyers from Destroyer Squadron 14. There was no change in the General Situation or the information given in Operation Plan No. 3-44. However, it was now made clear that Orange forces consisted of a major surface force of five battleships, seven to eight heavy cruisers, eight to ten light cruisers, and thirty to thirty-five destroyers, all reported to be anchored in Truk Harbor as of 8 September. There were also a reported twenty to twenty-five Orange submarines at Truk, and Orange aircraft strength was now reported as eighteen Catalinas, twenty Hellcat fighters, and ten Avenger and twelve Helldiver attack planes. Additionally, there were about seventeen Mariner naval patrol bombers, fourteen Catalina naval patrol bombers, and forty-three Flying Fortress heavy bombers in the Marshalls, Marianas, and Caroline Islands.[6]

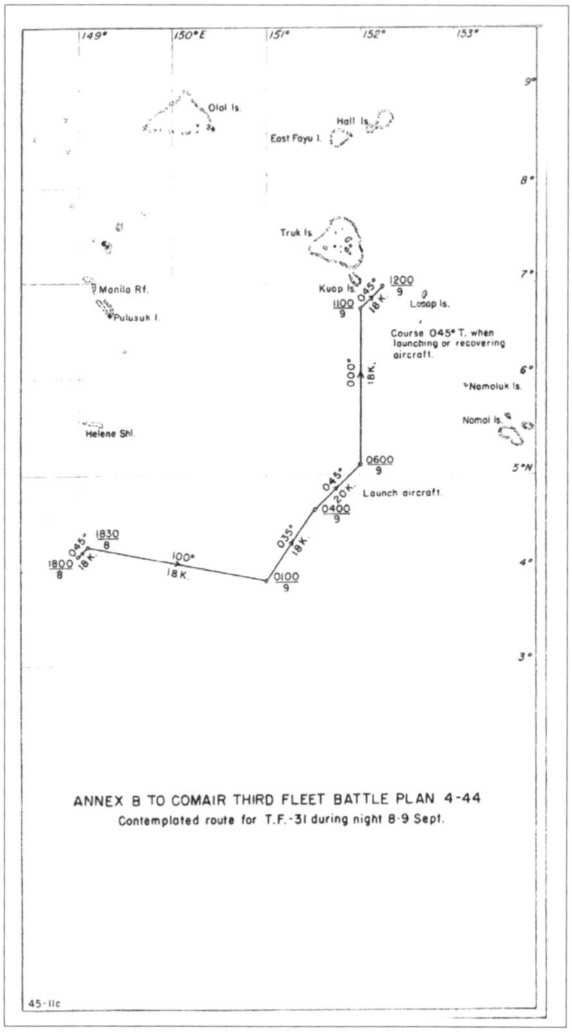

Fig. 59
Annex B to COMAIR 3rd Fleet Battle Plan No. 4-44

As to the disposition of his own forces, BF again pointed out that Blue submarines were operating in the vicinity of Truk and that Blue land-based aircraft would bomb Truk and scout for Orange surface forces prior to the main engagement. He assumed that Blue submarines would furnish contact reports on Orange surface forces upon the departure of those forces from Truk and their attempt to make night attrition attacks on Blue. Blue land-based aircraft would also report contacts with Orange surface forces as they bombed Truk, and they would do so through the night. He added an assumption that Orange light forces and submarines would attempt to mine the waters ahead of the Blue surface forces.[7]

His battle line would initially engage the Orange battle line at extreme ranges, close as rapidly as possible to ranges between twenty and twenty-two thousand yards, close to moderate ranges (under twenty thousand yards) when the rate and volume of enemy fire were reduced or made ineffective, and provide its own Air Spot and Battleline Inner Anti-Submarine Patrol. His Center, Right Flank, and Left Flank Forces were to defend the battle line against enemy light force attacks. Cruiser planes were to furnish Air Spot and Battleline Intermediate Anti-Submarine Patrol, while destroyers supported the cruisers with torpedo attacks against the Orange battle line when directed by the Officer in Tactical Command (OTC). The

Carrier Group was to make dawn air strikes against planes and air installations on Truk with a quarter of its fighters and half of its dive-bombers. It was to make dawn air strikes on the Orange battleships with the remainder of its planes, less those needed for antisubmarine and combat air patrols. The Carrier Group was ordered to make repeated air strikes on Orange air and surface forces throughout the day, furnish its own combat air patrol, as well as one for the battle line, and furnish its own Anti-Submarine Patrol. When the Blue battle line was engaged with its Orange counterpart, the Carrier Group was to operate on the disengaged side, out of range of the Orange main-battery guns.[8]

Planes were to report any drifting mines or submarines sighted. All planes armed with depth charges and all destroyers making contact with submarines were to attack "vigorously." All other ships were to follow the movements of the Carrier Group until deployed for surface action. Fighter direction officers were to be stationed in fleet carrier CV-10 and fast battleship BB-57. Cruising Dispositions 5 or 5N were to be used as signaled, followed by Approach Disposition 6 and Battle Disposition 7. Commander, Striking Group was again in BB-57; Rear Admiral B-1, Commander, Battleline, was designated the second in command. Battle Plan No. 4-44 was to become effective upon signal; Commander, Air, 3rd Fleet Communication Plan Dog was in effect. Annexes to the plan were included, as were instructions for distribution to the Commanders, Battleline, Carrier Group, and Center, Left Flank, and Right Flank Forces.[9]

Antiaircraft Fire Effect and Aircraft-Damage Maneuver Rules

A day later, additional Maneuver Rules were issued for Anti-Aircraft Fire Effect and Rules Governing Damage Inflicted by Aircraft. In terms of Aerial Combat, the Director of the Maneuver would, in general, assign aerial combat losses in accordance with Table A, making "due allowance" for the disposition of aircraft with respect to altitude, formation, and position. The aircraft initiating the attacks were designated "attackers" or N in number, while the aircraft being attacked were designated as "defenders" or M in number. For the purpose of determining the results of sustained attacks, a three-minute interval would take place between each attack. In addition, the losses due to attacks initiated by airplanes other than fighters and the results of attacks occurring at night or in fog would be decided after consideration of the conditions and circumstances under which each such engagement occurred. Each attack was to involve the expenditure of an eighth of the available machine-gun ammunition for the guns involved. Guided missiles carried to the vicinity of the target by parent aircraft would be destroyed if the parent aircraft was shot down prior to the launching of the guided missile. Guided missiles in free flight that were subject to interception by fighter aircraft could be shot down, but the loss incurred would only be half as much as if the guided missile were a defending fighter. The losses incurred by intercepted suicide aircraft would be the same as that listed in column 4 of Table A and the losses sustained by the aircraft making the interception in this case would be a fifth of that indicated in column 3 of Table A.[10]

When it came to Aircraft Losses Caused by Anti-Aircraft Fire, a battery of four antiaircraft guns of over 4.5-inch caliber was considered to be the standard unit of fire for the purpose of computing the effectiveness of antiaircraft fire. Guns of other calibers would be multiplied by the coefficients given in Table B to convert them to standard units of antiaircraft fire. Guns under two-inch calibers would be classified as antiaircraft machine guns.[11]

Attackers	Defenders	Attacker's Losses	Defender's Losses
(n)	(m) VF	m x 0.1	n x 0.15
(n)	(m) VO (s) single seat	m x 0.1	n x 0.2
(n)	(m) VO (s) other than single seat	m x 0.1	n x 0.6
(n)	(m) VA	m x 0.1	n x 0.5
(n)	(m) VP	m x 0.3	n x 0.3
(n)	(m) VP B-29 type	m x 0.9	n x 0.2

Fig. 60
Table A: Aerial Combat Losses

The number of antiaircraft gun standard units of fire and the number of antiaircraft machine gun standard units of fire that would bear on and have the attacking aircraft within effective range would be determined by the Director of the Maneuver. This decision would also take into consideration the type and direction from which the attack was made, the altitude of the aircraft involved, the disposition of the firing ships, and their antiaircraft commitments. The number of standard units of antiaircraft fire from any ship or ships that would bear on the attack might be reduced by the amount specified in Table C if the ship or ships concerned were affected by the conditions indicated in Table C. The basic percentage of aircraft losses to antiaircraft fire was taken from Table D. The basic percentage of loss to antiaircraft guns was obtained by using the number of effective standard antiaircraft gun units as an argument. The final percentage of airplane losses to antiaircraft guns and antiaircraft machine guns was determined by multiplying the basic percentages taken from Table D by Table E.[12]

Caliber of AA Guns	Coefficient
2"01 to 3"9	.15
3"91 to 4"5	.20
over 4"5	.25

Caliber of AA Machine Guns	Coefficient
under 0"50	.05
0"50 to 1"0	.10
1"01 to 2"0	.15

NOTE 20 mm = 0"8
40 mm = 1"6

Fig. 61
Table B: Antiaircraft Gun-Caliber Coefficients

At night or when the attacking aircraft could not be seen visually for more than two thousand yards, the percentage taken from Table D would be reduced by 50 percent for ships with fire control radar. Ships without fire control radar would not be able to fire under these conditions unless they could see the aircraft, in which case the percentage taken from Table D would be reduced by 90 percent. Aircraft losses to antiaircraft fire during the approach would be assumed to take place before the aircraft involved dropped their bombs or torpedoes or fired their rockets. Aircraft losses to antiaircraft fire during the retirement would be 20 percent of the losses incurred during the approach. When it came to the Accuracy of Aircraft Attacks, Table F was to be used as a guide in determining the percentage of hits

Condition	% Reduction
Firing ship under gunfire	20
Firing ship firing main or secondary battery	20
Firing ship strafed by at least 4 planes just prior to attack	35
Firing ship has 50% AW damage	100
Firing ship in rough sea	20

Fig. 62
Table C: Conditions Reducing Standard Units of Antiaircraft Fire

No. of A/C in Attack Group	NUMBER OF EFFECTIVE STANDARD UNITS OF AA FIRE BEARING ON ATTACK GROUP														
	1	2	3	4	5	6	7	8	9	10	11	12	13	14	15
5	30	35	40	45	50	55	60	65	70	75	80	85	90	100	100
10	25	30	35	40	45	50	55	60	65	70	75	80	85	100	100
15	20	25	30	35	40	45	50	55	60	65	70	75	80	100	100
20	15	20	25	30	35	40	45	50	55	60	65	70	75	100	100
30	10	15	20	25	30	35	40	45	50	55	60	65	70	75	100
40	8	10	12	15	20	25	30	35	40	45	50	55	60	65	100
60	6	8	10	12	15	20	25	30	35	40	45	50	55	60	100
80	4	6	8	10	12	15	20	25	30	35	40	45	50	55	60
100	2	4	6	8	10	12	15	20	25	30	35	40	45	50	55

Fig. 63
Table D: Basic Percentage of Airplane Losses to Antiaircraft Fire

Type of AA Fire	Aircraft Attack Method	Multiplier for Basic % Obtained from Table "D"
AA Guns	Minimum altitude bombing	1.5
	Horizontal bombing 300 - 3000 feet	1.3
	Horizontal bombing 3001 - 6000 feet	1.0
	Horizontal bombing 6001 - 9000 feet	0.9
	Horizontal bombing 9001 - 12000 feet	0.7
	Horizontal bombing 12001 15000 feet	0.5
	Horizontal bombing 15001 -18000 feet	0.3
	Horizontal bombing above 18000 feet	0.1
	Torpedo attack	0.4
	Dive or glide bombing	0.3
	Strafing or rocket attack	0.3
AA Machine Guns	Minimum altitude bombing	0.8
	Horizontal bombing 300 - 3000 feet	0.6
	Horizontal bombing 3001 - 5000 feet	0.1
	Horizontal bombing above 5000 feet	0.0
	Torpedo attack	0.4
	Dive or glide bombing	0.4
	Strafing or rocket attack	0.3

Fig. 64
Table E: Coefficients for Aircraft Losses

TARGET SIZE	BOMBS								Rockets	GUIDED MISSILES	
			Horizontal Bombing								
	Dive or Glide Bombing	Low altitude Bombing	3000 to 6000 feet	6000 to 90000 feet	9000 to 12000 feet	12000 to 15000 feet	15000 to 18000 feet	Above 18000 feet		Pilotless Aircraft	Suicide Aircraft
Large	20	32	12	10	9	7	4	3	35	25	45
Intermediate	12	20	7	6	5	4	2	2	30	18	40
Small	8	15	5	4	3	3	2	1	25	11	35
Destroyer	4	10	4	3	2	2	1	1	20	5	30

Fig. 65
Table F: Percentage of Hits by Bombs, Rockets, and Guided Missiles

Speed of Target	Rocket Attack	Dive or Glide Bombing	Low Altitude Bombing	Level Bombing
3 to 12 knots	1.0	1.0	1.0	0.9
12 to 25 knots	1.0	0.8	0.8	0.6
Over 25 knots	0.9	0.6	0.6	0.4

Fig. 66
Table G: Reduction Factor against Maneuvering Targets

Target	% Torpedo Hits		
	At Anchor	Underway	Manouvoring
BB	75	45	40
CVB, CV	75	50	45
CVL	65	45	35
CA	65	40	35
CL	50	35	25
DD	50	20	15
CVE	60	35	30
Aux or M/V	60	40	35

Fig. 67
Table H: Percentage of Aircraft Torpedo Hits

6255-5755
13/Aug/46 gq

TABLE I

BOMB, ROCKET AND TORPEDO DAMAGE IN PERCENT LOSS OF LIFE OF TARGET SHIP PER HIT

TARGET	BOMB TYPE																				ROCKET TYPE				
	2000# GP		1600# AP		1000# AP		1000# SAP		1000# GP		500# SAP		500# GP		250# GP		100# GP		11.75" AR		5" HVAR		Torpedo
	P	N	P	N	P	N	P	N	P	N	P	N	P	N	P	N	P	N	P	N	P	N	
BB IOWA	-	13	18	4	16	3	-	5	-	8	-	2	-	4	-	2	-	1	-	2	-	.1	13
BB SOUTH DAKOTA	-	13	18	4	16	3	-	5	-	8	-	2	-	4	-	2	-	1	-	2	-	.1	15
OBB	-	16	18	4	16	3	-	5	-	10	-	2	-	5	-	2	-	1	-	2	-	.1	20
CB ALASKA	-	18	18	5	16	4	-	6	-	11	-	2	-	6	-	2	-	1	-	2	-	.1	25
CVB MIDWAY	-	13	23	5	18	4	-	6	-	8	-	2	-	4	-	2	-	1	-	2	-	.1	5
CV ESSEX	-	22	25	8	19	4	14	8	-	12	-	3	-	7	-	3	-	2	-	3	-	.3	20
CA BALTIMORE	-	25	12	7	10	4	18	8	-	14	16	3	-	7	-	3	-	2	16	3	-	.3	30
CA SALT LAKE CITY	-	25	12	7	10	4	18	8	-	14	16	3	-	7	-	3	-	2	16	3	-	.3	30
CVL SAIPAN	-	27	12	8	10	5	20	10	-	18	17	4	-	9	-	4	-	2	17	4	-	.3	33
CL CLEVELAND	-	27	14	8	12	5	20	10	-	18	17	4	-	9	-	4	-	2	17	4	-	.3	30
CVL INDEPENDENCE	-	30	14	9	12	6	22	11	-	20	19	5	-	10	-	5	-	2	19	5	-	.3	33
CL RENO	50	35	20	10	18	8	25	13	30	22	19	6	20	10	-	7	-	3	19	6	-	.5	40
CVE ALL CLASSES	50	35	33	10	30	8	35	14	40	22	20	6	30	12	20	8	15	3	20	6	3	.5	45
DD (1630 Tons and above)	75	55	33	14	33	9	40	20	45	35	20	8	25	17	20	9	12	6	20	9	4	1.0	45
DD (under 1630 Tons)	85	60	40	17	40	12	60	23	70	40	25	10	33	20	25	10	15	6	22	10	5	1.0	75

Note: P = Penetrative hit
N = Non-Penetrative hit
When no value is indicated under "P" penetration to the vitals of the ship will not occur.

Fig. 68
Table I: Bomb, Rocket, and Torpedo Damage in Percent Loss of Target Life

obtained by aircraft making an attack with bombs or rockets as well as by guided missiles of the "pilotless" or suicide type. Bomb and rocket hits were determined to the nearest whole number after deducting aircraft losses to antiaircraft fire during the approach. The figures for the percentages of hits obtained by guided missiles were applied to the number remaining after interception by fighters and included losses to antiaircraft fire and misses. If the target maneuvered during the attack, the percentage of hits taken from Table F would be multiplied by the reduction factor taken from Table G to obtain the percentage of hits on a maneuvering target. In night or low-visibility attacks with planes equipped with radar or "suitable" radar accessories, the final percentage of hits taken from Table F would be reduced by 50 percent.[13]

Distance of ship in yards from explosion	Percent loss of life of ship
0 - 1000	100
1001 - 2000	90
2001 - 3000	80
3001 - 4000	70
4001 - 5000	60
5001 - 6000	50
6001 - 7000	40
7001 - 8000	30
8001 - 9000	20
9001 - 10,000	10
Above 10,000	0

Fig. 69
Table J: Damage Effect of One Atomic Bomb

When it came to aircraft torpedo hits on ships, torpedoes that ran hot, straight, and normal would have a probability of hitting various targets as indicated in Table H. The percentage of hits taken from Table H could be varied by the roll of dice. In this case, Roll 1 would decrease the percentage of hits by 20 percent, while Roll 2 would decrease it by 10 percent. Roll 3 would incur no change, while Roll 4 would increase the percentage of hits by 10 percent. Roll 5 would increase the percentage of hits by 20 percent; Roll 6 would incur no change.[14]

Table I was to be used as a guide in assessing the damage resulting from hits by various types of bombs, rockets, aerial torpedoes, and guided missiles on the target classes of ships indicated. The damage resulting from any one hit was to be expressed in "loss of life." For bombs and rockets, Table I gave the damage resulting from hits that penetrated the armored decks as well as hits that did not penetrate. The armor plate penetration curves in Appendix I would be used to determine whether a bomb hit was penetrative or nonpenetrative. Damage from guided missiles would equal that assigned to the type of bomb or rocket in Table I that most nearly approximated the type of warhead employed in the guided missile that was used. Damage from hits by suicide aircraft would be equal to that of a five-hundred-pound general-purpose (GP) bomb—assumed to be penetrative for ships with no armor—plus the damage resulting from the type of bomb or rocket carried. Table J was to be used as a guide in assessing damage resulting from the detonation of an atomic bomb within a target area.[15]

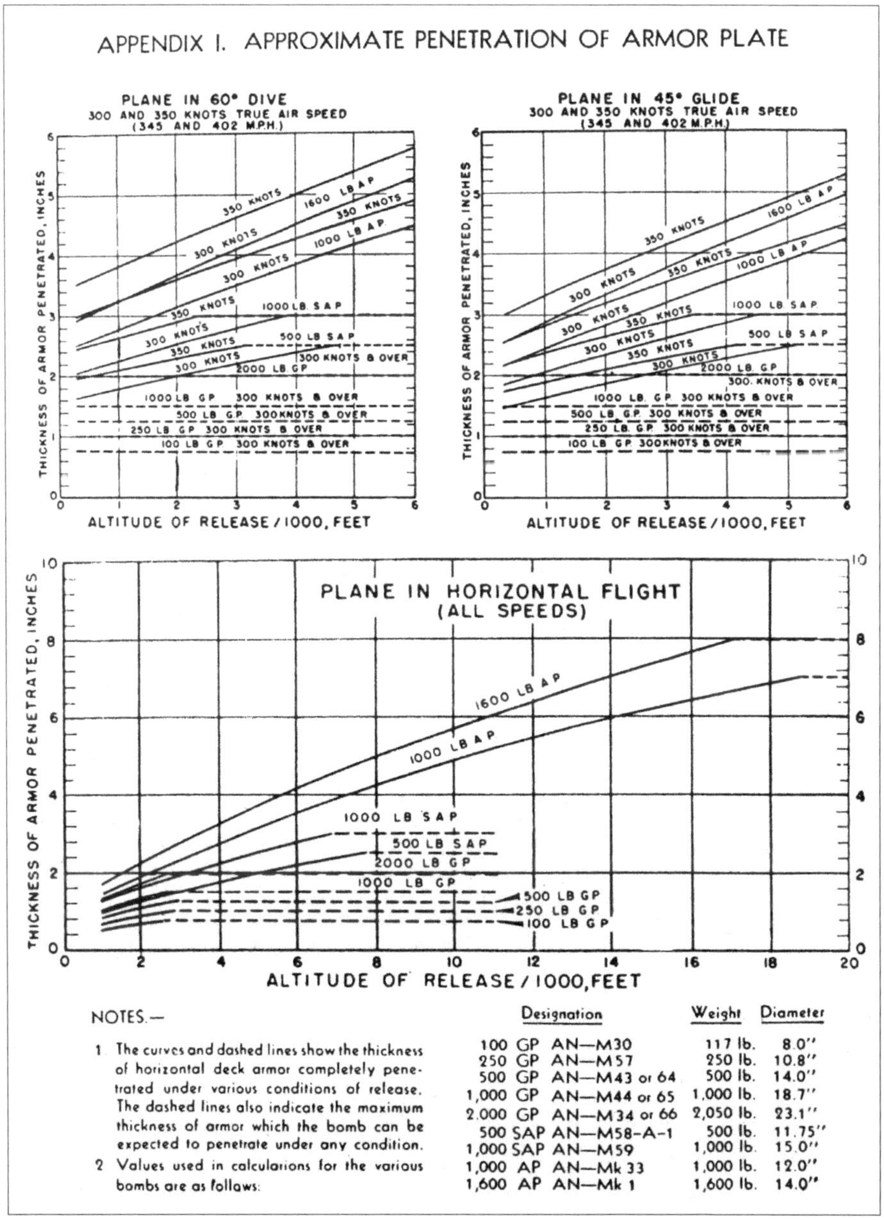

Fig. 70
Appendix I: Approximate Penetration of Armor Plate

Bomb or rocket damage was to be treated in the same manner as gunfire damage if it was desired to apply the element of chance in computing the amount of damage sustained by any particular target as outlined in the "Memorandum on the Application of the Chance Factor to the Scoring of Gunfire Damage on the Maneuver Board." If the bomb or rocket penetrated the armored decks of the target ship, as indicated in Table I, the "Chance of Sinking" sequence for the percent of life destroyed was entered and the first unused letter was noted. If this letter was S, the ship was considered "d," or sunk. If the letter was *N*, the ship was not sunk and the appropriate "Fire Effect Reduction" sequence was entered to obtain the loss of Fire Effect. If the damage sustained was the combined result of bombs or rockets that would penetrate and bombs or rockets that would not penetrate, then only the percent of loss of life suffered from the penetrative hits was to be used to enter the sinking sequence. In this case, if the ship was not sunk then the sum of loss of life inflicted by the penetrative and nonpenetrative hits would be used to enter the Fire Effect Reduction Sequence in order to obtain the resulting loss of Fire Effect. The Move Time employed for any one computation involving the Chance Factor, as described above, was to be limited to that required to complete the attack by the aircraft or the group of aircraft involved. The result of each additional attack would be figured separately.[16]

Target		Total Effective Deck Armor
BB Iowa		6.25"
BB South Dakota		6.25"
OBB		4.9"
CB Alaska		4.3"
CVB Midway	Flight deck 3.5"	4.25"
	Hangar deck 1.5"	
CV Essex	Hangar deck 2.5"	3.25"
	4th deck 1.5"	
CA Baltimore		2.5"
CA Salt Lake City		2.4"
CVL Saipan	Hangar deck 2.5"	2.5"
CL Cleveland		2.0"
CVL Independence	Hangar deck 2.0"	2.0"
CL Reno		1.25
CVE all classes		none
DD 1850 - 2200 Tons		none
DD 1500 - 1630 Tons		none

Fig. 71
Ship Deck-Armor Characteristics

NOTES

1 "Naval Communications: Operation Plan No. 3-44 and Battle Plan No. 4-44," 13 August 1946, pp. 1–2, folder 2584-K, box 135, RG 4, NHC. These exercises were obviously based on the U.S. operations against the Gilbert and Marshall Islands of late 1943 and early 1944, especially the raid against Truk in February 1944. For an account of those strikes in terms of wartime American combined-arms naval doctrine, see Trent Hone, "U.S. Navy Surface Battle Doctrine and Victory in the Pacific," *Naval War College Review* 62, no. 1 (Winter 2009), pp. 68–69, 76–81.

2 "Naval Communications," p. 2.

3 Ibid.

4 Ibid., pp. 2–3.

5 Ibid., p. 3.

6 Ibid., p. 4.

7 Ibid., pp. 4–5.

8 Ibid., p. 5.

9 Ibid., pp. 5–6.

10 "Anti-Aircraft Fire Effect and Rules Governing Damage Inflicted by Aircraft," 13 August 1946, pp. 1–2, folder 2577-J, box 133, RG 4, NHC.

11 Ibid., p. 2.

12 Ibid., pp. 2–4.

13 Ibid., pp. 4–6.

14 Ibid., p. 6.

15 Ibid., pp. 7, 9.

16 Ibid., pp. 9–10.

C-in-C.

VII *Board Maneuver Exercises, September 1946*

In early September, the junior class of June 1947 was issued instructions and parameters for six Board Maneuver Exercises that were to take place between 13 and 17 September. The object of Exercise 1 was to familiarize students with the sections of the Maneuver Rules that applied to the Board Maneuver, to furnish an exercise in the mechanics of ship movements on the board, and to acquaint students with the Move Sheet. Students were referred to various parts of Sections A–C of the Maneuver Rules, *Characteristics of the Blue Fleet,* and the Supplement to Maneuver Rules.

Exercise 1

For Exercise 1, students were divided into six groups, Section A consisting of Groups 1, 3, and 5 and Section B of Groups 2, 4, and 6. Members of the Naval War College Staff supervised. The exercise was divided into three phases, as indicated in figure 74. Odd-numbered groups were to maneuver on the northernmost part of the Maneuver Board and even-numbered groups on the southernmost. Each student prepared a Move Sheet for the forces to which he was assigned. The staff member supervising each group would examine each Move Sheet and notify the Director of the Maneuver when the group was ready to make the move on the board. Moves were to be made when the Director so announced; student officers made their moves in succession in the order of listing, the senior officer of each group going first.*

The General Situation for Exercise 1 had Blue units maneuvering independently. These units included fast battleships BB-58 and BB-59, organized as Battleship Division 3; heavy cruisers CA-68, -69, -70, and -71, as Cruiser Division 5; and Destroyer Squadron 19, divided into Destroyer Division 191, consisting of DD-697, -698, -699, and -700, and DESDIV 192, with DD-701, -702, -703, and -704. BATDIV 3 was steaming in column at a distance of a thousand yards (i.e., between ships) on a course of 000 degrees and an engine speed of sixteen knots. The ships of CRUDIV 5 were also in column but at a distance of six hundred yards,

* "Board Maneuver Exercises," 4 and 11 September 1946, pp. 1–3, folder 2587, box 135, RG 4, NHC. Subsequent page references in this chapter are to this source.

also on a course 000 but at fifteen knots. DESRON 19 was steaming in Squadron Formation No. 5M on a division line of bearing of 090 degrees, at intervals of five hundred yards, with each division in column. It was at a distance of three hundred yards on course 000 and with an engine speed of fifteen knots. The wind was from the northeast at Force 1, with seas smooth, visibility normal, and cloud ceiling ten thousand feet. (Pages 4–5.)

Group 1	Group 2
Col. S.E. Manzo, AC, USA	Col. H.J. Hawthorne, AC, USA
Comdr. G.S. Coleman, USN	Comdr. K. Loveland, USN
Comdr. A.M. Nibbs, USN	Comdr. C.A. O'Connell,Jr., USN
Lt.Col. S.W. Downey,Jr., CAV,USA	Comdr. A.T. Church,Jr., USN
Comdr. Baldridge, USN	Comdr. H.J.P. Foley,Jr.,SC,USN
Lt.Cdr. Aymond, USN	Lt.Cdr. W.Y. Howell, USN
Group 3	**Group 4**
Col. H.O. Edson, INF, USA	Col. R.J. Mason, AC, USA
Comdr. E.W. Bridewell, USN	Comdr. G.H. Miller, USN
Comdr. R.C. Morton, USN	Comdr. N.C. Gillette, Jr., USN
Lt.Col. Leary, AC, USA	Comdr. H. Marvin-Smith, USN
Comdr. Kirkpatrick, USN	Lt.Col. F.L. Street, FA, USA
Lt.Cdr. F.L. Taeusch, USN	Lt.Cdr. E.P. Bonner, USN
Group 5	**Group 6**
Lt.Col. L. Wallace, INF, USA	Lt.Col. M.H. Thompson, CE, USA
Comdr. C.D. Simonsen, USN	Comdr. J.E. Owers, USN
Comdr. W.K. Rogers, USN	Comdr. D.C. Richardson, USN
Comdr. G.F. Dalton, USN	Comdr. O.B. Lundgren, USN
Comdr. T.S. White, USN	Comdr. R.W. Rynd, USN
Comdr. R.P. Webber, SC, USN	Lt.Cdr. J.D. Ramage, USN

Fig. 72
Student groups, junior class of June 1947, Exercise 1, Board Maneuvers, September 1946

Phase I consisted of Moves 1, 2, and 3. In Move 1, Battleship Division 3, Cruiser Division 5, and Destroyer Squadron 19 continued on their courses and speeds, but in Move 2 BATDIV 3 performed a "column right" of forty-five degrees while maintaining engine speed. Cruiser Division 5 increased speed to twenty-four knots, and DESRON 19 performed a squadron change of course to 090 degrees at the same engine speed. Destroyer Division 192 then increased speed by six knots and changed its course to 090 degrees to maintain its intervals. In Move 3, BATDIV 3 reduced speed to twelve knots, while Cruiser Division 5's ships turned starboard to 045 and continued the increases of engine speeds that it had started in Move 2. DESRON 19 completed the maneuver it had started in Move 2, with DESDIV 191 continuing course and speed and DESDIV 192 slowing to fifteen knots. Upon completion of Move 3, when ordered by the Director, students shifted to their units for Phase II. They also wrote course and engine speed for Move 3 on the Maneuver Board for the group taking over. (Page 5.)

Director of the Maneuver - Captain D.S. Evans

Officers Assisting Groups

Group 1	Commander A.G. Pelling
Group 2	Commander J.W. Reed
Group 3	Commander D. Mayberry
Group 4	Lt.Cdr. V.G. Holzapfel
Group 5	Commander H.L. Haskell
Group 6	Lt.Cdr. J.H. Curran

Fig. 73
Staff members, Exercise 1, Board Maneuvers, September 1946

Groups	Phase 1	Phase 2	Phase 3
1 and 2	Batdiv 3	Crudiv 5	Desron 19
3 and 4	Crudiv 5	Desron 19	Batdiv 3
5 and 6	Desron 19	Batdiv 3	Crudiv 5

Fig. 74
Exercise phases

For Phase II (Moves 4, 5, and 6) the wind was still from the northeast, but this time at Force 4; the seas were now moderate; and visibility and ceiling were the same as in Phase I. In Move 4, BATDIV 3 continued on its course and decreased its engine speed, while Cruiser Division 5 turned to port fifteen degrees. Commander, DESRON (i.e., COMDESRON) 19 had his Division Commanders take charge of their divisions and act independently. Destroyer Division 191 changed course forty-five degrees to starboard and increased engine speed to twenty knots, while DESDIV 192 increased engine speed to twenty knots but maintained a course of 090 degrees. Move 5 began with Battleship Division 3 forming a line of bearing 135 degrees relative to the left of the Guide. Fast battleship BB-58 maintained course and engine speed, but BB-59 increased speed to 15.5 knots and changed course to 030. Cruiser Division 5 maintained course and speed, as did DESDIV 191, while DESDIV 192 turned port in column from 090 to 000. In Move 6, BB-58 maintained its course and speed, while BB-59 continued the maneuver it had started in Move 5. Cruiser Division 5's ships came left thirty degrees, and DESDIV 191 ships came left forty-five degrees while maintaining speed. Destroyer Division 192 ships continued their course and speed. Upon completion of Move 6, students again shifted, this time for Phase III, leaving course and speed information for the next groups. (Page 6.)

Phase III, or Moves 7, 8, and 9, simulated the same weather conditions as Phase II. In Move 7, BB-58 continued on a course of 045 degrees with a speed of twelve knots, while BB-59 reduced speed to twelve knots and changed course to 045. Cruiser Division 5 came right in column to 030 while continuing its speed of twenty-four knots. Destroyer Squadron 19's DESDIV 191 came right forty-five degrees and continued at twenty knots, while DESDIV 192 maintained course and speed. Move 8 saw BATDIV 3 ships turning right forty-five degrees and increasing speed to sixteen knots, while Cruiser Division 5 turned starboard sixty degrees and maintained engine speed. Within DESRON 19, DESDIV 191 turned port 135 degrees to course 000 and increased speed to twenty-five knots, while DESDIV 192 came right ninety degrees and increased to the same speed. Move 9 saw BATDIV 3 turning left in column ninety degrees and increased speed to the maximum; Cruiser Division 5, less heavy cruiser CA-71, turned starboard forty-five degrees and also increased engine speed to the maximum allowed. CA-71 left the formation and proceeded on course 180, also at maximum speed. DESRON 19's ships all increased speed to maximum and continued on the same courses. Move 9 ended Phase III as well as Exercise 1. (Page 7.)

Exercise 2

The object of Exercise 2 was to familiarize the student with the Gunfire Section of the Maneuver Rules, the Board Maneuver Rules, the Supplement to the Maneuver Rules, and the Gunfire Sheet used in a Board Maneuver. Students were referred to these sources, as well as to *Characteristics of the Blue Fleet*. Exercise 2 simulated an engagement between the Blue Force used in Exercise 1 and a Purple Force of the same number and types of ships. Student and staff assignments for Exercise 1 carried over to Exercise 2, with Groups 1, 3, and 5 controlling the Blue Force and Groups 2, 4, and 6 the Purple Force. Each student filled out a Move Sheet and a Gunfire Sheet for the unit to which he was assigned. The sheets were again inspected by a member of the staff, who notified the Director when the group was ready for the move. Again, moves were made on the board when announced by the Director; to reduce the time needed for the exercise, tracks were plotted prior to the exercise. The General Situation specified that the Purple Force comprised Battleship Division 1 (BB-1 and BB-2); Cruiser Division 1 (CA-1, -2, -3, and -4); and Destroyer Squadron 1, containing Destroyer Division 11 (DD-1, -2, -3, and -4) and DESDIV 12 (DD-5, -6, -7, and -8). The wind was again from the northeast at Force 4, the seas were moderate, visibility was normal, skies overcast, ceiling five thousand feet. (Pages 8–9.)

Section A BLUE Forces		Section B PURPLE Forces	
Group	Unit	Group	Unit
1	Batdiv 3	2	Batdiv 1
3	Crudiv 5	4	Crudiv 1
5	Desron 19	6	Desron 1
	Move Two		
5	Batdiv 3	6	Batdiv 1
1	Crudiv 5	2	Crudiv 1
3	Desron 19	4	Desron 1
	Move Three		
3	Batdiv 3	4	Batdiv 1
5	Crudiv 5	6	Crudiv 1
1	Desron 19	2	Desron 1

Fig. 75
Exercise 2 groups

The Blue Force had Battleship Division 3 in column at a distance of a thousand yards on course 000 degrees at an engine speed of twenty-three knots. BB-58 was serving as Guide, and one spotting plane from each ship was on station. Cruiser Division 5 was also in column, at a distance of six hundred yards on 045 and speed twenty-five knots. Heavy cruiser CA-68 was the Guide, and again each ship had a spotting plane on station. DESRON 19 too was in column, with intervals between divisions of 1,500 yards and a distance between ships within each division of three hundred yards. Its course was 045, its speed was twenty-five knots, and DESDIV 191 was in the lead. No Blue ships had received any previous damage. In Move 1, BATDIV 3 continued its course and speed; BB-58 opened fire with its main battery on Purple fast battleship BB-1 and with its secondary battery on Purple DESDIV 11. At the same time, Blue BB-59 opened fire with its main battery on Purple BB-2 and with its secondary battery on Purple Destroyer Division 12. Purple DESDIV 11 opened fire on Blue BB-58, and Purple DESDIV 12 opened fire on Blue BB-59. Cruiser Division 5 continued its course and speed; its heavy cruisers CA-68, -69, -70, and -71 opened main-battery fire on, respectively, Purple heavy cruisers CA-4, -3, -2, and -1. DESRON 19 continued its course and speed, with DESDIV 191 opening fire on Purple BB-1 and DESDIV 192 on Purple BB-2; the two Purple battleships returned the fire with their five-inch guns. (Page 9a.)

The Purple Force disposition had BATDIV 1 in column at a distance of a thousand yards on a course of 315 and at a speed of twenty-three knots. Fast battleship BB-1 was the Guide, and one spotting plane from each battleship was on station. Cruiser Division 1 was in column with a distance of six hundred yards on course 225 and speed twenty-five. Heavy cruiser CA-1 was the Guide, and the Purple cruisers likewise had one spotting plane each on station. DESRON 1 was in column with intervals between divisions of 1,500 yards at a distance between ships of three hundred yards on a course of 315 and at a speed of twenty-five knots; Destroyer Division 11 was in the lead. Purple ships had been undamaged prior to Move 1. In Move 1, Purple battleship BB-1 opened fire with its main battery on Blue battleship BB-58 and with its secondary battery on Blue DESDIV 191. Purple battleship BB-2's main battery fired on Blue battleship BB-59 and its secondary battery on Blue DESDIV 192. Purple Cruiser Division 1 continued its course and speed, and its heavy cruisers CA-1, -2, -3, and -4 opened up with main-battery fire on Blue heavy cruisers CA-71, -70, -69, and -68, respectively. Purple DESRON 1 also continued on course and speed, its DESDIV 11 opening fire on Blue BB-58 and its DESDIV 12 on Blue BB-59. (Page 10.)

In Move 2, Blue BATDIV 3 continued on course and speed; BB-58 continued its main-battery fire on Purple BB-1 and secondary-battery fire on Purple Destroyer Division 11. Blue BB-59 also continued shooting with its main battery at Purple BB-2 and with its secondary battery on Purple DESDIV 12, while Purple DESDIVs 11 and 12 continued firing on Blue BB-58 and BB-59, respectively. By this time, Blue fast battleship BB-58 had been damaged 20 percent. Blue Cruiser Division 5 continued on its course and speed, with heavy cruisers CA-68, -69, -70, and -71 continuing their main-battery fire on Purple heavy cruisers CA-4, -3, -2, and -1, respectively. By this time, Blue heavy cruisers CA-68 and CA-70 had each received 10 percent damage, CA-69 20 percent. Heavy cruiser CA-71 was still undamaged. Blue DESRON 19 maintained course and speed; DESDIV 191 continued to fire on Purple battleship BB-1 and DESDIV 192 on Purple battleship BB-2. Purple BB-1 and BB-2 were firing their secondary batteries at DESDIVs 191 and 192, respectively; DESDIV 191 had taken 20 percent damage by this time, and DESDIV 192 had been 10 percent damaged. (Page 11.)

Move 2 also saw Purple BATDIV 1 continuing on its course and speed, its fast battleship BB-1 continuing main-battery fire on Blue BB-58 and Purple BB-2 still shooting at Blue BB-59. Both Purple battleships were taking fire from their Blue counterparts and Blue Destroyer Divisions 191 and 192; Purple BB-1 had received 20 percent damage by this time, but Purple BB-2 was undamaged. Cruiser Division 1 continued on its course and speed; heavy cruisers CA-1, -2, -3, and -4 carried on firing at Blue heavy cruisers CA-71, -70, -69, and -68, respectively. By this time, the Purple heavy cruisers CA-1 and CA-3 had been damaged 10 percent, 20 percent for CA-2; CA-4 remained undamaged. Purple DESRON 1 was on its same course and speed, with its two divisions continuing their fire on the two Blue battleships and receiving secondary-battery fire in reply. By now Purple DESDIV 11 had been 20 percent damaged, and DESDIV 12 had been 10 percent damaged. (Page 12.)

Move 3 for Blue entailed Battleship Division 3 changing course to 060 degrees at the same speed. Blue BB-58 continued its main-battery fire on Purple BB-1 and its secondary-battery fire on Purple Destroyer Division 11, while Blue BB-59 continued its main-battery fire on Purple BB-2 and its secondary-battery fire on Purple DESDIV 12. There was a reduction of 0.2 of the Blue battleships' Fire Effect caused by the Purple battleships' changing course, and each Blue battleship had by now been damaged 20 percent. Cruiser Division 5 slowed by five knots and turned to 345; each Blue heavy cruiser carried on firing at its Purple counterpart from the previous moves. The Purple heavy cruisers too had changed course, causing a reduction of 0.2 in the Fire Effect of the Blue heavy cruisers; by this time, Blue heavy cruisers CA-68 and CA-69 had each been damaged 30 percent, CA-70 and CA-71

20 percent. Destroyer Squadron 19 did not change course or speed; Destroyer Divisions 191 and 192 continued their fire on Purple fast battleships BB-1 and BB-2, respectively. Their Fire Effect was reduced by 0.2 because of the change of course by the two Purple battleships, and they were still taking fire from the Purple battleships' secondary batteries; Blue DESDIV 191 had been 50 percent damaged by now, and Blue DESDIV 192 had taken 20 percent damage. (Page 13.)

Move 3 for the Purple Force saw BATDIV 1 changing course to 090 at the same speed, with BB-1 continuing its main-battery fire on Blue BB-58 and its secondary-battery fire on Blue DESDIV 191. Purple BB-2 continued its main-battery fire on Blue BB-59 and its secondary-battery fire on Blue Destroyer Division 192. As the Purple battleships received fire from the Blue DESDIVs, their Fire Effect was also reduced by 0.2 because of a change of course by the Blue battleships, and each Purple battleship by this time had been damaged 20 percent. Cruiser Division 1 had slowed by five knots and changed course to 000. Its ships were still firing on their Blue counterparts from the previous moves, and they too had seen their Fire Effects reduced by the change of course by the Blue heavy cruisers; also at this time, Purple heavy cruisers CA-1 and CA-2 were 30 percent damaged and CA-3 and CA-4 20 percent. Purple DESRON 1 was on its same course and speed, and its DESDIVs 11 and 12 were continuing to fire on Blue battleships BB-58 and BB-59, respectively. The Purple destroyer divisions were still taking fire from the Blue battleships, which had changed course sixty degrees to starboard. This change of course also saw the Purple DESDIVs' Fire Effect reduced by 0.2, DESDIV 11 damaged 50 percent, and DESDIV 12 20 percent damaged. Exercise 2 now ended. (Page 14.)

Exercises 3 and 4

Exercise 3 was for continued practice in the conduct of a Board Maneuver, in this case, another engagement between a Blue and a Purple force of equal strength. Again, there would be three phases,

Director of the Maneuver	Capt. D. S. Evans
Chief Damage Computer	Comdr. Pelling Comdr. Mayberry
Assist BLUE Batdiv 3 Staff	Comdr. Reed
BLUE Crudiv 5 Staff	Comdr. Reed
BLUE Desron 19 Staff	Comdr. Haskell
Assist PURPLE Batdiv 1 Staff	Lt. Cdr. Curran
PURPLE Crudiv 1 Staff	Lt. Cdr. Curran
PURPLE Desron 1 Staff	Lt. Cdr. Holzapfel

Fig. 76
Exercise 3 staff

BLUE	Phase 1	Phase 2	Phase 3	PURPLE
Group 1 Batdiv 3				**Group 2 Batdiv 1**
Col.S.E.Manzo,AC,USA	2	C	A	Col.H.J.Hawthorne,AC,USA
Comdr.G.S.Coleman,USN	3	A	2	Comdr.K.Loveland,USN
Comdr.A.M.Nibbs,USN	D	B	1	Comdr.C.A.O'Connell,Jr.USN
Lt.Col.Downey,Jr.CAV,USA	B	2	C	Comdr.A.T.Church,Jr.,USN
Comdr.Baldridge,USN	A	E	3	Comdr.H.J.P.Foley,Jr.SC,USN
Lt.Cdr.J.P.Aymond,USN	C	3	B	Lt.Comdr.W.Y.Howell,USN
Group 3 Crudiv 5				**Group 4 Crudiv 1**
Col.H.O.Edson,INF,USA	3	A	C	Col.R.J.Mason,AC,USA
Comdr.E.W.Bridewell,USN	A	C	2	Comdr.G.H.Miller,USN
Comdr.R.C.Morton,USN	B	D	3	Comdr.N.C.Gillette,Jr.USN
Lt.Col.R.E.Leary,AC,USA	2	3	B	Comdr.H.Marvin-Smith,USN
Comdr.Kirkpatrick,USN	1	2	E	Lt.Col.F.L.Street,FA,USA
Lt.Cdr.F.L.Taeusch,USN	C	B	A	Lt.Cdr.E.F.Bonner,USN
Group 5 Desron 19				**Group 6 Desron 1**
Lt.Col.L.Wallace,INF,USA	3	C	A	Lt.Col.M.H.Thompson,CE,USA
Comdr.C.D.Simonsen,USN	2	3	B	Comdr.J.E.Owers,USN
Comdr.W.K.Rogers,USN	A	2	C	Comdr.D.C.Richardson,USN
Comdr.G.F.Delton,USN	C	1	D	Comdr.O.B.Lundgren,USN
Comdr.T.S.White,USN	B	A	2	Comdr.R.W.Rynd,USN
Comdr.R.P.Webber,SC,USN	E	B	3	Lt.Cdr.J.D.Ramage,USN

Key: A - Unit Commander
B - Chief of Staff to A.
C - Move Plotter
D - OTC
E - Chief of Staff to D.

1 - Force Damage Recorder
2 - Damage Computer
3 - Move Umpire (for group only)

Fig. 77
Exercise 3 student groups and opposing forces

each of three moves; students were assigned to particular duties as each group controlled various forces in turn. Student Commanders were to maneuver their assigned units under the direction of an OTC, with the general mission being one of inflicting maximum damage on opposing forces while sustaining minimum damage to one's own. While torpedo fire was not simulated, each side's destroyers were maneuvered to torpedo-firing positions as directed by the OTC. (Pages 15–17.)

Exercise 4 was intended to provide practice in the use of the Radar Contact Report and Message Forms as well as in the plotting of radar contacts. A Staff Solution was issued for comparison with the student solutions; the reference was again the Maneuver Rules, along with Radar Bulletin No. 1A. In the Special Situation, Blue fast battleship BB-58 was steaming in company with a force of cruisers and destroyers, all organized as Task Group 45.2. They were in a Night Cruising Disposition and were assigned as the long-range surface radar guard (BB-58 was equipped with an SG-3 surface-search radar). The OTC was in one of the cruisers and considered contact with enemy surface units probable. The course of the formation was 000, speed was twenty-two knots. The wind was from the northeast at Force 1, the sea was smooth, visibility was normal, and radar conditions were normal.

Contact with three groups of unidentified surface vessels occurred sometime before 1900, at which time the OTC directed BB-58 to report the course, speed, and probable composition of each group at three-minute intervals. The students plotted the track of their own ships and of the contacts on Tactical Data Plotting Sheet 1925. The Radar Contact Forms presented BB-58 but not the plots of the other ships. Course, speed, and probable composition of each group for contacts at 1903, 1906, and 1909 were required, as was the text of a very-high-frequency (VHF) voice-radio message from BB-58 to the OTC at 1910. This radio message, written on a Message Form, reported the estimate made from the data obtained by 1909. In addition, the students were to provide a copy of a radio message from the OTC—Commander, TG 45.2—to Commander, Task Force 45, reporting the situation and the OTC's intended action. Commander, Task Force 45 was at a base five hundred miles distant. (Pages 18–19.)

Exercise 5

Exercise 5 was really two exercises in one, both being torpedo-firing problems designed to acquaint students with the use of Torpedo Fire Cards and the Torpedo Fire Form. Exercise 5 also familiarized students with the solution of torpedo-fire problems and with destroyer torpedo doctrine as given in the *Manual of Destroyer Torpedo Control,* Destroyer Tactical Bulletin No. 4-45. (Page 20.)

In Exercise 5(a), Vessel A was steaming on a course of 270 at twenty knots. A destroyer fired one Type G-2 torpedo at Vessel A when the latter's target angle was 330 degrees and was at a maximum effective range. No range allowance was made, and the wind was from the northeast at Force 2, seas smooth. The students

Form T-18
1944

Sheet 1

RADAR CONTACT REPORT

BLUE OPERATIONS PROBLEM B.M.Ex.4 (Tactical)

UNIT MAKING CONTACT

Exercise 4
Worksheet

Ship BB 58
Force BLUE
Student -
Room -

Date -
MOVE NO. -
MANEUVER TIME 1900
Latitude -
Longitude -

TARGET										
No	True Bearing	Range		Air or Surface	One Few Many	Approx.			Prev. Reported	
		Miles	I/D			Course	Speed	Alt.		
1	86	14.5	-	S	1	NW	25	-	No	
2	88	14.5	-	S	1	NW	25	-	No	
3	56	7.6	-	S	—	NW	25	-	No	Rng. & Brgs. Center of Grp.
4	61	7.8	-	S	—	NW	25	-	No	
5	45	15.1	-	S	4	SW	25	-	No	

P.P.I.

SCALES

1 Division = 3000 Yds.

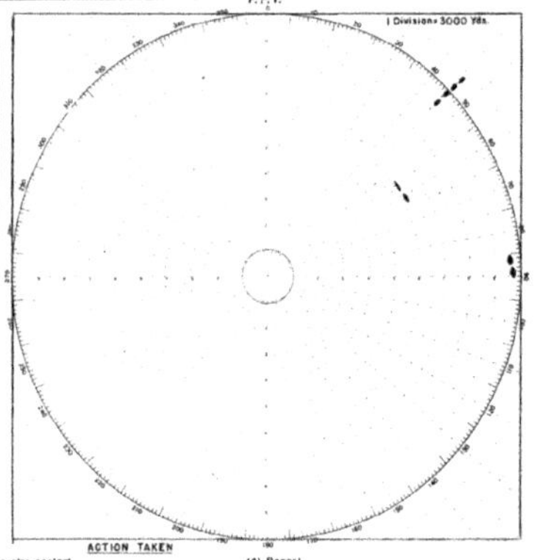

ACTION TAKEN

(1) Develop contact
(2) Carry out original search plans
(3) Hold contact
(4) Report
(5) Other action

Student's Signature

Form T-18
1944

Sheet 2

RADAR CONTACT REPORT

BLUE OPERATIONS PROBLEM B.M.Ex.4 (Tactical)

UNIT MAKING CONTACT

Exercise 4
Worksheet

Ship BB 58
Force BLUE
Student -
Room -

Date
MOVE NO. -
MANEUVER TIME 1903
Latitude -
Longitude -

TARGET										
No	True Bearing	Range		Air or Surface	One Few Many	Approx.			Prev. Reported	
		Miles	I/D			Course	Speed	Alt.		
1	87	13.7	D	S	1	NW	25	-	Yes	
2	89	13.7	D	S	1	NW	25	-	Yes	
3	53	6.7	D	S	—	NW	25	-	Yes	Rng. & Brgs. Center of Grp.
4	60	6.9	D	S	—	NW	25	-	Yes	
5	48½	13.1	D	S	4	SW	25	-	Yes	

P.P.I.

1 Division = 3000 Yds.

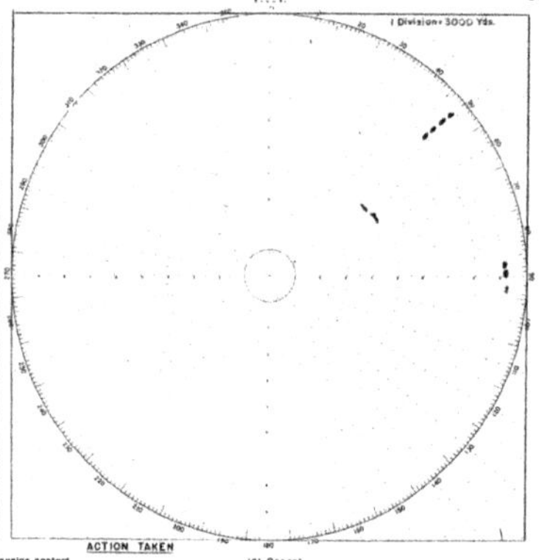

ACTION TAKEN

(1) Develop contact
(2) Carry out original search plans
(3) Hold contact
(4) Report
(5) Other action

Student's Signature

Required:
Course, Speed and probable
Composition of each group

Form T-18
1944

Sheet 3

RADAR CONTACT REPORT

BLUE OPERATIONS PROBLEM B.M.Ex.4 (Tactical)

UNIT MAKING CONTACT

Exercise 4
Worksheet

Ship BB 58
Force BLUE
Student -
Room -

Date -
MOVE NO. -
MANEUVER TIME 1906
Latitude -
Longitude -

TARGET										
No	True Bearing	Range		Air or Surface	One Few Many	Approx.			Prev. Reported	
		Miles	I/D			Course	Speed	Alt.		
1	88½	13	D	S	1	NW	25	-	Yes	
2	90½	13	D	S	1	NW	25	-	Yes	
3	50	5.8	D	S	--	NW	25	-	Yes	Rngs. & Brgs. Center of Grp.
4	58	6.	D	S	—	NW	25	-	Yes	
5	53	11.1	D	S	4	SW	25	-	Yes	

P.P.I.

SCALES

1 Division = 3000 Yds.

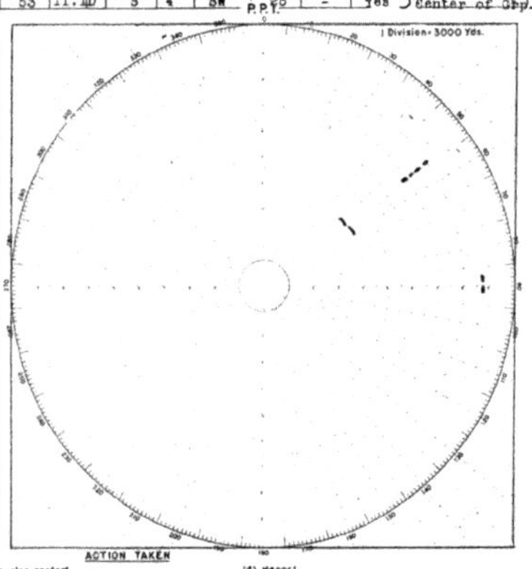

ACTION TAKEN

(1) Develop contact
(2) Carry out original search plans
(3) Hold contact
(4) Report
(5) Other action

Student's Signature

Required:
Course, Speed and probable
composition of each group

Form T-18
1944

Sheet 4

RADAR CONTACT REPORT

BLUE OPERATIONS PROBLEM B.M.Ex.4 (Tactical)

UNIT MAKING CONTACT

Exercise 4
Worksheet

Ship BB 58
Force BLUE
Student -
Room -

Date -
MOVE NO. -
MANEUVER TIME 1909
Latitude -
Longitude -

TARGET										
No	True Bearing	Range		Air or Surface	One Few Many	Approx.			Prev. Reported	
		Miles	I/D			Course	Speed	Alt.		
1	88½	12.4	D	S	1	Chng.		-	Yes	
2	90½	12.4	D	S	1	Chng.		-	Yes	
3	38	5	D	S	--	NW		-	Yes	Rng. & Brgs. Center of Grp.
4	47	5	D	S	--	NW		-	Yes	
5	54½	9.6	D	S	4	Chng.		-	Yes	

P.P.I.

SCALES

1 Division = 3000 Yds.

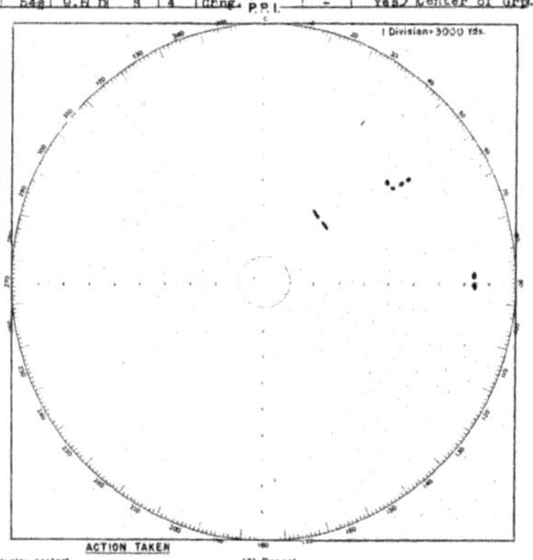

ACTION TAKEN

(1) Develop contact
(2) Carry out original search plans
(3) Hold contact
(4) Report
(5) Other action

Student's Signature

Required:
Course, Speed and Probable
Composition of each group.

Fig. 78
Radar Contact Report, Exercise 4

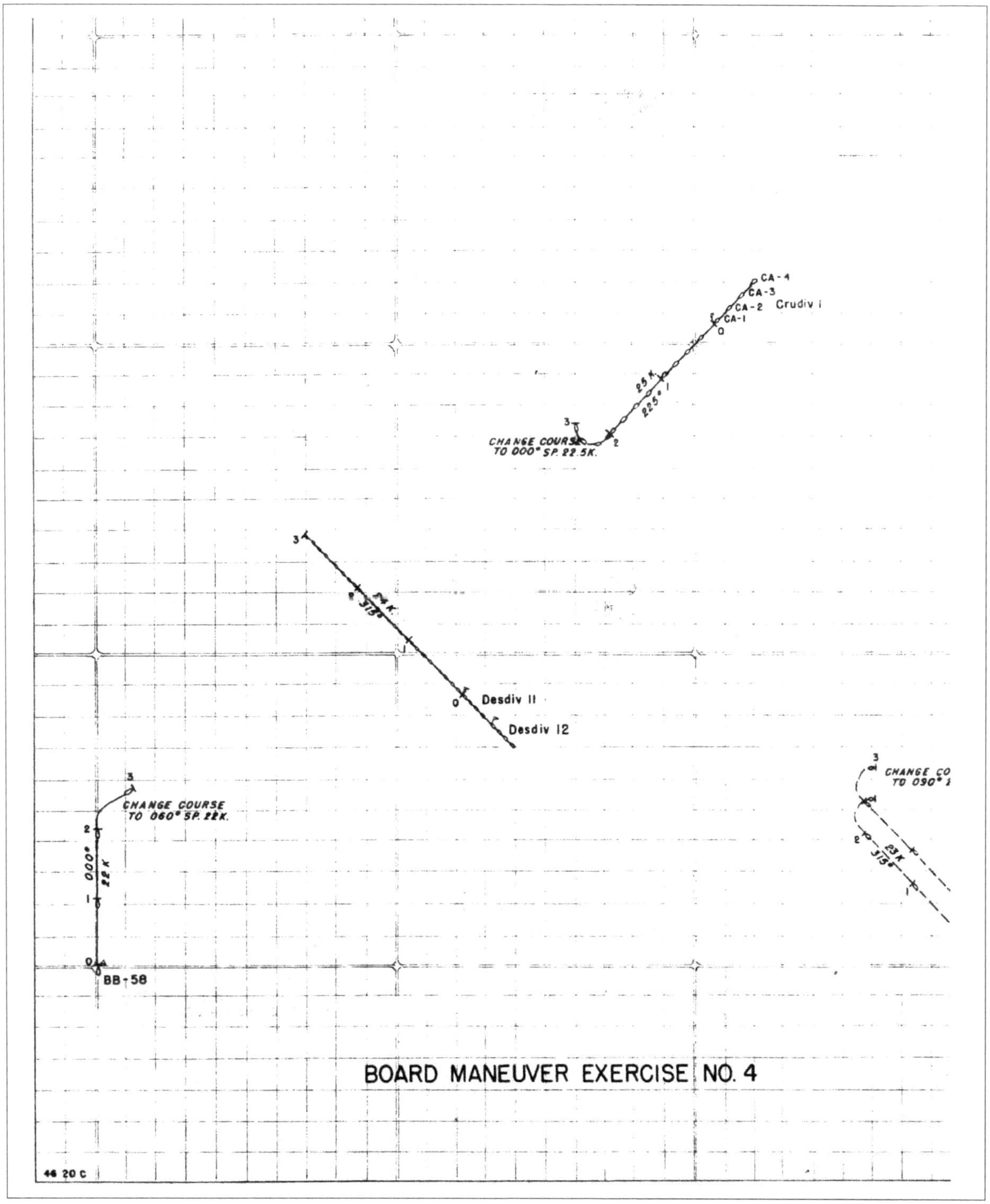

Fig. 79
Board Maneuver,
Exercise 4

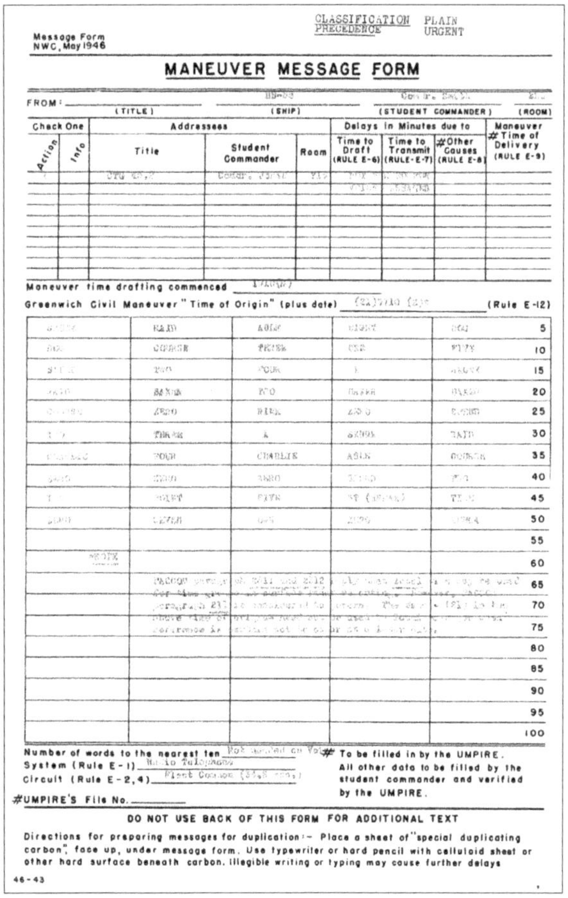

CLASSIFICATION PLAIN
PRECEDENCE URGENT

Message Form
NWC, May 1946

MANEUVER MESSAGE FORM

FROM: (TITLE) (SHIP) (STUDENT COMMANDER) (ROOM)

Check One: Action	Check One: Info	Addressees: Title	Addressees: Student Commander	Addressees: Room	Delays in Minutes due to: Time to Draft (RULE E-6)	Delays in Minutes due to: Time to Transmit (RULE-E-7)	Delays in Minutes due to: #Other Causes (RULE E-8)	Maneuver #Time of Delivery (RULE E-9)

Maneuver time drafting commenced

Greenwich Civil Maneuver "Time of Origin" (plus date) (Rule E-12)

Number of words to the nearest ten

System (Rule E-1) Radio Telephone

Circuit (Rule E-2,4)

#UMPIRE'S File No.

To be filled in by the UMPIRE. All other data to be filled by the student commander and verified by the UMPIRE.

DO NOT USE BACK OF THIS FORM FOR ADDITIONAL TEXT

Directions for preparing messages for duplication:– Place a sheet of "special duplicating carbon", face up, under message form. Use typewriter or hard pencil with celluloid sheet or other hard surface beneath carbon. Illegible writing or typing may cause further delays

46-43

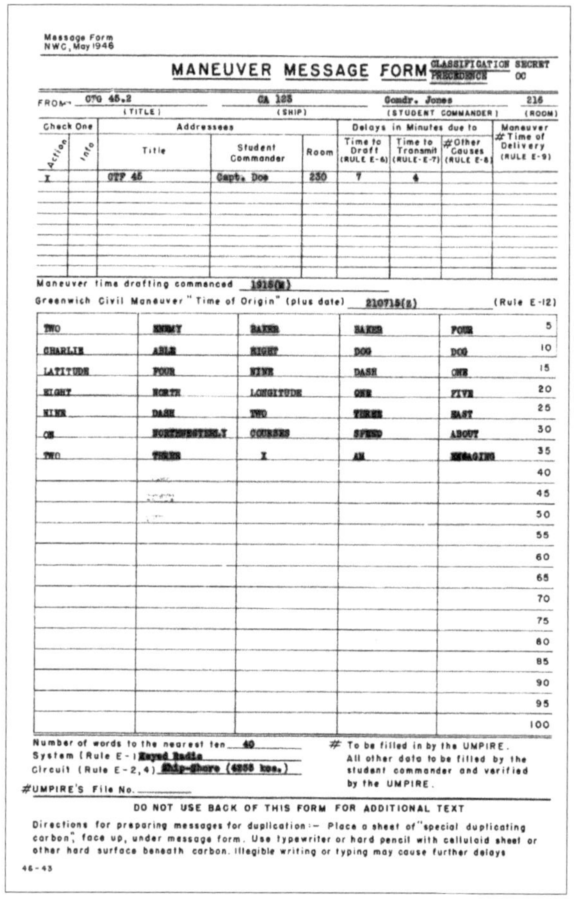

Message Form
NWC, May 1946

MANEUVER MESSAGE FORM

CLASSIFICATION SECRET
PRECEDENCE OO

FROM: CTG 45.2 (TITLE) CA 125 (SHIP) Comdr. Jones (STUDENT COMMANDER) 216 (ROOM)

Check One: Action	Check One: Info	Addressees: Title	Addressees: Student Commander	Addressees: Room	Delays in Minutes due to: Time to Draft (RULE E-6)	Delays in Minutes due to: Time to Transmit (RULE-E-7)	Delays in Minutes due to: #Other Causes (RULE E-8)	Maneuver #Time of Delivery (RULE E-9)
X		CTF 45	Capt. Doe	230	7	4		

Maneuver time drafting commenced 1915(Z)

Greenwich Civil Maneuver "Time of Origin" (plus date) 210715(Z) (Rule E-12)

TWO	ENEMY	BAKER	BAKER	FOUR	5
CHARLIE	ABLE	EIGHT	DOG	DOG	10
LATITUDE	FOUR	NINE	DASH	ONE	15
EIGHT	NORTH	LONGITUDE	ONE	FIVE	20
NINE	DASH	TWO	THREE	EAST	25
ON	NORTHWESTERLY	COURSES	SPEED	ABOUT	30
TWO	THREE	X	AM	ENGAGING	35
					40
					45
					50
					55
					60
					65
					70
					75
					80
					85
					90
					95
					100

Number of words to the nearest ten 40

System (Rule E-1) Keyed Radio

Circuit (Rule E-2,4) Ship-Shore (4235 kcs.)

#UMPIRE'S File No.

To be filled in by the UMPIRE. All other data to be filled by the student commander and verified by the UMPIRE.

DO NOT USE BACK OF THIS FORM FOR ADDITIONAL TEXT

Directions for preparing messages for duplication:– Place a sheet of "special duplicating carbon", face up, under message form. Use typewriter or hard pencil with celluloid sheet or other hard surface beneath carbon. Illegible writing or typing may cause further delays

46-43

Fig. 80
Maneuver Message Form, Exercise 4

computed the target track, the target angle, the line of sight, the track of the torpedo, the sight angle, the base torpedo course, and the track angle. All were to be shown in a diagram and properly labeled. In Exercise 5(b), Blue Destroyer Division 191, consisting of four *Sumner*-class destroyers, fired a spread of forty Type G-1 torpedoes at Purple BATDIV 1—battleships BB-1 and BB-2, identical to the U.S. *South Dakota* class. DESDIV 191, which was under effective fire from the Purple battleships' secondary batteries, was on course 225 at twenty-seven knots, before changing course to 180. The torpedoes were fired in the expectation that the targets would not maneuver, and the Base Torpedo Course Plan was used. The unit of spread was one degree, and the point of aim was the Purple battleships themselves.

Continuing Exercise 5(b), students now figured the torpedo depth setting, the torpedo speed to be used, and the base torpedo course. Using the Maneuver Rules, they also worked out the estimated number of hits on the Purple battleships and drew diagrams showing target track and torpedo spread as well as showing the limits of the division's spread. They were additionally to plot the position of the torpedoes at the end of each move and record a Torpedo Fire Form for DESDIV 191's attack by the end of the move. (Pages 20–21.)

Exercise 6 and the Concluding Exercise

Exercise 6 consisted of two problems in plotting smoke screens. In Exercise 6(a), a division of four destroyers in column at intervals of three hundred yards was steaming on a course of 270 at fifteen knots. The wind was from the southwest at Force 3, the seas were smooth, and smoke would lie. All ships commenced making smoke at the beginning of Move 1. They continued this operation in Moves 2, 3, 4, and 5 while the student plotted the track of the destroyer division and the smoke for the five moves. In Exercise 6(b), Purple DESDIV 12, four destroyers also

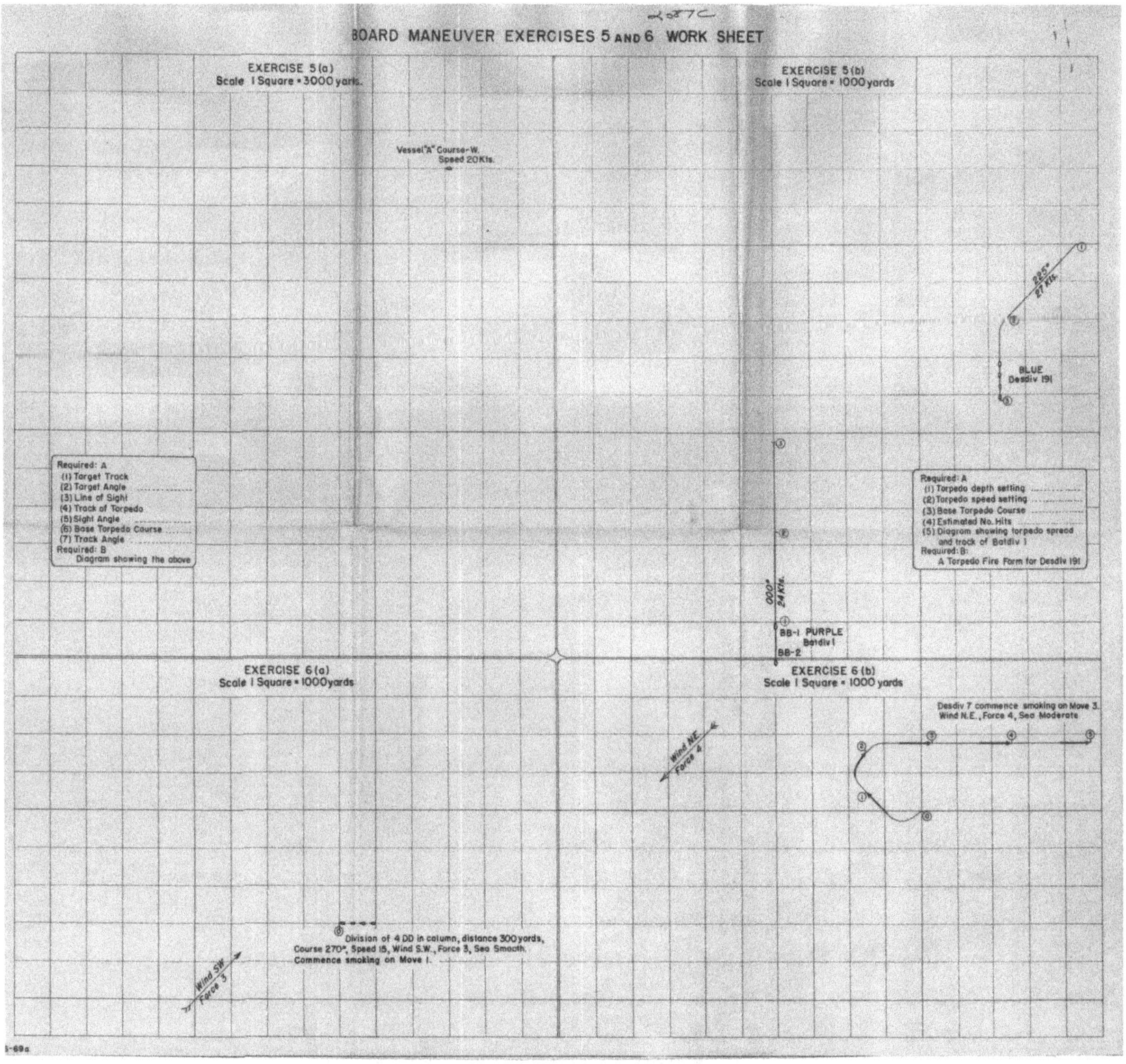

Fig. 81
Exercises 5 and 6 Work Sheet

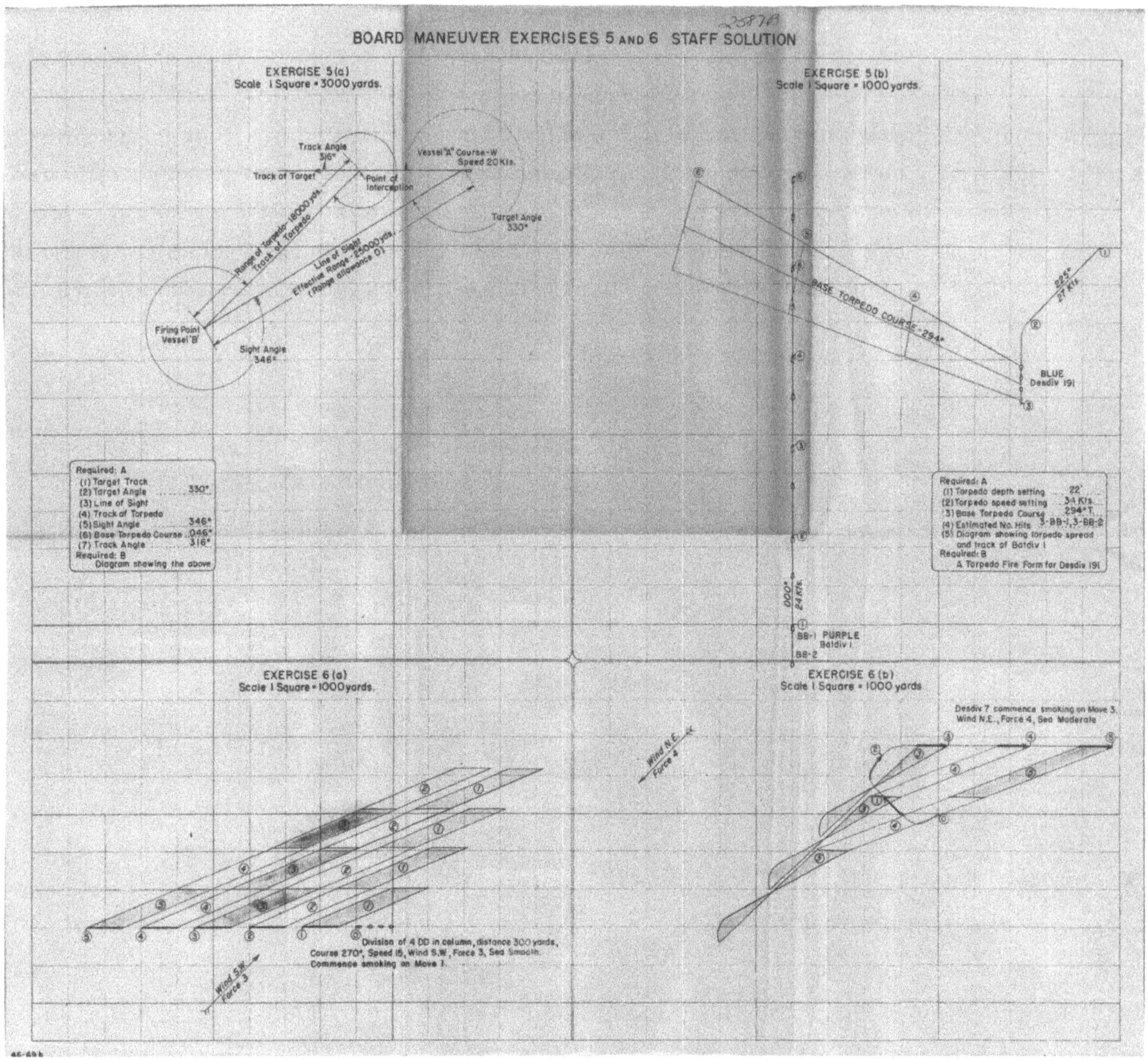

Fig. 82
Exercises 5 and 6 Staff Solutions

identical with those of the *Sumner* class, changed course to 315 in Move 1 and fired torpedoes. In Move 2, DESDIV 12 changed course to 045; in Move 3, it turned to 090 degrees and commenced making smoke, which it continued to do in Moves 4 and 5. The wind was now from the northeast at Force 4, the seas were now moderate, but smoke would still lie. Students plotted the smoke made by Purple Destroyer Division 12 for Moves 3, 4, and 5. The track of the division had been given to them on the Plotting Sheet; they added the curved shape of the smoke at the beginning of emission in Move 3. (Pages 21–22.)

About a month later, the senior class of June 1947 was posed the same set of exercises, with the same organization, reference materials, and purposes. There were some minor changes, however. In Exercise 2, the Director of the Maneuver decided that ships steaming in courses between 000 and 090 degrees would suffer reduction in speed because of sea conditions—battleships and cruisers by one knot, destroyers by three. The maximum speed allowed for destroyers was thirty-two knots. The Director also decided that ships steaming in the same course patterns would see their Fire Effect Multiplier, *M*, reduced 0.1 also because of sea conditions. Reductions other than that caused by Above Water Damage would be decided by the Director at the "appropriate" time.* With the conclusion of these exercises, the students moved on to the first Operations Problem of their academic year.

* "Board Maneuver Exercises," 4 October 1946, pp. 1–23, esp. p. 9, folder 2586, box 135, RG 4, NHC.

VIII *Operations Problem 1J*

Blue, September 1946

In early September 1946, the junior class of June 1947 was issued with Operations Problem 1J, a tactical Chart and Board Maneuver involving two opposing task forces in a theater of operations of "limited extent." The problem began on 20 September with student officers drawing the Statement of the Problem, going through an orientation, and attending a lecture in the auditorium in Pringle Hall. In accordance with Admiral Smith's instructions, the students worked on the problem until 27 September when, after another lecture, they began preparations for the maneuver the next day. The maneuver commenced on 30 September and continued until 8 October, when the students drew the Staff Solution and prepared for the Maneuver Critique, which took place on the 9th. The objectives of the problem were to have students make a tactical Estimate of the Situation, formulate a Battle Plan, and fight a day surface action. Another was to demonstrate the capabilities, limitations, and tactical employment of Blue ships and carrier-based aircraft and to bring out the hydrographic and meteorological features of the northern Pacific.[1]

The Blue Statement: General and Special Situations

Under the General Situation, Blue and Purple were at war, their offensive actions thus far consisting of offensive and reconnaissance submarine patrols and a "sustained" land-based-air offensive by Purple. There had been no change in the possession or occupation of Pacific bases. In the Special Situation, the Blue Pacific Fleet was based at Pearl Harbor. Fleet Admiral CD, Commander-in-Chief, Pacific Fleet (not played by a student officer), organized Task Force 21, or the Alaskan Defense Force, from the Pacific Fleet, under the command of Admiral BA, Commander, Battleship Squadron (COMBATRON) 1. TF 21 consisted of two fleet carriers under Rear Admiral BD, organized as Carrier Division 7. TF 21 also contained two fast battleships (Battleship Division 1, under Rear Admiral BX), two additional fast battleships, organized as Battleship Division 2 under Rear Admiral BE, and two battle cruisers under Rear Admiral BF (Cruiser Division 4).

There were additionally four light cruisers of CRUDIV 17 under Rear Admiral BG and nine destroyers under Captain B-2, Destroyer Squadron 4. This latter squadron was divided into a group of five destroyers as Destroyer Division 41 and

Admiral BA (Cdr. William Kirkpatrick, USNR)

Rear Admiral BD (Lt. Col. Richard Leary, USAAF)

Rear Admiral BX (Col. Hallett Edson, USA)

Rear Admiral BF (Cdr. George Dalton, USN)

four as DESDIV 42. Captain B-3 commanded an additional nine destroyers organized into DESRON 12, comprising five in DESDIV 121 and the other four in DESDIV 122. Another nine destroyers were under Captain B-4 as DESRON 21, within which five made up DESDIV 211 and four DESDIV 212. Blue destroyers had wakeless torpedoes with a range of forty thousand yards at thirty-two knots, or eighteen thousand yards at forty-six knots. (Pages 1, 6).

As to airpower, TF 21's fleet carriers each had fifty-seven fighter planes on board; CV-39's planes were forty-five F8F Bearcat fighters organized into two fighter squadrons; CV-37's air group consisted of forty-five F4U Corsair fighters, also organized into two squadrons. Each carrier also had four F6F Hellcat fighters as photographic reconnaissance planes and eight additional Hellcat night fighters attached to these fighter squadrons. There were also two attack squadrons on fleet carrier CV-39, with twenty-four SB2C Helldiver and nineteen TBM Avenger attack planes, five of the latter constituting a carrier tactical electronics warfare squadron (VA[Q]), another four a squadron of carrier airborne early warning aircraft. Fleet carrier CV-37 had the same number, types, and composition of attack planes, but all organized into just one attack squadron. Each of the battleship division's units also had four SC-1 observation planes, as did those of CRUDIV 4, while CRUDIV 17 had eight. (Page 7.)

Blue land-based air forces and airfields in the Aleutians and western Alaska had been neutralized by the Purple air offensive. In addition, Blue intelligence reports indicated that Purple had assembled a joint expeditionary force at Avacha Bay. It was apparently composed of two combat divisions and an amphibious force that included gunfire support groups made up of gunboats and rocket-firing ships. According to Blue intelligence, there was also a Purple Covering Force at Kashiwabara Bay, consisting of two to three fleet carriers of the *Essex* class, four to five fast battleships of the *South Dakota* and *North Carolina* classes, three to five heavy cruisers of the *Baltimore* class, three to four light cruisers of the *Cleveland* class, and three squadrons of destroyers comparable to Blue's latest types and equipped with long-range torpedoes. These reports were considered "excellent" in reliability.

Rear Admiral BG (Lt. Col. Stephen Downey, USA)

Captain B-2 (Cdr. Robert Webber, USN)

Captain B-3 (Cdr. Garrett Coleman, USN)

Captain B-4 (Lt. Cdr. Frederick Taeusch, USN)

Other "reliable" observation reports, however, indicated that Purple land-based air forces in the North Pacific had been reduced to "impotency" by the heavy losses incurred in their recent offensive. (Pages 1–2.)

Fleet Admiral CD had issued a Fleet Operation Plan, dated 20 July, that was to take effect upon receipt by Admiral BA. The Fleet Operation Plan assigned Admiral BA the tasks of preventing Purple from establishing forces ashore in the Aleutians and of exploiting every opportunity to bring Purple surface combatant forces in the northwest Pacific into "decisive" action. Admiral BA was aware that air reconnaissance from Pearl Harbor indicated that Purple submarines in unknown numbers were operating to the north of Hawaii. In addition, Admiral BA was told that neither carrier- nor land-based air replacements or surface reinforcements would be available to the Alaskan Defense Force for sixty days. Blue submarines were conducting reconnaissance patrols off Kamchatka and Attu and on the Cape Kamchatka–Cape Wrangell Line, but they were not present in sufficient strength to be used offensively, and their early reinforcement was "impracticable." However, it was not expected that Purple losses in land-based air strength could be made good for sixty days either. The rest of the Blue Pacific Fleet—less the Alaskan Defense Force—was to contain the Purple Pacific Fleet, less the Purple forces that were currently based in Kamchatka and the Kuriles. Finally, it was pointed out to Admiral BA that the Blue Pacific Fleet's Service Squadron 6, Task Unit 11.2.2, would provide logistical support to TF 21. Moreover, fuel and aviation gasoline sufficient for the Alaskan Defense Force were available at Dutch Harbor and Unalaska. (Pages 2–3.)

For purposes of the maneuver, it was reported that at 0600 on 28 July, TF 21 rendezvoused with TU 11.2.2 and fueled on a southwesterly course. Just prior to completing refueling, the Commander, Task Force 21 received a message from the Commander-in-Chief of the Pacific Fleet stating that Purple troops had landed on Attu but that the Purple Covering Force was not present. Upon completion of fueling at latitude 44° N, longitude 178°15′ E, CTF 21 ordered Commander, Task Unit (CTU) 11.2.2 to operate to the eastward of 180 degrees east and south of 42 degrees north while awaiting orders. TF 21 then proceeded at twenty knots to Point X-ray,

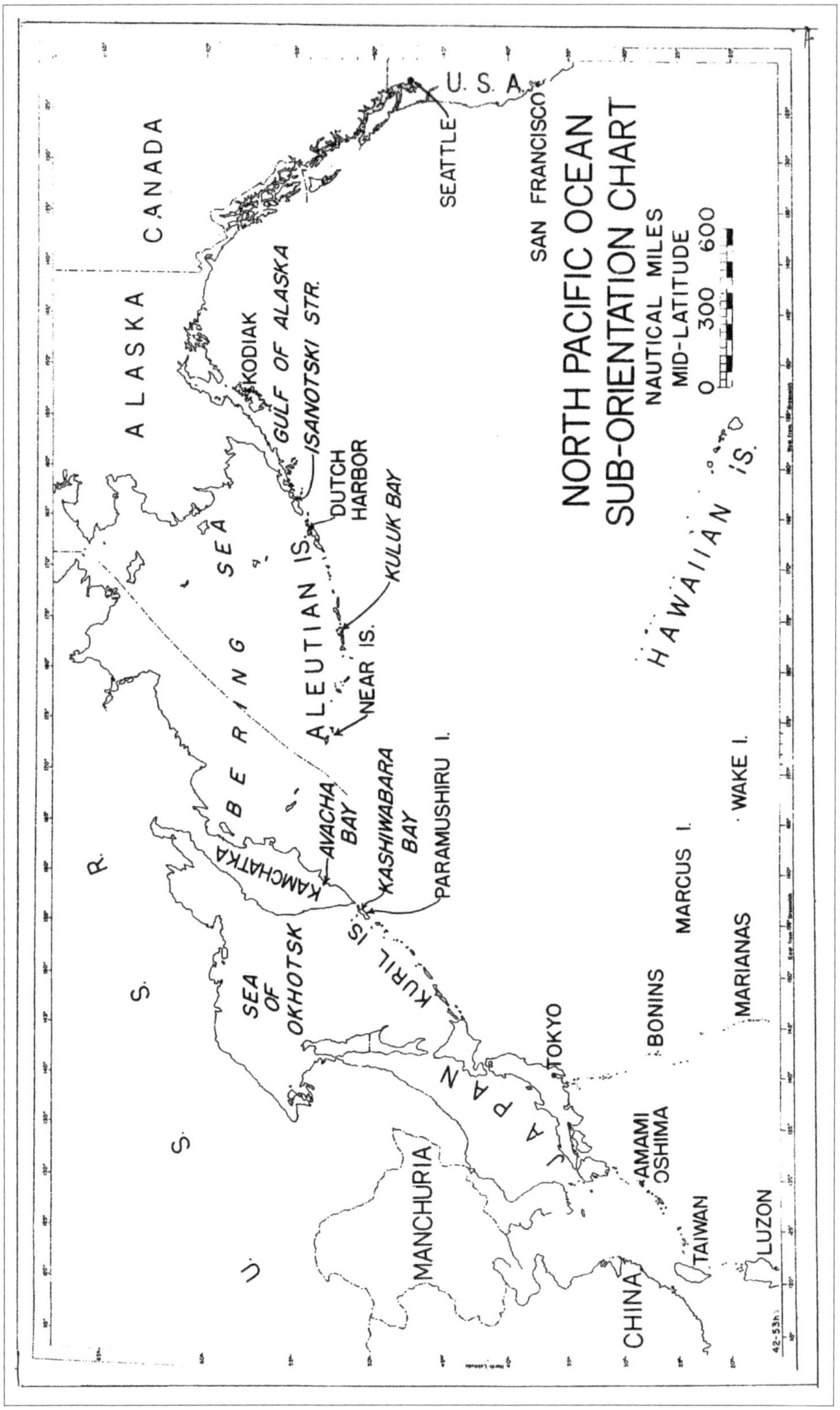

Map 5
North Pacific Ocean

latitude 47°31′ N, longitude 173°40′ E. At 0200 on 29 July at Point X-ray, CTF 21 launched a reinforced search with bearings limited to between 330 and 020 degrees true out to a distance of 335 miles for the purpose of discovering the Purple Covering Force. At 0300, CTF 21 launched a day combat air patrol of twelve planes and landed eight night fighters. (Page 3.)

At 0355, CTF 21 received a contact report from one of the search planes at 52°36′ N, 171°30′ E on an enemy force composed of two fleet carriers, four to five fast battleships, eight or nine cruisers, and about thirty destroyers. At 0400 on 29 July, when Task Force 21 was at 48°10′ N, 173°25′ E, the search plane reported that reduced visibility prevented amplification. The weather in the vicinity of TF 21 was wind from the southwest at Force 3, sea moderate from the southwest, and visibility variable from three to nine miles. Cloud cover was six-tenths, with the top of the cover at four thousand feet and the bottom at 1,500 feet. Sunrise had been at 0300 and sunset would be at 2100. The first move was to take place between 0400 and 0700 on 29 July. (Pages 3–4.)

As for Student Requirements, CTF 21 included an annex to his Operation Plan, laying down several battle plans covering various contingencies. One of these battle plans envisioned a situation approximating the one confronting him now. On that basis, the student officers submitted Admiral BA's Estimate of the Situation at 0400 on 29 July as well as their Solutions to the Problem. The students utilized data previously compiled, considered, and carried in the Running Estimate as well as all information received as of 0400 on 29 July. The students then provided a typewritten copy of CTF 21's modified Battle Plan as well as the Cruising, Approach, and Battle Dispositions for TF 21. They also supplied a Communication Plan written to specification from Section E of the Maneuver Rules, and the number and location of any airborne-early-warning planes and any radar pickets that were to be employed as of 0400 on 29 July. In addition, the student officers designated the day combat air patrol and night fighters landed at 0300 on 29 July, providing flimsies and Flight Forms covering all aircraft airborne by 0400 on 29 July. Flimsies covering the movement of surface forces and aircraft launched during Move 1 and associated Flight Forms were also prepared, using Strategic Plotting Chart No. 22. (Pages 4–5.)

The Blue Chart Maneuver

About two weeks later (in game time), Admiral BA had created a Task Organization for the Alaskan Defense Force. Rear Admiral BX's Battleship Divisions 1 and 2 would constitute the Battleline, while Rear Admiral BF's Cruiser Division 4 and Destroyer Division 42 would be the Center Force. Captain BGA's light cruisers CL-91 and CL-101, along with Destroyer Squadron 12, would constitute the Right Flank Force, and Rear Admiral BG's Cruiser Division 17 (less light cruisers CL-91 and CL-101) and Destroyer Squadron 21 would make up the Left Flank Force.

Captain BGA (Lt. Cdr. John Aymond, USN)

Rounding out the Task Organization would be Rear Admiral BD's Carrier Group, consisting of CARDIV 7 and DESRON 4 (less DESDIV 42).

After reviewing the General Situation and the disposition of enemy forces (including the possibility of a third Purple aircraft carrier at Kashiwabara Bay), Admiral BA noted his assumptions. He assumed that the Purple surface force that was reported southwest of Attu was engaged in supporting and covering the Purple landing force and that Purple would employ air search to the "maximum practicable extent" to locate and track the Alaskan Defense Force. He also assumed that Purple would employ long-range air strikes against his force to the maximum extent permitted by weather conditions. Admiral BA thought that Purple submarines might be encountered and that Purple would accept a surface engagement to protect its landing force at Attu. In regard to this surface engagement, Admiral BA took for granted that Purple would attempt to engage at under thirty thousand yards, that the action would take place during the day, that visibility would be low, and that aircraft operations would still be "practicable." BA asserted that his force would destroy the Purple Covering Force by successive air attacks followed by "major action" preparatory to destroying the Purple amphibious forces landing at Attu. Air and surface forces would ensure the full destruction of that amphibious force in order to prevent Purple from establishing forces ashore in the Aleutians.[2]

Admiral BA detailed what each component would accomplish. His battle line was initially to engage the enemy battle line at a maximum range of forty thousand yards. It was to close rapidly and maintain "extreme ranges" of thirty-four to thirty-five thousand yards until the rate and volume of enemy fire were reduced or ineffective. At this time, the Blue battle line would close rapidly to moderate ranges of twenty-one to twenty-five thousand yards. The battle line would provide its own Air Spot and battle-line Inner Anti-Submarine Patrol. There were no specific roles for the Center Force or the Right Flank Force, but the Left Flank Force was to protect the Blue battle line from Purple light-force attacks.[3] Prior to the engagement of the battle lines, the Blue battle cruisers would operate offensively against the Purple light forces at initial ranges of twenty-nine to thirty-one thousand yards and exploit "every opportunity" to destroy or damage them. Blue destroyers were also to make torpedo attacks on the Purple battle line when directed by the Officer in Tactical Command. The Left Flank Force too would furnish its own Air Spot and Inner Anti-Submarine Patrol. Finally, the Carrier Group would locate and track the Purple air and surface forces. It would endeavor to destroy these forces by "successive" air attacks, with the priority of targets being the Purple aircraft carriers, then the

Chief of Staff, Commander, Task Force 21 (Cdr. Thomas White, USN)

battleships, heavy cruisers, light cruisers, and destroyers. These air attacks would commence at the maximum radius for the Blue aircraft. If visibility prohibited attacking priority targets, the aircraft were to select and destroy targets of opportunity. The Carrier Group was also to furnish its own antisubmarine patrol as well as its own combat air patrol and the combat air patrol for the battle line. To ensure its safety, it was to operate on the disengaged side of the battle line upon deployment.[4]

In addition, all ships and planes were to report mines and submarines sighted. Destroyers and antisubmarine patrol planes were ordered to attack submarines "vigorously," continuing these attacks until threatened surface units were "well clear" of the danger zone or the submarines had been "positively" destroyed. All ships and planes were to be prepared to attack and destroy Purple amphibious forces in Massacre Bay of Attu Island when directed by the OTC. They were also told that prior to deployment they were to follow the movement of the Blue carriers during flight operations. The direction to deploy would be signaled, and the Operation Plan was to take effect upon that signal. The Communication Plan was Commander, Battleship Squadron 1 Communication Plan X-ray (which is detailed below). All units were to use Zone Mike (minus 12) time. Fighter direction responsibility was placed in fleet carrier CV-39 and fast battleship BB-61, with CV-37 and BB-63 as their Relief Fighter Direction ships, respectively. In the Cruising and Approach Dispositions, additional fighter direction was assigned to designated cruisers and picket destroyers. All units were to use Cruising Disposition 6-R, Approach Disposition 6-N (or 6-V, as signaled), and then Battle Disposition 6. CTF 21 was to be the Officer in Tactical Command in fast battleship BB-62; Rear Admiral BX, Commander, Battleship Division 1, in fast battleship BB-61, was second in command.[5] For purposes of the Chart Maneuver phase of Operations Problem 1J, TF 21 was broken down into the Blue component commanders: CTF 21, his Chief of Staff, and Commanders, Battleships, Cruisers, and Destroyers. There were also room assignments for the Commander, Carriers; his Chief of Staff (also Commander Coleman) and Operations Officer (also Lieutenant Commander Aymond); and Commanders, Carrier Air Groups (COMs CVG), fleet carriers CV-37 (also Commander Webber) and CV-39 (also Lieutenant Commander Taeusch). Per the Maneuver Rules, these assignments included separate rooms when the Commanders, Carrier Air Groups were in the air and when they were on board ship. For purposes of the Board Maneuver, commanders were also designated for the Battleline, the Center Force, the Right Flank and Left Flank Forces, and the Carrier Group.[6]

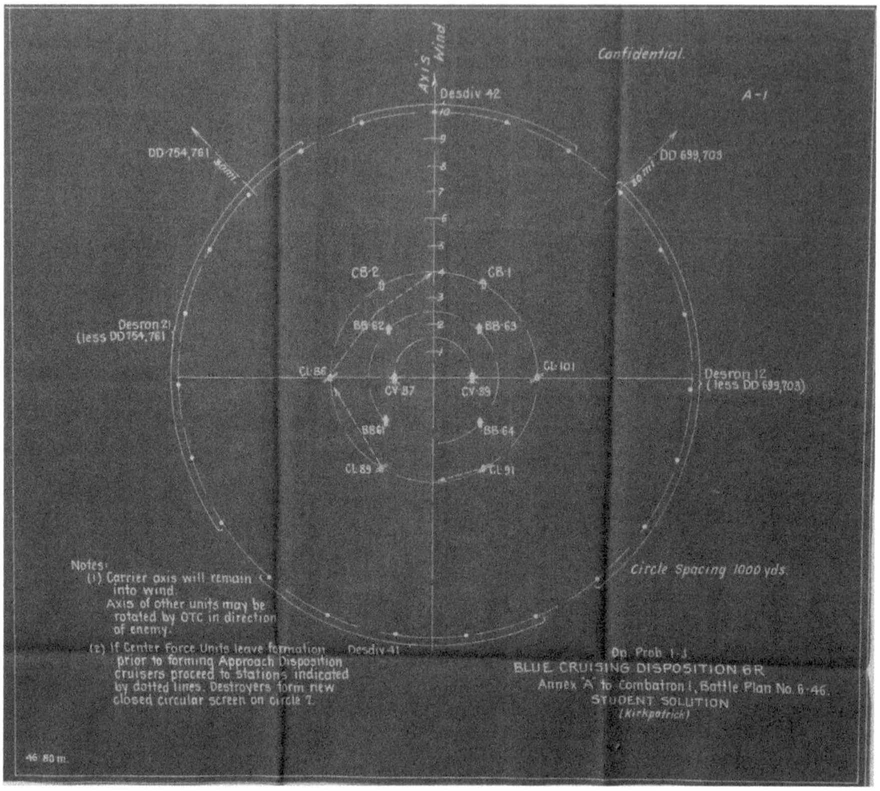

Fig. 83
Blue Cruising Disposition 6-R

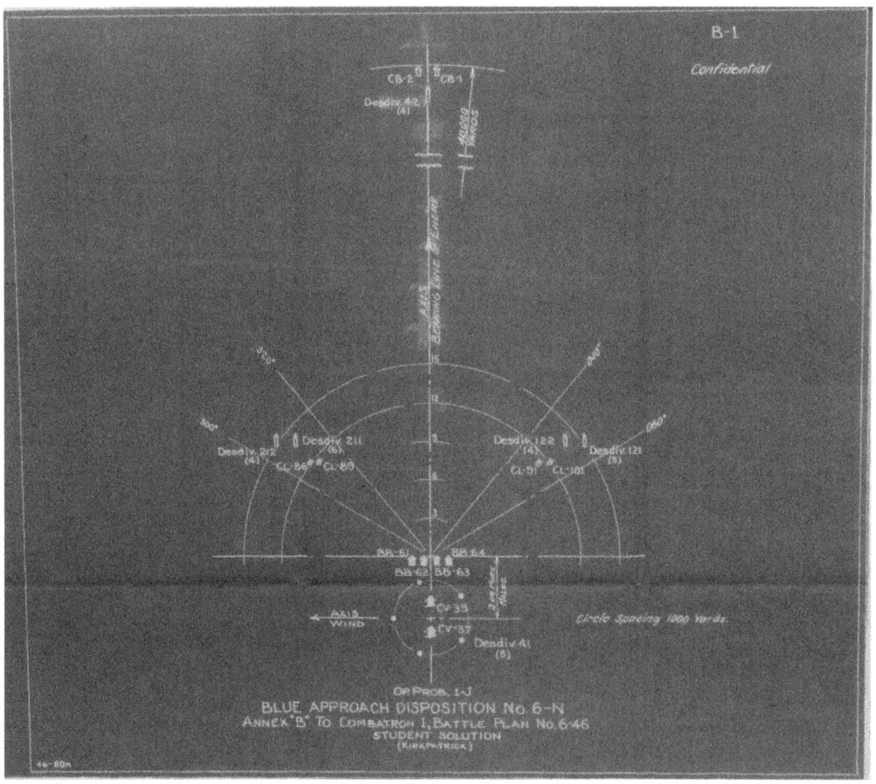

Fig. 84
Blue Approach Disposition 6-N

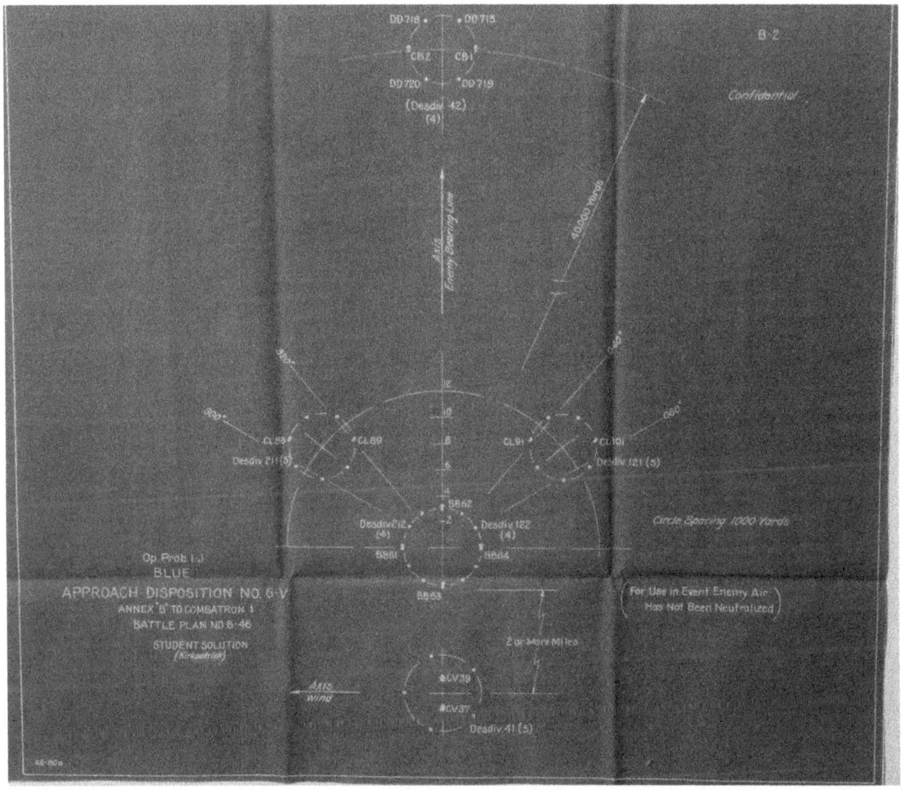

Fig. 85
Blue Approach Disposition 6-V

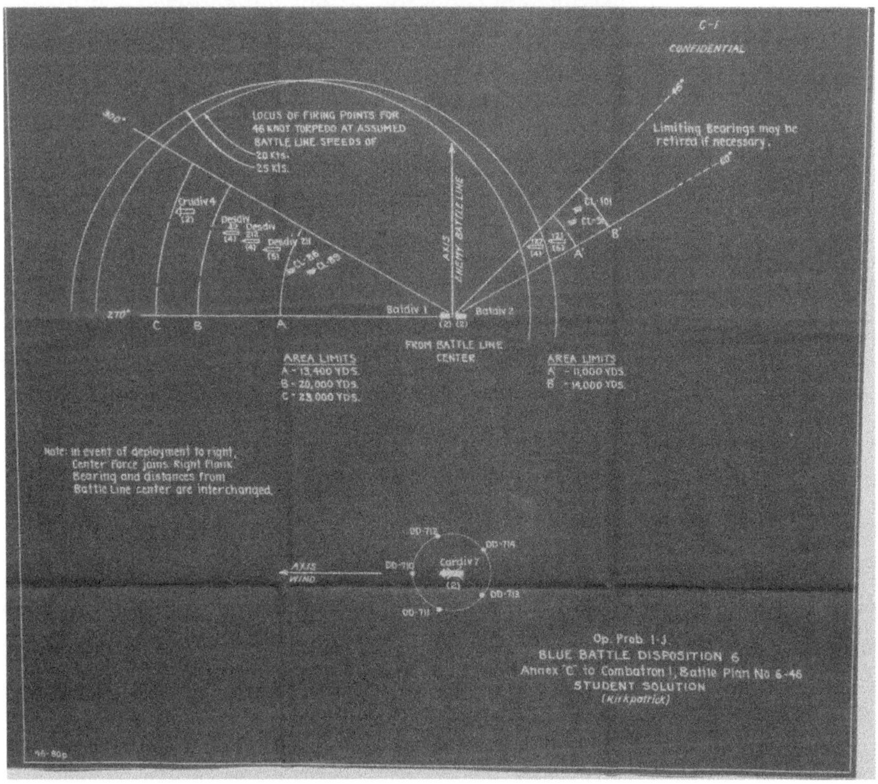

Fig. 86
Blue Battle Disposition 6

The Alaskan Defense Force's Communication Plan was in accordance with an actual Pacific Fleet plan, PAC-70 (B). As noted above, it was to take effect upon receipt by the component commands, and it entailed a "Fox Schedule" (or fleet broadcast, containing traffic variously addressed to all ships and commands assigned to "copy" it), derived from the 9th Fleet and designated NPG (F), and transmitted from Radio San Francisco. It directed radio silence until contact with the enemy and designated Tactical Radio frequencies, both Primary and Secondary Voice as well as Primary and Secondary Continuous Wave frequencies. There was also a General Warning Voice frequency, as well as an Air Spot frequency. The Air Search and Fighter Director frequencies also had Primary and Secondary designations; the Air Search frequency was continuous wave and the Fighter Direction frequency voice. The Air Attack Common and Task Force Combat Air Patrol Common frequencies were also both voice. Specific ships were designated for radar Identification Friend or Foe, radar countermeasures control, and silence. No condition of radar silence would be imposed, and all ships and aircraft were to have their Identification Friend or Foe transponders on. In addition, guard ships were to have their interrogator-responder units on and ready for actuation at all times. Radar intercept guard ships were designated for certain frequencies; the OTC was to control jamming. The OTC would also control radio deception and radio intercept; deception was to be conducted in various frequencies by selected ships. The Communication Plan was rounded out with codes for various standardized commands, voice call signs for each component commander, and authentication instructions for all units.[7]

Communication Plan X-ray followed these guidelines but added some additional details. X-ray was to become effective upon the execution of COMBATRON 1's Battle Plan No. 6-46. Functional circuits with radio frequencies were now given for Talk Between Ships, Emergency Tactical, and voice. There was also a Voice Command Channel for Task Group Commanders, Continuous Wave Dispatch and Maneuvering Signals for a Task Force Common frequency, and frequencies for Inter-Fighter Director and Aircraft Launching and Control nets. In addition, there were specific tactical and voice frequencies for the Battleline, Center Force, Right Flank Force, Left Flank Force, and Carrier Group. Also added were frequencies for airborne early warning control and strike control. When the Alaskan Defense Force formed into its Approach or Battle Dispositions, the senior officer of each task group could designate ships as radar, Identification Friend or Foe, or radar countermeasures control and silence guard ships. Also, when picket destroyers were stationed, the picket ship that was controlling fighter direction was to assume all air-search and identification guards (that is, responsibilities for monitoring circuits), and the standby destroyer would assume these duties for all surface-search and recognition guards. Moreover, it was ordered that nothing was to preclude any ship from ensuring that any unreported contact was brought to the attention of the Officer in Tactical Command. The rest of X-ray outlined in more detail the codes to be used, the voice calls for the component forces and their individual ship units, and some additional details about radar jamming. In particular, all ships equipped for this type of jamming were to maintain their equipment in Standby Condition and be ready to jam when so directed.[8]

The Blue Maneuver Staff

Rear Admiral Smith's instructions listed the personnel assigned to the Maneuver Staff, giving a concrete sense of how these maneuvers were organized. In addition to a Director of the Maneuver, there was an Assistant Director and Chief Damage Computer. An officer was detailed to perform the role of Assistant Director for Blue, another officer that of Assistant Director for Purple; there were also an Air Umpire, a Communication Umpire, and the Historian. Further, there were Move Umpires and Force Damage Recorders for both Blue and Purple and, in the Chart Maneuver, a number of deputies to the Air Umpire. These latter assistants comprised an Aerial Combat Loss Computer, a Hit Computer, an Anti-Aircraft Loss Computer, and two Status Board Recorders. There were also an Operational Loss Computer, two Radar Contact Recorders, and two civilian Plotters. Additionally there were four Standby Damage Computers to assist the Air Umpire if necessary. For the Board Maneuver, there were Damage Computers for the Battleline, the Detached Wing and Center, the Right Flank Cruisers and Right Flank Destroyers, and the Left Flank Cruisers and Left Flank Destroyers. A civilian Plotter and two civilian Assistant Plotters rounded out the Maneuver Staff for Operations Problem 1J.[9]

The Blue Staff Solution and Board Maneuver

In early October 1946, the Blue Staff Solution to Operations Problem 1J was issued. After repeating the General Situation about Blue and Purple losses in the area, the deployment of both forces, and the Purple movement toward Attu, it stated, in a Tactical Estimate of the Situation, that Admiral BA had been informed by Fleet Admiral CD, Commander-in-Chief of the Pacific Fleet, that Purple troops were landing on Attu but that the Purple Covering Force was not present. At 0355 on this day, Admiral BA received an aircraft contact report indicating the presence of a Purple combatant force 275 miles from his present position on a bearing of 344 degrees from him and interposed between the Alaskan Defense Force and Attu. Admiral BA concluded that this was the Purple Covering Force. He understood his mission as preventing Purple from establishing forces ashore in the Aleutians by exploiting every opportunity to bring Purple surface combatant forces to decisive action in the northwest Pacific. He understood that by defeating the Purple Covering Force—which was in itself his secondary assignment—he would be accomplishing his primary mission, since Purple, lacking its Covering Force, would not be able to get reinforcements and supplies to Attu from Kamchatka and the Kuriles. He deduced that he could interfere "most directly" by attacking Purple shipping in the transport area at Attu, but as long as the Purple Covering Force was deployed, it had to be his first target. If, for instance, he attacked the Purple Joint Expeditionary Force with his carrier aircraft without first disposing of the Purple Covering Force, he would lay his ships open to serious damage from Purple carrier aircraft, which in turn would prevent him from carrying out his subsequent tasks.[10]

In looking at the Situation and Courses of Action, Admiral BA analyzed the Objective of Enemy Forces, Relative Combat Power, the Characteristics of the Theater of Operations, Time, Space, and Logistics. Concerning the Objective of Enemy Forces—now that Purple had neutralized Blue Aleutian and Alaskan land-based air forces and had troops landing on Attu, and given the location of the Covering Force—its intention of seizing Attu either to use it as a base or to deny it to Blue was clear, as were his orders to prevent this. He then compared the combat power of the two forces. He had two sources of information on the Purple Covering Force: an intelligence report and the 0355 contact report. According to the intelligence report, Purple had two to three fleet carriers, four to five battleships, three to five heavy cruisers, three to four light cruisers, and three squadrons of destroyers. According to the contact report, there were two fleet carriers, four or five battleships, eight or nine cruisers, and "about" thirty destroyers. He took the best current estimate to be that of the contact report. Since the pilot could not make positive identification of all of the Purple ships, Admiral BA surmised that the Covering Force had thirteen large ships in addition to the two fleet carriers. These thirteen ships comprised either four battleships and nine cruisers or five battleships and eight cruisers. "To be on the safe side I will conclude that the enemy force contains 5 BBs and 8 cruisers. On the basis of the intelligence report I will divide these cruisers evenly into 4 CAs and 4 CLs." The intelligence report gave Purple a destroyer strength of three squadrons. Since a destroyer squadron normally consisted of two divisions of four destroyers each, plus a squadron leader, that implied a total of twenty-seven destroyers. The aircraft contact report gave "about thirty" destroyers, which compared favorably with the intelligence report. Accordingly, Admiral BA was going to base his plans on a Purple destroyer strength of thirty ships. (Pages 3–4.)

He could now compare his own two fleet carriers, four battleships, two battle cruisers, four light cruisers, and twenty-seven destroyers with Purple's, noting that he had no submarines for offensive use and was unsure on that point for Purple. In addition, he had very little information on Purple carrier aircraft, so he considered that Purple would have the same as he did—106 fighter planes, seventy attack planes, and twenty-six specialized aircraft. He knew, however, that Purple had twenty-six observation planes to his twenty. He also knew that neither he nor his Purple counterpart would get reinforcements in time to affect the existing situation. He took the combat efficiency of the two sides' personnel to be the same and then began comparing the combat efficiency of the various types of ships. (Pages 4–5.)

As for fleet carriers, each side had the same number, and he thought they were comparable in terms of class, maximum and sustained speeds, and armor. He also thought that the Purple carriers were the same age as his and had identical antiaircraft weaponry. Accordingly, he surmised that Blue and Purple carrier strength was the same. When it came to the battleships, however, he was inferior in numbers by a ratio of four to five. Still, his battleships were heavier and had a four-knot speed advantage, and he had two battle cruisers to help offset Purple's advantage. His battle cruisers were six knots faster than the Purple battleships, but Admiral BA knew he had only thirty-six sixteen-inch guns and thirteen twelve-inch guns to Purple's forty-five sixteen-inch guns. His battleships outranged Purple's by six thousand yards and his battle cruisers by one thousand yards, but the battle cruisers were lightly armored and no match, ship for ship, for the Purple battleships. Though penetrative ranges could vary by several thousand yards—depending on list, roll, yaw, and quality of armor—a consideration of theoretical penetrative ranges was of value in determining the "most probable" advantageous and disadvantageous ranges. He thought that his battleships were more resistant to penetration at ranges of eighteen to twenty-five thousand yards, given favorable target angles, while he thought that the Purple battleships were more resistant to penetration at ranges of twenty-three to twenty-eight thousand yards. His battle cruisers, he knew, were vulnerable to Purple battleship fire at all ranges while the Purple battleships were comparatively invulnerable to battle-cruiser fire except at ranges over thirty-one thousand yards or when presenting an end-on target angle at medium and short ranges. (Pages 5–7.)

Admiral BA had no heavy cruisers to match Purple's four, but Purple's heavy cruisers were, he asserted, no match for his two battle cruisers. For instance, his battle cruisers were one knot faster in maximum speed and their guns outranged those of the Purple heavy cruisers by seven thousand yards. With favorable target angles, penetration of his battle cruisers by Purple's heavy cruisers could theoretically occur only at the extreme range of the Purple eight-inch guns. Theoretical penetration of the Purple heavy cruisers by his battle cruisers, in contrast, could occur at all ranges in excess of sixteen thousand yards. Purple and Blue were equal in the characteristics and numbers of light cruisers. Still, the Purple heavy cruisers outranged his light cruisers by four thousand yards and were superior ship to ship. As for destroyers, and having no more definite information, he assumed that the Purple destroyers were comparable to his DD-692 class and had ten torpedo tubes each. Since he was not sure how many destroyers had this many tubes, he assumed that all of them did. He also surmised that Blue and Purple destroyers had the same characteristics in other respects, but since Purple was thought to have three destroyers more than Blue, Admiral BA took the Purple commander to have a total of seventy-five torpedoes at his disposal. The guns on Blue's light cruisers outranged those of the Purple destroyers by eight thousand yards. "Their fast shooting characteristics and range of gunfire make them vastly superior to a DD target. However the DDs have a 1.5 knot speed advantage over the CLs." (Pages 7–9.)

Summing up, Admiral BA noted that he had a speed advantage over Purple's battleships as well as a six-thousand-yard range advantage in main batteries. His battleships were almost three years younger and had superior antiaircraft suites. Purple, however, had one more battleship than he did and nine more main-battery guns. BA's battle cruisers' main batteries outranged those of the Purple battleships by a thousand yards and had a six-knot speed advantage but were vulnerable to the Purple battleships at nearly all ranges. The Blue battle cruisers, however, were superior to Purple's heavy cruisers in terms of speed, main-battery range, armor, "life" (i.e., age), and antiaircraft defense, but Purple had four heavy cruisers to none for Blue. In addition, Purple probably outnumbered Blue by three destroyers, and Blue was probably inferior in numbers of observation planes. Though inferior in numbers of battleships, Admiral BA took his speed advantage as allowing him to select the range at which the engagement was fought. Given the theoretical penetrative range bands and the effect of target angles, he thought that in a battle-line engagement he would be "wise" to exploit any advantage at moderate ranges under twenty-three thousand yards while presenting a forty-five-degree target angle.

To get that close to the Purple battle line, he would have to compensate for Purple's numerical superiority in battleships by air or light-force torpedo attack on Purple's battleships. He also thought it might be possible to exploit his gun-range advantage to inflict damage at extreme range before closing with Purple. He took his main advantage over Purple to be his two battle cruisers. He thought that these ships could "handle" all four of Purple's heavy cruisers and should be able to destroy Purple's light forces unless Purple opposed them with a "Detached Wing" of battleships. If his battle cruisers could destroy the Purple light forces, they would be able to bring their main batteries to bear on the Purple battle line. Moreover, even if the Purple battleships were not damaged by air or light-force torpedo attack, he thought, his battle cruisers could be maneuvered ahead of the Purple battle line in order to "cap the *T*." He could then close with his battleships and concentrate thirty-six sixteen-inch and eighteen twelve-inch guns on the enemy. Even if Purple opposed his battle cruisers with a Detached Wing of battleships, this would work to Admiral BA's advantage since his battle line would immediately become superior to Purple's and he could close to "decisive" range. Overall, then, and aside from

Purple's probable superiority in destroyers and torpedoes, Admiral BA took the two forces to be "about" equal. (Pages 10–11.)

Turning to the weather, Admiral BA asserted that conditions in the Aleutians were the most unpredictable in the world. Winds up to ninety miles an hour were commonplace, as were generally bad weather and "howling" storms. Temperature ranged from 5 to 80 degrees Fahrenheit, and overcast was persistent, with occasional breaks over isolated areas. There were also conditions of fog and low ceilings. Clear weather typically extended only over expanses of twenty miles; clear weather over larger areas was seldom encountered, though the northern coastlines had "far better" weather than the southern ones. He also noted that southerly winds indicated the approach of cyclones accompanied by low ceilings; northerly winds followed the cyclones. In July, there was fog 25 to 70 percent of the time. Fog and mist could persist for days, and the fog could become quite thick. Still, he said, fog rarely exceeded four thousand feet and was often thin, lying close to the water less than a hundred feet in height. Average cloud cover in the area was eight months out of the year, however; 60 percent of the time, the area experienced winds from the southwest at speeds up to Force 3. The average wind was ten knots, but there could be moderate gales of Force 7. There was also minimal rainfall ten to fifteen days a month; average air temperature was just over 48 degrees Fahrenheit, sea temperature just over 47 degrees. (Page 12.)

Given this weather, Admiral BA considered flying conditions hazardous and assumed that the weather would "greatly" hinder aerial search operations. As of 0400 on 29 July, however, the forecast indicated weather suitable for flying, though cloud cover was six-tenths, with cloud layer from 1,500 to four thousand feet and visibility variable from three to nine miles. "AEW [airborne early warning] planes will be of great value in tracking enemy movements if the visibility decreases." In addition, he took dive-bombing to be "feasible" only if his planes could find openings in the overcast. These attack aircraft, therefore, had to be prepared to employ glide-bombing techniques. In contrast, his torpedo planes would be suitable for the situation; in fact, he surmised, they could exploit the overcast cover prior to commencing their runs. Also, since he did not think that photographic missions had much chance of success, he was going to employ six of his eight photographic reconnaissance planes as fighters. He added that the use of radar by both ships and planes would be "essential." This meant, however, that the main-battery range superiority in his battleships would be negated by the need to employ Radar Spot. Plane Spot, he was hopeful, would help in this regard. Since Purple was subjected to these same radar limitations, Admiral BA saw the damaging of Purple battleship radars as an excellent objective for his carrier aircraft. (Page 13.)

Although both forces were equipped with long-range torpedoes, he did not think their longer ranges would be of much use, given visibility conditions. If the torpedoes were fired outside of eighteen thousand yards, the fire-control data, especially bearings, would only be approximate. At ranges in excess of visibility, fire control would have to be based on radar information and the effectiveness of torpedo fire would be greatly reduced. He took it as probable, therefore, that torpedoes would be fired at ranges under eighteen thousand yards and with high speed settings. (Page 13.)

Since sunrise was at 0300 and sunset at 2100, there would be "ample" time for a daylight engagement. At 0400 on 29 July, Admiral BA assessed, Purple forces bore 344 degrees true at a distance of 275 miles from his position and were approximately fifty miles from Attu. The course and speed of the enemy force, however, were unknown. From his present position, Attu bore 357 degrees and was 280 miles away, the extreme range of his carrier aircraft. His location with respect to Attu would force an action by Purple. Therefore, Admiral BA intended at 0405 to turn his force into the wind and launch aircraft. He anticipated that Purple would do the same and that the forces would then close each other at a relative speed of forty knots. This would place them within about forty thousand yards of each other by 1100. In view of the proximity of Attu to the Purple forces, he expected the island to be of use as an emergency landing field for aircraft in distress. Given the climate, he took it that his pilots would have a much better chance of surviving on Attu, in spite of the Purple landing, than if they ditched at sea. (Pages 13–14.)

In terms of the effectiveness of logistics support in relation to bases, facilities, and supplies, Admiral BA cited Avacha Bay and Kashiwabara Bay as locations that were 480 miles and six hundred miles, respectively, from Purple's 0400 reported position. Both of these Purple bases were understood to be well stocked and equipped for the replenishment of fuel and supplies and to be able to handle emergency repairs to battle-damaged ships. Admiral BA had TU 11.2.2 assigned to him for logistical support; it was currently operating east of 180 degrees east and south of 42 degrees north. The Blue base at Dutch Harbor had fuel and aviation gasoline sufficient for his needs, but it was approximately 850 miles from his 0400 position. Admiral BA, however, had refueled Task Force 21 on the morning of 28 July and had supplies in ample quantities. In addition, Kiska, which was 475 miles from his 0400 position, was equipped to handle emergency repairs; damaged ships were to be escorted to Kiska as the situation permitted. Given that logistics were sufficient on both sides, Admiral BA did not consider them to represent a critical factor in the present operation. (Pages 14–15.)

Admiral BA proposed to seize the initiative and set the pace rather than conform to Purple's. He therefore first addressed his own Course of Action and only then

Purple's capabilities that might prevent him from accomplishing his mission, rather than vice versa. His mission was the destruction of the reported Purple forces, and he had several ways to accomplish it. Purple forces could be destroyed by aircraft bombs, torpedoes, rockets, or guns, or by surface-ship gun and torpedo attacks. He considered that he was equal to Purple in carrier aircraft and probably in destroyer strength, as well as "somewhat" superior in heavy-ship gunfire, given his speed and range advantages. He also thought that he could get in the first air strike; if he was lucky enough to catch Purple with its planes still on board, he might, by repeated air attacks, complete the destruction of Purple by air alone. However, he did not think that he could surprise Purple to this extent and therefore assumed that an air strike would merely be a preliminary to a surface engagement. Also, since he could not deploy his destroyers with much hope of success if they were not supported by heavy gunfire, he surmised that his best Course of Action was to damage Purple by air and complete the destruction of the Purple Covering Force by heavy gunfire and destroyer torpedo attacks. (Pages 15–16.)

Turning to enemy capabilities, he took the Purple Covering Force's mission to be preventing his interference with the Purple Joint Expeditionary Force's landing on Attu. The Purple Covering Force might retire from the area. It could alternatively retire during daylight and return to attempt to disable the Blue force in a night action, then drive it off in a daylight air and surface action. Finally, it could try to destroy Blue by a daylight combined air and surface action. Admiral BA did not think his aggressive enemy likely to simply retire. Nor did he think that Purple would fight a retiring engagement, given the position of Blue and Purple forces relative to Attu, since that would expose the Purple amphibious force. As for the last possibility, Admiral BA did not intend to be driven off by Purple and inferred that Purple would consider the attempt not very feasible. Given that destroying the Blue Alaskan Defense Force was Purple's most suitable method for achieving its objective, Admiral BA took it as the most probable enemy action. (Page 16.)

In light of these probable courses of action by his force and Purple's, Admiral BA set out a detailed analysis, first looking at air strike capability, while again assuming that weather would permit flight operations. He had to ensure that the loading of his aircraft would permit them to reach the target and return. Helldiver bombers could carry torpedoes, but it was doubtful that they could make a round-trip so loaded. Therefore, he was going to arm his forty-eight Helldiver bombers with bombs and his twenty-two Avenger bombers with torpedoes. He had three "systems" of air strikes available. One was a combined, coordinated strike comprising all of his carrier attack aircraft, one was a two-strike attack, and one was a series of repeated strikes. Whereas it would take approximately ninety minutes for strikes to reach Purple and the possibility of surprising the Purple force was remote, he did not think there was any advantage in sending off "deck loads" of aircraft (i.e., the fraction of his aircraft that could be spotted on flight decks at once) as soon as they were launched in the hope of finding Purple with some of his planes on deck, as was the method in a two-strike system. Admiral BA considered it better to wait until he had all of his attack planes in the air so they could proceed in one strike; just before arriving at the target, however, they would divide into two groups and attack from two different directions in order to saturate Purple's air and ship defenses, as well as to achieve maximum damage. Accordingly, he did not consider repeated strikes applicable in this situation. However, his returning planes could refuel, rearm, and strike again, and in this way he hoped to maintain "unrelenting" pressure on Purple's forces, depending on the condition of his carriers following an enemy air strike. (Pages 17–18.)

Admiral BA listed his priority of targets as the Purple fleet carriers, then damage to at least one Purple fast battleship, followed by damage to the Purple battle line's radar capability. He had to be certain that the Purple carriers had been so damaged that they were incapable of operating aircraft before his attack planes shifted to another target. The complete destruction of the fleet carriers was not considered necessary, only that they be made nonoperational and set up for complete destruction later. For this purpose, he thought that thousand-pound GP bombs would be the most suitable weapon. To damage a Purple battleship, he thought his best weapon the aerial torpedo, especially as the weather would favor its use. Thousand-pound bombs not needed to damage the carrier flight decks would be used to destroy the Purple battle line's radar; the Air Coordinator would decide when to shift the attack planes from the Purple carriers to the battleships. Once successful in these tasks, Admiral BA would order his attack planes to select targets of opportunity. He also surmised that radar-jamming planes could be used in conjunction while his planes were at the target; jamming was not to be employed, however, unless specifically ordered by him. Conversely to all this, BA thought that Purple's logical targets for its carrier attack aircraft were the Blue fleet carriers, then the battleships, and finally the battle cruisers. He anticipated that the Purple commander too would execute one large strike, possibly preceded by a "fast bomber" strike. To counter this, Admiral BA planned to assume a defensive antiaircraft disposition with his surface ships as well as employ high speeds and "radical" maneuvers as the situation dictated. By compromising on the number of fighter planes detailed to accompany his air strike, he thought he could maintain a strong combat air patrol. In addition, by making maximum use of his airborne early-warning planes and radar pickets, Admiral BA expected to detect incoming raids in time to take defensive actions. (Pages 18–19.)

At this point, BA explored the advantages of day versus night actions. All previous night engagements with which he was familiar appeared to him to have relied on the "chance factor" to an extent out of all proportion to the strength of the forces involved. Since he considered his present force at least equal and possibly superior to Purple's, he did not want to gamble on a night action, of which the results could be so unpredictable. He had concluded that Purple too was unlikely to wait for a night action if the Blue Alaskan Defense Force proceeded toward Attu. Should Purple, however, decide on a night action, Admiral BA assumed he would be able to turn to avoid general action during the night and engage the Purple Covering Force during daylight the following day. In the meantime, BA would seize the opportunity to destroy the Purple Joint Expeditionary Force at Attu. His expectation, therefore, was to fight a day surface action. In order to exploit his speed advantage, he needed to select, with respect to wind and sea, a deployment course that did not impair this speed advantage. Such a course would have to permit him to steam downwind and allow his light forces on both flanks to execute attacks with the wind and sea on the quarter or astern. Such a course was not suitable for carrier flight operations, but it could be taken as soon as the air phase of the engagement was over.

He foresaw that his fleet carriers, if not disabled by then, would be able to render assistance at some distance from the battle lines. With the wind and sea from the southwest, he would deploy on a northerly course. He would then have the choice between a "normal action" and a "reverse action" (i.e., the battle lines steaming in the same direction or in opposite directions, respectively). Purple would be forced to deploy in the same direction to prevent being outflanked by Blue and Blue's getting at the Purple transports at Attu. A normal action would also best allow him to get in a light-force torpedo attack with his Van. However, he did not want to be bound to one plan. If he was able to force an early Purple deployment, as he hoped, he could dispose of his Center Forces to the Van or to the Rear, or both—whichever course he considered most advantageous—and accept either a reverse or a normal action. (Pages 19–20.)

Having damaged the Purple battle line to the extent of destroying its radar, by either air attack or destroyer torpedo attack, Admiral BA would close with his battleships. The other possibility for offsetting the Purple battleships' superiority in number of ships—aside, that is, from exploiting the Blue battleships' speed and gun range—was deployment of the Blue battle cruisers in his battle line. He did not, however, consider this proper since the battle cruisers were so vulnerable to battleship fire; in addition, his light forces would be deprived of an element that could handle the Purple heavy cruisers. He hoped later in the engagement, however, to use the battle cruisers' main batteries against the Purple battleships. Once he closed with the Purple battle line following the air strikes, he would engage its battleships at extreme range. Once his destroyers carried out their torpedo attack, he assumed he would be able to close with his battle line to moderate ranges under twenty-three thousand yards and destroy Purple. (Page 20.)

As to light forces, he had twenty-seven destroyers, all but nine of them identical. These nine differed from the rest only in that they carried five torpedo tubes instead of ten. It was logical, therefore, to assign these latter ships to defensive tasks and the others to the Van for attack. Some destroyers, however, would have to be detailed to protect the carriers when they proceeded to the disengaged side. He saw his light cruisers as best deployed defensively, in support of his destroyers. Purple, of course, had thirty destroyers, though these too would have to be divided among screening the Purple fleet carriers, defending the Rear and the Van, and constituting an attack element. He knew that the Purple commander would need at least four destroyers to screen his fleet carriers and thought he would then probably dispose of two-thirds of the remaining twenty-six in the Van and one-third in the Rear. Purple could evenly divide its destroyers between Van and Rear, but such a deployment would not provide sufficient force in either for an effective attack force. Another possibility would be for Purple to put all of his destroyers in the Van, but this would leave his Rear too dangerously exposed to an attack by Blue Rear forces. Such a deployment could also be an attraction for a reverse action by Blue. Since Admiral BA did not think a reverse action probable, he saw Purple deploying four destroyers to his carriers, nine more to his Rear, and the remaining seventeen to his Van. (Page 21.)

Admiral BA surmised that the Purple heavy cruisers would probably be used as a counter to the Blue battle cruisers and would lead the Purple light forces during the approach. On deployment, he assumed, the Purple heavy cruisers would be sent to the Van and the four light cruisers used as defensive support, with two in the Van and two in the Rear. No other disposition of the Purple cruisers on deployment appeared likely to Admiral BA since Purple would not achieve concentration of force in either Van or Rear but weaken both. BA considered that his own battle cruisers would prove decisive in the surface engagement. He therefore thought about exploiting their capabilities—to outrange and outlast any Purple ships except the battleships—to the maximum by designating them as a Detached Wing to operate with the Center, help drive back the Purple light forces, and force Purple to deploy prematurely. If Purple tried to forestall this move by putting one or more of his battleships in his Center Force, Blue could seize this opportunity to press home an early destroyer torpedo attack and damage one or more of the Purple battleships that Admiral BA considered so important. (Pages 21–22.)

To do so, Admiral BA would have to place one squadron and one division of destroyers in the Center during the approach. He also needed to be prepared to shift one destroyer division to the Rear upon deployment. This latter decision would depend on Purple's battle disposition. If this action was required, BA assumed, he could direct the move by signal. Following the Blue deployment, BA's battle cruisers would destroy enemy heavy units in Purple's Van and support the Blue destroyer attack on the battleships. He saw this use of the battle cruisers as "feasible" unless Purple placed a Detached Wing of one or more battleships in his Van, in which case, as already noted, the Purple Force would at the same time be negating its numerical superiority in the battle line. The result would be that BA could immediately close with the Blue battle line and engage at gun ranges under twenty-three thousand yards.

Following the destruction of the Purple Van heavy units, the Blue battle cruisers could operate ahead of the Purple battle line and concentrate fire on the battleships. Admiral BA could then detach his battle cruisers to attack the Purple transports at Attu, should the situation warrant. From the above comparison of alternatives and the means for accomplishing them, Admiral BA considered his Course of Action to be suitable and feasible. He expected major losses but also considered such losses acceptable if he could destroy the Purple Covering Force. Having summarized that the Alaskan Defense Force would destroy the Purple Covering Force by early and repeated air attacks, a daylight surface engagement at moderate ranges under twenty-three thousand yards, and light-force torpedo attacks to assist in preventing Purple from establishing forces ashore in the Aleutians, Admiral BA proceeded to prepare his task organization and assign the various tasks for his Battle Plan. (Pages 22–23.)

Admiral BA focused first on the protection of his force when in the Cruising Disposition. He knew that Purple submarines of an unknown number were at sea conducting offensive reconnaissance patrols; he saw these as a constant threat. He also saw air attack as "imminent." Therefore, while the Alaskan Defense Force was in Cruising Disposition, it needed to defend itself from air attack and enemy snoopers by means of night fighters, a low combat air patrol, and a radar-picket combat air patrol. In addition, the Alaskan Defense Force had to locate and report the enemy's location, movements, and composition by means of aircraft search, using a combination of day fighters, night fighters, and AEW aircraft. Destroyers —eight in number—would also fulfill roles as radar pickets and "radar guards" to detect and report on enemy air and surface units; the other nineteen destroyers would be employed in an antisubmarine screen, and fighter or attack planes would carry out antisubmarine patrols. (Page 24.)

The next set of tasks entailed his forty-eight Helldiver bombers destroying the flight decks of Purple's fleet carriers and then destroying the radar capability of the Purple battle line. At the same time, his twenty-two Avenger bombers armed with torpedoes would disable at least one Purple battleship, while his ten electronic-warfare planes would jam Purple radars when directed by the OTC. Forty-eight of his Bearcat and Corsair fighters would provide a high-cover force of sixteen

CAP Launched at 0300

AIRCRAFT FLIGHT FORM

FORCE T.F. 21 — STUDENT -- — ROOM NO. --

MOVE NO. 1 — MANEUVER TIMES OF MOVE 0400-0700/29 July — WEATHER AREA --

List of Task Groups: Base Ship or Station	Squadron	Plane Numbers	Time of Start of Flight	Time of End of Flight	Duration of Flight	Air Speed, Knots	Altitude, Feet	Fuel in Planes at Start of Flight or Move	Bombs, Torpedoes or Special Equipment	Orders to Aircraft Task Group Commanders and Radio Frequency.
CV-37	VF-372	1 to 12	0300	0700	4	150	See Note	384	M.G. Wing Tank	Note: When all VF are launched for maximum protection of own formation.
REINFORCEMENT FOR 0300 CAP										LOCAP-8VF-1,500 RAPCAP for two advanced picket stations 4V each at 15,000 Total 8 VF
CV-37	VF-371	21 to 24	0430	0830	4	150	See note	384	M.G. Wing Tank	RADCAP for two rear picket stations 2 VF each at 15,000 Total 4 VF
CV-39	VF-392	21 to 28	0445	0830	4	140	See Note	333	M.G. External Tank	CAP - 12 VF at 10,000 12 at 15,000
ADDITIONAL VF CONDITION TEN										4 VF at 20,000
CV-39	VF-391	21 to 24	-							
CV-39	VF-392	1 to 20	-							

0400 AIR STRIKE

AIRCRAFT FLIGHT FORM

FORCE T.F.21 — STUDENT -- — ROOM NO. --

MOVE NO. — MANEUVER TIMES OF MOVE 0400-0700/29 July — WEATHER AREA -

Time of Start of Flight (all rows): 0401 commence catapulting 18 planes from each CV, free launch remainder by 0422

Time of End of Flight (all rows): Commence recovery about 0730

List of Task Groups: Base Ship or Station	Squadron	Plane Numbers	Time of Start of Flight	Time of End of Flight	Duration of Flight	Air Speed, Knots	Altitude, Feet	Fuel in Planes at Start of Flight or Move	Bombs, Torpedoes or Special Equipment	Orders to Aircraft Task Group Commanders and Radio Frequency.
CV-37	37-F-1				3.5	180	15,000	384	M.G. Belly Tank	Air Coordinator
CV-39	39-F-1				3.5	180	15,000	333	M.G. Belly Tank 1-1000# G.P.	Relief Air Coordinator Escort & Cover Strikes
CV-37	VF-371	1 to 20 25 to 28			3.5	180	15,000	384	Wing Tank 2-1000# G.P.	High Cover 16 VF Bomber Escort 16 VF Torpedo Escort 16 VF
CV-39	VF-391	1 to 20 25 to 28			3.5	180	15,000	333	Belly Tank 1-1000# G.P.	VF-371-25 and VF-391-25 take damage assessment photos. Bomb Carrier Flight Decks
CV-37	VA-371	3 to 24			3.5	180	15,000	453	Aux.Tank 1-1000# G.P.	Coordinated attack on carrier flight decks.
CV-39	VA-391	1 to 24			3.5	180	15,000	453	Aux.Tank	

0400 AIR STRIKE (cont.)

AIRCRAFT FLIGHT FORM

FORCE T.F. 21 — STUDENT -- — ROOM NO. -

MOVE NO. 1 — MANEUVER TIMES OF MOVE 0400-0700/29 July — WEATHER AREA --

Time of Start of Flight: Same as preceding form

Time of End of Flight: Same as preceding form

List of Task Groups: Base Ship or Station	Squadron	Plane Numbers	Time of Start of Flight	Time of End of Flight	Duration of Flight	Air Speed, Knots	Altitude, Feet	Fuel in Planes at Start of Flight or Move	Bombs, Torpedoes or Special Equipment	Orders to Aircraft Task Group Commanders and Radio Frequency.
CV-37	372-A	1 to 11			3.5	180	10,000	535	1 Torpedo 2 Tanks	Disable one Battleship Coordinated torpedo attack on one enemy battleship.
CV-39	392-A	1 to 11			3.5	180	10,000	535	1 Torpedo 2 Tanks	
CV-37	372-A	12, 13			3.5	125	15,000	335	RCM	RCM Jam enemy air search and fire control radar during attack as ordered by OTC
CV-39	392-A	12, 13			3.5	125	15,000	335	RCM	
CV-39	372-A	17			125	10,000	335	AEW		AEW Avoid action, track enemy surface ship movements, at a distance of 60 miles from formation
CV-39	372-A	18			3.5	125	10,000	335	AEW	

Fig. 87
Blue Staff Aircraft Flight Form, 29 July

fighters, a bomber escort of another sixteen fighters, and a torpedo escort with the final sixteen fighter planes. As noted earlier, these planes would proceed as one combined strike, dividing near the target. (Pages 24–25.)

Admiral BA concluded that one of the requirements for Blue success was damage to or destruction of one Purple battleship before the two battle lines engaged at decisive ranges. To this end, he assigned his battle cruisers and light forces certain tasks. His Center—the two battle cruisers of CRUDIV 4 and the thirteen destroyers of DESRON 4 and DESDIV 212, all under Rear Admiral BF—would drive back the Purple Center Force and force an early deployment. It would then attack the Purple battleships with torpedoes, if the battleships were present in the Purple Center Force. On deployment, Admiral BF's battle cruisers would form the Detached Wing and the rest of his forces would join the Blue Alaskan Defense Force in the Van. The Right Flank Group—the two light cruisers from Cruiser Division 17 and the five destroyers of DESRON 12 (less DESDIV 122), all under Rear Admiral BG—would attack the Purple battle line with torpedoes when a favorable opportunity presented itself (this was later changed to attack upon order from the OTC). The Right Flank Group was also to defend the Blue battle line from Purple light-force torpedo attacks. The Left Flank Group—the two remaining light cruisers from CRUDIV 17 and the five destroyers of DESRON 21 (less DESDIV 212), under Captain C-9—would also participate in these torpedo attacks and the defense of the Blue battle line. When the Carrier Group—the two fleet carriers of CARDIV 7, under Rear Admiral BD—detached to continue providing defense against Purple air attack and support the light forces and battleships on the disengaged side with Air Spot, it was to be escorted by the four destroyers of Destroyer Division 122. (Pages 25 and 27–28.)

When the Detached Wing of two battle cruisers deployed, it was to support the Blue Van forces by engaging the heavy units in the Purple Van. Upon the destruction of these heavy units, the Detached Wing was to concentrate its main-battery fire on the Purple battle line and then prepare to attack the Purple Joint Expeditionary Force at Attu. Admiral BA's battle line of four battleships from Battleship Divisions 1 and 2, under Rear Admiral BX, would then destroy the Purple battle line by engaging initially at extreme ranges. When directed by the Officer in Tactical Command and upon the completion of the light-force torpedo attacks, the Blue battle line would close rapidly and engage at moderate ranges under twenty-three thousand yards. All units were to use Zone Mike time for communications; CTF 21 was the OTC in fast battleship BB-62; Rear Admiral BX, in fast battleship BB-61, was the second in command. The Blue Alaskan Defense Force was to employ Cruising Disposition 21-R, Approach Disposition 21-AD, and Battle Disposition 21-M. (Pages 25–28.)

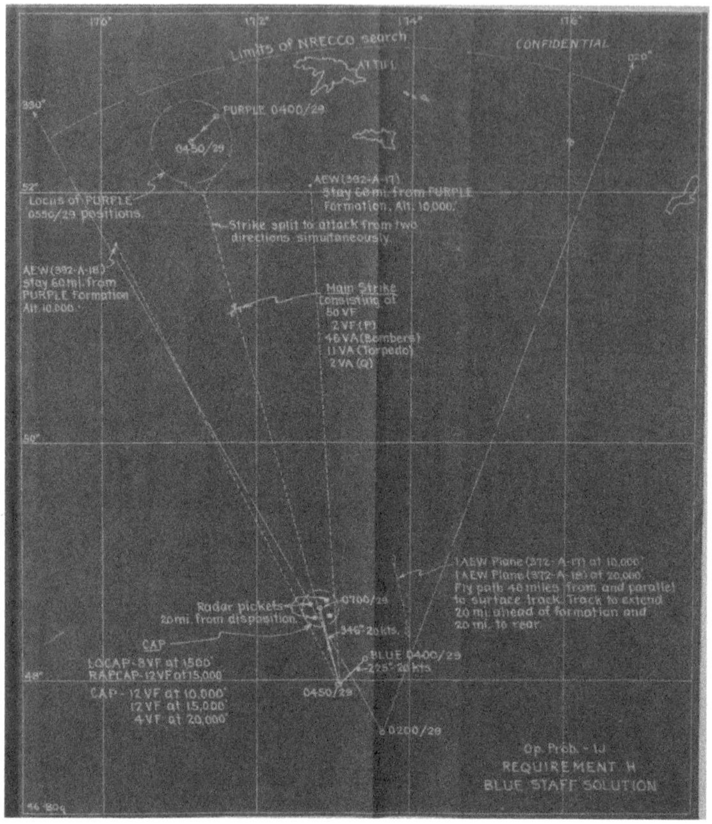

Fig. 88
Blue Staff Solution, Air Operations

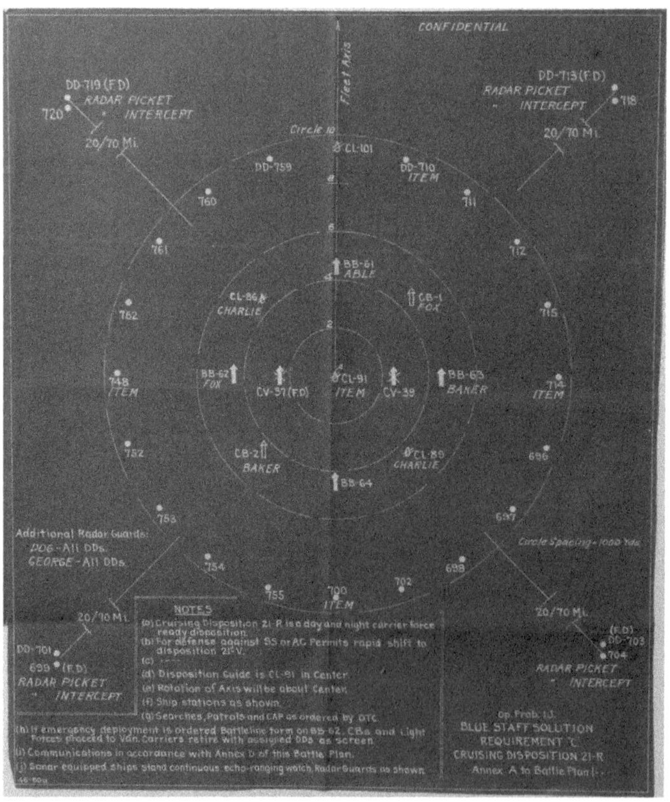

Fig. 89
Blue Staff Solution, Cruising Disposition 21-R

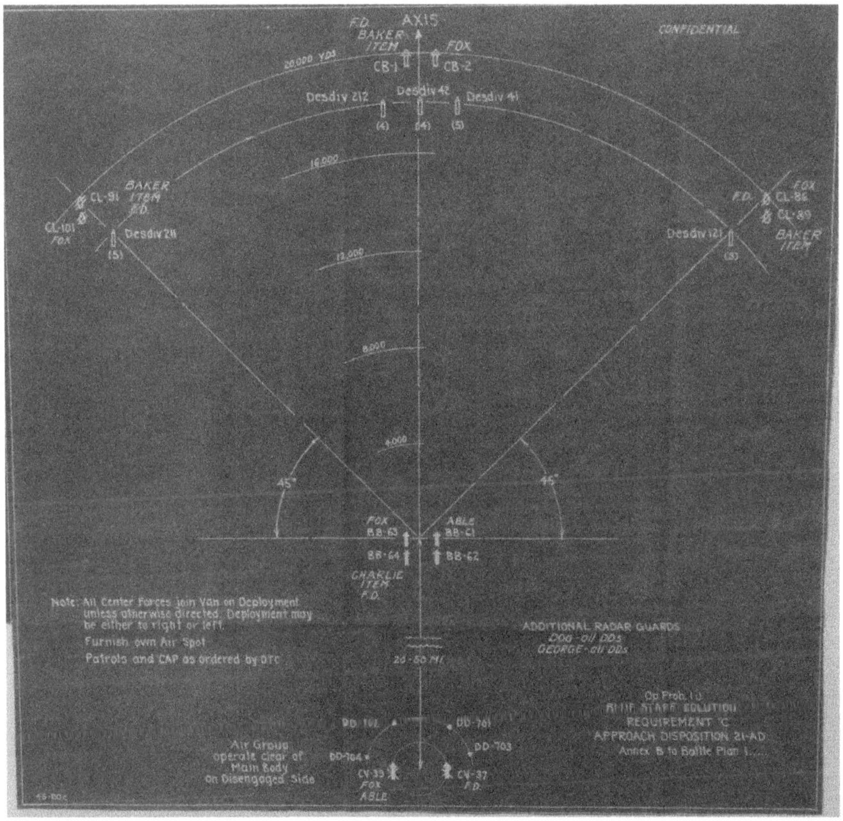

Fig. 90
Blue Staff Solution, Approach Disposition 21-AD

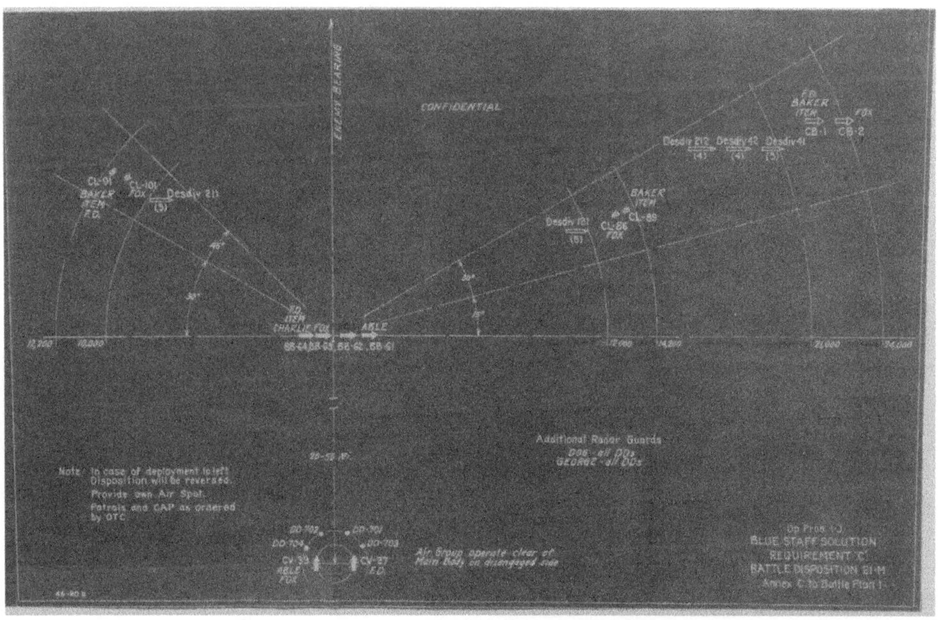

Fig. 91
Blue Staff Solution, Battle Disposition 21-M

Just as his Solution to the Problem was nearly identical to the Blue Statement of the Problem, so Admiral BA's Communication Plan was virtually the same, but with some details added.[11] The communications were now in accordance with the Blue Pacific Command instead of a particular Pacific Fleet communication plan. In addition, radio silence and radio security were to include the use of visual communications in lieu of radio whenever possible. Radar was unrestricted now, but all ships and aircraft were to keep their Identification Friend or Foe transponders energized at all times and restrict their interrogators to emergency use, unless they were recognition guard ships. Further, all ships with two Talk Between Ships systems were to have both the Task Force Common and Task Group systems set up. Task Group Commanders were to relay information received on the Task Force Common Circuit to ships in their task groups guarding only the Task Group Circuit; and all ships were to guard the Very High Frequency Fleet Common Circuit. There were also now special code words for such actions as "Attack," "Attack with Torpedoes," "Aircraft," "Retire," "Affirmative," and "Negative," as well as voice calls for the Officer in Tactical Command and the various commanders.[12]

NOTES

1 "Operations Problem 1J: The Blue Statement," 10 September 1946, p. 1, Schedule of Events, folder 2602, box 138, RG 4, NHC. Subsequent page references, until the next endnote, are to this source.

2 "Operations Problem 1J: Blue—Section A," 27 September 1946, pp. 1–2, folder 2602, box 138, RG 4, NHC.

3 Light surface forces consisted of light cruisers, destroyers, and destroyer escorts, as opposed to heavy surface forces, which comprised battleships, battle cruisers, and heavy cruisers.

4 "Operations Problem 1J: Blue—Section A," p. 2.

5 Ibid., p. 3.

6 "Blue: Section A, Chart Maneuver—Operations Problem 1J," 28 September 1946, folder 2602-B; "Blue: Section A, Board Maneuver—Operations Problem 1J," 28 September 1946, folder 2602-C; both box 138, RG 4, NHC.

7 "Annex C: Communication Plan," 25 September 1946, pp. 1–5, folder 2602-H, box 138, RG 4, NHC.

8 "Annex D: Communication Plan XRAY," 28 September 1946, pp. 1–5, folder 2602-E, box 138, RG 4, NHC.

9 "Maneuver Staff, Operations Problem 1J," 30 September 1946, folder 2602-J, box 138, RG 4, NHC.

10 "Operations Problem 1J: The Blue Solution," 2 October 1946, pp. 1–2, folder 2602-P, box 138, RG 4, NHC. Subsequent page references, until the next endnote, are to this source.

11 The Blue Solution to the Problem was not very different—aside from the absence of two American carriers—from that of another North Pacific scenario, one that had actually occurred during the war. In the battle of Midway, the Japanese Aleutians Force under Vice Adm. Hosogaya Moshiro attacked U.S. positions in the Aleutians with the two light carriers of the 2nd Carrier Striking Force. Rear Adm. Robert Theobald, Commander, Task Force 8, sought to defeat Hosogaya by, first, having the land-based aircraft of Brig. Gen. William Butler, USAAF, strike the Japanese carriers and damage as many surface ships as possible. Theobald planned then to use his five cruisers and four destroyers to engage the remaining Japanese surface forces with gunfire. The plan did not work, however, because of the weather and the lack of antiship strike capability among the Army pilots; Craig Symonds, *The Battle of Midway* (New York: Oxford Univ. Press, 2011), pp. 199, 201–202. A reverse action, with battle lines moving in opposite directions and chasing the enemies' tails, was believed to offer significant advantages for the U.S. Navy, because the Japanese were expected to put fast and powerful forces in the van. If the battle moved away from them, their effectiveness would be reduced; Trent Hone, "U.S. Navy Surface Battle Doctrine and Victory in the Pacific," *NWCR* 62, no. 1 (Winter 2009), pp. 70–71.

12 "Operations Problem 1J: The Blue Solution," pp. 29–32.

NORTH

IX *Operations Problem 1J*

Purple, October 1946

The Purple Statement began as had the Blue Statement, with the Schedule of Events related by Admiral Smith. The Scope of the Problem was also obviously the same, as were the Objectives and the General Situation. In the Special Situation, the Purple 8th Fleet, under Admiral PQ (no designated student officer), was based on the Kamchatka Peninsula and in the Kuriles. Its major components were its Amphibious Force at Avacha Bay and its Covering Force at Kashiwabara Bay. The latter, designated Task Force 81, was under the command of Vice Admiral PS, who was also Commander, Battleship Squadron 2.

The Purple Statement and Situation

Task Force 81 was composed of two fleet carriers under Rear Admiral PV, organized as Carrier Division 5; three fast battleships organized into Battleship Division 3 and commanded by Rear Admiral PL; and two additional fast battleships organized into BATDIV 4, under Rear Admiral PR. TF 81 also had four heavy cruisers under Rear Admiral PT organized as Cruiser Division 2, four light cruisers under Rear Admiral PZ organized as Cruiser Division 14, and twenty-seven destroyers. Nine of these destroyers, those of Destroyer Squadron 7 under Captain P-5, were broken down into Destroyer Division 71 (five destroyers) and DESDIV 72 (four). Another nine destroyers were in DESRON 9 under Captain P-6, with five ships in DESDIV 91 and the other four in DESDIV 92. The final nine destroyers made up DESRON 17 under Captain P-8, five destroyers in Destroyer Division 171 and four in DESDIV 172. Purple destroyers had torpedoes with ranges of forty thousand yards at thirty-two knots or eighteen thousand yards at forty-six knots. Like the Blue destroyers' torpedoes, Purple destroyer torpedoes did not leave wakes. Purple airpower matched Blue's almost to a tee, except that Purple had four attack squadrons between its two carriers instead of three and twenty-six observation planes on its battleships and cruisers to Blue's twenty.[1]

Vice Admiral PS acknowledged that Purple land-based air forces had been reduced to impotence by their recent air offensive. He also understood that the Blue Pacific Fleet, based at Pearl Harbor, had organized a task force for operations in the northwest Pacific. This force was apparently composed of two to three fleet carriers

Vice Admiral PS (Cdr. David Richardson, USN)

Rear Admiral PV (Col. Robert Mason, USAAF)

Rear Admiral PL (Lt. Col. Miles Thompson, USA)

Rear Admiral PR (Lt. Cdr. Emmett Bonner, USN)

Rear Admiral PT (Cdr. Robert Rynd, USN)

Rear Admiral PZ (Cdr. Harry Marvin-Smith, USNR)

Captain P-5 (Cdr. George Miller, USN)

Captain P-6 (Cdr. Harry Foley, USN)

of the *Essex* class. In addition, there were five to seven fast battleships and battle cruisers, of which three ships were definitely *Iowa*-class battleships and two were *Alaska*-class battle cruisers. The Blue force also had three to four light cruisers of the *Cleveland* class, and there were believed to be three squadrons of destroyers of the latest fleet type, which could carry long-range torpedoes. Admiral PS further understood that Blue land-based air forces and airfields in the Aleutians and western Alaska had been completely neutralized by the recent Purple air offensive.[2]

Commander, Purple 8th Fleet issued a Fleet Operation Plan, dated 10 July and to take effect upon receipt. The Purple 8th Fleet was to seize, occupy, defend, and develop Attu as an advanced air base for further operations against Blue bases in the eastern Aleutians. TF 81, the Covering Force, was to prevent Blue naval forces from interfering with the landing operations of the Purple Joint Expeditionary Force on Attu. All Purple forces were also to exploit every opportunity to destroy Blue naval forces encountered in the northwest Pacific, north of 40 degrees north latitude and west of 170 degrees west longitude. D-Day was designated as 28 July. Commander, 8th Fleet's Operation Plan included additional information for Vice Admiral PS: that Attu was weakly defended by a Blue garrison of about five thousand troops; that there were now no Blue air bases operational in the western Aleutians; that the Blue Pacific Fleet was at Pearl Harbor as of 9 July; and that Blue submarines in unknown strength were operating off Kamchatka and the western Aleutians.

In addition, Purple submarines were on reconnaissance patrols on the line of 45 degrees north latitude between longitudes 165 east and 175 west, as well as in the Hawaiian area, but they were not in sufficient strength for offensive use. Admiral PS was told that neither carrier- nor land-based air replacements would be available for sixty days. Admiral PS assumed, therefore, that there would be no strong land-based air opposition prior to or during the Assault Phase against Attu. PS also assumed (correctly) that Blue land-based air losses in Alaska and the Aleutians could not be replaced for sixty days and that the greater part of the Blue submarine force would continue to operate against Purple supply lines in the Far East. He further assumed that the Blue Pacific Fleet would continue to operate south of 40 degrees north latitude and was insufficiently powerful to assure timely reinforcement of the Blue Northwest Pacific Force.[3]

At 0030, 29 July—the 8th Fleet, including TF 81, having sortied from its bases on 26 July in accordance with the Operation Plan and the Joint Expeditionary Force having landed troops on Attu at 0510 on 28 July—while cruising to the southwest of Attu, Commander, Task Force 81 intercepted a contact report from a Purple submarine on a large enemy group of ships, including aircraft carriers, other large vessels, and destroyers, proceeding at twenty knots on a northerly course at latitude 46°53′ N, longitude 173°57′ E. At 0200, at 52°54′ N, 171°00′ E, CTF 81 launched a reinforced search between bearings 140 and 180 out to 340 miles to locate the Blue force. Admiral PS also launched twelve day-combat planes at 0300 as he landed eight night fighters. Fifty-five minutes later, he received another contact report, this time from one of the search planes, that an enemy force of two fleet carriers, six fast battleships or battle cruisers, four cruisers, and about thirty destroyers had been spotted at 48°10′ N, 173°25′ E. Five minutes later, when TF 81 was at 52°36′ N, 171°30′ E, the same plane reported that reduced visibility prevented amplification of the initial contact report.

Meteorological conditions, of course, were the same for the Blue force, as was information given about the situation at the time of the first move. Student Requirements for Purple were the same as for Blue: an Estimate of the Situation as of 0400 on 29 July; the Solution to the Problem; and CTF 81's typed Battle Plan. The latter also supplied Cruising, Approach, and Battle Dispositions; a Communication Plan; the location of his airborne warning aircraft and radar pickets; and his day-combat air patrol and night-fighter dispositions. Just as for Blue, flimsies and Aircraft Flight Forms were to cover all surface and air force movements during Move 1.[4]

The Purple Chart Maneuver

In preparation for the Chart Maneuver, students were informed that Cruising Disposition 5-R was a "normal" carrier-force, day or night, high-visibility disposition. It was considered suitable for defense against air attack while affording "reasonable" protection against submarine attack. The Guide was to be in the battleship at the center, about which any rotation of the axis was to be conducted. Station assignments could be varied by the Officer in Tactical Command to suit existing conditions, and the screening ships would be stationed in accordance with Part III of USF 10B (a real-world reference). Plane-guard destroyers were to be assigned by the Screen Commander when ordered by the OTC. Searches, patrols, and offensive air operations would also be ordered by the OTC. If emergency deployments were ordered, the heavy ships would be deployed in the Van, the aircraft carriers would retire to the disengaged side, and the Screen would divide as designated by the Screen Commander. All sonar-equipped ships, except for fighter-direction destroyers, were to stand continuous echo-ranging watches unless otherwise directed. These ships were to use acoustic equipment in accordance with current doctrine. Battleship- and cruiser-based aircraft would be kept in Condition II readiness, loaded with depth bombs. Exceptions would be planes designated as lifeguards. Finally, antisubmarine patrols would be launched only on the instructions of the Officer in Tactical Command.[5]

At the same time, Vice Admiral PS developed his Battle Plan. Admiral PS's task organization was slightly different from Blue's in that he separated out his fleet flagship, fast battleship BB-60, under Captain PI (also Lieutenant Colonel Thompson, USA, junior class of 1946–47), from the rest of the Covering Force. His battle line, under Rear Admiral PL, consisted of Battleship Division 3 (less BB-60) and light cruisers CL-55 and CL-57. His Center Force, under Rear Admiral PR, was made up of BATDIV 4 and light cruisers CL-56 and CL-58, while his Right Flank Force, under Rear Admiral PT, consisted of heavy cruisers CA-122 and CA-123, DESRON 7 (less DD-725 and DD-770), and DESDIV 171. The Left Flank Force, under Captain LM, comprised heavy cruisers CA-124 and CA-132, DESRON 9 (less DD-730 and DD-731), and DESDIV 172. The Carrier Group, under Rear Admiral PV, was made up of CARDIV 5 and destroyers DD-725, -730, -731, and -770. In his General Situation, Admiral PS noted that "nearly equal" Blue forces were approaching Attu in "order to foil Purple Amphibious landing operations in progress." Blue had two to three *Essex*-class carriers in Hawaiian waters, along with five to seven battleships and battle cruisers—two of which he was certain were of the *Iowa* class and two definitely of the *Alaska* class. He noted three to four *Cleveland*-class light cruisers and the estimated three squadrons of modern destroyers carrying long-range torpedoes. Since the Purple Joint Expeditionary Force was scheduled to capture Attu on D-Day, 28 July, he understood his task was to have his Covering Force destroy or drive off the Blue "sea forces" and reduce enemy air forces by "early and repeated" carrier air attacks as well as by daylight surface engagement. The ultimate objective of this action was to destroy or drive off Blue forces attempting to interfere with the Attu operation.[6]

Captain LM (Lt. Cdr. Willard Howell, USN)

His instructions as to precisely how his units were to operate began with his fleet flagship, which would join BATDIV 3 in the battle line for the fleet engagement. His battle line was to engage the Blue battle line at extreme ranges, which meant between twenty-two and twenty-five thousand yards. The Purple battle line was to close as rapidly as possible to this range, continuing in to under twenty-two thousand yards when the rate and volume of Blue fire had been reduced. As with the Blue Alaskan Defense Force, the Purple battle line was to supply its own Air Spot and Anti-Submarine Warfare Patrol. The Center Force was similarly to engage the Blue Center Force at extreme ranges, in order to force Blue to deploy, moving to the flank toward which the heavy units of Blue's Center Force would have to deploy. The Purple Center Force was then to fall back on the Purple battle line and take position in case the Blue battle line was encountered. Purple's light cruisers were to provide protection for the Center from Blue destroyer attacks. Following deployment and when ordered by Admiral PS, the Center was to join the light cruisers of the battle line and protect the Purple battle line from destroyer attacks.[7] While the Right Flank held its position, the Left Flank was to have its destroyers—supported by its cruisers—make torpedo attacks against the enemy battle line as well as against the Blue battle cruisers and heavy cruisers when directed to do so by the OTC.

The Carrier Group was given the most detailed orders of the component forces. It was to destroy enemy air and surface forces by repeated air strikes, employing maximum forces with weapon loads sized to ensure they reached the enemy main force. The Carrier Group was to make "concentrated" air attacks against enemy aircraft carriers and destroyers until Blue carrier flight operations had ceased. It was then to deliver repeated air strikes against Blue ships, the priorities being the fleet carriers and destroyers and then the battleships and light cruisers. Simultaneously, the Carrier Group was to furnish TF 81 with combat air patrols. During the surface engagement, the Carrier Group was to take station on the disengaged side of the formation with its own destroyers, when directed to do so by the OTC, and coordinate torpedo and bombing attacks on the Blue battle line. The Carrier Group was to provide its own combat air patrol and AEW aircraft and keep the OTC informed of the situation in the target area.[8]

The direction of deployment of Task Force 81 would be as signaled. It was further ordered that all planes that were armed with depth charges and all destroyers making contact with submarines attack their targets "vigorously." Prior to forming for the Approach Disposition, all ships were to follow the movements of the carriers during flight operations. The Purple Battle Plan, like its Blue counterpart, was also to be considered in effect upon signal. Additionally, damaged ships that were incapable of further action were to proceed to Kashiwabara Bay. For communications, all units were to use TF 81 Communication Plan Able and use Zone Mike time (minus 12).

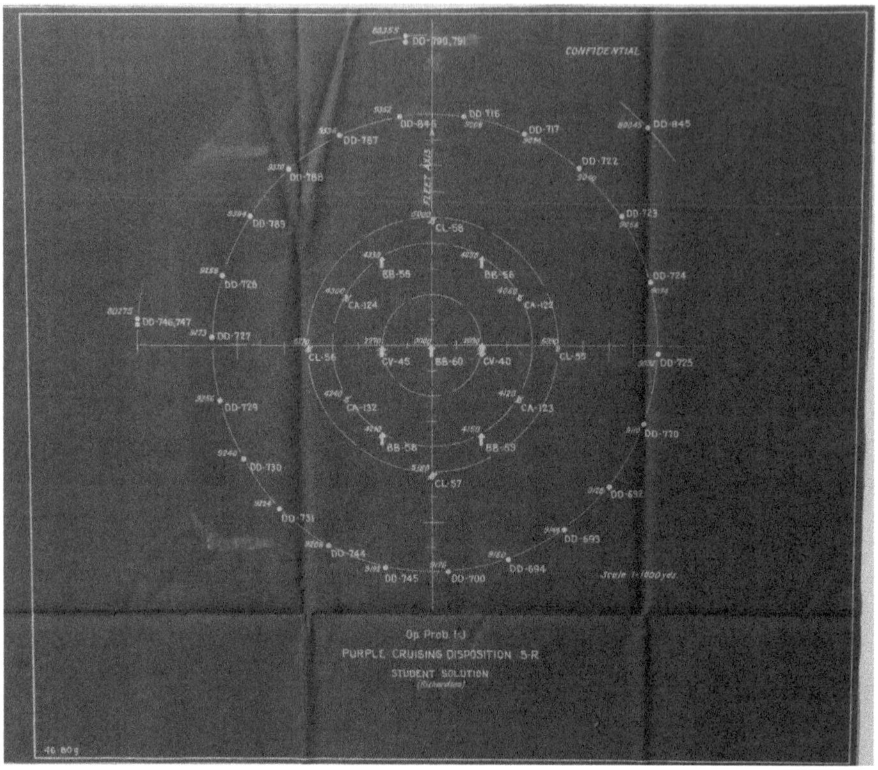

Fig. 92
Purple Cruising Disposition 5-R

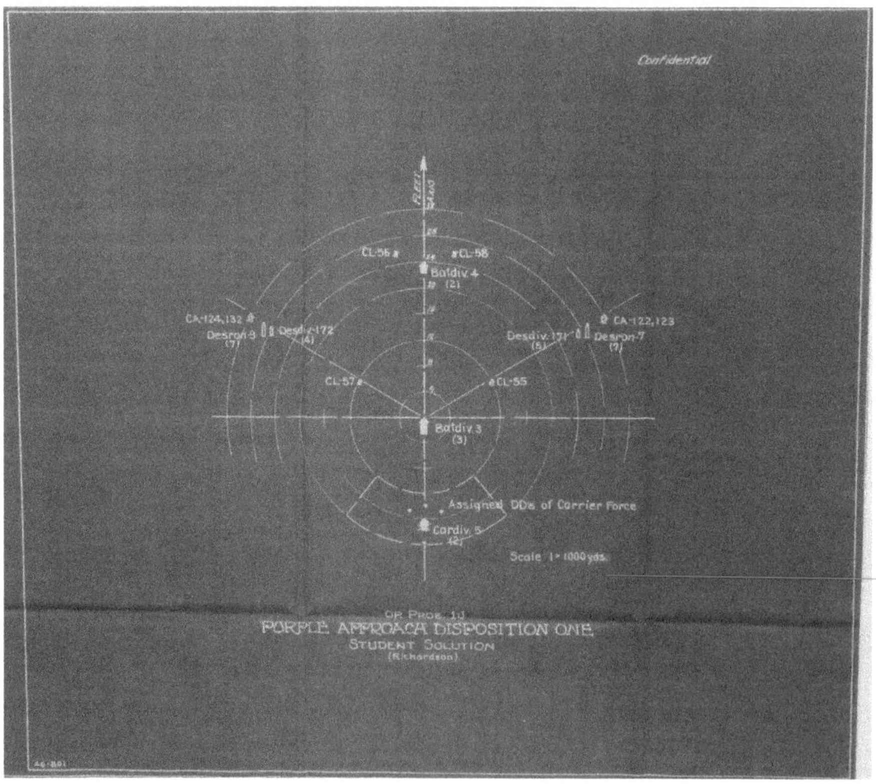

Fig. 93
Purple Approach Disposition 1

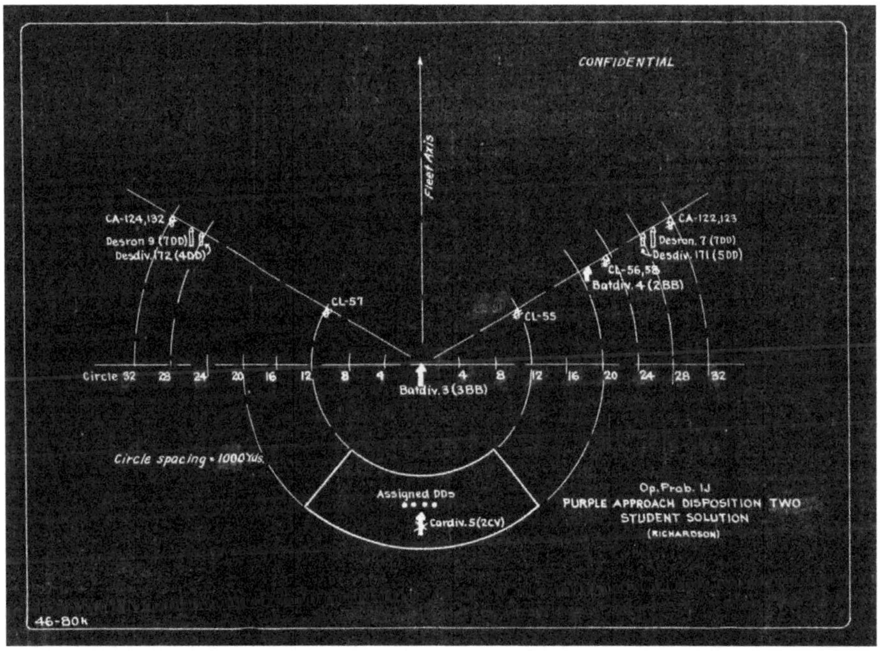

Fig. 94
Purple Approach Disposition 2

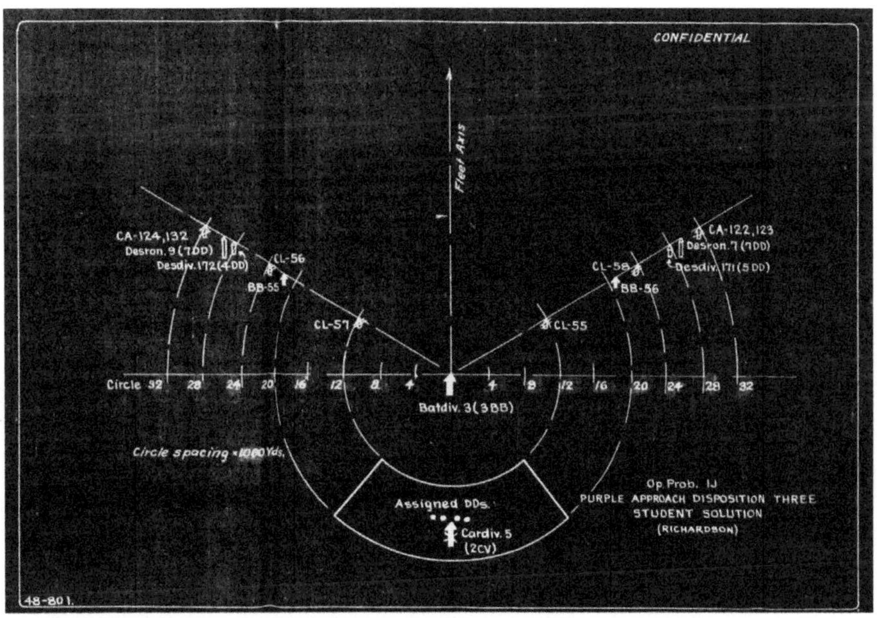

Fig. 95
Purple Approach Disposition 3

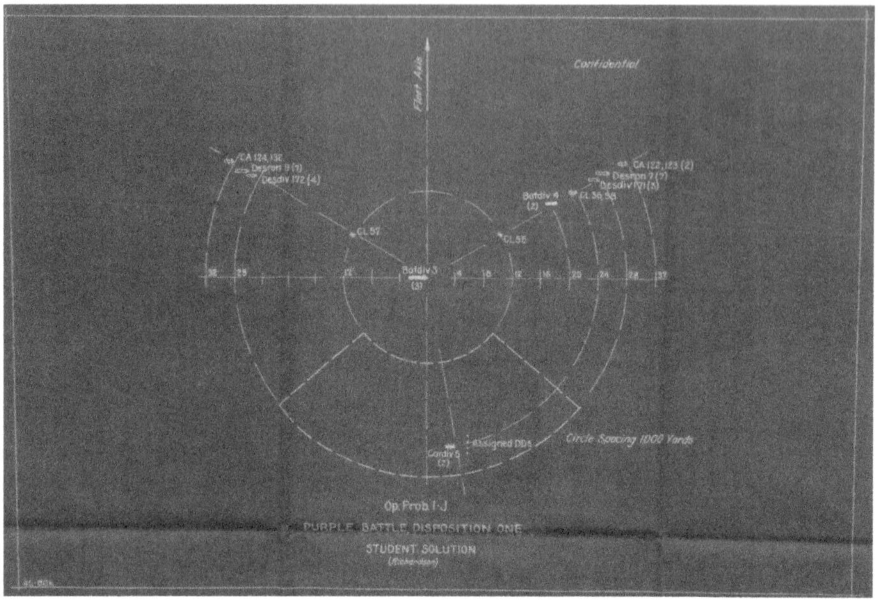

Fig. 96
Purple Battle Disposition 1

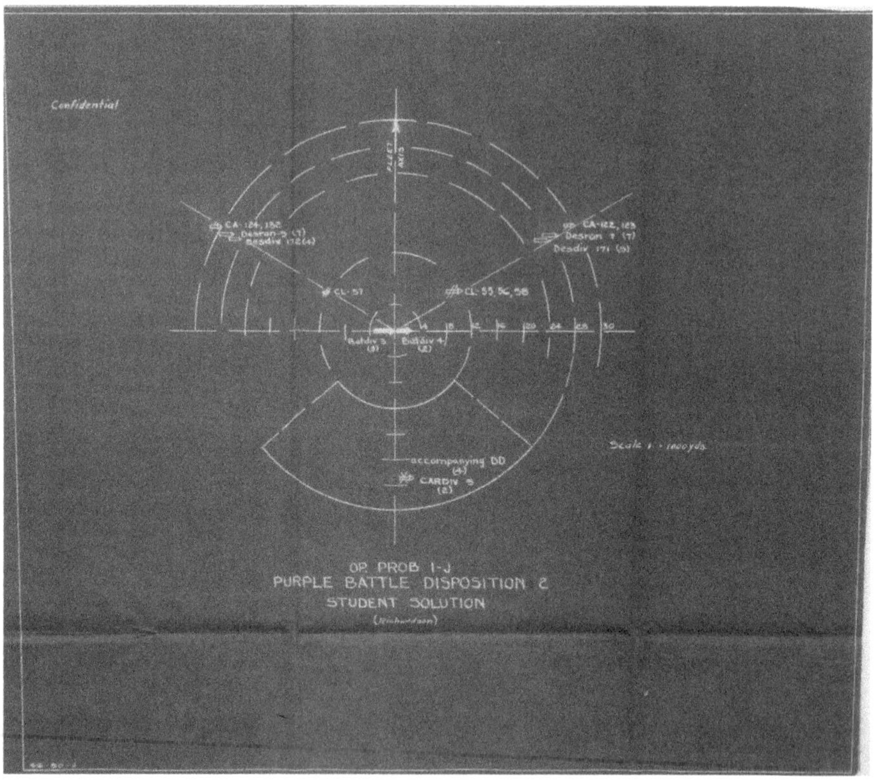

Fig. 97
Purple Battle Disposition 2

Cruising Disposition 5-R was to be employed, as was either Approach Disposition 1, 2, or 3, upon signal. Battle Disposition 1, 2, or 3 would also be signaled. The Officer in Tactical Command was Commander, Task Force 81 in fast battleship BB-60. Rear Admiral PL, Commander, Battleline, was designated second in command.[9]

For purposes of the Board Maneuver, TF 81 was broken down into its component commands, just as the Blue Alaskan Defense Force had been. Student officers and members of the Maneuver Staff were assigned to play the roles of Commander, Task Force 81; his Chief of Staff; the Commander, Purple Battleline; Commanders, Battleline Battleships and Cruisers; Commanders, Center, Right Flank, Left Flank, and Carrier Groups; and their subordinate commanders. In the Chart Maneuver, component forces were broken down somewhat differently; roles were played as Commander, TF 81; his Chief of Staff; Commanders, Battleships, Cruisers, and Destroyers; and the Commander, Carriers, along with his Chief of Staff (also Commander Rynd), his Operations Officer (also Lieutenant Commander Bonner), and the two Air Group Commanders (also Commanders Foley and Howell). Just as with the Blue Air Group Commanders, the Purple Commanders, Carrier Air Groups would be in different rooms when they were in the air versus when they were on their carriers.[10]

Cdr. Albert Church, USN

Also not surprisingly, the Purple Communication Plan Able was an almost exact match with Blue's, with the exception that the Fox Schedule was transmitted from Radio Amchatka. Rules concerning radio silence and categories of radio frequencies were the same, though the radio call signs were based more on ships' hull numbers than were Blue's. Responsibilities for radar Identification Friend or Foe, Radar Countermeasures Control, and silence were also assigned to certain ships and radar guard ships, fighter-direction ships, and radar intercept guard ships were listed. Ships were also designated for radio intercept and deception; radar jamming and radio deception were again under the control of the OTC. Guard ships, like their Blue counterparts, were to have interrogator-responder units on, and all ships and aircraft were also to have their Identification Friend or Foe transponders on. In addition, there was an authentication table and system.[11]

The Purple Maneuver Staff

In addition to the Director of the Maneuver, who was played by Capt. Donald Evans (see chapter 1), an instructor for the junior class, there was an Assistant Director and Chief Damage Computer—Cdr. Dale Mayberry, an instructor for the same class (see chapter 1)—as well as Assistant Directors for Blue and Purple, Cdr. James Reed (chapter 1) and Lt. Cdr. Valenti Holzapfel (chapter 1), respectively, both also instructors for the junior class. The Assistant Director for Purple was also the Air Umpire; there was a Communication Umpire and a Historian, respectively Lt. Cdr.

Cdr. Edward Baldridge, USN

Cdr. Edward Bridewell, USN

Cdr. Norman Gillette, USN

Col. Harry Hawthorne, USAAF

Cdr. Raleigh Kirkpatrick, USN

Cdr. Kenneth Loveland, USN

Cdr. Oscar Lundgren, USNR

Col. Salvatore Manzo, USAAF

Cdr. Robert Morton, USN

Cdr. Alan Nibbs, USN

Cdr. George O'Connell, USN

Cdr. James Owers, USN

Lt. Cdr. James Ramage, USN

Cdr. William Rogers, USN

Cdr. Carl Simonsen, USN

Lt. Col. Frank Street, USA

Lt. Col. Lee Wallace, USA

James Curran, a reservist who was an instructor in the Department of Intelligence (chapter 1), and Cdr. Raleigh Kirkpatrick, another instructor for the junior class. Additionally, there were Move Umpires for Blue and Purple—Cdrs. Alan Nibbs and Kenneth Loveland, respectively, both junior-class students—and Force Damage Recorders for Blue and Purple, Lt. Col. Lee Wallace and Cdr. James Owers, respectively, also both junior-class students.

In Phase I, the Chart Maneuver, there were Assistants to the Air Umpire. These consisted of an Aerial Combat Loss Computer, played by Col. Harry Hawthorne; a Hit Computer, a role filled by Col. Salvatore Manzo; an Anti-Aircraft Loss Computer, played by Cdr. Robert Morton; two Status Board Recorders, Cdr. Carl Simonsen for Blue and Cdr. Norman Gillette for Purple; and the Operational Loss Computer, Lt. Cdr. James Ramage.

Further, there were two Radar Contact Recorders, Cdr. Edward Baldridge for Blue and Cdr. Oscar Lundgren for Purple; two civilian draftsmen serving as War Gaming Plotters from the Department of Tactics, John Lawton and Frederick Wagner; and four Standby Damage Computers, Cdr. Edward Bridewell, Cdr. George O'Connell, Cdr. William Rogers, and Lt. Col. Frank Street.

With the exception of the civilians, all of these were junior-class students. Phase II, the Board Maneuver, saw Blue and Purple Damage Computers for the battle lines, the Detached Wings and Centers, the Right Flank cruisers, the Right Flank destroyers, the Left Flank cruisers, and the Left Flank destroyers played by the officers and civilians noted above, with the exception that in Phase II draftsman Adele DeMarco from the Department of Tactics served as an Assistant Plotter.[12]

The Purple Staff Solution and Board Maneuver

In the Purple Staff Solution to the Problem, Vice Admiral PS cited his orders from Commander, 8th Fleet to prevent Blue naval forces from interfering with the landing operations of the Purple Joint Expeditionary Force on Attu and to exploit "every opportunity to destroy Blue naval forces encountered in the Northwest Pacific north of latitude 40 degrees North and west of longitude 170 degrees West." He understood Commander, 8th Fleet's General Plan intent as to seize, occupy, defend, and develop Attu in order to provide an advanced air base for further operations against Blue bases in the eastern Aleutians. He knew that a Blue force comparable in size to Purple TF 81 had been located 275 miles southeast and was in a position to strike Purple amphibious forces at Attu by air within two hours. This enemy force was within air-striking distance of his force as well, and intelligence indicated that it contained all the combatant units organized by Blue for its operations. Any interference with Purple landing operations would come from this force; even if there was a second Blue force in the area, this larger one would be his major target. By destroying the Blue force just sighted, he would be able to prevent Blue interference with Purple landing operations and destroy the Blue naval forces encountered.[13]

Vice Admiral PS began his Estimate of the Situation with a general summary of events since 26 July.[14] Admiral PS confirmed that TF 81 consisted, aside from assigned aircraft, of two fleet carriers, five fast battleships, four heavy and four light cruisers, and twenty-seven destroyers, all steaming in Circular Disposition 81-R. The Purple 8th Fleet Joint Expeditionary Force was in the process of seizing Attu and Purple submarines were on reconnaissance patrols on the line of 45 degrees north latitude between longitudes 165 east and 175 west, in the Hawaiian area. They were not, however, in numbers that could be used offensively, and Admiral PS could not expect air or surface reinforcements. Based on intelligence estimates and the contact report he had just received, Admiral PS assumed that the Blue force consisted of two *Essex*-class fleet carriers, three *Iowa*-class fast battleships, two *Alaska*-class battle cruisers, another battleship or battle cruiser, four *Cleveland*-class light cruisers, and twenty-seven destroyers of either the *Sumner* or *Gearing* class. Purple had designated this force the Blue Northwest Pacific Force and knew that the Blue Pacific Fleet itself was still in Hawaiian waters on 9 July. Blue submarines in unknown numbers were operating off Kamchatka and in the western Aleutians, but no other Blue force was believed to be in the area. Commander, Purple 8th Fleet had also made the assumption that the Blue Pacific Fleet would not be able to reinforce quickly the Blue Northwest Pacific Fleet. It was further assumed that Blue submarines operating in the area had detected and reported the movements of the 8th Fleet, including Task Force 81 and the assault on Attu. Since the 8th Fleet was invading Blue territory and since the Blue force's composition was unsuited for seizing a base, Admiral PS deduced that the mission of the Blue Northwest Pacific Fleet was to destroy the Purple naval forces engaged in the amphibious assault on Attu and prevent the establishment of a Purple base on that island. (Pages 2–3.)

Like Admiral BA, Admiral PS compared the relative combat power of the two forces. He compared carriers, battleships, battle cruisers, heavy cruisers, light cruisers, destroyers, and carrier aircraft by such characteristics as number, maximum and sustained speeds, age, main battery and range, and automatic (i.e., 40 mm and 20 mm) weapons. Numbers of planes were compared on the carriers and other surface ships, as were the torpedoes on the destroyers. Since each side had two fleet carriers of comparable classes, Admiral PS thought it probable that Blue aircraft strength was equal to his, which was a total of 202 carrier and twenty-six observation planes. Also, since he did not have any information to the contrary, he also assumed that Blue personnel were equal to Purple's in terms of combat efficiency. Neither side could rely on land-based airpower, and he saw the two sides evenly matched in fleet carriers, light cruisers, and destroyers. He knew that Blue battleships and battle cruisers had a greater gun range than his battleships. He also knew that the Blue ships had a four- or five-knot speed advantage and could command somewhere between twenty-seven and thirty-six sixteen-inch guns, as well as somewhere between eighteen and twenty-seven twelve-inch guns. Aside from the weather, operating ability would greatly depend on "radar efficiency." (Pages 4–7.)

PS thought Blue's disadvantage of one or two battleships was offset by his two, possibly three, battle cruisers. In the worst case, Admiral PS thought, Blue might have four battleships and two battle cruisers. He also noted that the Blue battleships were heavier and faster than his own but had no firepower advantage. Admiral PS had forty-five sixteen-inch guns to match the potential mixes the Blue commander had. Though Blue could outrange Purple by six thousand yards, PS did not think the Blue guns would be very accurate at extreme ranges. In addition, the Blue battle cruisers were "comparatively" lightly armored and no match for his battleships, ship for ship. Though penetrative ranges might vary by several thousand yards, depending on list, roll, yaw, and the quality of armor, he felt that theoretical penetrative ranges would be of value in determining probable advantageous ranges. Accordingly, he argued that his ships were more resistant to penetration at favorable target angles of twenty-three to twenty-eight thousand yards and that Blue's battleships were more resistant at eighteen to twenty-five thousand. He also argued that Blue battle cruisers were vulnerable to Purple battleship fire at "nearly all" ranges but could seldom be penetrated by eight-inch gun at extreme ranges, at which eight-inch fire was "relatively" ineffective. Therefore, he saw Purple's battleships as "comparatively" invulnerable to the twelve-inch guns on Blue's battle cruisers, except at ranges over thirty-one thousand yards or when presenting end-on or broadside target angles at medium or short ranges. (Page 7.)

While noting that he had four heavy cruisers to Blue having none, Admiral PS admitted that his heavy cruisers would be "relatively ineffective" against the Blue battleships and battle cruisers, albeit "extremely" effective against Blue's light cruisers. He saw his four light cruisers as more resistant than Blue's to penetration at ranges of thirteen to nineteen thousand yards. He also knew, however, that Blue's battle cruisers would "somewhat" offset his superiority in heavy cruisers. As for destroyers, though he did not know the exact number Blue had, he thought that the destroyer forces appeared evenly matched. He also argued that in spite of the four-to-five-knot speed advantage in Blue's battleships, Blue might have difficulty maintaining that high speed for sustained periods because destroyers would be unable to maintain station. He did not know how many torpedoes the Blue destroyers carried, but he knew he was inferior in battleship speed and maximum range. Still, he saw his battleships and heavy cruisers as either superior, equal, or in the worst case only "slightly" inferior to Blue's heavy forces. He assumed that Blue would use his battle cruisers as a Detached Wing, to which he could oppose only his "inferior" heavy cruisers—and he would have to, or see his forces on that flank driven back on the Purple battle line for support. In general, though, as he wrote, "If I can come to grips with Blue I should be superior or at worst equal in over-all fighting strength." His best ranges for battle-line engagement were twenty-five to twenty-eight thousand yards; he knew that Blue's speed advantage had to be overcome by some means. (Pages 7–8.)

Given the positions of the two forces by this time, Admiral PS deduced that the Blue Northwest Pacific Fleet was within two hours' steaming of air-strike distance against Attu. Since neither side could receive reinforcements and the Blue force was 2,500 miles from Pearl Harbor, PS thought, he could use Attu as a refuge for damaged ships and an emergency ditching area for aircraft. He assumed that Blue had refueled at sea, given his distance from his bases, and that the Blue commander could similarly use Kiska as a ditching area for his planes. Tying in with all of this, of course, was weather. Like Admiral BA, he noted how poor the weather generally was and that it could "influence the coming action considerably." He thought that the wind would be favorable to him, however, since Blue would be forced to steam away from Attu to conduct flight operations, although Blue aircraft would have a tailwind en route to their objective during an initial air strike. Admiral PS's force would have a headwind on their initial strike and would face the additional problem of having to cover the amphibious landings on Attu. Given the overcast, fog, and unpredictable conditions, PS also decided to use his observation planes to report on weather in addition to their tactical scouting functions. Admiral PS concluded that in order to ensure victory, he had to cut down the speed of the Blue battle line and reduce the effectiveness of Blue's battle cruisers. (Pages 8–10.)

As for Blue's possible Courses of Action, Admiral PS argued that the only known restrictions were that the Blue force could not be reinforced and that Blue land-based air had been neutralized in the Aleutians. Blue's possible actions were, first, to engage the Purple Covering Force with aircraft and surface ships in an effort to destroy or drive it off and then destroy the Purple Joint Expeditionary Force at Attu. Blue could also engage his force with aircraft and surface ships while the Blue Detached Wing evaded TF 81 and destroyed the Purple amphibious forces. The Blue commander could likewise try to avoid the Purple Covering Force, use a strong combat air patrol for defense, and use his remaining aircraft for an air strike on the Purple Joint Expeditionary Force. Finally, Blue might strike the Purple carriers and avoid a surface action upon completion of his air attacks on Attu while preparing for a night surface action on 29 July or a day surface action on the 30th. Admiral PS noted the weather, the task of covering the Joint Expeditionary Force, and his own lack of reinforcement as the constraints on his own Courses of Action. (Page 11.)

One Course of Action was destroying the Blue Northwest Pacific Fleet by air while avoiding a surface action, but he did not think that it would be possible for him to destroy it completely, given his air strength. It would be possible by air strikes, he thought, only to knock out the Blue fleet carriers, reduce the speed of

Blue's battleships, and damage its battle cruisers, so he rejected this Course of Action as infeasible. The next possibility was to destroy the enemy force by air attacks and daylight surface engagement—"Preliminary destruction by air strikes followed by a daylight surface engagement appears to be suitable and feasible." Last was the idea of attacking the enemy force by day and then avoiding a surface engagement until night. He did not think, however, that there was any advantage in a night surface engagement, especially since it gave Blue time to close in on Attu and strike the Joint Expeditionary Force by air; this Course of Action was also rejected. The only Course of Action, therefore, was to destroy the Blue Northwest Pacific Fleet by a combination of air strikes and daylight surface battle. (Pages 11–12.)

Admiral PS asserted that he had tested his chosen Course of Action against the enemy's capabilities and still saw his decision as suitable, feasible, and acceptable as to costs. He thought this because with "adequate" warning he could detach part of his surface or air forces to destroy Blue units attempting an end run. Some of the enemy alternatives entailed divided air strength; Admiral PS assumed that such a situation would make his task of destroying the Blue Northwest Pacific Fleet easier. If, however, Blue decided to fight a night surface engagement, he knew he would have to revise his Estimate of the Situation. Accordingly, his decision was that TF 81 would destroy the Blue Northwest Pacific Fleet by "early and repeated" air attacks and a daylight surface engagement in order to prevent its interference with landing operations by the Purple Joint Expeditionary Force on Attu. (Page 12.)

Since his "Complementary Military Problem" was to destroy the Blue Northwest Pacific Fleet, Vice Admiral PS looked again at the assumptions he had made on the basis of relative combat power. He now had to consider the most advantageous use of his ships and aircraft. When it came to his aircraft, his first consideration was to ensure that he hit Blue first. He thought he could best ensure that by a fighter-bomber attack designed to put the enemy carrier decks out of action. Since he considered the coming air battle to be an "all out" engagement between the two carrier task forces, he was going to follow his fighter-bomber strike with a full-air-group strike. He opted for this (instead of a deck-load strike) so as to be sure none of his own planes remained on board while his carriers were under attack. He also thought that the use of a full-air-group strike would saturate Blue's antiaircraft fire. With these advantages, however, he had to accept the disadvantage of steaming for a longer period into the wind, thus decreasing his opportunities for defensive maneuvering should his carriers come under attack. His actions after the initial air strikes would depend on how much damage his carriers had received and also on the damage his planes inflicted on Blue's carriers. (Pages 13–14.)

In the air strike, radar-countermeasures aircraft would precede so as to simulate a raiding force and draw off enemy fighters while the air group approached from another direction. Radar-countermeasures aircraft would also accompany the air group to jam Blue fire-control radars. These aircraft were also going to be used to track the Blue force, should conditions for visual or radar tracking be unfavorable. Admiral PS also wanted the radar-countermeasures aircraft to jam enemy surface force radars as he directed. Additionally, photo planes would accompany the strike to obtain damage-assessment photographs and AEW aircraft would be on station as well. The best tactical use of the latter appeared to be stationing one of them over the Purple Covering Force, one sixty miles out in the probable direction of attack by the Blue aircraft, and one or more to track the Blue Northwest Pacific Fleet itself. Given the high peak power of its radar, the airborne early-warning aircraft, he surmised, should be able to track the Blue force out of range of Blue fighter direction and report both enemy positions and movement by continuous-wave radio. Admiral PS knew he had to provide "sufficient" protection for his carriers, with a strong combat air patrol. He also had to schedule planes for antisubmarine patrols as well as Air Spot for his surface forces. Since weather could be an important factor, he would station weather-reporting planes in the probable areas of operation. In addition, all planes on station were ordered to make periodic weather reports. As of 0400 on 29 July, he had twelve fighters aloft as a day combat air patrol, as well as six search planes and two airborne early-warning aircraft in the air. In addition, his carrier decks were spotted with armed and fueled planes. (Pages 14–15.)

As for the Blue battleships and battle cruisers, Admiral PS assumed the worst case of Blue having four battleships and two battle cruisers instead of three of each. The most probable use for the battle cruisers was, he thought, in the Center of the Blue formation during the approach and in the Van, or operating as a Detached Wing after Blue's deployment. The only way he thought he could counter these positions and moves was to use one or two of his battleships against the Blue battle cruisers until the latter's Fire Effect was reduced. As soon as his assigned battleships had dealt with the Blue battle cruisers, they would rejoin his battle line and concentrate their fire on the leading Blue battleship. While cruising or when enemy aircraft were attacking, his battleships would be used for antiaircraft defense. Since his own heavy cruisers could not stand up against Blue's battle cruisers, it was also going to be necessary to reinforce them with one or two battleships. "Two CAs concentrating on one CB should work out successfully." Tentatively, he would also use two heavy cruisers in the Center Force as well as one each in the Right Flank and Left Flank Forces. If Blue attempted a reverse action, it would be advantageous for him to have one heavy cruiser in the Rear. The heavy cruisers in the Van and the Rear would assist the light forces when they attacked. (Page 15.)

The light forces he found evenly matched. By splitting his light forces three ways, Admiral PS thought, he could be prepared for any allocation of light forces that Blue might make. His light forces would be given freedom to exploit any favorable opportunity for a torpedo attack but no such attack would be made until the Blue battle line was under fire from his own battleships. Also, the light forces that were assigned as defense elements would be used to defend the Purple battle line against Blue light force attacks as well as be prepared to exchange stations with the destroyers assigned to offensive tasks after those ships had expended all their torpedoes. The light forces could also be used for feint attacks on the Blue battle line. While cruising, light cruisers and destroyers were to form the antiaircraft screen; the destroyers would also be employed as an antisubmarine screen as well as radar pickets. After deployment, the 710-class destroyers would be assigned to escort the fleet carriers as necessary, depending on the operational state of the carriers by that time. (Pages 15–16.)

In his consideration of surface tactics to destroy the Blue Northwest Pacific Fleet, Vice Admiral PS first had to ensure that he kept TF 81 interposed between the Blue force and Attu so as to prevent the Blue force from making an end run with its superior speed. He also had to take into account the effect of wind and sea on the speed of the Blue battleships, the use of aircraft, and the employment of light forces. Additionally, he needed to assess Blue's speed advantage, use of his battle cruisers, visibility, and maximum-range advantage. As to the type of action to be fought, he had to choose between normal, reverse, and retirement actions. Since Admiral PS desired to close the range, a retirement action would not be to his advantage: "I must not let the enemy maneuver into position so that he can fight a retirement action towards Attu." With superior speed, a Blue reverse action might enable the Blue Northwest Pacific Fleet to circle the rear of Purple Task Force 81. In order to prevent Blue from fighting a reverse action, Admiral PS surmised, he had to make his Rear strong enough to make the option "unattractive" to Blue. Since he had a stronger battle line and he desired a decisive action, it appeared to him that a normal action would be best. Still, he thought he could fight a reverse action for short periods of time and then reverse to a normal action in time to prevent Blue from getting at Attu. (Page 16.)

Given the necessity of keeping his force interposed between Attu and the Blue Northwest Pacific Fleet, Admiral PS thought that, in general, he had to conform to Blue's movements as to deployment course. The best deployment course for TF 81 would be into the wind and sea, as this would reduce the enemy's maximum useful speed, take Blue away from Attu, and facilitate Purple air operations. If he deployed to the south or southwest, there was no assurance that Blue would follow. If the general bearing line of the two forces on the approach was roughly north–south, Admiral PS would favor deployment to the west, as this would also reduce the enemy's maximum useful speed. PS wanted to attempt to gain the windward position, since his light forces would then have both the wind and sea with them in attacking.

He thought that these factors would allow those forces to get into their attack positions more quickly and that the windward position would permit a more effective use of smoke screens. (Pages 16–17.)

Admiral PS thought that he should close to twenty-five to thirty thousand yards for the best results, avoiding the eighteen-to-twenty-three-thousand-yard zone. He did not think that Blue would desire to remain at extreme ranges because of the poor return on ammunition expended and the poor visibility, necessitating Radar Spot or indirect fire with Plane Spot. Either of these would give lower Fire Effect than would closer ranges. He knew, however, that he would be able to close the range only if he was able to slow down the Blue battle line with an air attack, the sea reduced Blue's maximum speed, or the Blue commander decided himself to close the range. The presence of Blue battle cruisers made the use of a Purple Detached Wing "almost" mandatory. Admiral PS knew he had to use a battleship to form a Detached Wing on the approach and station this unit so as to counter a Blue Detached Wing. He planned on forming his Detached Wing with one battleship and two heavy cruisers, though he was not going to "tie down" the heavy cruisers by having them operate in formation with the battleship. That, he argued, would deprive them of full use of their capabilities.

While his battleship engaged one Blue battle cruiser with six of its guns and the other with its remaining three, the second battle cruiser could be taken under fire by the heavy cruisers, thus giving the second Blue battle cruiser a "multiplicity" of targets. As to his light forces, he thought that the best distribution in a normal action would be with two-thirds of these forces in the Van and one-third in the Rear. He knew that the distribution of Blue's light forces might change the allocation of the Purple light forces, but he concluded that a normal action with a Detached Wing on a deployment course within the southern semicircle, northwest to southeast, at a range of twenty-five to thirty thousand yards was his best tactical option. (Pages 17–18.)

PS focused on four factors that would influence his air tactics. One was the relative distance between the Blue and Purple forces, while another was the movement of the two forces after they launched their aircraft. In addition, the combat radii of the aircraft mattered, as did weather conditions. As of 0400 on 29 July, with the two forces 275 miles apart, Admiral PS assumed that Blue had detected his force and had been in contact since at least TF 81 had discovered the Blue Northwest Pacific Fleet. As for aircraft combat radii, the Bearcat fighter had a radius of three hundred miles if loaded with a belly tank and two thousand-pound bombs. The Helldiver bomber had a range of 230 miles if loaded with a torpedo and 245 miles with one two-thousand-pound or two thousand-pound bombs. If a Helldiver bomber had one thousand-pound bomb and a drop tank, its range was 335 miles; the Avenger bomber had a range of 425 miles with two wing tanks and one torpedo. If the Blue force headed south after launching aircraft, it would be out of range of Admiral PS's Helldivers that were loaded with torpedoes or two-thousand-pound bombs. Blue had to advance to the north to accomplish his mission, but Admiral PS thought that the Blue commander could momentarily head south until certain that the Purple air attacks had failed. The same situation also applied to Blue if Admiral PS headed north. (Page 18.)

However, Admiral PS had available AEW aircraft that he thought would provide "excellent" information as to enemy movements. Although he would like to have loaded all of his Helldivers with torpedoes, he could not, because he needed some attack planes armed with bombs to be carrying out strikes at the same time as the attack planes armed with torpedoes as a way to saturate Blue's antiaircraft defenses. None of the above, moreover, would affect his initial fighter-bomber strike with Bearcat fighters loaded with two thousand-pound bombs each. A compromise load for the attack-plane strike, however, had to be devised. There would be thirty Helldiver bombers loaded with one thousand-pound bomb each, in addition to drop tanks; eighteen more Helldiver bombers loaded with torpedoes; twenty-two Avenger bombers with torpedoes and wing tanks; and twenty-one escorting fighters. Should information from tracking aircraft indicate that Blue had turned away, the eighteen Helldiver bombers with torpedoes would be recalled in order to prevent their loss. The remaining aircraft, he thought, had sufficient combat radii to cover that contingency. He also surmised that his TF 81, after launching two strikes, would turn away and maintain a distance of 275 miles from Blue until the enemy carriers were hit. This would prevent Blue from utilizing the maximum loading on his planes. Admiral PS's subsequent maneuvers would depend on the damage suffered by Blue and received by Purple. He further noted that weather conditions in Blue's vicinity might preclude dive- and glide bombing, so Purple planes had to be prepared to make the best attack possible under the existing conditions. (Pages 18–19.)

For the "Concept of the Battle Plan" section of his analysis, Vice Admiral PS reiterated that the enemy force just sighted was formidable, had equality in air and light forces, and was superior in the speed and gun range of its heavy forces. In order to ensure victory, Admiral PS knew he had to gain control of the air first; he saw a high-speed strike to put Blue's carrier decks out of action as essential for gaining air control. A heavy attack plane strike would finish off the fighter-bombers' strike mission if necessary; then the Blue battleships could be slowed with an aerial torpedo strike. The third priority for his air group was the destruction of Blue's battle cruisers. (Page 19.)

He was then going to approach Blue with a strong Center Force, including one battleship and two heavy cruisers to deal with Blue's battle cruisers should they be in the Blue Center on the approach. The Purple Center would attempt to "hurry,

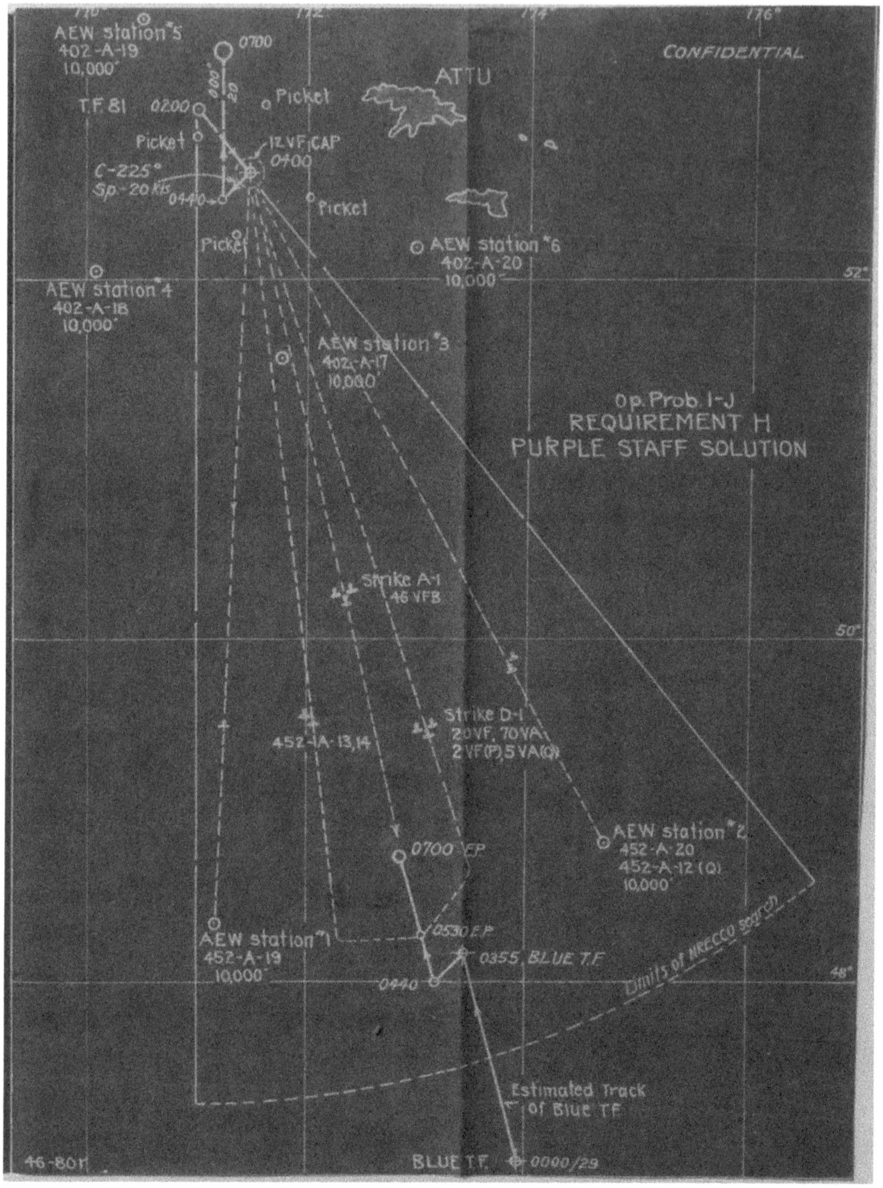

Fig. 98
Purple Staff Solution Air Operations

hamper or force the deployment of the enemy." Since his battleship in the Detached Wing would have no speed advantage over the Blue battle line, he thought it might be necessary upon deployment for the Purple battle line to slow or reverse course in order to permit his Detached Wing to attain its station in his Van. At this point, the Purple battle line would close in for a normal action at ranges between twenty-five and thirty thousand yards to destroy the Blue battle line by gunfire. Again, he thought that his best deployment course was within the southern semicircle, between northwest and southeast, so as to keep between the Blue Northwest Pacific Fleet and Attu. He assumed Blue would employ its battle cruisers as a Detached Wing in its Center on the approach and then in the Van upon deployment. Since his own heavy cruisers were no match for the Blue battle cruisers, two of the heavy cruisers would be accompanied by the battleship in the Purple Detached Wing. His Detached Wing would also be in his Center on the approach and would maneuver thereafter as necessary to counter the operations of the Blue Detached Wing. If Blue did not deploy a Detached Wing, the Purple battleship would join the Purple battle line upon deployment, while the heavy cruisers would join the Van. (Pages 19–20.)

Vice Admiral PS was going to station a heavy cruiser, two light cruisers, and five destroyers on each flank during the approach. On deployment, the heavy and light cruisers in the Van would attack simultaneously with the destroyers, on signal. If the Detached Wing was operating in the Van, the Van would consist of one battleship, three heavy cruisers, two light cruisers, and fourteen destroyers. The light forces would then seize opportunities that might present themselves for torpedo attacks after the Blue battle line came under Purple main-battery fire. If the air strikes had slowed the enemy force and the gunfire action was progressing "favorably," Admiral PS thought that the light-force attack might not even be necessary. He stressed, however, that every effort had to be made to deny information about the Purple force to Blue: "This is all-important to the ultimate victory." He thought this aim could be accomplished by taking advantage of local weather conditions, such as keeping TF 81 in areas of bad weather during air attacks and while being shadowed. In addition, covertness could be accomplished by destroying Blue snoopers and spotting planes, employing radar jamming on known frequencies, and observing strict communications discipline. Damage inflicted on the enemy by Purple air attacks and on Purple TF 81 by Blue's air attacks might also necessitate a change in the Battle Plan. (Page 20.)

In devising the tasks for his Battle Plan and creating his task organization, Admiral PS divided "Cruising Tasks" and "Battle Tasks" by unit type. His carrier, or "Air," forces, comprising the two fleet carriers of CARDIV 5 and the four destroyers of Destroyer Division 172, were under the command of Rear Admiral PV. The carrier aircraft were to defend Task Force 81 from enemy air attack and provide information on the enemy as well as furnish a thirty-plane combat air patrol. A further seven specialized planes would take part in airborne early-warning, search, and contact-scouting operations. In addition, nine fighter planes and thirty-six fighter-bombers would put the Blue carriers out of action. Another twenty-one fighters and seventy attack planes would inflict maximum damage on the Blue surface force, slowing its battle line with torpedo attacks and destroying its battle cruisers. The priority of targets here was to be the battleships, then the battle cruisers, and finally the light cruisers. Air was also to obtain and maintain air control before contact with the enemy and then support the light-force attacks, on deployment, by striking Blue light forces with rocket attacks, as directed by the OTC. The carrier aircraft were additionally to protect Purple spotting planes and destroy Blue's spotting planes, as well as have smoke tanks ready for use. (Pages 21–26.)

Admiral PS's observation planes were to furnish antisubmarine patrols and weather planes. Eight of these planes would scout, another thirteen would provide Air Spot, and the remaining five would provide the antisubmarine patrols. The battleships and cruisers that launched these planes would provide antiaircraft screens for the carriers and radar guards as assigned, while the destroyers would also provide part of the antiaircraft screen, serve as radar guards and pickets, and provide antisubmarine sound screens. The battleships would be organized into the battle line, four battleships from Battleship Divisions 3 and 4 under the command of Rear Admiral PL. The battle line was ordered to destroy the Blue battle line by normal action between twenty-five and thirty thousand yards in which it provided its own Air Spot. Its secondary mission was to support the light forces engaged in the torpedo attacks on the Blue battle line. The Detached Wing of one battleship and two heavy cruisers was under the command of Rear Admiral PR. On the approach, it was ordered to take station in the Center and drive back the Blue Center Force, thereby forcing a Blue deployment. It was also to coordinate the operations of the Center and upon deployment to destroy the Blue Detached Wing and then support the Purple light forces. (Pages 21–26.)

The Center was commanded by Captain P-5 and consisted of thirteen destroyers, from Destroyer Squadron 7 plus DESDIV 92. This force was to defend the area ahead between the forces on its own flanks against attacks by Blue light forces, aircraft, and submarines, as directed by the commander of the Detached Wing. On deployment, DESRON 7 would join the Van, while DESDIV 92 would join the Rear. The Left Flank was under Rear Admiral PZ and consisted of one heavy cruiser, two light cruisers, and the five destroyers of Destroyer Division 171. It had no particular tasks assigned. The Right Flank, under Rear Admiral PT, however, was to attack the Blue battle line

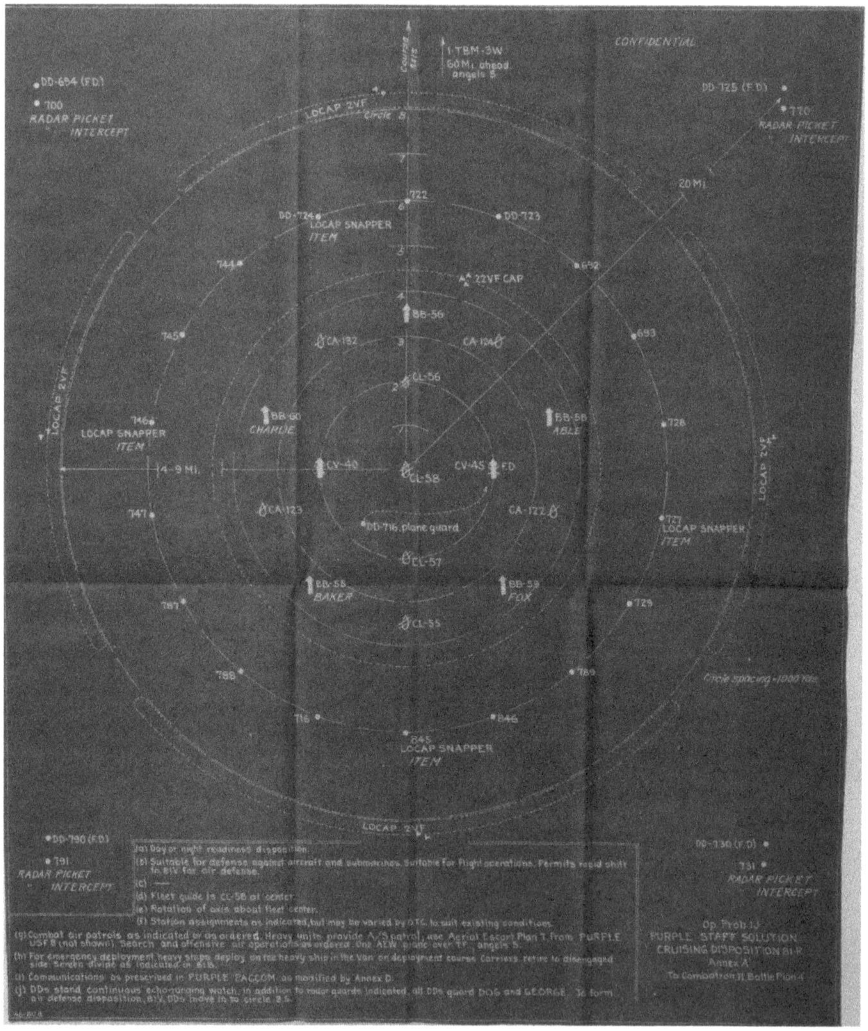

Fig. 99
Purple Staff Solution Cruising Disposition 81-R

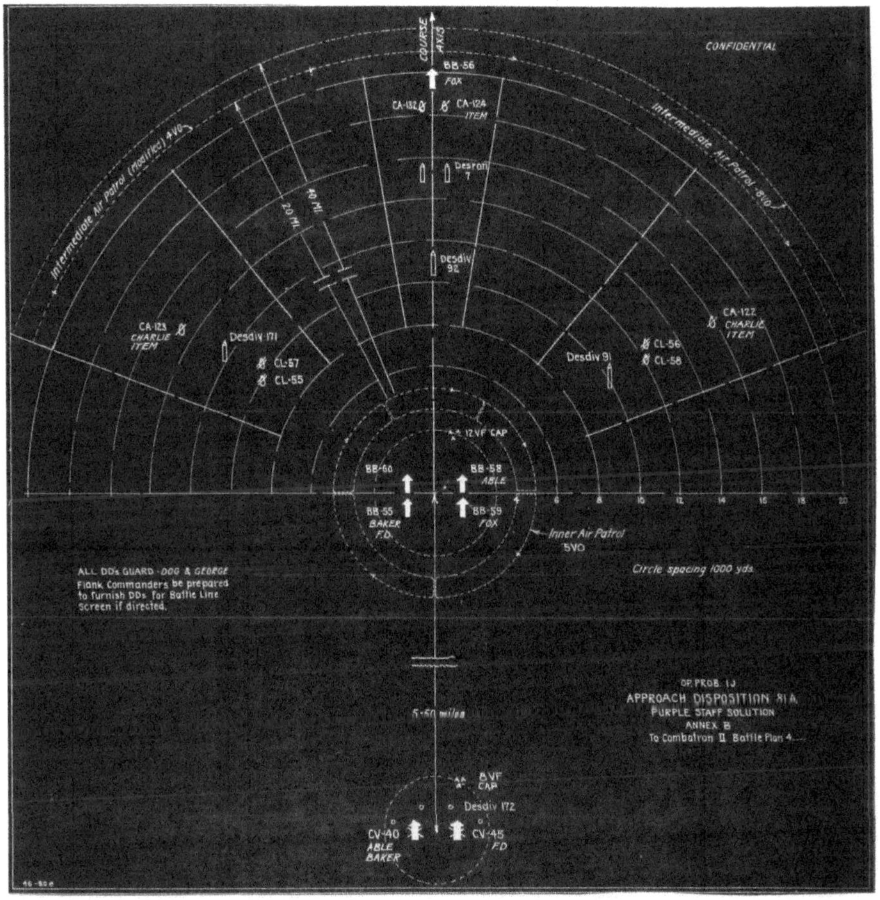

Fig. 100
Purple Staff Solution Approach Disposition 81-A

with torpedoes when a favorable opportunity arose and at the discretion of the flank commanders. The forces in the Rear were also to feint torpedo attacks as ordered by the Van commander. The Right Flank Force was then to defend the Purple battle line from Blue light-force torpedo attacks, furnish its own Air Spot, and make smoke as directed by the flank commanders. To carry out its tasks, the Right Flank had been given a heavy cruiser, a light cruiser, and the five destroyers from DESDIV 91. (Pages 21–26.)

As to his dispositions, Admiral PS's primary consideration in the Cruising Disposition was defense against air attack, with submarines a secondary consideration. He also wanted his Cruising Disposition to facilitate air operations in order to launch his air strikes on Blue. Since Cruising Disposition 81-R best met these requirements, Admiral PS's orders were for TF 81 to employ that formation during its early approach. When it came to the Approach Disposition, he would signal this formation only when "reasonably" secure from enemy air attack. Due to variable visibility and the fact that Blue and Purple deployment directions were uncertain, he thought it was best to split his light forces three ways, keeping his destroyer squadrons and divisions together as much as possible. Since Approach Disposition 81-A fit this tactical situation, he would use it, with modifications if conditions warranted. He was anticipating all this when he cautioned his units to be ready to go straight from their Cruising to Battle Dispositions, specifically Battle Disposition 81-B, which best fit a normal action (to be consistent with earlier renditions of this) with a preponderance of light forces in the Van plus the employment of a Detached Wing, as necessary or desirable. The Battle Plan was to go in effect upon receipt, with CTF 81 in fast battleship BB-60 as the OTC and Rear Admiral PL in BB-58

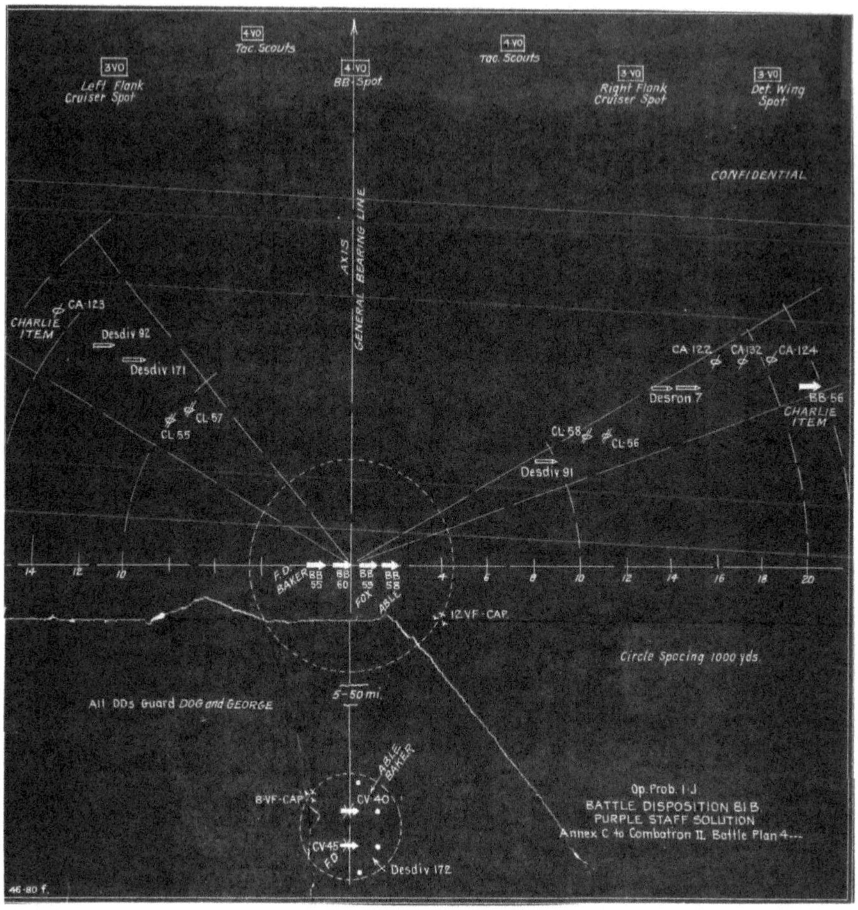

Fig. 101
Purple Staff Solution Battle Disposition 81-B

as the second in command. (Pages 21–26.) The Purple Communication Plan mirrored Blue's exactly in terms of categories, effective date, frequencies and circuits, code words, and voice calls. (Pages 27–30.)

The Critique

Several days later, a Critique of Operations Problem 1J was held in the Maneuver Room of Luce Hall for both Blue and Purple. While the dialogue of the Critique was not recorded, its format and organization provide some sense of what the Naval War College Staff thought was important for the student officers in the lessons of the exercises and maneuvers. Student officers gave five-minute presentations on the indicated portions of their own and the staff's solutions. Staff comment followed "when and as appropriate." Students for both Blue and Purple were assigned to talk about the "Estimate of the Mission," with Commander Bridewell presenting for Blue and Commander Lundgren presenting for Purple. In addition, there was an "Estimate of the Situation." These latter discussions were further broken down into Blue and Purple presentations on "Relative Combat Power Considerations," presented by Commander Simonsen for Blue and Commander Gillette for Purple; "Other than Relative Combat Power," with Colonel Manzo presenting for Blue and Colonel Hawthorne presenting for Purple; "Enemy Capabilities," presented by Commander Morton for Blue and Lieutenant Commander Ramage for Purple; and "Own Courses of Action," with Lieutenant Colonel Wallace for Blue and Commander Owers for Purple. Then were the presentations "Analysis of Opposing Courses of Action," by Commander Baldridge and Lieutenant Commander Taeusch for Blue and Commander O'Connell for Purple, and "Comparisons of Own Courses of Action and the Decision," by Commander Rogers for Blue and Lieutenant Colonel Street for Purple. These were followed by "Complimentary Problems," with Commander Nibbs and Lieutenant Colonel Downey presenting for Blue and Commanders Loveland, Marvin-Smith, and Miller for Purple; "Battle Plan," delivered by Commander White for Blue and Commander Church for Purple; "Dispositions," by Commander Rogers for Blue and Lieutenant Colonel Street for Purple; and "Communication Plans," analyzed by Lieutenant Commander Curran (see chapter 1). A summary by Captain Evans (chapter 1) was also included.[15]

The Critique included an analysis of the Air Action, which comprised the Chart Maneuver. Here, the analysis included presentations on the "OTC's Plan and the General Execution ofthe Air Action," with Commander Kirkpatrick presenting for Blue and Commander Richardson for Purple. There was also a category on air operations in general, delivered by Lieutenant Colonel Leary for Blue and Colonel Mason for Purple, and a summary by the Air Umpire (Lieutenant Commander Holzapfel) and his assistants on the "Use of Search Planes, the Use of Airborne Early Warning Planes and Pickets, Strikes, Interceptions, Anti-Aircraft Fire, Damage to Ships, and Operational Losses." Officers involved here included Colonels Hawthorne and Manzo, Commanders Morton, Gillette, Simonsen, Lundgren, Baldridge, Bridewell, O'Connell, and Rogers, Lieutenant Colonel Street, and Lieutenant Commander Ramage, along with Mr. Lawton and Mr. Wagner. This phase also entailed a discussion of communications by Lieutenant Commander Curran as well as a conclusion by Captain Evans and Commander Raleigh Kirkpatrick before presentations on the surface action—which constituted the Board Maneuver—began.[16]

Here, the categories included the "Officer in Tactical Command's Plans"—analyzed by Commander Kirkpatrick for Blue and Commander Richardson for Purple—and then the action. This latter category broke down into "Approach" and "Deployment," explored by Kirkpatrick and Richardson again; "Center Force Action," critiqued by Commander Dalton for Blue and Lieutenant Commander Bonner for Purple; "Battle Line Action," carried out by Colonel Edson for Blue and Lieutenant Colonel Thompson for Purple; "Detached Wing Action," analyzed by Dalton and Bonner again; "Light Forces in the Van," delivered by Commander Coleman, Commander Webber, and Lieutenant Commander Aymond for Blue, with Commander Foley, Commander Marvin-Smith, and Lieutenant Commander Howell looking at Purple's role; and "Light Forces in the Rear," explored by Lieutenant Colonel Downey and Lieutenant Commander Taeusch for Blue, and Commanders Rynd and Miller for Purple. Thereafter was a talk on the "Officer in Tactical Command's Future Intentions," with Commanders Kirkpatrick and Richardson again fulfilling this role, and then a summary that focused on "Damage," "Fire Control," and "Radar," analyzed by Commanders Mayberry and Owers with Lieutenant Colonel Wallace, respectively; "Ship Handling," explored by Commanders Loveland and Nibbs; "Communications," by Lieutenant Commander Curran; "Blue

Operations," looked at by Commander Reed (see chapter 1); and "Purple Operations," looked at by Lieutenant Commander Holzapfel, followed by "Conclusion to the Critique" by Captain Evans.[17]

NOTES

1 Junior Class of June 1947, "Operations Problem 1J: The Purple Statement," 10 September 1946, pp. 1, 6–7, Schedule of Events, folder 2602, box 138, RG 4, NHC.

2 Ibid., pp. 1–2.

3 Ibid., pp. 2–3.

4 Ibid., pp. 4–5.

5 Junior Class of June 1947, "Operations Problem 1J: Purple—Section B," 27 September 1946, p. 1, folder 2602-J, box 138, RG 4, NHC.

6 Junior Class of June 1947, "Operations Problem 1J: Purple—Section B," 28 September 1946, p. 1, folder 2602-K, box 138, RG 4, NHC.

7 Ibid., pp. 1–2.

8 Ibid., p. 2. This scenario is not radically different from another Midway episode. Even after it became his last remaining operational carrier, the commander of the 1st Carrier Strike Force, Vice Adm. Chuichi Nagumo, still hoped that *Hiryu* could damage, if not destroy, one and perhaps two of the American carriers. Nagumo planned then to engage the American force at night with his battleships, cruisers, and destroyers, reinforced by an even larger force of the same types from Vice Adm. Nabutake Kondo's Midway Invasion Force; see Craig Symonds, *The Battle of Midway* (New York: Oxford Univ. Press, 2011), pp. 310–12.

9 Junior Class of June 1947, "Operations Problem 1J: Purple—Section B," 28 September 1946, pp. 2–3.

10 "Purple: Section B, Board Maneuver—Operations Problem 1J," 28 September 1946, folder 2602-F, and "Purple: Section B, Chart Maneuver—Operations Problem 1J," 28 September 1946, folder 2602-G, both box 138, RG 4, NHC.

11 "Annex A: Communication Plan Able," 28 September 1946, pp. 1–4, folder 2602-L, box 138, RG 4, NHC.

12 Junior Class of June 1947, "Operations Problem 1J: Maneuver Staff," 30 September 1946, p. 1, folder 2602-J, box 138, RG 4, NHC.

13 Junior Class of June 1947, "Operations Problem 1J: The Blue Solution," 2 October 1946, pp. 1–2, folder 2602-P, box 138, RG 4, NHC.

14 Junior Class of June 1947, "Operations Problem 1J: The Purple Solution," 3 October 1946, pp. 1–2, folder 2602-Q, box 138, RG 4, NHC. Subsequent page references, until the next endnote, are to this source.

15 Junior Class of June 1947, "Operations Problem 1J: The Critique," 8–9 October 1946, p. 1, folder 2602-O, box 138, RG 4, NHC.

16 Ibid., pp. 1–2.

17 Ibid.

X *Operations Problem 2*

Purple, September–October 1946

Operations Problem 2 was the next maneuver. Its purpose for the senior class of June 1947 was to give practice to students in making an Estimate of the Situation, to prepare directives from that Estimate, and to study the characteristics of the North Pacific theater of operations. In addition, students would get practice in dealing with different types of aircraft and ships. Finally, students would execute a planned action by means of a Chart Maneuver.[1] On 26 September, Rear Admiral Smith provided students with the Detail of the Maneuver Staff—that is, staff personnel assignments. For Operations Problem 2, there was to be a Director of the Maneuver, Capt. Einar Johnson, an instructor in the Naval War College's Department of Strategy. Two other instructors from the same department, Capt. Harry McIllhenny and Cdr. Frank Acker, were the Assistant Directors and Plotting Officers for odd- and even-numbered moves, respectively. An additional instructor from the Department of Strategy, Capt. Paul Crosley (see chapter 1), was the Move Umpire for Blue and Purple, with Capt. Richard Dole (an instructor in the Naval War College's Department of Correspondence Courses) as the Assistant Move Umpire for Blue and Capt. Dashiell Madeira (a student in the June 1947 senior class), as the same for Purple. Cdr. Henry Haskell, a reservist and an instructor in the Naval War College's Department of Intelligence, was the Communication Umpire for Blue and Purple. Commander Haskell's Assistant Communication Umpire for Blue was Capt. Edwin Taylor, a student in the June 1947 senior class, and the Assistant Communication Umpire for Purple was another senior student, Capt. William Outerbridge. The Air Umpire was Capt. George Montgomery, an instructor in the Naval War College's Department of Strategy. He had two Assistant Air Umpires, Cdr. Lauren Johnson, an instructor in the Naval War College's Department of Tactics (see chapter 1), and Capt. George Van Deurs, a student in the June 1947 senior class. Finally, the Maneuver Staff comprised a Damage Umpire, Captain Johnson; a Plotter, draftsman Charles Ward; and three additional draftsmen from the same section and department, Mr. Wilson and Philip Gaudet as the Assistant Plotters, and one Mr. Scannevin as the Historian.[2]

Capt. Einar Johnson, USN

Capt. Harry Mc-Illhenny, USN

Cdr. Frank Acker, USN

Capt. Richard Dole, USN

Capt. Dashiell Madeira, USN

Cdr. Henry Haskell, USNR

Capt. Edwin Taylor, USN

Capt. William Outer-bridge, USN

Capt. George Mont-gomery, USN

Capt. George Van Deurs, USN

The Purple Statement: The General Situation

In the General Situation, relations between Purple and Blue had deteriorated "considerably" since the defeat of Germany and Japan in 1945, and there was little hope of bettering those relations. Purple's aim was "naturally" to better its position in world affairs, provide for its "natural" expansion, and ensure its future security. Understanding that it would be looked on with suspicion by other powers, Purple was nevertheless willing to accept war to advance and protect its interests: "Blue and Purple have not declared war, but the implications are plain." The General Situation also illustrated that Blue was now in possession of the Japanese home islands, the Ryukyus, and Korea south of 38 degrees north latitude. Purple, of course, was occupying Sakhalin, the Kuriles, and Korea above 38 degrees north. Purple had withdrawn from Manchuria, but its effective encirclement gave it "virtual" control of that province. (Page 3.)

Blue naval forces in the Pacific had been strengthened, and the Blue Army had been brought to full strength. In addition, Blue had activated its principal bases in the Aleutians as well as its minor bases, and Blue land-based airplanes had been seen over the Kurile Islands. Blue bases in the Aleutians were "adequately" equipped with the latest types of radars, antiaircraft batteries, communications equipment, ground-control equipment, and "FIDO" (fog intensive disposal, or dispersal, equipment, for clearing fog from in front of moving equipment and vessels). Blue bases in Alaska and the Aleutians included a naval air station at Sitka that could accommodate thirty carrier-type planes. The base had only a limited seaplane capability, but seaplanes could be tended at nearby Juneau. There was also an airfield and seaplane area at Skagway, a naval auxiliary air facility at Yakutat, and another naval auxiliary air facility at Annette Island. Kodiak Naval Air Station also had a naval operating base and a submarine base. The naval air station housed three airstrips between five and six thousand feet long, all of them able to handle the heaviest aircraft. These fields could accommodate thirty fighters and twenty naval patrol planes in Normal Conditions, while 280 fighters, 144 naval patrol planes, and ten very-long-range aircraft could be accommodated under Emergency Conditions. For purposes of the maneuver, there would be ten fighters, sixteen naval patrol bombers, and five very-long-range aircraft at Kodiak under Normal Conditions, 136 fighters and seventy-six naval patrol bombers under Emergency Conditions. The submarine base had complete facilities, and the harbor had facilities of twelve berths for destroyers or ships under five hundred feet length, or three for cruisers or ships over five hundred feet. (Pages 1–2.)

At Cold Bay, Alaska, Purple noted, Blue had an airstrip of 7,500 feet and another of five thousand, the larger of which could handle B-29 Superfortress very heavy bombers. These facilities could also park fifteen Navy transport planes, ten fighters, and five naval patrol planes. Cold Bay had no seaplane facilities, and its only harbor facilities were five mooring buoys, but seaplanes could take off and land there. Dutch Harbor had an airstrip of 4,300 feet, but local conditions restricted the size of the planes that could operate there to PBY Catalina naval patrol bombers and R4D transport planes. The field could, however, take twenty-four fighters under Normal Conditions and forty-eight under Emergency Conditions, and it could accommodate twenty-four PBM Mariners. Its harbor could also handle battleships, heavy and light cruisers, and destroyers. For Operations Problem 2, Dutch Harbor would have nine fighters and nine naval patrol bombers during Normal Conditions and eighteen of each type under Emergency Conditions. (Pages 2–3.)

The Blue base on Unalaska also had a Navy fueling base on nearby Akutan Island, a Navy auxiliary air facility at Otter Point, a seaplane anchorage at St. Lawrence, and an emergency land-based and seaplane area at Sand Point. Adak had a naval air station with four airstrips, varying in length from 5,277 to 7,800 feet, that

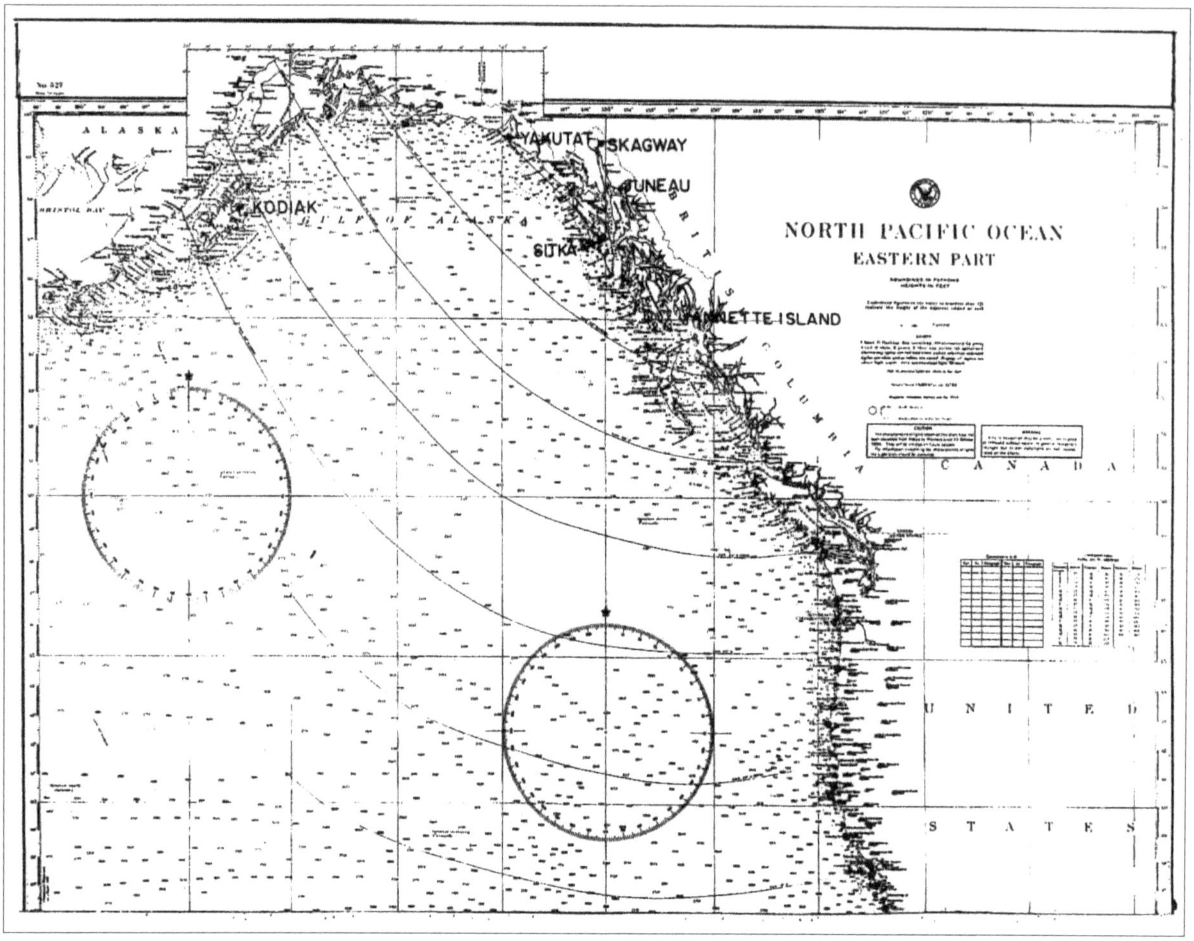

Map 6
North Pacific Ocean: Eastern Part

could handle Catalinas, R5D transport planes (larger than the R4D), and Superfortresses. It also had limited seaplane facilities for six naval patrol bombers and could handle between 435 and 612 fighters during Normal and Emergency Conditions, respectively, 275 to 340 naval patrol bombers, and between ninety-five and 116 very-long-range aircraft. During the maneuver, however, there would be 210 fighters, 130 naval patrol planes, and thirty-six very-long-range aircraft during Normal Conditions. During Emergency Conditions, Adak would have 315 fighters, 172 naval patrol bombers, and fifty-four very-long-range aircraft. Adak had unlimited anchorage for all classes of ships, while Amchitka only had buoys for three cruisers. Amchitka had airfields between 5,255 and 10,220 feet in length that could take Superfortresses or other planes up to forty-five thousand pounds in weight, as well as between 150 and 250 naval patrol bombers (Normal and Emergency Conditions, respectively). For purposes of Operations Problem 2, though, Amchitka would have fifty fighters, fifty naval patrol bombers, and a hundred very-long-range aircraft during Normal Conditions, a hundred fighters, a hundred naval patrol bombers, and 150 very-long-range aircraft in emergencies. (Pages 3–4.)

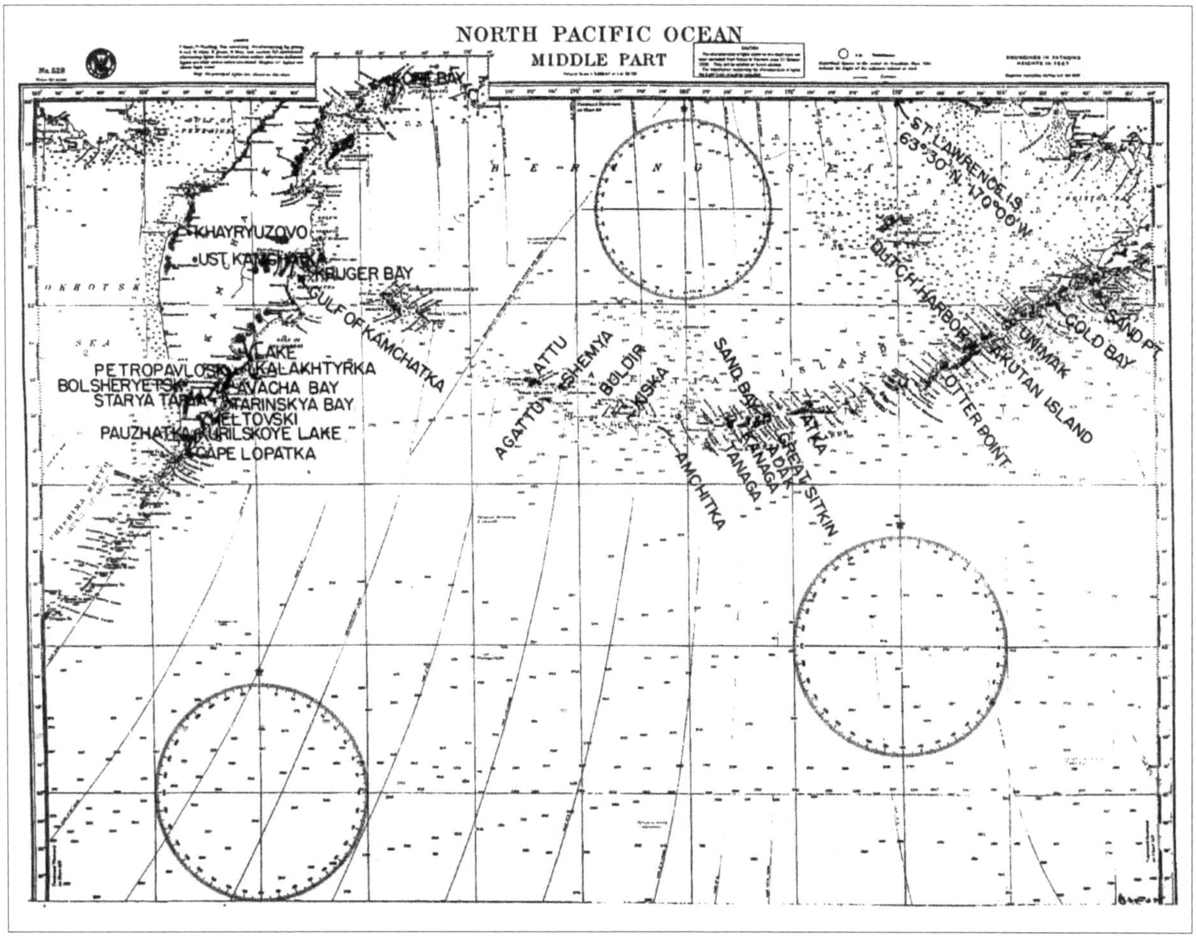

Map 7
North Pacific Ocean: Middle Part

The Blue base at Tanaga had one airstrip of five-thousand-foot length that could handle aircraft up to sixty thousand pounds in weight but had limited service facilities. It could take five Catalinas or five PB4Y-2 Privateer naval patrol bombers, but it only had limited harbor facilities, usable for small craft of less than twenty-foot draft. Sand Bay had unlimited seaplane tender facilities and four square miles of anchorage for all classes of ships but no land-plane area. There was also a Navy auxiliary air facility at Atka, a Navy fuel base and emergency seaplane area at Great Sitkin; a weather station and emergency seaplane area at Kanaga; and a ship anchorage, emergency landing field, and emergency seaplane area at Kiska. Attu had a naval air station and a naval submarine base. Its airstrips numbered two, both 6,500 feet in length, that could take airplanes weighing sixty-four thousand pounds. Attu had a seaplane area in Casco Bay, though seaplanes larger than PBM Mariners could not be accommodated. Attu could, however, handle 192 fighters and ninety-six naval patrol bombers under Normal Conditions, 480 fighters and 240 naval patrol bombers under Emergency Conditions. For the Chart Maneuver, the strength at the air base would be forty-eight fighters and seventy-two naval patrol bombers under Normal Conditions, 120 fighters and 180 naval patrol bombers in emergency. Its harbor facilities were also significant, with seven cruiser buoys, thirty-five destroyer anchorages, and twenty-one additional berths for battleships or cruisers along with aviation gasoline, fuel oil, and diesel oil available from shore tanks, which delivered the fuel by barge. (Pages 4–5.)

Shemya also had airstrips, three between 5,400 and ten thousand feet in length that could handle Superfortresses, though the shortest runway could do so only in emergencies and the second-shortest only for staging purposes. Shemya could handle between 108 and 150 fighters, seventy-two to a hundred naval patrol planes, and ninety-six to a hundred very-long-range aircraft, the numbers again depending on Normal or Emergency Conditions. During the maneuver, air strength at Shemya would be twenty-seven fighters, eighteen naval patrol bombers, and twenty-four very-long-range aircraft under Normal Conditions. Under Emergency Conditions, this air strength would be fifty-four fighters, thirty-six naval patrol bombers, and forty-eight very-long-range aircraft. Its harbor facilities could accommodate any ship that could operate in waters between eleven and fourteen fathoms deep. There was an emergency seaplane area at Aggatu and both an emergency seaplane area and weather station at Buldir. Purple's summation of Blue's facilities ended with LORAN (long-range radio aid to navigation) stations at Dutch Harbor, Adak, Amchitka, Attu, and Unimak. There were also direction-finding radio stations at Sitka, Kodiak, Dutch Harbor, Adak, Amchitka, and Attu, as well as high-power radio stations at Kodiak, Dutch Harbor, and Adak and radar stations at Sitka, Kodiak, Dutch Harbor, Adak, Amchitka, Attu, Shemya, Cold Bay, Tanaga, and Sand Bay. (Page 5.)

Purple had a "very good" naval and air base under construction at Petropavlovsk (today Petropavlovsk-Kamchatka), on Avacha Bay, but it was not yet ready to support operations. Until then, protection of the area rested on afloat forces. There was, however, at Vladivostok an "excellent" air and naval base that was capable of supporting major operations. Vladivostok was the base of the Purple 3rd Fleet, which was operating in the waters adjacent to the Kurile Islands. As of 1 June, the 3rd Fleet had been concentrated around Vladivostok for logistics and training exercises in anticipation of extended operations, and all of its ships were on forty-eight-hour readiness. When Admiral PA assumed command of the Purple Pacific Fleet, one of his missions was to maintain control of the sea approaches to Purple territory north of 44 degrees north latitude and west of 170 east longitude. This control would assist in the protection of Purple "Asiatic" territory and islands; it had been approved by the Chief of the Purple Naval Staff (not played by a student officer). To date, Admiral PA's immediate tasks were to patrol the waters adjacent to the Kuriles, Kamchatka, Sakhalin, and the Siberian coast against Blue submarine and air reconnaissance as well as Blue surprise attacks by air and sea. (Page 6.)

Admiral PA (Capt. Russell Smith, USN)

The Special Situation

As of 1 June, the Purple Special Situation amounted to the receipt by Admiral PA of a dispatch from the Chief of the Purple Naval Staff informing him that "fairly reliable" sources indicated a Blue expeditionary force was being organized in Puget Sound. This force included about twenty attack transports (APAs) and twenty attack cargo ships (AKAs), to be escorted by two prewar, modernized battleships of the *Maryland* class, two light cruisers, two or three escort carriers of the *Block Island* class, about twenty-five destroyers, and several minesweepers. This force was reported as making preparations to depart on "distant service" early in June; its destination, according to their crews (presumably by way of undercover agents ashore), was somewhere in the Aleutians. Some of the attack cargo ships might be tankers or service ships. Purple had arranged to get "timely" information on the departure of the force, and it was known that the Blue Commander-in-Chief had constituted a new task force from units of the Blue Fleet. This latter force probably comprised four to five heavy cruisers, three light cruisers, five or more fleet carriers, and twenty-five destroyers, all of the latest type. This force, Admiral PA was told, had not yet been formed and its units were still operating within task force organizations to which they had been previously assigned. However, the disposition of the Blue Fleet in the Central Pacific was such that this task force could concentrate and be ready for distant operations on very short notice if it took on fuel and supplies at Pearl Harbor before departing. Admiral PA's specific orders were now to prevent the establishment of a Blue advanced base west of 170 east and above 44 degrees north, as well as to continue his current operations until he was informed of the departure of the Blue forces or otherwise directed. (Pages 6–7.)

Purple's Pacific Fleet was significant. Admiral PA flew his flag in a heavy cruiser. Rear Admiral PB had command of two battle cruisers, organized as Cruiser Division 1. Rear Admiral PC commanded CRUDIV 2, three heavy cruisers that included Admiral PA's flagship. Another three heavy cruisers were organized as CRUDIV 3, with Rear Admiral PD in command; CRUDIV 4 consisted of four light cruisers under Rear Admiral PE (Rear Admiral PE, Commander, Cruiser Division [COMCRUDIV] 4 [played by the same student as Admiral PD]). There were also two light and two escort carriers under Rear Admiral PF, as Carrier Division 1. The light carriers each had two fighter squadrons and one attack-plane squadron, with one of the fighter squadrons consisting of eighteen Bearcat fighters and the other of twelve Hellcats. There were also six Hellcat night fighters assigned to each light carrier. Each light carrier's attack-plane squadron consisted of six Helldivers and six Avenger night-attack planes. The escort carriers each had one fighter squadron and one attack-plane squadron, the fighters consisting of eight Fireballs and eight Hellcat night fighters, while the attack-plane squadrons each had twelve Avenger bombers. The cruiser divisions also carried observation planes, four in CRUDIV 1, six each in CRUDIVs 2 and 3, and eight in CRUDIV 4. Carrier-based plane and crew replacements were available at any Kurile Island Air Base by ferry flights from Vladivostok within thirty-six hours after receipt of request by Commander, Far East Naval District in Vladivostok. Admiral PA was warned, however, that the total number of planes in service at any one time could not be increased. (Pages 8–9.)

Commodore PG was to command Destroyer Squadron 1, a light-cruiser flagship and fifteen destroyers organized into three divisions of five ships each. Commodore PH commanded Destroyer Squadron 2, with an identical composition and organization. Commodore PI led Task Force 39, the Service Force, which consisted of Commodore PL's Service Squadron 1, six oilers evenly divided as Service Divisions 2 and 3. There was also Mine Squadron 1, commanded by Commodore PR, consisting of one light minelayer (DM) as the flagship, four destroyer-minesweepers (as Mine Division 1), and six minesweepers (Mine Divisions 2 and 3). Concerning the Purple surface and air forces, student officers were given comparisons of Purple ship types to Blue's—for instance, Purple battle cruisers were similar to the Blue *Alaska*-class battle cruisers, while Purple's heavy cruisers were similar to Blue's *Baltimore*-class units. Purple light cruisers were comparable to the Blue *Cleveland* class, and Purple light carriers were similar in capabilities to the Blue *Saipan* class. In the same way, Purple's escort carriers compared to Blue's *Commencement Bay* class, and its destroyers were like Blue's *Gearing* class. Purple oilers were seen as similar to Blue's *Cimarron* class, its submarine tenders (ASs) to Blue's *Fulton* class, its light minelayers to Blue's *Robert H. Smith* class, its destroyer-minesweepers to Blue's *Ellyson* class, and its minesweepers to Blue's *Heed* class. (Page 9.)

The Purple Submarine Force was commanded by Rear Admiral PJ. At his disposal were two submarine squadrons, each of which had four submarines organized into two divisions per squadron; there was also a submarine tender for each squadron flagship. As of 1 June, Admiral PJ was at Fleet Headquarters in Vladivostok, as was the flagship of Submarine Squadron 1. Submarine Squadron 2's flagship was at Kashiwabara Bay, while the boats of Submarine Division 1 were in overhaul at Vladivostok and were not scheduled to depart to patrol in the Aleutians until 10 June. SUBDIV 2 was already on patrol in the Aleutians but was due for three weeks of maintenance in Vladivostok when relieved by SUBDIV 1. SUBDIV 3 was training in the vicinity of Kashiwabara Bay and was scheduled to relieve SUBDIV 1 on 10 July; SUBDIV 4, which was also training in the vicinity of Kashiwabara Bay and had completed overhauls on 31 May, would relieve SUBDIV 3 on 10 August. All of these submarines were capable of speeds of six, nine, twenty,

Rear Admiral PB (Capt. Harry Hummer, USN)

Rear Admiral PC (Col. Anthony Grossetta, USAAF)

Rear Admiral PD (Col. Allen Koonce, USMC)

Rear Admiral PF (Col. George Privett, USA)

Rear Admiral PJ (Cdr. Jack Roudebush, USN)

Commodore PG (Capt. William Hoffheins, USN)

Commodore PH (Lt. Col. John Lane, USA)

Commodore PI (Capt. Emmett Sullivan, USN)

Commodore PL (Col. Von Shores, USAAF)

Commodore PR (Capt. Charles Heberton, USN)

and twenty-five knots on diesel engines with *schnorchel*s and were able to operate submerged on batteries or hydrogen-peroxide engines, and on the surface with diesel engines. Endurance using these four types of propulsion was two, one, six, and two hundred hours, respectively, and cruising radius was twelve, ninety, ninety, and thirty thousand miles, respectively. These boats had six torpedo tubes in the bow, carried eighteen torpedoes, and could substitute two mines for each torpedo. Use of the *schnorchel* obliged them to have to operate at periscope depth. The submarines undergoing maintenance or training were on forty-eight hours' notice for operations. Therefore, any plan made by Admiral PA had to consider this time lag between issuance of his orders and the actual beginning of the operation. (Page 10.)

Rear Admiral PK was Commander, North Pacific Air Command, with Headquarters, Fleet Air at Kashiwabara. At Vladivostok, Commodore PL led Fleet Air Wing 10, three seaplane tenders with six naval patrol plane squadrons, four of the squadrons each having nine Mariners and the other two squadrons having six Catalinas. At Matsuwa, Commodore PM led Fleet Air Wing 20, four naval patrol plane squadrons and two fighter squadrons. The patrol squadrons were equipped with two squadrons of fifteen Privateers each, a squadron of twelve Harpoon naval patrol bombers, and a squadron of six Catalinas. The fighters consisted of thirty Hellcats, twenty-four of them day fighters and six night fighters. At Paramushiru, Commodore PN led Fleet Air Wing 30. This unit had two squadrons of eighteen Privateers, one squadron of eighteen Harpoons, and a squadron of six Catalinas. It also had two squadrons of thirty-six Hellcat day fighters, one squadron of twenty-four Hellcat day fighters, and two squadrons of Hellcat night fighters, one with twenty-four planes and the other with thirty-six. (Page 11.)

The North Pacific Air Command consisted of three additional air wings. Fleet Air Wing 40 was based at Shimushu and was commanded by Commodore PO. It consisted of fifty-four Harpoons in three squadrons of eighteen each, one squadron of six Catalinas, a squadron of thirty-six Hellcat day fighters, and two squadrons of Hellcat night fighters, with twenty-four planes in each squadron. Fleet Air Wing 50 was commanded by Commodore PP and was based at Shikuka. It consisted of two squadrons of Privateers with eighteen planes between them, one squadron of six Catalinas, one squadron of twenty-four Hellcat day fighters, and one squadron of twelve Hellcat night fighters. The final air wing, Fleet Air Wing 60, based in Etorofu, was commanded by Commodore PQ and had a squadron of twelve Privateers, one of six Catalinas, one of twelve Hellcat day fighters, and one of six Hellcat night fighters. These aircraft were at the above locations as of 1 June and could be redeployed on that date as desired. All of these fields were equipped with radar and communications and were "well-defended" by antiaircraft batteries. In addition, the airfields were capable of servicing planes up to their operating capacities, except in landing-only and emergency-landing areas; shore-based air facilities at Avacha Bay, however, were not going to be ready for operations on 1 June. Land-based planes—like the carrier-based planes—and their crews could be replaced by ferry flights from Vladivostok within thirty-six hours after the receipt of request by the Commander, Far East Naval District, but the total number of planes in service at any one time could not be increased. (Pages 11–12.)

Purple air and naval bases were extensive, in addition to the construction still going on at Vladivostok. Purple had in the Kuriles and Sakhalin naval bases and stations that were well equipped and stocked with fuel and supplies for any ship that could use their harbors. Kamchatka's bases and stations were now less well equipped and supplied to support ships but they would be fully operational after

Rear Admiral PK (Capt. Eugene May, USN)

Commodore PM (Capt. James Hardin, USN)

Commodore PN (Cdr. John Munholland, USN)

Commodore PO (Cdr. Terrell Nisewaner, USN)

Commodore PP (Col. Robert Erlenkotter, USA)

Commodore PQ (Cdr. Richard Andrews, USN)

1 August, as would their airfields. The Kuriles and Sakhalin were well defended by coastal and antiaircraft batteries and had radars and direction finders. In the Kuriles, on Shimushu Island were five airfields, including an emergency dry-weather landing field at Bettobu and a seaplane base at the same field; Shimushu could service fifteen naval patrol bombers, but only in the summer. Imai Saki also had an emergency dry-weather landing field, Kataoka had two runways that could handle forty-five fighters and sixty naval patrol bombers, and Miyoshino had two runways at an Army-controlled field that could service forty fighters and thirty naval patrol bombers. Paramushiru Island had four airfields. Kashiwabara had a runway and facilities that could handle sixty fighters and forty naval patrol planes, while thirty fighters and twenty naval patrol planes could be serviced at Suribachi. Suribachi Bay also had a seaplane base for tender-based planes but no shore facilities; Kurabu had two runways that could accommodate forty-five fighters and sixty naval patrol planes. A final airfield on Paramushiru, not mentioned above, could handle forty fighters and thirty naval patrol planes. (Pages 12–13.)

On Matsuwa Island were two runways; the island served as the primary air defense station for the central Kuriles and as the shuttle point for aircraft from the south and west. It had a capacity of thirty fighters and forty naval patrol bombers. Etorofu Island also had two airfields, Tennei (with two airstrips that could handle thirty-six fighters and twenty-four naval patrol bombers) and Rubetsu (an emergency-only landing field). Kunashiri and Shikotan Islands were also emergency landing fields, but on Kamchatka there were extensive air facilities. Kalakhtyrka Lake had a seaplane base, hangars, aprons, and ramps for eighteen flying boats. Petropavlovsk had a seaplane base, for eighteen flying boats, as did Tarinski Bay. That field also had a land-plane base, three airstrips that could take sixty naval patrol bombers. Another field, Bolsheretsk, could take forty-eight fighters, as could Khayryuzovo and Staraya Tarya. The air base at Cape Lopatka could handle thirty-six fighters, as could the bases at Pauzhatka and Zheltovski. The base at Kurilskoye Lake could take forty-eight naval patrol planes. There were incomplete facilities at Ust Kamchatka and Korf Bay, each of which would be able to take eighteen flying boats when finished. (Pages 13–15.)

Sakhalin Island's air facilities were also extensive, at least in number. The base at Alexandrovsk had an emergency landing field, a seaplane base for nine small flying boats, and a capacity of forty-eight fighters. In addition, the air base at Baydukov Island had a seaplane base—this one for six small flying boats—as well as

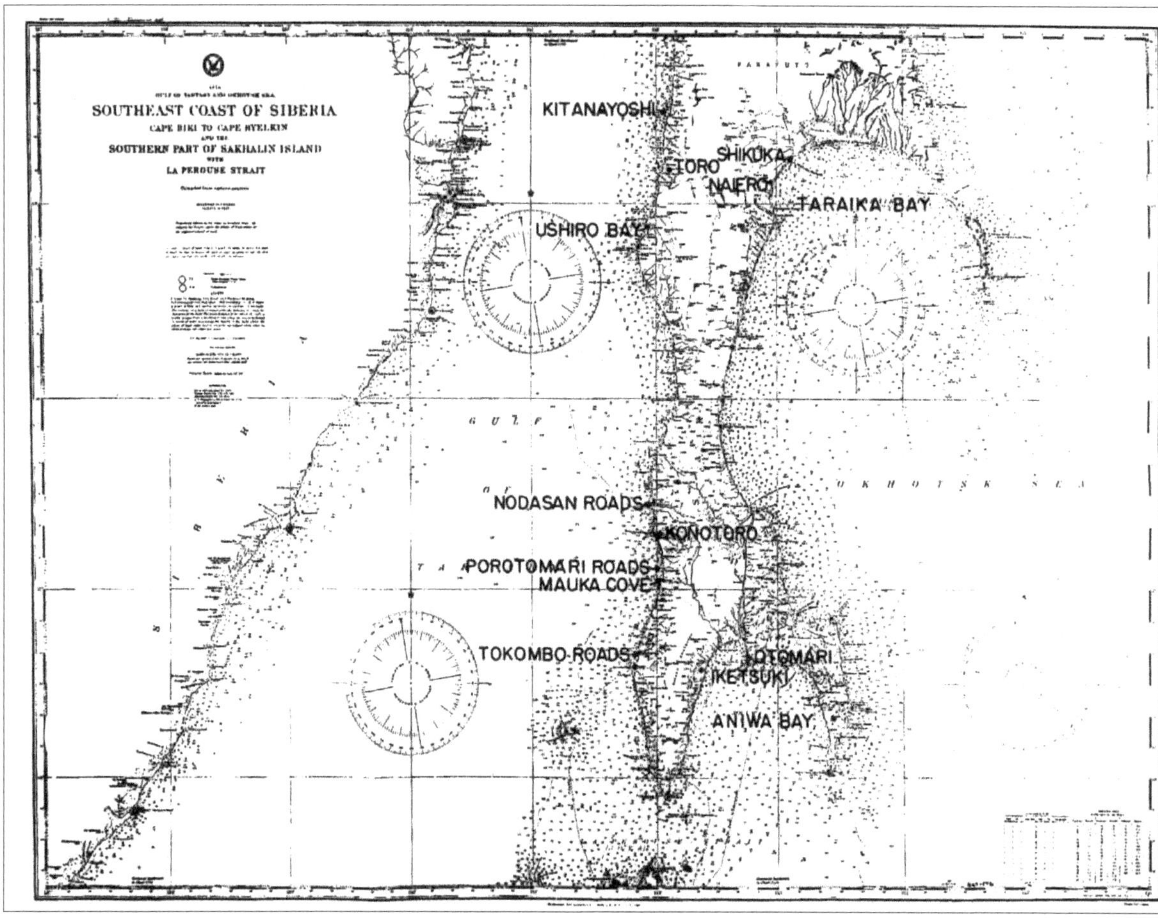

Map 8
Southeast Coast of Siberia, southern portion

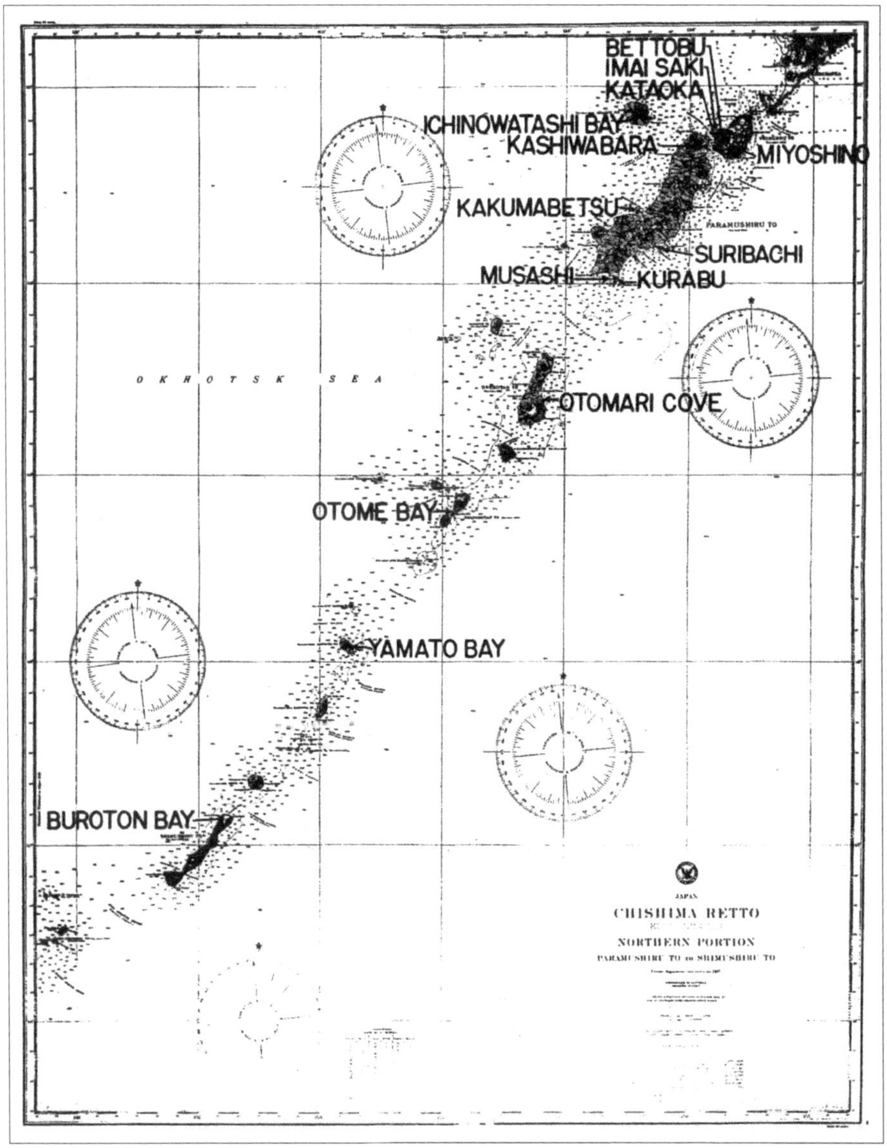

Map 9
Chishima Retto: Northern Portion

an emergency landing field. At Chaivo Bay, the seaplane base could take fifteen flying boats, while the large facility at Moskalvo had a seaplane base that could take twenty-five flying boats and a large airfield that could accommodate sixty naval patrol bombers. The base at Nogliki was merely an emergency landing base, but the two fields at Okha could handle sixty-six fighters between them. Okha also had a seaplane base that could service twenty flying boats; Pilevo and Pogranichnoye were both emergency landing fields. Viakhtu could accommodate eighteen fighters and twenty flying boats. There were also air bases at Iketsuki, Konotoro, Nairo, Otomari, Shikuka, and Toro with very modest facilities. (Pages 15–16.)

There were extensive anchorages in the Kuriles as well. Though Kataoka Bay had a limited capacity, Kashiwabara Bay had a pipeline leading to storage tanks ashore, extensive warehouses, minor repair facilities, and nine piers for ships of lengths from a hundred to five hundred feet. There was also an eight-hundred-yard berth and two five-hundred-yard berths; the whole complex, along with Kataoka, was considered a major fleet anchorage. Kakumabetsu Bay had warehouses and boat-repair facilities ashore, was well protected, and had a five-hundred-yard berth and two of eight hundred yards. Musashi Bay had nine thousand-yard berths, two eight-hundred-yard ones, and another nine berths that were five hundred yards long. Suribachi Bay had two eight-hundred-yard berths and five of five hundred yards, as well as four large piers and numerous warehouses and underground storage facilities. Araido Island had a five-hundred-yard berth, a small-boat anchorage, and boat-repair facilities. Onnekotan Island had a small-vessel anchorage at Otomari Cove, but Otome Bay on Shasukotan Island had berths of a thousand yards long, two eight hundred yards, and two five hundred yards. Additionally, Matsuwa Island had a thousand-yard berth and two eight-hundred-yard berths, along with three of five hundred yards, oil storage and warehouses, and two piers, all at Yamato Bay. Shimushiru Island's Buroton Bay could take only small vessels up to a thousand tons, but Tsurigane Bay on Uruppu Island had nine thousand-yard, three eight-hundred-yard, and seven five-hundred-yard berths, constituting a destroyer, submarine, and seaplane base, respectively. Futagoshima Bay and Tokotan Bay both had good anchorages but were exposed to weather. (Pages 17–18.)

Etorofu Island's Bira Sentai Bay had an unlimited anchorage but was open to the north. On the same island, Hitokappu Bay—where the Japanese First Air Fleet had

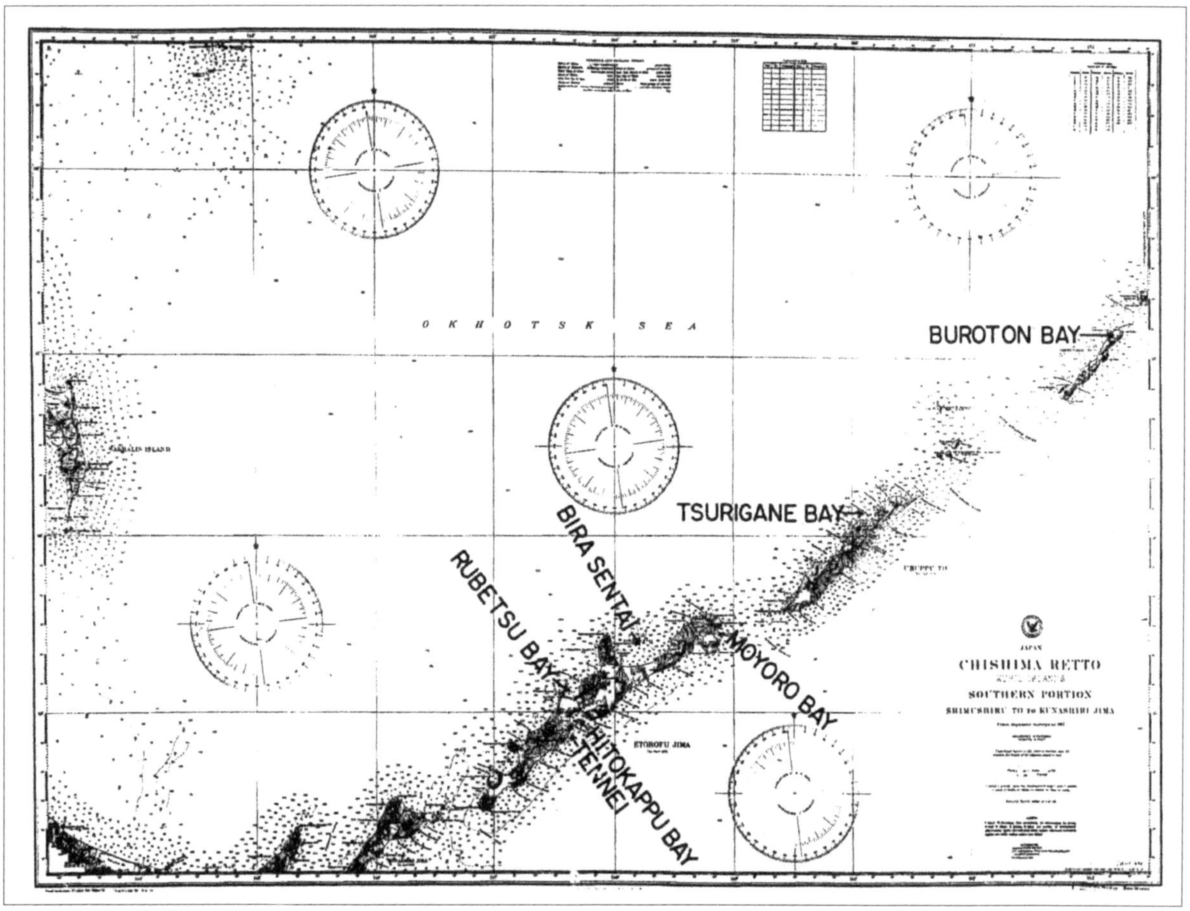

Map 10
Chishima Retto: Southern Portion

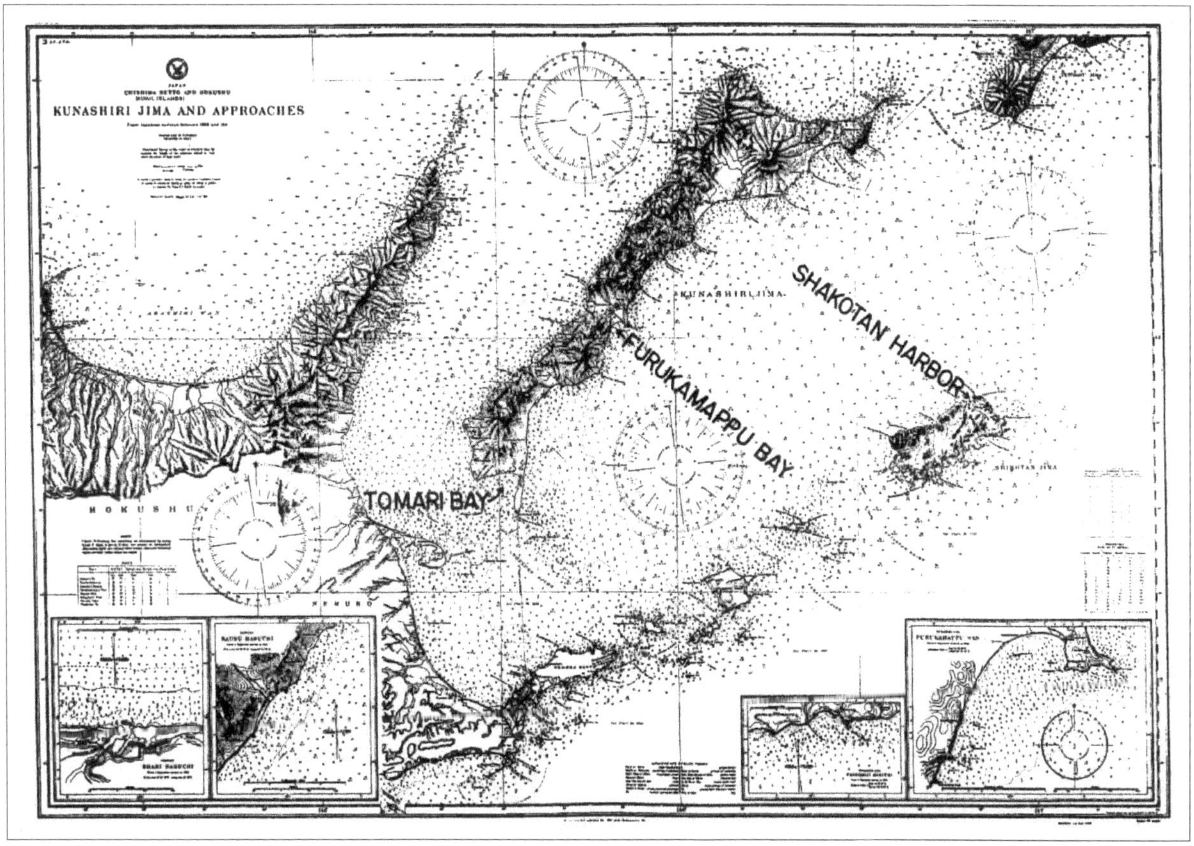

Map 11
Kunashiri Jima and Approaches

deployed for its attack on Pearl Harbor in 1941—was a major fleet anchorage for battleships, carriers, cruisers, destroyers, and all classes of auxiliaries, with twelve thousand-yard and three eight-hundred-yard berths. Also on Etorofu, Moyoro Bay had nine thousand-yard berths, and Rubetsu Bay had forty-three thousand-yard, five eight-hundred-yard, and eighteen five-hundred-yard berths. Kunashiri Island's Tomari Bay and Furukamappu Bay could accommodate only small craft, but Shikotan Island's sheltered Shakotan Harbor had a 450-yard berth in the inner harbor, as well as a thousand-yard berth and an eight-hundred-yard berth in the outer harbor. Matsugahama Bay had a small but good harbor that boasted a submarine base.

On Kamchatka, Avacha Bay had a large harbor for all classes of ships. Korf Bay had a modest but good harbor and was "well sheltered" for seaplanes and submarines operating from tenders. Ust Kamchatka had a small harbor, but it was unprotected; Kruger Bay had three five-hundred and six four-hundred-yard berths for ships. Sakhalin, Alexandrovsk, Moskalvo, Viakhtu, and Rogati Roads all had open roadsteads. Mosiya Bay was a partially sheltered anchorage with limited space, while Kazakevicha Bay and Agnevo Roads were both open roadsteads. Pilevo Bay, partially sheltered, was for small craft; Kitanayoshi Anchorage, Ushiro Bay, Nodasan Roads, Porotomari Roads, and Tokombo Roads were open roadsteads. There were a small, partially sheltered harbor for small craft at Mauka Cove, an unlimited anchorage at Otomari (which was part of Aniwa Bay, on the southern end of Sakhalin), and unlimited anchorages in the northern part of Taraika Bay. Sakhalin further offered unprotected anchorages on its east coast at Luniski Roads, Nabiliski Roads, Niski Roads, Chaivo and Kyakrvo Anchorages, Urkt Roads, and Kuegda Roads. (Pages 18–21.)

Admiral PA was supplied with the characteristics of Blue atomic bombs. These, he was told, could be handled like any other ammunition but carried only by very-long-range heavy bombers like the Superfortress. It was assumed that a Blue atomic bomb would cause 100 percent damage out to five hundred yards, thereafter less damage in increments down to 10 percent at a radius of 4,500 to five thousand yards, and no damage outside five thousand.

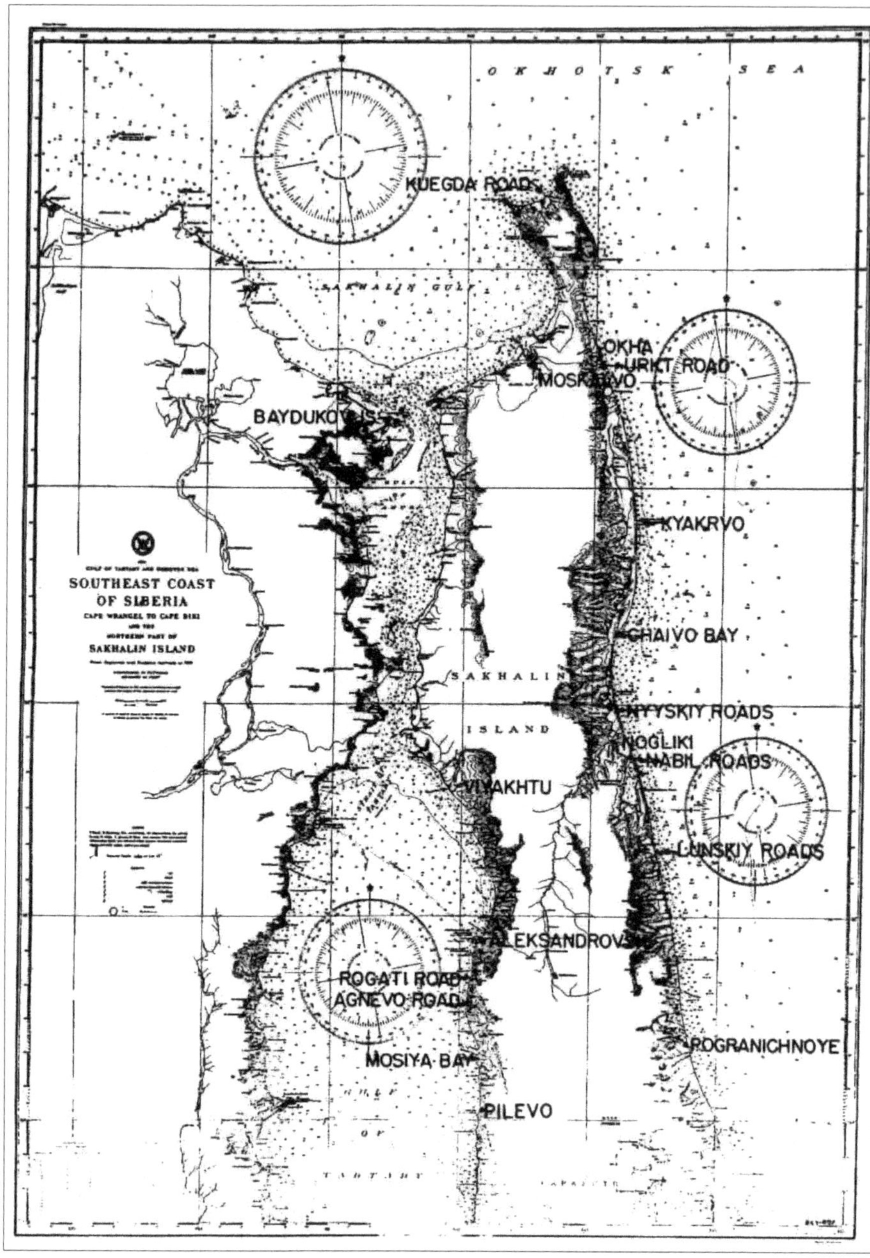

Map 12
Southeast Coast of Siberia, northern portion

Student officers completed Admiral PA's Estimate of the Situation and Completion of the Plan. They also had to prepare—in typescript—Admiral PA's directive as well as submit, in pencil, various annexes to the Operation Plan. These included a Movement Plan, including routes and rendezvous; an Intelligence Plan; a Search Plan, both air and surface; a Communication Plan; a Logistics Plan for fuel oil

LOCAL CIVIL TIME OF SUNRISE, SUNSET, MOONRISE and MOONSET in the Theatre.
(for use in this Problem only)

Date	SUN Lat.50°N Rises	SUN Lat.50°N Sets	MOON Lat.45 Rises	MOON Lat.45 Sets	MOON Lat.50 Rises	MOON Lat.50 Sets	MOON Lat.55 Rises	MOON Lat.55 Sets
June 10	0351	2007	1107	0007	1059	0016	1050	0027
13	0350	2009	1416	0129	1419	0127	1423	0125
16	0350	2011	1745	0304	1759	0252	1818	0234
19	0350	2012	2105	0535	2120	0517	2145	0454
22	0350	2013	2319	0908	2328	0857	2338	0843
25	0351	2013	0022	1241	0023	1242	0024	1242
28	0352	2013	0149	1556	0140	1608	0128	1622
July 1	0354	2013	0537	1851	0320	1909	0259	1932
4	0356	2012	0617	2105	0604	2120	0526	2140
7	0358	2010	0857	2238	0840	2245	0837	2254
10	0401	2009	1201	2358	1202	2355	1203	2351
13	0404	2006	1522	0059	1534	0049	1550	0036
16	0407	2004	1846	0316	1905	0257	1928	0234
19	0410	2001	2116	0646	2126	0632	2139	0616
22	0414	1958	2254	1028	2253	1027	2252	1026
25	0418	1954	----	1349	----	1359	----	1412
28	0422	1950	0137	1647	0121	1704	0100	1726
31	0426	1946	0359	1905	0342	1921	0320	1942

In Lat. 45° sunrise is approximately 20 minutes later than in Lat. 50° and sunset is approximately 20 minutes earlier than in Lat. 50°.

In Lat. 55° sunrise is approximately 30 minutes earlier than in Lat. 50° and sunset is approximately 30 minutes later than in Lat. 50°.

Twilight lasts all night in Lat. 50° from 2 June to 12 July.

Fig. 102
Local Civil Time of Sunrise, Sunset, Moonrise, and Moonset in the Theater

and aviation gasoline; and a Mine Plan. Student officers also provided typed copies of directives covering the preliminary deployment of forces. (Pages 21–24.) The students drew the Statement of the Problem on 17 September, completed the Estimate of the Situation and the Decision by Friday, 20 September, and completed the details of the Operation Plan by Tuesday the 24th.

The rough draft of the Operation Plan was completed the next day, and the day after that a "smooth" copy was delivered to Room 206. Students then drew copies of the Operation Plans and were detailed as Subordinate Commanders for the Chart Maneuver. On 27 September, the Estimate of the Situation was conducted from the perspective of the Subordinate Commander and necessary plans would be prepared from the Estimate. By 30 September, a smooth copy of the Subordinate Commander's Operation Plan was delivered to Room 206. The first day of October was used for Chart Maneuver preparations, while the maneuver itself was conducted from 2 October until the morning of 8 October. The afternoon of the 8th, as well as 9 October, was for the Critiques of the Plans and the Chart Maneuver. (Pages 24–25.)

The Purple Staff Solution

In early October, the Purple Staff Solution to Operations Problem 2 was issued. In the Estimate of the Situation, Admiral PA cited his "incentive to action" as a dispatch from the Chief of the Purple Naval Staff, received on 1 June, that ordered him to prevent the establishment of a Blue advanced base in Kamchatka as well as maintain control of the sea approaches to Purple's territory and protect its Asiatic territory and islands. He was also to continue his current operations until informed of the departure of Blue forces. He was not to concern himself with either Purple or Blue operations in Japan, the Ryukyus, or the Asian continent. He reviewed the situation: Blue naval forces were being strengthened in the Pacific, the Blue Army was being brought up to strength in preparation for war, Blue bases were being activated in the Aleutians, Blue land-based planes were patrolling over the Kuriles, and there was a probability of Blue submarines patrolling the waters he was ordered to control. He additionally reviewed the concentration of the Blue Expeditionary Force at Puget Sound, the Blue Covering Force at Pearl Harbor, the probable strength of each, and his anticipation of "timely" information on the embarkation dates for these forces. Finally, he noted that the Purple Pacific Fleet at Vladivostok had just completed a long training cycle—with the exception of one submarine division, on patrol in the Aleutians—and could be ready for operations in forty-eight hours.[3]

Admiral PA then sought to deduce the Blue mission in terms of effect on his own potential Courses of Action. Admiral PA surmised from the Blue preparations that the enemy was preparing for a major amphibious landing and that Purple was the obvious target but that the remoteness of Blue's primary bases from the Purple bases that it was planning to hit was a disadvantage. Therefore, the establishment of a Blue advanced base west of the Aleutians was the probable objective of the Blue Expeditionary Force at the initiation of hostilities; PA's orders were to prevent this extension of Blue control in the western Pacific. Admiral PA next assumed that the Blue Expeditionary Force would depart sometime in early June in conjunction with the Blue Covering Force in the Central Pacific. He also assumed that "Blue will seek early advantage through the employment of atomic bombs." (Pages 2–4.)

In addition to Blue's Expeditionary Force and Covering Force, he knew, Blue had submarines operating in his area, in unknown numbers; a chain of both strong and minor naval and air bases in the Aleutians; land-based air assets that also included Superfortress heavy bombers, in unknown numbers; and an unknown number of atomic bombs. Reviewing his own forces and bases in the Kuriles–Sakhalin–Vladivostok area, he noted that his bases on Kamchatka—particularly the one at Petropavlovsk—would not be fully operational or protected until 1 August. In a detailed comparison of Purple and Blue Forces, he had 324 fighters (not shipborne), compared with his estimate of Blue's 750; 258 naval patrol bombers, compared with five hundred; and no very-long-range or transport aircraft, compared with Blue's 250 and one hundred, respectively. These numbers represented what Admiral PA thought would be maximums for Blue throughout its Aleutian bases. Though Blue's land-based planes were formidable and superior to his North Pacific Air Forces, PA thought, most were for defense, patrol, and search. Moreover, located so far from Purple bases as they were, he did not consider them decisive in attacks on his bases, except for the very-long-range aircraft. He was concerned, however, that the Blue planes would seriously impede the ability of his own planes to strike Blue's bases, especially given the large number of Blue night fighters, which could operate in poor visibility. (Pages 4–5.)

He was also concerned that Blue very-long-range aircraft could strike his bases, probably in flights of up to two hundred against the Kuriles–Kamchatka area and up to fifty against the Sakhalin area. He was especially concerned that these Blue planes could carry atomic bombs: "They offer an extremely serious threat to my forces in that their logical targets are the bases I depend upon for my logistic support." In spite of his fighter and ground defenses, Admiral PA did not expect to achieve "measurable" success against Blue's raids on the Purple air bases. At the very least, he knew, he had to keep his forces dispersed, both among airfields and at each base. He also knew that since he could not expect the total number of aircraft to increase, any operation had to be planned with the present numbers. Still, he thought that he could locate Blue's surface forces somewhere between eight hundred and a thousand miles from his bases and that Blue—once within the combat radius of his fighters—would suffer "heavily." Accordingly, he intended to subject Blue to fighter attacks at the same time that he was undertaking surface action against the Blue Covering and Expeditionary Forces. (Pages 5–6.)

As to fleet-borne air forces, Admiral PA noted Purple's seventy-six fighters to Blue's 175, twenty-eight night fighters to an estimated one hundred, thirty-six attack planes to three hundred, twelve night-attack planes to an estimated 130, and twenty-eight observation planes to thirty-nine. Admiral PA was assuming that Blue had six fleet carriers and three escort carriers. Given Blue superiority in, especially, night fighters, Admiral PA further assumed he had to apply attrition tactics as soon as Blue came within range of his shore-based fighters. With his fleet and land-based planes combined, however, he had 230 fighters compared with Blue's 175, 148 night fighters to Blue's estimated hundred, thirty-six attack planes to Blue's three hundred, two hundred night-attack planes to Blue's estimated 130, and twenty-eight observation planes to Blue's thirty-nine. His two hundred night-attack planes included 188 naval patrol bombers that had radar and could be loaded for attack missions. He took into account that not all of his fighters would be employable, since they were land based and might not be able to get at the Blue naval forces. He further acknowledged that his observation planes—on his battleships and cruisers—were not valuable except as spotters. He did have the advantage of secure bases and superiority in numbers, if facing Blue's only fleet-based air forces. Blue, however, had the advantages of greater numbers of attack planes as well as high mobility and degree of concentration. Nevertheless, Blue's fleet carriers were highly vulnerable to Admiral PA, and their destruction was "clearly" "vitally" important. (Pages 6–7.)

From the tabulation (figure 103) of Blue and Purple surface combatant characteristics, it was evident to Admiral PA that Blue held decided advantage in carrier strength and that this strength had to be reduced as early as possible. PA thought that his submarines might be the solution here. In addition, while Purple had two battle cruisers, it had no battleships, while Blue had no battle cruisers but had two "battlewagons." In cruiser strength, Purple had superiority in numbers of ships and in gun power, except for automatic weapons, where he thought the two sides were evenly matched. Destroyer strength also favored Blue, in a five-to-three ratio, but

6277-5777-PSS
10/Sep/46rg

- 9 -

CONFIDENTIAL

SURFACE COMBATANT SHIPS

TYPE	SPECIAL DESIGNATION (BLUE ONLY)	NUMBER PURPLE	NUMBER BLUE	LIFE PURPLE	LIFE BLUE	Max. Speed PURPLE	Max. Speed BLUE	Radius Econ. Speed PURPLE	Radius Econ. Speed BLUE	Total No/Cal. Guns PURPLE	Total No/Cal. Guns BLUE	Range Thous. Yards PURPLE	Range Thous. Yards BLUE	Total Automatic Weap. A.A. Battery PURPLE	Total Automatic Weap. A.A. Battery BLUE	Total Torpedoes PURPLE	Total Torpedoes BLUE	Total Planes # PURPLE	Total Planes # BLUE	Total [illegible] or [illegible] PURPLE	Total [illegible] or [illegible] BLUE
CV	C	--	5 **	--	10.5	--	33	--	17670	--	72:5"/38	--	16	-- --	432:40mm 420:20mm	--	324: E-1	-- --	222 VF 390 VA	--	[illegible] D.C.
CVL		2	--	5.5	--	33	--	11000 Est.	--	--	--	--	--	80:40mm 64:20mm	-- --	36 MK 13	--	72 VF 24 VA	-- --	144 DC	--
CVE	E	2	3 *	4.6	4.6	18.5	18.5	30000	30000	4:5"/38	6:5"/38	16	16	72:40mm 40:20mm	108:40mm 60:20mm	-- 24: E-1	-- 36: E-1	32 VF 24 VA	64 VF 36 VA	192 DC	[illegible] DC
BB	E	--	2	--	18.5	--	19.5	--	15000	--	16:16"/45 32:5"/38	-- --	33 16	--	80:40mm 88:20mm	--	--	--	4 VO	--	12 DC
CB		2	--	10.7	--	33	--	11500	--	18:12"/50 24:5"/38	-- --	35 16	-- --	112:40mm 68:20mm	-- --	--	--	4 VO	--	48 DC	--
CA	C	6	5 *	5.5	5.5	32	33	9400	9000	54:8"/55 72:5"/38	45:8"/55 60:5"/38	28 16	30 16	288:40mm 132:20mm	240:40mm 100:20mm	-- --	-- --	12 VO	20 VO	144 DC	120 DC
CL	E	6	2	4.6	3.3	32.5	32.5	10500	8370	72:6"/47 72:5"/38	24:5"/38 --	24 16	16 --	168:40mm 60:20mm	48:40mm 32:20mm	72 EST.	-- --	12 VO	6 VO	72 DC	24 DC
CL (New)	C	--	3	--	4.6	--	32.0	--	9000 EST.	--	36:6"/47 D.P.	--	24	--	144:40mm 60:20mm	--	--	--	9 VO	--	36 DC
DD	25 E 25 C	30	50	1.6	1.6	34	34	7800	7800	180:5"/38	300:5"/38	16	16	360:40mm 600:20mm	600:40mm 100:20mm	150 G-1	250 G-1	--	--	1980 DC	3330 DC

TOTAL LIFE: 150.2 242.3

"E" signifies unit of BLUE Expeditionary Force.
"C" signifies unit of BLUE Striking Force.

* may be one less than is indicated in this table.
** may be one less or unknown number more than is indicated in this table.

EST.= Estimated.

\# Total plane figures this table may differ slightly from preceding aircraft comparison. Latter is rounded to even figures for simplicity. In this table, VF and VF(N) are lumped under VF, as are VA and VA(N) under VA. See preceding aircraft comparisons for breakdown.

Fig. 103
Blue and Purple surface combatant characteristics, Operations Problem 2

Blue also had to project and protect an expeditionary force. Blue additionally had a six-to-one superiority in aerial torpedoes, as well as what PA thought was a slight edge in surface-combatant torpedo strength. (Pages 7, 10.)

As for submarines, Admiral PA found it difficult to compare the two sides' characteristics. He had sixteen modern, high-speed submarines and two tenders. Lacking better information, he assumed that Blue had the same, as well as up to two hundred World War II–vintage submarines. He therefore had to take every possible precaution against submarine attack. He knew that Blue transport and cargo ships, like his own, had heavy antiaircraft firepower and that aircraft attacking them had to expect opposition. The presence of these ships in the Blue Expeditionary Force, however, should restrict the speed of the Expeditionary Force to about twelve knots, and Blue had to protect them from submarine, air, and surface attack. He estimated that the Blue Expeditionary Force could embark about thirty-eight thousand troops, either at Puget Sound, in the Aleutians, or at a Central Pacific base. From a survey of Purple bases and territory, he surmised that these troops would be superior in number and equipment to the forces at any one of the bases that he thought might be an immediate Blue objective. "However, this superiority cannot be asserted unless they are landed—it is my job to prevent their landing and thus eliminate any question of superiority in troops." (Pages 10–11.)

Operating Condition	Max. Sustained Speed (kts.)	Endurance (Hrs. at Max. Sust. Speed)	Cruising Radius
Surface-Diesel Engines	25	200	30,000(10 kts.)
Submerged-Hydrogen-peroxide Engines	20	6	--
Submerged-Batteries	9	1	90(2 kts.)
Submerged-Diesel-Schnorchel	6	2000	12,000(6 kts.)

Torpedo Tubes: 6 Bow, None Aft.
Torpedoes : 18 Forward
Mines : 2 Mines can be substituted for each torpedo

Fig. 104
Blue and Purple submarine characteristics, Operations Problem 2

Admiral PA looked at the influence of relative distances on combat power and operations. Purple forces were located with the Pacific Fleet—less its submarines—at or near Vladivostok, while the Purple Pacific Submarine Force had four boats in overhaul at Vladivostok, along with one submarine tender. Another submarine tender was at Kashiwabara Bay, as were the eight boats training, and four boats were on Aleutian patrol. The North Pacific Air Forces had four wings of planes in the Kuriles, one at Vladivostok, and one at Shikuka. Blue forces included the Expeditionary Force in the Puget Sound area and the Striking Force at or near Pearl Harbor. He did not know the location of the Blue Submarine Force but assumed that Blue boats were on patrol in the Kuriles–Kamchatka area. Blue, he knew, also had very-long-range planes, up to fifty based at Shemya and Adak each, up to 150 at Amchitka, and up to ten at Kodiak. Atomic bombs were "probably" stored at some or all of these bases. From the considerable distances by which the opposing forces were separated, he deduced that Blue objectives for bases had to be in the Kuriles–Sakhalin–Kamchatka area, including potentially the Komandorskis.

One of Blue's major problems, however, PA asserted, was that of distance itself—its Expeditionary Force had to travel between 1,700 and two thousand miles before even reaching an Aleutian base. Once there, it would still be 1,200 to 1,400 miles from its probable base objectives, while the Blue Striking Force at Pearl Harbor was some three thousand miles from them. In contrast, his forces, though scattered, were within easy concentration range, with the surface forces at Vladivostok and the majority of his submarines and air wings in or near the Kuriles. All of this indicated to him the value of early detection of Blue's movements as well as the prolonged period he would have in which to conduct repeated attrition attacks by aircraft and submarine. These alone, he hoped, could

6277-5777-BSS
10/Sej/45rf

- 12 -

CONFIDENTIAL

RELATIVE DISTANCES OPERATIONS PROBLEM TWO

	PUGET SOUND	PEARL	KODIAK	COLD BAY	DUTCH HARBOR	ADAK	ATTU	KORP BAY KAMCHATKA	UST KAMCHATKA	AVACHA BAY	KASHIWABARA BAY	MATSUWA	ETOROFU	BUROTON BAY	ALEXANDROVSK	SHIKUKA	OTOMARI	VLADIVOSTOK	AMCHITKA	SHEMYA	KOMANDORSI ISLANDS
PUGET SOUND	--	2350	1250	1500	1700	2000	2400	2800	2800	3000	3000	3100	3300	3150	--	--	--	--	--	--	--
PEARL	2350	--	2200	2000	2000	2030	2350	3000	2800	2760	2800	2800	3000	2800	--	--	--	--	--	--	--
KODIAK	1250	2200	--	450	570	1050	1300	1750	1850	--	2000	--	--	--	--	--	--	--	--	--	--
COLD BAY	1500	2000	450	--	150	570	1050	--	--	--	--	--	--	--	--	--	--	2800 (AIR)	--	--	--
DUTCH HARBOR	1700	2000	570	150	--	470	750	1050	1100	1300	1400	--	1800	--	--	--	--	--	600	--	1050 (AIR)
ADAK	2000	2030	1050	570	470	--	400	800	800	920	1000	--	--	--	--	--	--	2300 (AIR)	220	--	650
ATTU	2400	2350	1300	1050	750	400	--	500	450	530	650	800	1200	900	1200 (AIR)	1150	1200 (AIR)	1850 (AIR)	250	--	300
KORP BAY KAMCHATKA	2800	3000	1750	--	1050	800	500	--	350	600	780	--	--	--	--	--	--	--	--	--	300
UST KAMCHATKA	2800	2800	1850	--	1100	800	450	350	--	300	450	--	--	--	--	--	--	--	--	--	150
AVACHA BAY	3000	2760	--	--	1300	920	530	600	300	--	200	350	700	--	--	--	--	1350 (1200) (AIR)	800	600	300
KASHIWABARA BAY	3000	2800	2000	--	1400	1000	700	780	450	200	--	200	550	300	--	500	600	1100 (AIR)	870	700	450
MATSUWA	3120	2800	--	--	--	--	800	--	--	350	200	--	300	70	--	350	480	700	--	--	650
ETOROFU	3300	3000	--	--	1800	--	--	--	--	700	550	300	--	230	--	300	280	700 (AIR)	1300	1060	800
BUROTON BAY	3150	2900	--	--	--	--	900	--	--	--	300	70	230	--	--	--	--	900 (AIR)	--	--	--
ALEXANDROVSK	--	--	--	--	--	--	1200 (AIR)	--	--	--	--	--	--	--	--	--	400	--	--	--	--
SHIKUKA	--	--	--	--	--	--	1150 (AIR)	--	--	--	500	350	300	--	--	--	330	--	--	--	--
OTOMARI	--	--	--	--	--	--	1200 (AIR)	--	--	700	600	480	280	--	400	330	--	--	--	--	1010
VLADIVOSTOK	--	--	--	2800 (AIR)	--	2300 (AIR)	1850 (AIR)	--	--	1350 (1200) (AIR)	1100 (AIR)	700 (AIR)	700 (AIR)	900 (AIR)	--	--	--	--	2115 (AIR)	1930 (AIR)	1600
AMCHITKA	--	--	--	--	600	220	250	--	--	800	870	--	1300	--	--	--	--	2115 (AIR)	--	--	520
SHEMYA	--	--	--	--				--	--	600	700	--	1060	--	--	--	--	1930 (AIR)	520	--	330
KOMANDORSI ISLANDS	--	--	--	--	1050 (AIR)	650	300	300	150	300	450	650	--	--	--	--	--	1600	--	330	--

Fig. 105
Relative Distances, Operations Problem 2

be serious enough to cause Blue to return to base and assume the defensive. (Pages 11–14.)

Admiral PA next sought to analyze in greater detail the probable Blue objectives, offering a "reasoned estimate" rather than any degree of certainty, a probability pattern by which to position his forces. Given what he saw as the "comparatively small" size of the Blue Expeditionary Force, he concluded that it was not designed to seize an objective near a very large enemy base—such as Vladivostok—from where Purple land-based air could continually mount air raids of such strength as to nullify the value of any base seized. Further, it appeared to him that Blue could not risk forcing the Sea of Okhotsk without first securing a base in either Kamchatka, the Komandorskis, or the Kuriles, to which he therefore narrowed his consideration. Kamchatka had only one initial area that seemed worth seizing for development as an operating base—Avacha Bay, where Purple was developing the Petropavlovsk base. The only other potential base sites, Ust Kamchatka and Korf Bay, he found unacceptable for several reasons. First, Korf Bay in the extreme north represented a loss in distance relative to logical Purple targets as compared with Attu, which was already in Blue's possession. In addition, from November to May, it was frozen over almost entirely. Ust Kamchatka in the Gulf of Kamchatka was closer to logical Purple targets than Attu but only by some two hundred miles. Ice was again a problem from November to May, and in May there was drifting ice from the Bering Sea. (Pages 14–15.)

Avacha Bay would not be well defended until 1 August, and Blue could hit it well before then. It also offered the best harbor facilities in Kamchatka and had suitable landing beaches if the harbor facilities were destroyed. Further, it contained Petropavlovsk, the political, economic, and military center of the Kamchatka Peninsula. Its sheltered harbor had a deepwater entrance and the adjacent level terrain was suitable for air operations. The climate was not what he would call attractive but was less severe than elsewhere in Kamchatka; summer temperatures ranged from 50 to 60 degrees Fahrenheit, average night temperatures in the lower forties. While Petropavlovsk was "somewhat" icebound from December to April, the anchorage was rarely frozen over completely and icebreakers got through in February and March. Unlike most of Kamchatka, Petropavlovsk offered conditions favorable for air operations during most of the year. In general, PA thought, Avacha Bay was the most important airplane, seaplane, and naval base on Kamchatka. It was also, however, halfway between Dutch Harbor and Vladivostok, within fighter range of Purple's base at Kashiwabara Bay and the numerous airstrips and facilities at Paramushiru and Shimushu, and within fighter range of the Komandorskis. Admiral PA also knew, however, that Avacha Bay was only just within fighter range of Matsuwa, Purple's primary air defense station for the central Kuriles, and outside fighter range from the other major Purple air bases in the Kuriles and Sakhalin.

Still, Blue possession of Avacha Bay would isolate and neutralize the Komandorskis and could put Blue four hundred miles closer than did Attu to the Purple targets it wanted to hit. (Pages 15–16.)

As for the Komandorskis, Bering and Copper Islands were large but mountainous. They had several small anchorages but these were poor and storms would force ships out to sea. However, these islands were ice-free the year round, except for some drift ice in January and February. The climate was tolerable, albeit marked by heavy snow in the winter and fog and rain in the summer. At Nikolski, on the northwest coast of Bering Island, there was a fair anchorage for about twenty ships, but westerly winds caused heavy swells. Two level sections ashore appeared "acceptable" to modest air-base development, so Bering Island was a potential target for Blue. About 1,100 miles from Dutch Harbor and 1,600 from Vladivostok, it was within fighter range of Blue's Attu base and of Purple's Korf Bay, Ust Kamchatka, and Avacha Bay bases. If Purple held the base on Bering Island, it and Blue's Attu would oppose each other on a one-to-one basis. A Blue base on Bering, though, would face land-based fighter attack from possibly three directions and have to distribute its own efforts over two or three simultaneous targets, depending on the weather. Bering Island was outside the fighter range of the Purple bases in the Kuriles–Sakhalin area and occupied a commanding position with respect to the southwestern entrance to the Bering Sea. On average and compared with Attu, the Komandorskis represented an advance for Blue of about two hundred miles relative to its future Purple targets. For these reasons, he found the Komandorskis to be a possible Blue target but not as probable as Avacha Bay. (Pages 16–17.)

In the Kuriles, Admiral PA knew, Blue would not see many worthwhile objectives. Onnekotan, Shasukotan, and Matsuwa Islands had beaches suitable for landing operations but were "extremely limited" by topography as sites for airfields. Matsuwa was already a Purple base but only had two airfields, neither of which could handle Blue's Superfortresses; between them, they could handle only a total of seventy fighter and naval patrol bombers. Onnekotan had only a small-ship anchorage, and Shasukotan, like Matsuwa, had a harbor with only one large, two medium, and two small berths for ships. Rejecting these three, he next looked at Shimushiru and Uruppu, which had suitable landing beaches but were likewise limited by topography in terms of air-base development. The former had an anchorage in Buroton Bay, but only for ships up to a thousand tons. The latter had nine large, three medium, and seven small berths but had proved practical for Purple only for destroyers and smaller ships. Rejecting these as well, he turned to Kunashiri and Shikotan. The former had several potential land-plane base sites but an anchorage usable only for small craft; the latter could anchor no more than five large ships and offered what he saw as "meager" potential as an aircraft base. He rejected these locations as probable Blue objectives. (Pages 17–18.)

That left two major possibilities in the Kuriles, Kashiwabara Bay and Etorofu. Admiral PA knew that Kashiwabara, a well sheltered bay, offered a major fleet anchorage and had been heavily developed by the Japanese. Paramushiru, flanking Kashiwabara Bay, had a number of beaches accessible to attack by sea, and both Paramushiru and Shimushu offered ready-built airfields, air bases, and seaplane bases. A strong Blue base there would "practically" control the entrance to the Okhotsk Sea. It did have "considerable" drift ice through January, February, and March and dense fog from April to September. Mean temperatures ranged from 26 to 52 degrees Fahrenheit in August; snow and rain had to be expected about two-thirds of the year. Still, Kashiwabara Bay was equidistant by air from Dutch Harbor and Vladivostok, within fighter range of Avacha Bay, Matsuwa, and Buroton Bay, and outside the fighter range of southern Kurile and Sakhalin bases. If successfully established as a Blue base, it could dominate the northern and central Kuriles and could even prevent Purple from getting Avacha Bay onto a "useful" footing. Moreover, Kashiwabara Bay bypassed the Komandorskis, which Blue could then neutralize from Attu. On average, compared with Attu, Kashiwabara Bay would mean an advance of some six hundred miles relative to future Purple targets, the most of all the bases he had analyzed so far. (Pages 18–19.)

Etorofu also had a "suitable" major fleet anchorage. A large island about 125 miles long and twenty-five miles wide at its widest point, Etorofu had excellent landing beaches; airfields had already been built there, and a southern portion of the island was adaptable to constructing more. The island also possessed a lake large enough to accommodate seaplanes. As in the northern Kuriles, fast ice formed in most bays and sheltered areas from January to March or April. Fog was even more prevalent than in the northern Kuriles, and temperatures were low, about like those at Paramushiru. Again, it had the "weather handicap" that was common to the entire Kamchatka–Kurile region but this problem did not entirely eliminate it as a potential base site for Blue. It was 1,800 miles from Dutch Harbor but only seven hundred by air from Vladivostok. It was within fighter range of Matsuwa and Buroton Bay, as well as of Shikuku, and Otomari on Sakhalin Island. Additionally, it was the only base he had tentatively accepted as worth analysis that was within fighter range of the Sakhalin bases; it was liable to attack from four Purple bases whose bombers could enjoy fighter escort. Blue's lines of communication and supply to this base would be flanked by Purple forces in the northern Kuriles, the Komandorskis, and Kamchatka. Etorofu, however, also dominated the southern and central Kuriles and was outside fighter range of the northern Kuriles. Compared with Attu, Etorofu would furnish Blue an advance of seven to eight hundred miles relative to various Purple targets, though relative distance was actually unchanged vis-à-vis a few Purple bases to its north. (Pages 19–20.)

Before trying to project Blue's probable Aleutian destination, he first wanted to assure himself that it was reasonable to expect Blue to stage in the Aleutians as a preliminary to seizing an advanced base at all. Blue, in fact, had several possible alternatives. The Blue commander might embark all troops and supplies at Puget Sound and proceed directly to the ultimate objective. Admiral PA, however, did not think that he would do so because of the distances and logistical complications. Alternatively, Blue might sweep southward and pick up troops in the Central Pacific; in this scenario, it would be using, in a cold-weather operation, troops trained in and acclimatized to warm weather. Still, he wanted to devise a search plan that took both of these contingencies into account. As for Blue's likely Aleutian objective or destination, he limited his analysis to bases that offered adequate anchorages, ample logistical facilities, and air protection. Aside from their inferiority compared with Dutch Harbor, both Kodiak and Adak were not close enough to Blue's ultimate objective to be useful as stopovers. He also rejected Cold Bay because it had not been developed as a major base, and he rejected Amchitka because of limited harbor facilities. Attu, another major Blue base, had adequate anchorage and logistics but he assigned it a low probability as Blue's destination because as Blue's westernmost existing base, it would have poorer air protection than one nearer the central Aleutians and a ship assemblage that would be more liable to detection by Purple patrol planes. In addition, the Blue commander, who might have atomic bombs on board, might hesitate to concentrate his shipping in harbor needlessly to westward when he could just as easily choose a safer base and get the same services for his force. (Pages 20–21.)

Adak offered the Blue commander all necessary facilities and was centrally located in the Aleutians chain. It had its own "considerable" air strength and could count on protection from Amchitka, Attu, and Dutch Harbor; Amchitka, in particular, was only about two hundred miles distant and could provide strong support. PA thought accordingly that Adak had a high probability as the Blue "stop" in the Aleutians. Dutch Harbor remained for consideration, however, since it was Blue's primary base in the eastern Aleutians. It had ample anchorage and logistical facilities but its land-based air force was inferior to Adak's, and it was more remote from supporting air bases than was Adak. Compared with Adak, it increased by three to four hundred miles the distance from the Aleutians to the final objective. He therefore saw Dutch Harbor as only a "medium" probability as the Blue stopover in the Aleutians. He thought, however, that Blue might split his force between two Aleutians bases and reconcentrate after departing the chain. (Pages 21–22.)

For the critical leg from the Aleutians to the final objective, Admiral PA saw the Blue Expeditionary Force leaving from Adak and Dutch Harbor to head north of the Aleutians and directly to its destination. The Blue forces could also head south of the Aleutians before turning to their objective or they could follow an evasive course either north or south. He thought the latter the least probable because it put additional logistical demands on the Blue Expeditionary Force and it would mean keeping Blue's troops at sea in—probably—poor weather longer than necessary, thus deteriorating their health and vigor for the impending assault. However, Admiral PA also noted that if headed for Avacha Bay, Blue might send his convoy far to the north and so attempt to avoid detection. His searches, therefore, had to cover this contingency as well. PA thought the first and second possibilities of routes were equally satisfactory in imposing minimal logistical problems for the Blue Expeditionary Force. The first promised more effective weather cover in the Bering Sea. Both permitted land-based air cover against submarines all the way to the target and land-based fighter aid as well, except for the last three or four hundred miles, depending on the final Blue destination. (Pages 22–23.)

From the intelligence available, PA thought that the Blue Striking Force would assemble at Pearl Harbor and be definitively committed to the operation as soon as it left there. He now considered the final Blue objective to be any Purple bases in and about the Kuriles–Kamchatka area that could oppose the Blue Expeditionary Force from the time it left a Blue base until it completed its amphibious landing. Given all of this analysis, he considered that the Blue Striking Force might take any of five tracks from Pearl Harbor. First, he thought it might proceed directly to an Aleutian base from Pearl Harbor, then head north of the Aleutians to the final objective; this route might entail a far northerly sweep and an approach to the final objective from the northeast. Second, the Blue Striking Force might head from Pearl Harbor to an Aleutian base and then pass south of the Aleutians to the final objective. Third, it might steam from Pearl straight through the Aleutians before heading to the final objective; this route would again lend itself to a sweep to the far north and an approach on the objective from the northeast. Fourth would be a route west from Pearl Harbor then to the final objective from the south; the fifth and last possibility he analyzed was directly from Pearl Harbor to the final objective itself. (Pages 23–24.)

In comparing these possible routes, he felt that the element of surprise would be a major factor influencing Blue's selection, since attack by the Blue Striking Force might be the first blow Purple received and it could be made devastating. Either of the first two possibilities listed would be desirable from a logistical perspective since Blue would have to steam about a thousand miles from the last base before it reached striking distance of Purple. Of these possibilities, Admiral PA chose Dutch Harbor as the most logical. Adak was a "fair possibility," but Dutch Harbor was far enough east to contribute to Blue's achieving surprise. The first course offered better possibilities for weather cover but both permitted land-based air cover—not very important for the Blue Striking Force but critical for the Expeditionary

Force. In this phase, however, Admiral PA assumed that the Blue Striking Force commander would want complete mobility and minimal ties to land-based air. He did not think that the third course was very probable since Blue would then arrive within striking distance of Purple having steamed over three thousand miles from its last base and would have been subjected to the same intensity of Purple searches as in the first two possibilities. The fourth possible course to the objective would mean arrival within striking distance of Purple after having steamed about 3,500 miles; it therefore appeared even more unlikely to Admiral PA. He retained it in his analysis, however, with a low probability since Blue might try this for deception. The fifth course would entail combat after Blue steamed about 2,400 miles since its last replenishment. This was logistically preferable to the fourth potential course but it would not offer as much chance of surprise since Blue had to assume Purple would know about its departure from Pearl Harbor. Also, neither of the final two possibilities took advantage of weather cover to the north of the Aleutians so he thought both had a low probability of adoption by Blue. Accordingly, Admiral PA bet on Blue's pursuing either of the first two possible courses to the final objective, with the first—proceeding directly to an Aleutian base from Pearl Harbor, then heading north of the Aleutians to the final objective—being more probable. (Pages 24–25.)

Admiral PA's next analysis was the "Combat Efficiency of Personnel." He thought that both forces were about equal in that respect, with a "high order" of personnel efficiency. Blue had the advantage of "invaluable" large-scale combat training and operational experience in the Pacific theater of operations during the Second World War but Admiral PA thought that this might make Blue overconfident, gauging Purple's personnel as "second-rate." Purple, he thought, had built up from its victory in the war and the communist indoctrination of its armed forces gave it a "virile" cult of invincibility, a culture that he argued was "superior to the easy-going easy-living way of the capitalist society." Of vital importance to him was that his forces were hardened and "thoroughly acclimatized" to the rigors of cold-weather operations, while most of Blue's were not. As for material, Purple, he argued, should enjoy superiority since every item had been tested and proven over long periods of time in cold weather. In addition, Purple would operate from its own bases while Blue would have a long voyage on which afloat repairs would be limited by weather and available repair facilities. (Pages 25–26.)

Admiral PA did not anticipate any Purple reinforcements beyond carrier- and land-based planes on a thirty-six-hour-delay basis. Moreover, he expected these reinforcements to be able only to bring his forces up to their original strength. In addition, he could replace aircraft losses at one base only by drawing from other bases. Once Blue left his Aleutian bases, however, it would have no reinforcements of any kind in the operation. In terms of logistics, Purple would have close at hand well-stocked bases that were, in turn, supplied by interior lines to the Purple homeland. Six oilers, two submarine tenders, and three seaplane tenders supplemented the base facilities available to Purple. In contrast, Blue had to carry everything along needed for combat power, equipment, and sustenance. Replenishments, to reach Blue, had to travel a long supply line that was exposed to submarine and air attack. Damaged Purple ships had a good chance of getting back to base, while Blue cripples would be in an "extremely precarious" position. Admiral PA summarized "Political Considerations" by asserting that Blue and Purple were mutually suspicious of each other's intentions and sincerity. The Purple masses, he argued, were convinced that capitalism had to be eliminated and would support any military action by their government. The Blue populace he saw as difficult to lead into supporting war; he thought that an initial victory by Purple forces might cause "serious dissension" among Blue political factions. He further pointed out that Purple agents in Blue territory had gained power in political, labor, and publicity organizations and could aggravate any internal strife that might follow an early Blue defeat. However, in view of Blue's success in intelligence gathering during the world war, and despite Purple countermeasures, PA assumed that Blue would have "reasonably" accurate information on Purple strengths and dispositions at the beginning of the operation. (Pages 26–27.)

As to the characteristics of the theater of operations—which he took to be between 40 degrees and 61 degrees north latitude, bounded on the west by the Kurile–Kamchatka chain and extending east about a thousand miles—Admiral PA did not think that its hydrography had any unusual aspects. The open sea was free of dangers to navigation, charts were adequate, and the general movements of water masses were controlled by the Kamchatka Current. This stream flowed from its departure from the Japan Current at about 38 degrees north latitude in a north-northeasterly direction. One part of it continued easterly along the southern side of the Kuriles and just before reaching the American coast divided into a branch toward the north (into the Gulf of Alaska) and a branch to the south (along the west coast of the United States, where it became the California Current). The second part of the Kamchatka Current turned north and entered the Bering Sea, followed along the northern side of the Aleutians, and circled to the Bering Sea counterclockwise. In the Bering Sea, this current was further cooled and flowed south to reach the northern islands of Japan. He thought it noteworthy that while some of the water that flowed into the Bering Sea around the Aleutians contributed in the summer to the current that flowed north through the Bering Strait, the major flow turned around and ran back south along the coast of Kamchatka. Velocities were in the range of one-quarter to one knot, depending on the force and direction of the wind. The currents through the passes in the Kuriles and the Aleutians, while influenced by the winds, were predominantly tidal and could reach eight knots or more, especially in narrows. (Pages 27–28.)

In terms of topography, the theater was predominantly mountainous. There were many areas suitable for airfields but no extensive plains or plateaus. Rocky coastlines were well protected by such natural defenses as steep cliffs, outlying rocks, and the kelp that fouled so much of the Kuriles; they did not, however, have to be considered for purposes of this operation and he had already discussed potential base sites. Weather, of course, was vital, given its potential to restrict operations. In general, the weather in the Kurile–Kamchatka–Aleutian area was bad and unpredictable. Rain, mist, fog, and falling snow were to be found in all seasons of the year, with fog predominating in the summer months and snow in the winter. The weather conditions at one spot were no evidence of what to expect for another spot just a few miles away. Summer weather meant temperatures of about 50 degrees Fahrenheit but there would be fog 35 to 75 percent of the time. In the Kuriles, over half of the days of June, July, and August were foggy. The fog originated to the south and east of Kamchatka through the movement of moisture-laden air from the south into contact with the cold air moving out from the Siberian High and with prevailing southwesterly winds, which carried it to the Aleutians. Several days of north or northwesterly winds were required to drive the fog south and dissipate it, a condition that he said seldom occurred. (Pages 28–29.) Rarely did sky covered by clouds become less than 70 percent in the Kuriles; the probability of a clear sky was only about 20 percent, though clear periods lasting eight to ten hours might come at any time. Cloud ceilings were low, from about one to two thousand feet, higher in the winter. Serious icing occurred at flying levels at all times of the year, less in winter than in summer. Gales during the summer were uncommon but not unknown. Smooth seas were to be anticipated except in local storm areas.

PA foresaw several specific impacts on operations because of weather. First, air operations would be "materially hindered" by fog, generally poor visibility, cloudiness, low ceilings, icing, and rugged topography. This was the general standard that he expected but there would be certain local exceptions; at Petropavlovsk, for instance, conditions were "comparatively favorable." Low clouds and poor visibility obviously offered the Blue forces better prospects of achieving surprise in the summer than in any other season. Parachute operations, however, would be extremely hazardous and Blue had to refuel its ships in the open sea. He noted that there was no "truly dark" period; twilight lasted from sunset to sunrise at 50 degrees north latitude from 2 June to 12 July. (Pages 29–30.)

In reference to facilities and fixed defenses, Admiral PA pointed out that Blue air bases in the Aleutians were "adequately" equipped with the latest types of radars, antiaircraft batteries, communications equipment, ground-controlled-approach equipment, and fog disposal/dispersal equipment. The Aleutian bases also had LORAN, direction-finding equipment, high-power radio equipment, and radar stations. Purple, for its part, had a major and well defended operating base at Vladivostok. In addition, Purple's naval bases and stations in the Kuriles and Sakhalin were well equipped and stocked as well as defended by coastal and antiaircraft batteries, radar, and direction finders. Until 1 August, however, Kamchatka bases and stations would have to depend on forces afloat for their own protection. Kamchatka airfields could be used but they would not be able to service planes before 1 August. In the related area of communications, primary reliance was being placed on radio by both Purple and Blue. Aircraft delivery could also be employed by both sides but was subject to the weather. Blue had radio stations at Kodiak, Dutch Harbor, Adak, and Pearl Harbor, and in the continental United States. Purple had similarly "satisfactory" facilities, extending from Vladivostok through major bases in the Kuriles and Sakhalin. Communications, he thought, were satisfactory for both sides. (Pages 30–31.)

NOTES

1 Senior Class of June 1947, "Operations Problem 2: The Purple Statement, Section A," 13 September 1946, p. 1, folder 2603, box 138, RG 4, NHC.

2 Senior Class of June 1947, "Operations Problem 2: Detail of the Maneuver Staff," 26 September 1946, p. 1, folder 2603-G, box 138, RG 4, NHC. Subsequent page references in this chapter, until the next endnote, are to this source.

3 Senior Class of June 1947, "Operations Problem 2: The Purple Staff Solution," 8 October 1946, pp. 1–2, folder 2603-B, box 138, RG 4, NHC. Remaining page references in this chapter are to this source.

XI *Operations Problem 2*

More Purple, September–October 1946

Admiral PA saw as Blue's "Strength Factors" the ability to choose the objective and the route to the objective. Blue could also choose the time of the operation and had available its Aleutian base complex. In addition, because of the characteristic low clouds, poor visibility, and frequent fog, Blue had a good chance of achieving surprise.

The Purple Staff Solution: Strength and Weakness Factors

Admiral PA assumed that Blue had high-speed submarines, atomic bombs, and very long range bombers. Its mobility and concentration of airpower were great because so much of Blue's airpower was carrier based and because of its superiority in attack planes that could be committed at the critical period of the operation. Moreover, Blue had a six-to-one superiority in planes that could launch aerial torpedoes and a slight torpedo superiority on its surface ships. This was combined with its two battleships and a five-to-three superiority in destroyer numbers, guns, and torpedoes. Blue's "Weakness Factors," however, were the possibility that poor weather would nullify its air superiority at critical periods; the vulnerability of much of its airpower, because it was carrier based; and a restriction on the Blue Covering Force's freedom of action, which had to protect the slow Blue Expeditionary Force. Blue also had the disadvantages of long lines of supply and the unavailability of immediate reinforcement.*

Purple's Strength Factors included freedom to choose the method of attack on the Blue Force and the ability to remain concentrated. In addition, Admiral PA's forces were operating from "comparatively" short, interior lines of supply, had aircraft replacements available within thirty-six hours, and had personnel and material that were better acclimated than Blue's to the operating area. He also saw advantages in the nightlong twilight, which would favor searches and attrition tactics if Blue was located early enough, and in superiority in fighter strength that could be engaged at the critical stage. Admiral PA further noted his own high-speed submarines, his two battle cruisers, and his superiority in cruiser numbers and

* Senior Class of June 1947, "Operations Problem 2: The Purple Staff Solution," 8 October 1946, pp. 32–33, folder 2603-B, box 138, RG 4, NHC. All subsequent page references in this chapter are to this source.

weapons—with the exception of automatic weapons. Moreover, he did not have a convoy to protect, and his ships, if damaged, had a better chance of getting back to a base than did Blue's. His Weakness Factors were the possibility that poor weather would curtail the employment of his air strength at critical times and the fact that Blue land-based air forces could seriously damage the search aircraft he needed to locate the Blue force. He also could not expect reinforcements, except replacement aircraft. Of all of these factors, the one that really stood out to him was the weather. He intended never to lose sight of its central importance as he considered his mission, the relative combat power of the two forces, the other conditions in the theater of action, the acceptability of costs, and the ability of planes to fly with operational losses. While he could not answer these questions until operations actually began, he was certain that his forces had to be "alert at all times to take advantage of the vagaries of the weather, and to anticipate similar action by Blue." (Pages 32–33.)

Admiral PA thought that his assigned mission narrowed his consideration of Blue's possible objective to the area west of 170 degrees east longitude and north of 44 degrees north latitude. He did not think that he was restricted to this operating area, but other factors limited Blue's capabilities. First, Vladivostok was remote from Blue's bases and so strongly held that it might be considered invulnerable. He also did not think that Blue would attempt to seize a base west of the Kuriles–Kamchatka line, since such a base would be highly vulnerable to Purple attack and too difficult to supply. Again, the Blue Covering Force was limited by its obligation to protect the Blue Expeditionary Force and Blue's unavailability to reinforcement would probably restrict the Blue commander as to the types of risks he would be willing to take. Given all of this, Admiral PA thought that Blue could have its Expeditionary Force evade major Purple forces and seize its objective while the Blue Striking Force (the Covering Force combined with Blue's land-based air assets that were within range) acted independently to contain any threatening major Purple forces. Blue could also have its Expeditionary Force and Striking Force concentrate prior to leaving Blue-controlled waters, make the passage together, and seize the objective while destroying or driving off the Purple forces they encountered. The Blue Striking Force could, alternatively, isolate the objective by fast carrier- and land-based air while the Expeditionary Force seized the objective and the Striking Force suppressed Purple's perimeter air bases. Finally, Admiral PA thought, Blue could attempt a combination of these tactics. (Pages 33–34.)

Courses of Action

To Admiral PA, the first potential Blue option entailed several "questionable" features. First, Blue would need "extremely" good luck and prolonged bad weather to evade the searches that PA intended to maintain. The division of Blue forces might permit Purple, in fact, to place its entire strength against the Blue Expeditionary Force, especially in the bad weather, which might restrict or preclude flight operations. Given these risks, Admiral PA assumed, Blue would not pursue the first option. The second option had the "seeming" advantage of maximum concentration of striking power and on-the-spot protection of the transports. In addition, losses per attack were likely to be less when forces were concentrated and more strength was on hand to protect cripples. The disadvantage, however, was that the Blue Expeditionary Force would have to accept the full brunt of Purple's strength, as would the Blue Striking Force during the entire passage to the objective: "I doubt that the Expeditionary Force could survive it." In addition, Blue would be using its Striking Force as a defensive guard rather than the offensive force that it was meant to be. Admiral PA rejected the second option as unviable. (Pages 34–35.)

The third option—to isolate the objective by fast carrier- and land-based air while the Expeditionary Force seized the objective and the Striking Force suppressed Purple's perimeter air bases—he saw as conforming to a pattern that Blue had found to be of "immense value" in World War II. It permitted the Blue Striking Force constant readiness for action, "extreme" mobility, and a great degree of freedom of action from the Expeditionary Force's movements. The Striking Force, under this option, could hit where and when it wanted to without concern for the safety of the Expeditionary Force. The latter force would be out of the zone of active operations and under the protection of Blue land-based air forces. When the Expeditionary Force finally came to the objective, Purple's strength would have been considerably reduced, presumably enough that Blue could get its transports through. Admiral PA saw Blue expecting to isolate the objective by destructive air attacks on the objective and bases within supporting range of it. Admiral PA also assumed that Blue knew Purple air forces could be provided with replacements but that this resource could be nullified by rendering their transit bases inoperable. He thought this third option highly likely, perhaps combined with elements of the other two. For instance, the Blue Expeditionary Force might or might not follow evasive courses once it left the Aleutians. Also, the Blue Striking Force might concentrate enough to provide close air support to the Expeditionary Force between the latter's Aleutian bases and the objective. In general, though, he saw Blue pursuing the third option. (Pages 35–36.)

Given Blue's need to seize and establish a base that would allow it to extend its control in the western Pacific, the forces this would take, and the need to get those forces to the objective with acceptable losses in a Purple-dominated area, Blue needed a powerful enough striking force to thrust itself into the area prior to the Expeditionary Force's arrival and destroy or neutralize the bulk of Purple's forces. Therefore, Admiral PA took the Blue Striking Force to be his own first objective; it would venture into his operational area first. Moreover, if he was able to defeat it decisively, he assumed, he would not have to deal with the Expeditionary Force. If, however, Blue was able to neutralize Purple airpower and bring the Expeditionary

Force into the operational area, that latter would become his primary objective, and he would engage the Blue Striking Force only as needed to get at the Expeditionary Force.

His orders called for a "vigorous" offensive, but he took into account the limitations on his capabilities. Since his forces did not include amphibious units, he could not attack Blue's territory so as to throw Blue back on the defensive. His land-based air forces were also limited by their locations and the geographic nature of the operational area. His surface forces were of such limited strength that he had to keep them concentrated for mutual support when action became probable. If, for instance, he dispatched surface scouts, he could not let them get far from his main force. In addition, he could not lose sight of the dependence of the Avacha Bay base on his forces for protection until 1 August. Still, he argued that the possibility of offensive air strikes on Blue bases had to be exploited. However, since Purple and Blue were not yet at war, there was a question about how far into Blue waters Admiral PA could go in an initial offensive action. In any case, until he received orders to execute his plan, he could take no offensive action. Nevertheless, he could station submarines off Pearl Harbor to get him accurate data on the Blue Striking Force. If the Blue Expeditionary Force departed Puget Sound before the Striking Force left Hawaii, he thought, his plan could be executed "in time to permit our sinking Blue carriers departing from Pearl Harbor." Given all the above restrictions, his Course of Action was going to be to reduce Blue strength by repeated attrition attacks on the Striking and Expeditionary Forces; to destroy the Blue Striking Force by air, surface, and submarine attacks; to do the same to the Blue Expeditionary Force; and to reduce Blue's strategic western Aleutian bases by land-based and fleet air attacks. (Pages 36–38.)

Looking now at the practicality of his Course of Action, Admiral PA asserted that attrition attacks on the two Blue forces would depend on early success in locating them as well as on "well reasoned" or lucky mining and stationing of his submarines. Attrition attacks, he knew, would have to be primarily delivered by air and submarine, but he saw them as a practicable, if partial, solution to his problem. As for destroying the Blue Striking Force, he thought it vital that he attempt this destruction when his land-based air forces could be effective since this was his major weapon. Entirely aside from the distances involved, he felt that World War II was so recent that he could not "Pearl Harbor" the Blue Striking Force at one of Blue's Aleutian bases (i.e., attack it without warning before the declaration of hostilities), and he could not risk his surface forces attacking Blue beyond coverage from his land-based air forces. He knew that the farther east he got, the poorer his chances of success, because he would have available less power from his land-based air forces, less search and attrition capability from his six fleet air wings, reduced time for his submarines to carry out attrition strikes, and longer supply lines for his surface forces, all of which would give Blue air and submarine forces greater opportunities. However, it was Blue that had to venture forth, and these handicaps would accrue to it; Admiral PA thought Blue's moves would permit him to use his full land-based air strength. He would keep his surface forces close to their bases—one of which was Blue's probable objective—and concentrated. He would also strengthen the air defenses and air warning systems at his bases. He saw all of these measures as within his capabilities. (Pages 38–39.)

If he had to engage the Blue Expeditionary Force, its destruction would accomplish his mission. If he destroyed the Blue Striking Force, he would not have to engage the Expeditionary Force, which would withdraw. If, however, he failed to destroy the Striking Force and Blue thought him sufficiently damaged and neutralized, Blue might think that its Expeditionary Force had a reasonably good chance of accomplishing its mission. His own best bet would be to conduct attrition attacks on the Blue Expeditionary Force when it was closest to his bases; all would depend on the amount of damage the Blue Striking Force inflicted on him. He saw this part of his mission as practical but foresaw that it would inflict a heavy cost. Also practical was the reduction of Blue's western Aleutian bases by air attacks; but strikes on the eastern Aleutian bases would be less so given their distance from his air bases. Enemy opposition he saw as likely to be "intense" but not enough to prevent him from inflicting damage. If he was able to reduce Blue's bases, that would degrade Blue's search and strike capability available to support the Striking and Expeditionary Forces. (Page 40.)

Admiral PA now focused on his submarine dispositions. He ordered Rear Admiral PJ, Commander, Pacific Fleet Submarine Force, to maintain two submarines on patrol in the Aleutians off the approaches to Dutch Harbor and to detach the other two currently on patrol there. Upon arrival at their base, these submarines were to take on maximum supplies and personnel, even to the extent of cannibalizing parts from other boats. They were then to proceed to stations off Pearl Harbor to observe the Blue carrier forces departing Hawaii. The submarines were to avoid detection, merely observing and reporting unless attacked. They were then to trail the fast carrier forces. (Pages 40–41.)

In any attempt to reduce Blue strength by attrition attacks, PA's primary weapons would be aircraft and submarines. His land-based air forces were distributed over a number of "well separated" bases and could search as far as a thousand miles. He expected heavy air losses in his attrition attacks given the strong air elements in the Blue Striking Force, but he accepted these losses as necessary. He recognized that his fleet-based air was weak and had to be tied to his land-based air, except for surprise strikes on Blue bases. In view of the weakness of his fleet-based air, he did not plan on using it in attrition attacks until the Blue Striking Force had been considerably "cut down." Also, his submarines were few in number. They were fast, however; had "tremendous" endurance; would be comparatively safe from destruction given their long submergence times; and had the capacity to destroy as many as six major targets each. They would be decisive, provided that they were positioned so as to intercept the Blue forces when located: "Since they will be observing off Pearl Harbor, Dutch Harbor, and Adak, I hope for immediate warning of enemy sortie and extremely early attrition." (Pages 42–43.)

His primary targets would be the Blue fleet carriers, and he was hoping to destroy two or three of them before the Blue Striking Force got into a position to launch fighter-escorted strikes on the Purple air bases. He did not think that much attrition could be inflicted on the Blue Expeditionary Force before it departed the Aleutians but after that he expected to inflict heavy damage on that force, knowing at that point where it was departing from. Purple submarines with high speed would also have great advantages in tracking and attacking these units, especially the transports, his main targets in the Expeditionary Force. He expected to destroy at least half the transports before the Expeditionary Force came within fighter range of his bases. The key, however, was either to destroy or to drive off the Blue Striking Force first. Though this meant great risk, the survival of his force and the accomplishment of his mission meant forcing an early confrontation with this force and, if not destroying it, damaging it to such an extent that it became impotent and was forced to retire. (Page 43.)

He did not expect an opportunity to execute this task until and unless his attempt to destroy the Blue Striking Force had failed. At this point, he assumed, he would have suffered heavy damage and his air forces would be, temporarily at least, "extremely weak." He would then have to depend chiefly on submarine attacks and long-range, unescorted, small-scale air attacks. In this case, Blue might reach the landing phase before Purple could inflict decisive damage, unless the Blue Striking Force's air assets had been "tremendously" reduced and Purple's fleet units remained "relatively" undamaged. He saw his strikes on the western Aleutian air bases by his land-based and fleet air forces as putting "highly desirable" restrictions on Blue, in that the latter depended so heavily on Adak for the replenishment of its forces. Blue sea forces also depended heavily on the western Aleutian bases for air support, especially those at Attu, Shemya, Amchitka, and Tanaga. Admiral PA further thought that his search and observation would be more satisfactory as he weakened Blue's base of air support. However, he also knew that his land-based air forces would have to conduct many missions without fighter escort, though his sea forces could give some fighter and attack support to the land-based air strikes against Attu and Shemya. These sea forces could remain under Purple land-based fighter cover, but he was under no illusion about losses in planes and crews. He nevertheless thought the results would justify them. (Page 44.)

PA now contemplated the complementary problems that had to be solved—a Search Plan, the Combat Employment of Air and Surface Units, a Movement Plan, a Mining Plan, a Logistics Plan, the Task Organization and Tasks, and a Communication Plan. On the basis of his assumption that Blue would continue to send reconnaissance planes over his bases and that Blue needed to be denied information about them, Admiral PA ordered Rear Admiral PK, Commander, North Pacific Air Command, to require effectively and immediately any foreign planes detected over Purple bases or territory to land for examination. Planes unwilling to comply were to be shot down. Commander, North Pacific Air Command was also to implement "suitable" defense and examination measures, while units at sea were to report all suspect plane contacts. The exceptions were Purple submarines at sea, which had more specific orders to carry out. (Pages 44–46.)

The Solution of the Problem

Admiral PA first considered his Search Plan, which would determine such other aspects of the campaign as the Movement Plan and the Logistics Plan. His operations would comprise two phases, Search and Destruction. They would overlap, and the second phase would not be successful unless the first was. His Search Phase would provide him with the intelligence needed to destroy the Blue force. His signal to execute his Operation Plan would be receipt of information that the Blue Striking and Expeditionary Forces had left their bases. Because this information might come from his submarines, he wanted his Communication Plan to pay particular attention to rapid contact with all units—especially the submarines—when it came time to broadcast the order to execute the Operation Plan. In addition, he understood that his search had to obtain for him information on the enemy force's strength, disposition, course, speed, and destination, not merely to locate it. The Search Plan was also predicated on the idea that the Blue Striking Force would enter the operational area first, while the Expeditionary Force would hold back under the air cover from the Aleutian bases until the Striking Force had considerably weakened Purple's bases and forces. (Page 47.)

Admiral PA estimated that the Striking Force's most probable route from Pearl Harbor to the Aleutian bases was via Dutch Harbor or Adak (stopping briefly) and then north of the chain to the Blue objective. This might include a "far northerly" sweep ending with an approach to the final objective from the northeastward. Another possible route for the Blue Striking Force would be from Hawaii to the Aleutians and then south of the archipelago to the final objective, again with a brief stop at one of the Aleutian Islands (see below for the Komandorskis). The Striking Force, of course, might also steam straight to its objective from Pearl Harbor or leave Hawaii on a westerly course and approach the objective from the south. The most probable route of the Expeditionary Force would be from Puget Sound to the Aleutians, then to the objective, passing north or south of the Aleutians chain. The Expeditionary Force alternatively might head from Puget Sound to the Central Pacific and then on to its objective, or even directly from Puget Sound to the objective. He did not think, however, the last two routes very likely. He saw the Blue Striking Force then attempting to reduce or neutralize the Purple bases at Paramushiru, Shimushu, and Avacha Bay as well as the air bases supporting them. The Expeditionary Force, he thought, would have Paramushiru and Shimushu as its first objective, then Avacha Bay, Etorofu, and the Komandorski Islands, in that order. He did not, however, think that the last two locations were likely objectives. (Pages 47–48.)

Since the Striking Force was so powerful, locating it early was critical. At about twenty knots, the Striking Force could reach either the Aleutians or PA's thousand-mile operational area in about four days; the Expeditionary Force would require about six or seven days. To search for these forces, he had 248 naval patrol planes, two light and two escort carriers, two battle cruisers, six heavy and six light cruisers, thirty destroyers, and sixteen high-speed submarines. His naval patrol planes were all radar equipped, able to search with "fair effectiveness" in any weather in which they could fly. Since the weather would be generally poor, he knew, these planes would usually be able to fly and search by radar alone. His thirty-six Mariner naval patrol bombers could be either land or tender based and had a search radius of 1,050 miles, while his forty-two Catalinas, also land or tender based, had a radius of about 880 miles. His ninety-six Privateers had a search radius of 970 miles but were only land based, as were his eighty-four Harpoons, which had a radius of 750 miles. He could operate from his three seaplane tenders only the fifty-four planes they could service; his land-based planes would have to take on four missions to every one that was carried out by the tender-based planes. (Pages 49–50.)

PA found the Gulf of Kamchatka to be desirable for bases for seaplane-tender operations for a number of reasons. It would permit him to conduct extended searches of the northern part of the area he had to cover, something not possible from his land bases. In addition, for bombing purposes, planes from the gulf would have a trip to Adak of eight hundred miles, as compared with over nine hundred from Avacha Bay and a thousand from the Kashiwabara Bay area. Furthermore, the Gulf of Kamchatka was usable in the summer. It would also allow these units to have antisubmarine as well as fighter protection (all to be marked on a chart in an appendix to the Search Plan). There were risks, however, to exposing the tenders and their support units. He thought, though, that these ships could be provided fighter cover; also, if he detected the Blue forces early, the tenders could retire to Avacha Bay, where they could continue their searches under fighter cover from the Paramushiru area. Or he could at that point retire the seaplane tenders to rear areas and continue his searches with land-based planes only. Meanwhile, Fleet Air

Wing 10 was the logical choice for this assignment; he ordered it to a tender-based group in the Gulf of Kamchatka, warning it of the probability that the tender would be pulled back once the Blue Striking Force was located. He left Commodore PL, COMFAIRWING 10, as much freedom of action as possible in searching the areas he had delineated. Search-area boundaries had been selected so as to maximize ranges in the northern sectors, where he saw the weather offering the maximum concealment for surprise strikes. To compensate, however, for the possibility of a "high speed" strike by the Blue Striking Force coming directly from Pearl Harbor or from the south, he established a submarine line. (Pages 50–51.)

As for his surface forces, the Purple escort carriers (built on auxiliary- or merchant-type hulls) were slow, so Admiral PA would detach them as a Service Force Air Support Group. His light carriers, however (having cruiser-type hulls and propulsion plants), would participate as a source of combat air cover, as an anti-submarine patrol force, and as a messenger service for the Seaplane Tender Group. Because the weather would probably handicap air operations, he was going to divide his surface search assets into two groups, HAMMER and SICKLE, leaving their organization, stationing, and other details to Commander, Fleet Attack Force, Rear Admiral PB, who was going to command the bulk of PA's fleet units. PB would manage the critical matter of concentrating the combatant surface forces and, through him, PA could order any surface searches desired by PA himself or his Air Group Commander. However, he did not want his two battle cruisers or their screening units employed in surface searches; these units were to be held in Kashiwabara Bay, under continuous land-based air cover, until it was necessary to concentrate the Purple Pacific Fleet for action. He further ordered his Surface Force Commander to furnish fighter cover to the naval patrol planes when needed. (Page 51.)

SUBMARINE PATROL LINES

1. Potential submarine patrol lines are delimited as follows:

Name of Line	Description of Line
HOPS	Line between latitudes 52°N and 50°N along longitude 170°W.
CASTOR	Line between latitudes 56°N and 54°N along longitude 170°W.
SUMAC	Line between latitudes 50°N and 49°N along longitude 177°E.
HELLBROTH	Line between latitudes 51°N and 50°N along longitude 180°.
HEMLOCK	Line between latitudes 52°-10'N and 51°N along longitude 173°E.
POPPY	Line between latitudes 54°N and 53°N along longitude 180°.
BROOM	Line between latitudes 55°N and 54°N along longitude 176°E.
OLIVE	Point of origin latitude 54°N longitude 170°E. Line runs through point of origin in direction 120°-300° true, extending in each direction to within ten miles of MEDNI ISLAND and ATTU ISLAND respectively.
POISON	Point of origin latitude 35°N longitude 170°E. Line extends two hundred miles from point of origin in direction 210° true.

Fig. 106
Submarine Patrol Lines, Purple Staff Solution

Turning to his submarines, he had four on patrol in the Aleutians. Two of these had orders to proceed to Dutch Harbor, and the other two were being diverted to Pearl Harbor. Four more submarines were in overhaul until 10 June, and eight additional were training around Kashiwabara Bay; he had worked out a rotation schedule for their employment. Each of his four submarine divisions had four submarines. The two boats of Submarine Division 1 that were overhauling in Vladivostok would depart on the 10th, top off fuel from tenders in the Kuriles, and then proceed to their patrol stations, arriving off Dutch Harbor and Adak, respectively, by 17 June. The four units of SUBDIV 2 that were off Dutch Harbor and en route to Pearl on 4 June would return to Vladivostok by 17 June for refit. The eight boats of SUBDIVs 3 and 4 would continue training at Kashiwabara Bay until "Jump Day," the day that Admiral PA's Operation Plan was executed. On Jump Day, SUBDIVs 3 and 4 would proceed to their patrol and picket stations. (Pages 51–52.)

6277-5777-PSS
19/Sept/46gq

- 52 -

CONFIDENTIAL

DATE	SUBDIV 1		SUBDIV 2		SUBDIV 3	SUBDIV 4
	Two Units	Two Units	Two Units	Two Units		
4 June	OVERHAUL VLADIVOSTOK		OFF DUTCH HARBOR	ENROUTE PEARL	TRAINING KASHIWABARA	TRAINING KASHIWABARA
10 June #	DEPART FOR PATROL			OFF PEARL	TRAINING KASHIWABARA	TRAINING KASHIWABARA
17 June	OFF ADAK	OFF DUTCH HARBOR	RETURN FOR REFIT		TRAINING KASHIWABARA	TRAINING KASHIWABARA
JUMP Plus 2* DAY	JUMP-DAY DATE UNKNOWN		JUMP-DAY DATE UNKNOWN		PROCEED PATROL AND PICKET STATIONS	PROCEED PATROL AND PICKET STATIONS

#On departure VLADIVOSTOK proceed at maximum sustained speed to KURILE based tender; top off and proceed immediately to patrol stations. SUBDIV 1 may anticipate the 10 June departure date from VLADIVOSTOK.

*JUMP-DAY is day of execution of my operation plan. Patrol stations for SUBDIVS will be indicated on the chart appendix to search plan. Picket stations will be discussed later.

Fig. 107
Purple Submarine Force rotation schedule, Operations Problem 2

All six fleet air wings—comprising both land- and tender-based air units—were under Rear Admiral PK as part of the Air Strike Force. The Air Strike Force, in turn, would be divided into the Land-Based Air Group and the Kamchatka Group. The Land-Based Air Group was to consist of Fleet Air Wings 20 through 60, inclusive, as well as two submarines for radar-picket duty; it would be under Commodore PM. The Kamchatka Group, under Commodore PL, would comprise the three seaplane tenders, one escort carrier and three destroyers (escort carrier CVL-2 and destroyers DD-4, -5, and -6), and Fleet Air Wing 10, as well as one light carrier and three destroyers from the Fleet Attack Force. (The latter units would not return to the Fleet Attack Force until Admiral PA was certain that a surface engagement was imminent.) In general, the tasks of the Air Strike Force would be to search and patrol in accordance with the Search Plan; report, track, and develop (i.e., collect maximum possible information about) any contacts; and provide search and rescue services. It was also to establish and maintain a seaplane tender base at the Gulf of Kamchatka as well as conduct photoreconnaissance of bases designated in the Search Plan. (Page 52.)

The Submarine Force would be under the command of Rear Admiral PJ. Comprising fourteen submarines and two submarine tenders after two boats were detached as radar pickets, the Submarine Force was to patrol off Pearl Harbor, Dutch Harbor, and Adak to cover probable approach routes and detect and report on Blue forces. Its primary initial mission was to make contact reports, but once this was accomplished the mission became the destruction of Blue forces encountered. The submarines were to patrol along patrol lines designated above, making amplifying reports when able to do so without jeopardizing their primary mission. One submarine tender would be based at Vladivostok for refits, the other at Kashiwabara to conduct repairs and training. However, when Submarine Divisions 3 and 4 departed for their stations two days after Jump Day (Jump Day+2), the second tender would proceed to Taraika Bay in Sakhalin and serve as a base for refits and repairs. Admiral PA designated the target priorities for the submarines as carriers, transports and cargo ships, battleships, tankers, and then other combatant ships. (Page 53.)

When possible, submarine stations would be "staggered" in depth to increase the possibility of contacts. The patrol lines for SUBDIVs 3 and 4—less one boat each for radar-picket duty—would be CASTOR and POISON; radar-picket stations would be assigned by Commander, Air Strike Force. One likely patrol line was between the Gulf of Kamchatka and Attu, three hundred miles long. Another would be between Kashiwabara and Attu, four to five hundred miles, positioned to furnish aircraft warnings and intercept ships. The submarine protecting Kashiwabara could be backed up if desired by radar-picket destroyers under land-based fighter cover. (Pages 53–54.)

The Surface Search Groups would patrol their assigned stations, conduct the searches ordered, and maintain readiness for rapid concentration on the Battle Cruiser Group, requesting fighter cover for the naval patrol planes.

Admiral PA understood that all of these search measures would put his command under "strenuous use," but he also knew these actions would provide early and accurate intelligence on Blue that was so vital to the destruction of those forces. His own headquarters would be ashore at Kashiwabara Bay, where he would have communication facilities and be closer to the hub of search operations than if he were at sea. This arrangement would also allow him to be in immediate contact with Commander, Air Strike Force. (Page 54.)

Admiral PA's 250-odd planes were more than enough to carry out searches, leaving plenty for strike operations. He planned initial air strikes on Attu, Shemya, Amchitka, and Adak–Tanaga. These strikes would commence as soon as the Blue forces had been located by his search units in accordance with a schedule devised by Commander, Air Strike Force, Admiral PA's second in command for the operation. In general, air strikes on Blue's bases would be without fighter cover, though fighters could be furnished from carriers if that did not mean exposing the carriers too much. Regardless, however, Attu and Shemya were his primary targets since these bases were so close to his proposed tender base. Once the Blue Striking Force was within effective range, Purple's land-based air units would concentrate on destroying it with bombs, torpedoes, and rockets. Details of this plan would be prepared by Commander, Air Strike Force, who was to allocate his land-based fighters; set up and maintain emergency capacity air defenses at Paramushiru, Shimushu, Avacha Bay, Matsuwa, Etorofu, and Sakhalin Island; set up a replacement system for planes and personnel; and be responsible for the air defense of the Purple bases. (Page 55.)

Commander, Air Strike Force would be assisted in the latter task by the two submarines assigned to radar-picket duty. In addition, the Kamchatka Group would have about fifty fighter planes in addition to those fighters from light and escort carriers, a total of about a hundred planes, which Admiral PA thought adequate given that the Seaplane Tender Group would retire to Avacha Bay once the Blue Striking Force had been located, since the enemy could then be tracked from land bases. He understood that the Tender Group could have a strong fighter cover from Kashiwabara, but that was six hours' flight time distant, during which the tenders would be under their original land-based fighters as well as those from the Purple carriers; he was willing to accept this risk to his naval patrol planes in order to obtain the maximum possible warning of the Blue Striking Force's approach, if from the north. Early detection would allow him to exploit the capabilities of his high-speed submarines and ideally eliminate half or more of the Blue carriers. His Air Strike Force, after completing its searches, would destroy the Blue Striking Force with bombs, torpedoes, rockets, and guns at the earliest opportunity. It was then to attack the Blue Expeditionary Force in the same manner, destroy any Blue submarines and aircraft encountered, and defend Purple's bases in the Kuriles, Kamchatka, and Sakhalin from air attack. The Air Strike Force was to control and coordinate all the Purple air strikes against the Aleutian bases and provide antisubmarine protection for the Gulf of Kamchatka. (Pages 55–57.)

Surface forces were organized into the Fleet Attack Force, under Rear Admiral PB—two light carriers, two battle cruisers, six heavy and six light cruisers, and twenty-four destroyers. This "fast, strong" group, with its own air cover that was to be supplemented by land-based air, would be supplemented further by the Kamchatka Group (one escort carrier and three destroyers), a Service Force (whose six oilers were protected by one escort carrier and three destroyers), and the one light carrier and three destroyers now detailed to the Air Strike Force. Admiral PA thought that a surface engagement was not inevitable but that an early air action was.

Strikes on the Blue Aleutian bases would allow him to relieve Commander, Fleet Attack Force of the task of the air strikes on these bases by Jump Day+4 and ensure the latter force's readiness to act against Blue surface forces. He expected the Fleet Attack Force to face a Blue Striking Force of six fleet carriers, five heavy cruisers, three light cruisers, and twenty-five destroyers, as well as three escort carriers, two battleships, two light cruisers, and twenty-five more destroyers in the Blue Expeditionary Force. Before Purple Fleet Attack Force and the Blue Striking Force had closed to gun range, he expected his attrition strikes to have reduced Blue air strength by half or more. But if the Blue Striking Force reduced Purple land-based air considerably, a surface engagement was probable. In it, he expected Purple to have a great advantage in speed but this meant that the Fleet Attack Force had to be ready for a surface engagement with all or part of the Blue Striking Force, all or part of the Blue Expeditionary Force, or some combination of both forces, minus only the destroyers protecting the Blue Expeditionary Force's transports. (Pages 57–58.) One solution was to tie in the two Surface Search Groups as flank forces when the Fleet Attack Force concentrated for its surface engagement.

Commander, Fleet Attack Force was to be supported by Commander, Air Strike Force, who was to provide fighter cover to surface forces within 250 miles of the fleet air wing shore bases. In fact, any commander needing shore-based fighter cover was to request it from the Air Strike Force. The Fleet Attack Force would provide antisubmarine surface patrols to the Tender Group at Kashiwabara Bay. (Pages 58–59.)

The operation could be implemented any time after 10 June. It was necessary to get his Seaplane Tender Group and its fighter support as near to the Gulf of Kamchatka as possible to take the forty-eight-hour time lag into account. This could be done by redeploying Fleet Air Wing 10 to Paramushiru and basing its necessary support units there for training instead of at Vladivostok. He therefore ordered Commander, North Pacific Air Command; Commander, Carrier Division (COMCARDIV) 1; COMDESRON 1; and Commander, Mine Squadron 1 to carry out the movements involved. Commander, North Pacific Air Command was to send Fleet Air Wing 10 to Paramushiru and two seaplane tenders to Suribachi Bay before 10 June and be prepared for extended operations. Commander, North Pacific Air Command, at his discretion, could also redeploy the naval patrol bombers of Fleet Air Wing 30. COMCARDIV 1 was to ready and train his two light carriers and three escort carriers for extended operations in the Okhotsk Sea south of latitude 53. After 10 June, COMCARDIV 1 was to base his unit at Kashiwabara Bay and await further orders. COMDESRON 1 was to detach six of his destroyers, half to Commander, North Pacific Air Command and half to COMCARDIV 1. Commander, Mine Squadron 1 was to ready one light minelayer and two destroyer-minesweepers for extended operations and have them report to

Commander, Fleet Air Wing 10 at Paramushiru no later than 10 June for training. (Pages 60–61.)

Admiral PA was concerned about the dispersal of forces against a major surprise attack by Blue before he executed his plan. For that reason, he wanted to retire his submarines in training and their tender to a base such as Uruppu, over three hundred miles farther from the Blue air bases than Kashiwabara. Such a move, however, would add a day's run to Patrol Line CASTOR, a cost that he considered unacceptable in view of the fact that the Blue Striking Force—if bound for Dutch Harbor—would arrive in four to five and a half days. Purple submarines from Kashiwabara, restricted by their two-day readiness notice, could reach their patrol line only on Jump Day+5 or +5½. Therefore, he decided to leave his submarines and their tender at Kashiwabara until Jump Day+2 in a bay well separated from the submarine group and under strong land-based air cover. (Page 61.)

By Jump Day, Commander, Air Strike Force would have issued his Search Plan and effected the necessary redeployment of his aircraft. Tender-based air was to be based by Jump Day—with its tenders, supporting carriers, destroyers, and minesweepers—at Paramushiru. Seaplane tenders and supporting units were given forty-eight hours' notice to leave Vladivostok for Paramushiru, ready for extended operations upon arrival. PA's Movement Plan called for their departure on Jump Day+1 from Paramushiru to the Gulf of Kamchatka, a move he did not expect to take more than a day and a half at fifteen knots.

On Jump Day, he would have eight submarines training at Kashiwabara Bay and another eight either at or en route to or from their patrol stations. The eight submarines training at Kashiwabara would be dispatched for patrol on Jump Day+2, at which time he would retire their tender and its escorts to Taraika Bay, near Shikuka on Sakhalin Island. Use of this bay for refits would save his boats over a five-hundred-mile transit each way as compared with Vladivostok. Taraika, in fact, was over a thousand miles from the nearest Blue base, would have local fighter protection from Alexandrovsk, and would enjoy the entire Kuriles chain for perimeter defense and warning. The tender would, however, need antisubmarine protection, for which he was considering minesweepers. However, submarines needing extended refits because of battle damage or operational casualties would still have to be repaired back at Vladivostok. These plans would have to be modified if the Operation Plan was not executed prior to SUBDIV 3's departure for patrol in early July. (Pages 61–62.)

The Fleet Attack Force would by this time have its Surface Search Groups at sea, with the remainder at Kashiwabara Bay. The Fleet Attack Force was to be ready for departure from Vladivostok on Jump Day+2 for Kashiwabara Bay, a three-day run at fifteen knots. It would be accompanied by minesweeping and possibly logistical units, but the light carrier and three destroyers under Commander, Fleet Attack

Force would be detached for operations in the Okhotsk Sea. Admiral PA was going to have the Service Force (one escort carrier, three destroyers, and six oilers under Commodore PI) brought forward from Vladivostok to Aniwa Bay, Sakhalin, except for units detached to travel with the Fleet Attack Force to Kashiwabara or independently to the Gulf of Kamchatka. Aniwa Bay was about 1,200 miles from the nearest Blue air base and could get fighter protection from Etorofu or Alexandrovsk. In addition, its position west of the Kuriles, he thought, would give "ample" warning for impending air raids. The Mine Force, comprising the light minelayer, four destroyer-minesweepers, and twelve minesweepers, would also be ready to depart from Vladivostok on Jump Day+2. The light minelayer and two destroyer-minesweepers had already been assigned to the Seaplane Tender Group, while the other units would be routed to Kashiwabara Bay, Taraika Bay, and Aniwa Bay. Finally, the Movement Plan required Commander, Fleet Attack Force to consult with Commander, Air Strike Force about rendezvous points, operating areas, fueling areas, routes between bases, and other details, and then issue orders. (Page 62–63.)

Admiral PA's Mine Plan was considered under two headings, minelaying and minesweeping. Minelaying fell into two categories as well, offensive and defensive. His potential minelaying facilities were the six fleet air wings, the two light and two escort carriers, the sixteen submarines, and the light minelayer. He did not plan on using his submarines for minelaying, since there were so few and their high speed gave them more offensive value carrying torpedoes. Moreover, torpedoes gave his submarines a defensive capability that mines did not. He also did not plan on using his surface units for offensive mining, because they would almost certainly be destroyed before they could reach offensive mining positions. That left his fleet air wings and carrier air groups. The ninety-six Privateer and thirty-six Mariner naval patrol bombers appeared best suited to offensive mining. Each could carry as many as eight thousand-pound or four two-thousand-pound mines. At a "reasonable" hazard in endurance, these planes could lay mines out to a 650-mile radius, which would allow them to reach Attu, Shemya, and Kiska. With only four mines per plane, the naval patrol bombers could reach Amchitka and Adak, though the cost would be excessive because of the heavy defenses there. Precise timing would allow for some carrier support in operations against Attu and Shemya, but at considerable risk. (Pages 64–65.)

To mine the Blue bases at Adak and Amchitka, he would have to resort to carrier air, but on a day with good visibility and using both light carriers, his fighters would be outnumbered as much as five to one. In poor visibility, he would be limited to twelve night fighters that could lay a total of twelve mines, which he did not think worthwhile. He speculated about using jet-assisted takeoff to launch naval patrol bombers from his carriers so as to extend their range, but he did not think that was technically feasible. In the end, he thought, a compromise was necessary between the cover afforded by poor visibility and the good visibility required for accurate mining. Admiral PA concluded that offensive minelaying near and at the extreme western Aleutian bases was practical for the fleet air wings, and he assigned this mission to Commander, Air Strike Force. For defensive mining and minesweeping, he decided to employ his light minelayer for the former and all of his destroyer-minesweepers and minesweepers for the latter. The light minelayer was a high-speed unit that could lay a full load of mines accurately in one day, as well as take on board another full load. These operations would require a mine base ashore, which would be Kashiwabara Bay. (Page 65.)

In view of the numerous entrances through the Kuriles into the Okhotsk Sea—some of them over a thousand fathoms deep—as well as the limited forces he could use for mining and patrolling access points that were not mined, Admiral PA decided to limit defensive mining to his ship bases as protection against submarines. The light minelayer and two of the destroyer-minesweepers at Paramushiru could, upon execution of the plan, make a high-speed run to the Gulf of Kamchatka, sweep and buoy a channel, and then plant a defensive minefield. The light minelayer could then return to Kashiwabara and plant defensive mines there, in conjunction with the sweeping and buoying of channels by the two destroyer-minesweepers arriving from Vladivostok with or preceding the Fleet Attack Force. The light minelayer could then plant defensive minefields at Aniwa Bay and Taraika Bay, where PA thought immediate protection was less urgent. Commodore PR, Commander, Mine Force, whose command would comprise Mine Squadron 1, also had to allot his minesweepers among the four bases employed in the operation and to issue the necessary movement plans for passage from Vladivostok. He would place his headquarters at Kashiwabara Bay so as to be in close contact with Admiral PA. Finally, Commander, Mine Force would evaluate all reports on mines and keep all commands informed. (Page 66.)

Admiral PA's Logistics Plan focused on fuel and aviation gasoline. He did not have a critical problem, because his bases would be so near his operations and he had short, interior lines of supply. He did, however, have to make do with what he now had. The Kurile and Sakhalin naval bases and stations, especially Kashiwabara Bay and Hitokappu Bay, were well equipped and stocked with fuel and supplies for any ship that could use the harbors there. Three of his task forces could be eliminated as potentially posing logistics problems. The Submarine Force would take all services from tenders, which in turn could exchange places with the Vladivostok tender. When required, fuel could also be sent via tanker. The Service Force was obviously self-sustaining. The Mine Force, being supported by a variety of bases, would fuel where located and would cause little drain on the local facilities. The other task forces, the Air Strike Force and the Fleet Attack Force, would both be fully loaded on arrival at Kashiwabara Bay and the Gulf of Kamchatka, respectively.

The Air Strike Force would have to replenish itself periodically at or near the Gulf of Kamchatka, especially with fuel for the Kamchatka Group's surface units. There would also be a need to replenish aviation gasoline used by the tender-based and carrier-based planes as well as by the land-based fighters at Khayryuzovo on the Kamchatka Peninsula. Diesel fuel for minesweepers would have to be taken into account, though he thought it could be supplied by the seaplane tenders in the Gulf of Kamchatka. (Page 67.)

Admiral PA assumed that a tanker arriving at the Gulf of Kamchatka every fifteen days could take care of "average expected" needs; the schedule could be adjusted readily as conditions required since he had short, well-covered lines of supply. He would use two tankers, to allow even more frequent arrivals, about every ten or twelve days. Whatever fuel Kamchatka could not take would be off-loaded at Kuriles bases as feasible for returning to Vladivostok. This arrangement would leave him with four tankers to care for the Fleet Attack Force and the Fleet Air Wing units at the Kurile–Sakhalin bases, supplemented by partial loads from tankers en route from Kamchatka to Vladivostok. This arrangement appeared ample since fleet units would not all be operating twenty-four hours a day and the tankers would have shorter runs to the Kurile bases than to the Gulf of Kamchatka. The Service Force, even when fueling at sea, could expect to be within land-based fighter range at all times, though he had assigned one escort carrier and three destroyers to protect it and, if circumstances required, he would assign additional escort units. His Movement Plan directed Commander, Service Force to sail from Vladivostok on Jump Day+2 and establish a base at Aniwa Bay, Sakhalin, shifting his flag (i.e., his headquarters on board ship) as necessary. In the event of tanker losses and a lack of replacements, Admiral PA would have well stocked bases to draw on for a considerable period by the rotation of units to port. (Pages 68–69.) All large ships were to be prepared to refuel destroyers and other small ships. On or about 1 August, when logistics became available at Avacha Bay, units would base there.

APPENDIX ONE - LOGISTICS PLAN - AVGAS

6277-5777-PSS
15 Sept/45 gq

Ship or Base	No. and Type Planes	Estimated Average No. Plane Missions Per Day	Estimated Gallons Expenditure per Mission	Total Expenditure Gallons
CVE	16 VF	30 CAP - COMBAT	300	9000
CVE	12 VA	5 PATROL - COMBAT	500	2500
CVL	36 VF	60 CAP - COMBAT	300	18000
CVL	12 VA	5 PATROL - COMBAT	500	2500
TENDERS	50 VP	9 SEARCH - COMBAT	2150 Weighted average for two tender base VP types	19350
GALS. PER DAY				51350
TANKERS PER DAY		TANKER LOADING "ABLE"		1/16

POSSIBLE TANKER LOADS (BBLS.)

Key Name	FUEL	DIESEL	AVGAS
ABLE	82300	--	19,600 BBLS. 823,000 GALS.
BAKER	68000	--	34,000 BBLS. 1,422,000 GALS.
CHARLIE	68000	16200	19,600 BBLS. 823,000 GALS.
DOG	97000	--	--

TENDER AVGAS CAPACITIES

Type	No.	Capacity Gals.	Total Capacity
AV	1	279,000	
AVP	1	84,300	
AVP	1	84,300	
			447,600

KAMCHATKA GROUP - 70 - CONFIDENTIAL

Fig. 108
Appendix 1, Logistics Plan, Purple Staff Solution, Operations Problem 2

APPENDIX TWO – LOGISTICS PLAN – FUEL

Ship Type	No.	Barrels Stowage Per Ship	Bbls/hr. 16 kts.	Bbls/hr 18 kts.	Bbls/hr. 20 kts.	Total For Type 16 kts.	Total for Type 18 kts.	Total for Type 20 kts.	Port Use Bbls/Day	Total Port Use For Type Bbls/Day
CVE	1	30338	23	32	—	23	32	Max. 32	63	63
#AY	1	14544	20	27.5	—	20	27.5	Max. 27.5	60	60
*# AVP	2	1776	6.5	—	—	13	Max. 13	Max. 13	30	60
DD	3	3324	13	17	21	39	51	63	14	42
CVL	1	12400	33	42	53	33	42	53	65	65
DD	3	3324	13	17	21	39	51	63	14	42
BBLS/HR						167	217	252		
BBLS/DAY						4008	5208	6048		332
TANKERS PER DAY						1/20	1/15	1/13		
CVL	2	12400	33	42	53	66	84	106	65	130
CB	2	24143	34	42	54	68	84	108	70	140
CA	6	14839	34	41	51	204	246	306	63	370
CL	6	13850	30	36	44	180	216	264	63	378
DD	24	3324	13	17	21	312	408	504	14	336
BBLS/HR						830	1038	1288		
BBLS/DAY						19920	24912	30912		1362
TANKERS PER DAY	TANKER LOADING ABLE OR DOG					1/4 (ABLE)	1/4 (DOG)	1/3 (DOG)		

*Indicates fuel is Diesel. Since AVPs will normally not be under way, these figures will care for Diesel expenditures of Ams based at GULF of KAMCHATKA.
#Will not be under way normally.
NOTE: (1) DM, DMS and AM are not included, since they will be apportioned among the bases, will fuel accordingly and have small consumptions.
(2) All fuel figures are in barrels.

6277-5777PSS
18/Sep/46
KAMCHATKA GROUP
- 71 -
FLEET ATTACK GROUP

Fig. 109
Appendix 2, Logistics Plan, Purple Staff Solution, Operations Problem 2

Kamchatka fighters and Catalinas were to fly from Paramushiru to Khayryuzovo prior to Jump Day with small maintenance crews and necessary equipment. Some of these supplies and equipment could be dropped by parachute, while the rest could be landed since naval patrol bombers could use its three-thousand-foot-long runways. On Jump Day, forty-eight fighters would be flown there from Paramushiru; the 6,700 gallons of aviation gasoline they required could be carried by four to seven naval patrol bombers. Refueling would be difficult up to 1 August but the added fighter protection for the Kamchatka Group would be valuable. In addition, he planned to have personnel relieved often. This plan required setting an additional task for the Air Strike Force, which was to establish and supply by air a Fighter Plane Group at Khayryuzovo. (Page 69.)

Admiral PA had only a "skeleton" outline for his task organization so as to allow task force commanders maximum flexibility. About 5 June, members of his staff would fly to Kashiwabara Bay, organize the base facilities there to fit the pattern he wanted, and handle preliminary coordination with Commander, North Pacific Air. On Jump Day, he would transfer to Kashiwabara as his base. Commander, Air Strike Force was designated as his second in command since that officer would control the major search instrument and combat weapon, the fleet air wings. By remaining at Kashiwabara with the Air Strike Force Commander, Admiral PA hoped to attain optimum flexibility in the coordination of his various groups. However, Commander, Air Strike Force would remain "well separated" from him to reduce the possibility of an air raid eliminating both of them at once. He code-named his operation BLUEPILL. (Page 72.) Admiral PA now repeated and summarized the tasks, detailed above, of his major subordinates. (Pages 73–74.)

Whenever Commander, Air Strike Force and Commander, Fleet Attack Force were engaged in coordinated action, the former would control. Surface force commanders were to be ready to take defensive measures at sea against atomic bomb attacks and to effect maximum practicable dispersal in harbor. Commander, Air Strike Force would request fighter support for the naval patrol bombers from Commander, Fleet Attack Force but only when the forces assigned to the former were unable to provide the needed support. Commander, Purple Pacific Fleet would also be involved in these cases. Task force commanders desiring shore-based fighter cover for surface and submarine units at sea would notify Commander, Air Strike Force twenty-four hours in advance. Shore-based and fleet-based aircraft that were lost could be replaced within thirty-six hours by dispatch request to Commander, Far East Naval District. (Pages 74–75.)

All forces were to be alert to take advantage of the weather and anticipate similar efforts by Blue. Related to this, task forces, groups, units, and aircraft were ordered to break radio silence and report by priority-precedence message indications of gale-force winds, heavy swells, or storms that could affect operations. Units were not to break radio silence, however, if that would impair operations in progress. Admiral PA would establish a major radio-intelligence center at Kashiwabara and assign additional communication personnel to it. In view of the possibility of heavy damage to the Kashiwabara facilities, a relief organization at Etorofu would be capable of taking over the entire communications system. Special provisions would be made for submarine broadcasts that repeated messages day and night. (Pages 75–76.)

A special "repeat-back" schedule was also going to be run between Kashiwabara and Vladivostok to permit the reception of messages by all units at sea without the need to receipt for them. PA attempted to ensure that procedures for contact and amplifying reports were "absolutely" clear. He prioritized traffic related to aircraft replacement, weather data, and submarines. Also of concern was radio silence for ships at sea, the assignment of zone times in messages, and the use of planes to deliver messages so that radio transmissions could be avoided. (Page 76.)

For the Board Maneuver, Admiral PA divided Rear Admiral PK's Air Strike Force into the Land-Based Air Group under Commodore PM and the Kamchatka Group under Commodore PL. The Land-Based Air Group comprised Fleet Air Wings 20, 30, 40, 50, and 60, with 210 naval patrol bombers, 193 day fighters, 126 night fighters, and two submarines among them. The Kamchatka Group was to comprise one of the light carriers, one of the escort carriers, all three seaplane tenders, six destroyers, and the forty-eight naval patrol bombers of Fleet Air Wing 10. Rear Admiral PJ's Submarine Force was to consist of the other fourteen submarines assigned to the Purple Pacific Fleet, along with their two submarine tenders. Commodore PR's Mine Force was the light minelayer, four destroyer-minesweepers,

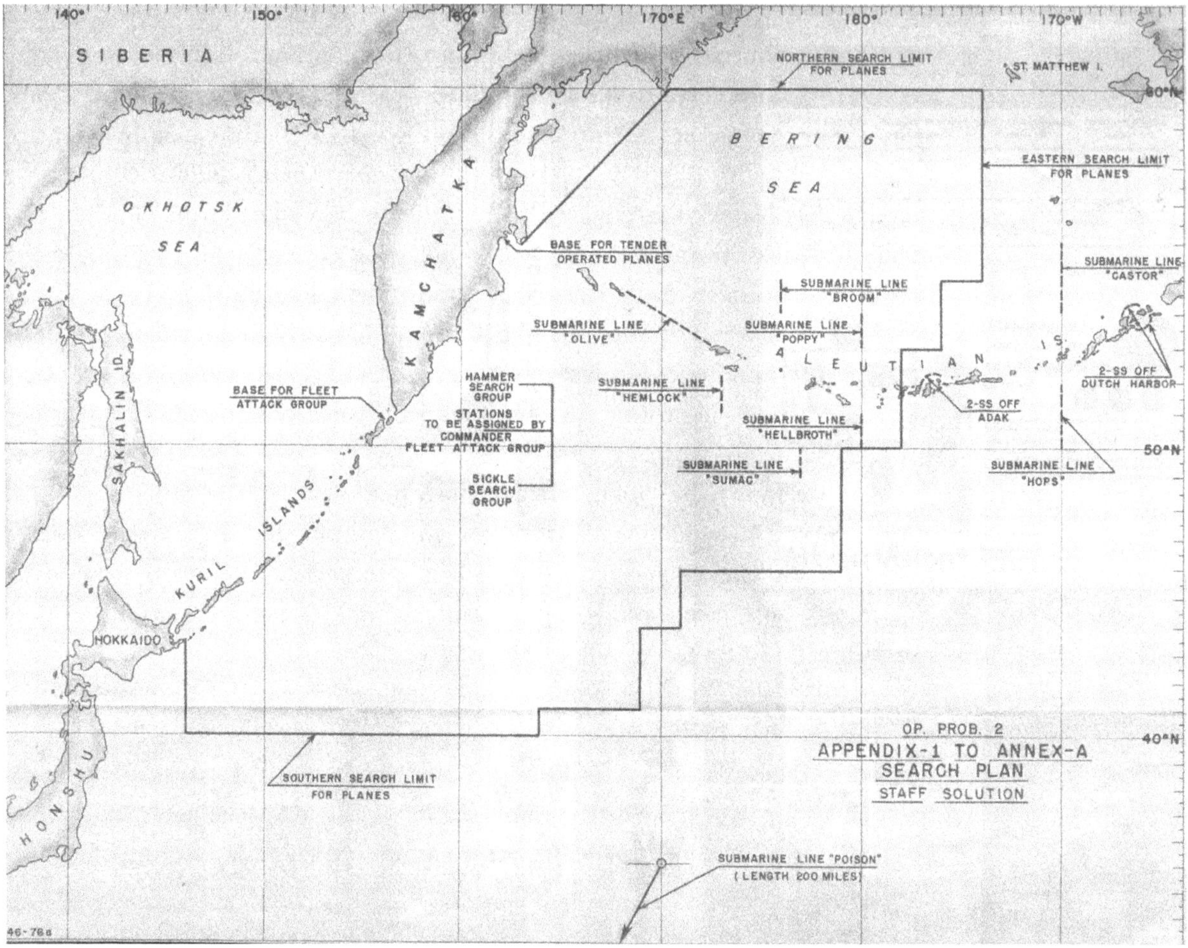

Fig. 110
Search Plan, Purple Staff Solution, Operations Problem 2

and twelve minesweepers noted above in connection with the Mine Plan. Commodore PI's Service Force comprised the second of Admiral PA's escort carriers, three additional destroyers, and the six oilers. The main body was Rear Admiral PB's Fleet Attack Force of two light carriers, two battle cruisers, six heavy and six light cruisers, and twenty-four destroyers. (Page 77.)

As of 5 June, there were no known Blue mines in Purple areas of operation. Given the Purple assumption that Blue would see an early advantage in the use of atomic bombs, Admiral PA also warned ships to effect maximum practical dispersal while in harbor. (Pages 78–80.)

As soon as was practicable after receipt of the Operation Plan, the Air Strike Force would effect preliminary redeployment of aircraft and equipment, including establishing its air groups in southern and central Kamchatka. Commencing on Jump Day, it was to search twice daily in the areas delimited in Appendix 1 to Annex A of the Operation Plan. When necessary on strike days or for other reasons—even when there was authorization to reduce the extent and number of searches—Commander, Purple Pacific Fleet was to be advised of any such reductions.

Whenever circumstances permitted, search ranges were to be extended as delineated in Appendix 1. Single-plane searches were to be made twice daily of the Komandorski Islands, once daily up the east coast of Kamchatka south of 57 degrees north, and once daily of the Kurile Islands. Negative results were not to be reported. Daily photoreconnaissance of the Aleutians as far east as 175 degrees west longitude was to be undertaken as feasible, with priority targets in the order of Adak, Amchitka, Attu, and Shemya. These results were to be reported. (Page 81.)

Upon execution of the Operation Plan, Submarine Division 1 would depart for patrol—on 10 June, or sooner if able. Two units were to proceed to patrol off Dutch Harbor, relieving the two boats already there. On Jump Day+2, SUBDIVs 3 and 4—each less one boat—would proceed to Patrol Lines CASTOR and POISON, with three submarines on each line. On Jump Day+2, the two radar-picket submarines would also report to Commander, Air Strike Force for duty. The Fleet Attack Force would initially operate to the east of the Kurile Islands between latitudes 52° and 49° N and within 250 miles from 50°30′ N, 156° E. It would also organize Surface Search Groups HAMMER and SICKLE and be prepared to conduct searches as ordered by Commander, Purple Pacific Fleet. All units were to advise Commander, Purple Pacific Fleet whenever it was considered that the searches conducted were providing less than 75 percent probability of detection of the large Blue surface forces. Upon initial contact with the Blue Striking Force, Commander, Air Strike Force was to be prepared to retire the Kamchatka Group to Avacha Bay or to the rear and from there continue his searches to the maximum feasible ranges. Appendix 2 of the Operation Plan contained the geographical limits of the potential submarine patrol lines. Task force commanders were to submit their implementing plans by 8 June. (Pages 81–82.)

Admiral PA directed no unit movements prior to Operation BLUEPILL except those noted in the Search Plan. He did, however, detail that when SUBDIVs 3 and 4 departed for their patrol and picket duties, submarine tender AS-3, escorted by one destroyer, would leave for Taraika Bay on Sakhalin to carry out submarine refits until further orders. Submarine tender AS-2 would remain at Vladivostok until further orders. The Fleet Attack Force would base such units as it deemed advisable at Kashiwabara Bay under "maximum practicable readiness" for getting under way. Prior to departure from Vladivostok, and not later than Jump Day+1, it would direct COMCARDIV 1 to detach escort carrier CVE-4 and COMDESRON 2 to detach destroyers DD-16, -17, and -18 to Commander, Service Force. The Service Force—less the units detached for logistical purposes—was to proceed on Jump Day+2 from Vladivostok for Aniwa Bay, Sakhalin, staying there until it received further orders. The Mine Force, less its units detached for implementing the Mine Plan, was to proceed on Jump Day+2 to Kashiwabara Bay and remain until further orders. (Page 84.)

On Jump Day, all task force commanders would commence final loading and preparations for "extended war service." On the same day, they would complete any aircraft preparation needed for extended war service. If practicable on the same day—but not later than Jump Day+1—Commander, Air Strike Force would dispatch light minelayer DM-1 and his two destroyer-minesweepers from temporary duty at Paramushiru to the Gulf of Kamchatka at maximum sustained speed with orders to buoy and sweep a channel as well as plant the minefields detailed in the Mine Plan. At this point, DM-1 would return to Kashiwabara Bay and report to Commander, Kamchatka Group. Also on Jump Day+1, Commander, Air Strike Force would detach one destroyer from the Kamchatka Group to the commanding officer of submarine tender AS-3 at Kashiwabara Bay to escort the submarine tender to Taraika Bay. Afterward, the destroyer would resume its duty with the Kamchatka Group. Not later than Jump Day+1, Commander, Mine Force and Commander, Service Force would report to Commander, Fleet Attack Force at Vladivostok to coordinate the passages of the three forces from Vladivostok to their forward areas and to ensure the adequacy of sweeping services. (Page 85.)

After consultation with and the approval of Commander, Air Strike Force, Commander, Fleet Attack Force would define and promulgate to all task force commanders and to Commander, Purple Pacific Fleet appropriate rendezvous points, operating and fueling areas, routes between bases, and other necessary details relating to movements, no later than 8 June. Purple Pacific Fleet Sortie Plan No. 3-4 remained effective; changes about swept channels would be broadcast by Commander, Mine Force. During Bluepill, Commander, Air Strike Force would be at Fleet Air Headquarters at Kashiwabara, while Commander, Submarine Force would be at Vladivostok. The Deputy Commander, Submarine Force would be at Purple Pacific Fleet Headquarters at Kashiwabara; Commander, Fleet Attack Force would be with his units at sea. Commander, Service Force would be at Aniwa Bay upon the initial arrival of the Service Force units from Vladivostok. Commander, Mine Force would be at the Purple Pacific Fleet Headquarters at Kashiwabara Bay upon the arrival of his units from Vladivostok. (Page 85.)

In his Logistics Plan, Admiral PA reiterated that Commander, Service Force's mission was to deliver fuel, diesel, and aviation gasoline to the forces engaged in Bluepill; he was also to prepare and promulgate schedules of delivery as soon as possible after the receipt of estimates (which were to submitted to Admiral PA by 7 June, with fueling times and rendezvous points, for the period of Jump Day to Jump Day+30). Task force commanders were to keep the Service Force informed of all changes in those requirements and of any requirements beyond Jump Day+30.

Task force commanders were as far as practicable to maintain fuel, diesel, and aviation-gasoline levels near full capacity. Sixty percent of full capacity was the lowest authorized level, except for tender aviation gasoline, which could drop to 40 percent. All forces engaged in Operation BLUEPILL were authorized to draw on any available base facilities as necessary, but no unit was to take fuel, diesel, or aviation gasoline unnecessarily. (Page 86.)

Offensive mining was to be limited to aircraft under Commander, Air Strike Force. Defensive mining would be confined to the protection of ships at bases against submarine attack. Initially, four bases—the Gulf of Kamchatka, Kashiwabara Bay, Aniwa Bay, and Taraika Bay—would be so mined. The senior Mine Force officer at each base would provide antisubmarine surface patrol units as directed by the Senior Officer Present Afloat (SOPA). Commander, Mine Force was also to submit by 8 June to Commander, Purple Pacific Fleet detailed charts of the proposed minefields to be planted by his units. Commander, Purple Pacific Fleet would disseminate these charts to all task force commanders; Commander, Mine Force would communicate any deviations from these charts to Commander, Purple Pacific Fleet as well as to the task force commanders. (Page 87.)

Upon completion of planting the minefields at Kamchatka and Kashiwabara, the Mine Force would supplement the Kamchatka minefields, proceed to plant fields at Aniwa Bay and Taraika Bay, and then provide minesweeping and channel buoy services at these two bays prior to the arrival of fleet units. The Senior Officer Present Afloat at each base would also control the frequency and limits of mine sweeps, but there would be at the minimum sweeps preceding the entry or sortie of all fleet units larger than destroyers as well as one daily sweep whether or not entries or sorties occurred. Upon request, Commander, Mine Force would also furnish minesweeping service to all task force commanders during Operation BLUEPILL. Commander, Purple Pacific Fleet would process the charts of mine plants submitted by Commander, Mine Force and distribute them by 10 June to all task force commanders. (Pages 87–88.)

Mining itself was to take place in specific ways. In the Kamchatka area, mines were to be laid in the channel to the anchorage area in two legs and 1,200 yards apart. The median line of the inner leg was to run 350 degrees true from Point OBOE, which was 56°03′ N, 162°14′ E, and terminate at Point ROGER, 56°07.5′ N, 162°13′ E. The median line of the outer leg was to run 110 degrees true from Point OBOE and terminate at Point X-RAY—at 56°02′ N, 162°21′ E—and at the channel entrance to seaward. Minefields were to be planted on each side of the channel as well. In the eastern sector, the minefield was to be bounded on the southeast by a line running from the northeast corner of the outer channel leg in a direction 080 degrees true and extending to the five-fathom curve, or about six hundred yards offshore where the five-fathom curve was closer than that. The minefield was to be bounded on the northwest by a line parallel to the southeast line and three miles distant from it to the northeast. This line was also to extend from the eastern side of the channel to the five-fathom curve, six hundred yards offshore. The minefield was to be bounded on the southwest by that portion of the right-hand limit of the outer channel as viewed from seaward, which was intercepted by the two minefield lines noted above. On the northeast, the minefield was to be bounded by a line joining the northeastern extremities of the first two minefield lines. (Page 89.)

The western sector of the Kamchatka area minefield was to be bounded on the north by a line running from the southwest corner of the outer leg channel in the direction 270 degrees true to the five-fathom curve, or a point about six hundred yards offshore. This minefield was going to be bounded on the northeast by the entire length of the left-hand limit of the outer-channel leg as viewed from seaward. It was also bounded on the southwest by a line running from the southeast corner of the outer channel leg in the direction 200 degrees true for a distance of three miles. It was bounded on the south and southwest by a broken line paralleling the above minefields, also at a distance of three miles from them to the south and southwest. The minefields in both the eastern and western sectors were to be planted in two stages, the first as soon as possible after the execution of BLUEPILL using one full load of a light minelayer, the second as soon as possible after the completion of the Kashiwabara plant. (Pages 89–90.)

In the Kashiwabara area, the minefield was to be laid in the channel to the anchorage area, also in two legs and about 1,200 yards apart. The median line of the inner leg was to run 310 degrees true from Point PETER, which was 50°30′ N, 156°27′ E, and terminate at Point WILLIAM, 50°33.9′ N, 156°19.8′ E. The median line of the outer leg of the minefield was to run 180 degrees true from PETER and terminate at Point X-RAY—50°26′ N, 156°27′ E, at the channel entrance from seaward. The northern entrance to the anchorage was also to be mined in the vicinity of 50°45′ N, 156°10′ E in order to effect complete closure. Minefields on the channel sides were also to be planted. The eastern sector of this minefield was to be bounded on the south by a line running from the eastern corner of the channel entrance—six hundred yards east of Point X-RAY—in the direction of 090 degrees true for three miles. On the east, the minefield was to be bounded by a line running from the eastern extremity of the above line of mines in the direction of 000 degrees true for twelve miles. The minefield was also to be bounded on the north by a line running from the northern extremity of the second line of mines 310 degrees true up to the five-fathom curve, or a point about six hundred yards offshore. On the west, the minefield was to be bounded by a broken line paralleling the above lines of mines, also at a distance of three miles from them in the channel direction. (Pages 90–91.)

The western sector of this minefield was to be bounded on the south by a line running from the western corner of the channel entrance—six hundred yards west of Point X-RAY—in the direction 270 degrees true for a distance of twelve miles. It was to be bounded on the west by a line running from the western extremity of the line above in the direction 000 degrees true to the five-fathom curve or about six hundred yards offshore. The minefield was bounded in the north by a line running from the northern extremity of the line above in the direction 090 degrees true to the intersection with the western limit of the outer-channel leg of mines. On the east, the minefield was to be bounded by that portion of the western limit of the outer channel that was included in the eastern extremities of the lines noted above. All mine plants were to be made in accordance with standard doctrine for protection against submarines. Also—as concerned the Aniwa Bay and Taraika Bay areas—Commander, Mine Force, upon receipt of the Operation Plan, was to confer with Commander, Service Force and Commander, Submarine Force to ascertain their desires for the respective anchorages, prepare mine plans for them, and submit those plans not later than 10 June to Commander, Purple Pacific Fleet for approval and promulgation. (Pages 91–92.)

Admiral PA's Communication Plan was to become effective upon receipt and was in accordance with Purple Standard Fleet Communication Plan PAC-70 (B). All commands were to preserve radio silence below sixty megacycles per second at staging and assembly points, except for warnings or defensive operations against the enemy. Visual communications were to be employed whenever possible; for purposes of communications security, all bases involved in BLUEPILL were to be considered as "advance bases and staging areas." Fox (broadcast) Schedules were also established for the fleet, submarines, and Radio Kashiwabara, the last to be responsible for requesting corrections and repetitions. Authentications were provided and task force commanders and Officers in Tactical Command were to prescribe the conditions for radio silence for the forces at sea. There was also a Port frequency provided as well as Talk Between Ships frequency assignments at both fleet and force levels for Task Forces 31, 32, 33, 34, and 35. All ships were accordingly to have two sets of Talk Between Ships equipment but would use the Fleet frequency when not in company with other forces within the fleet. Additionally, there were task force, task group, and task unit circuits and there were frequencies for air-search and reconnaissance planes. Aircraft contact reports would be broadcast by shore stations to the fleet on its Fox Schedule and on the Operational Intelligence Circuit. Aircraft required to communicate with independent early warning submarines were to use the Air-Sea Rescue frequencies. (Pages 93–94.)

Finally, submarine contact reports were to be made on specific channels, one each to Radio Vladivostok, Radio Etorofu, Radio Matsuwa, and Radio Kashiwabara.

Appendix (1) to Annex (E)

RADIO CALLS

Com THIRD Fleet	ABDUCT
CTF 30	BARLEY
THIRD Fleet	CRONY
CTF 31	ABDUL
TF 31	CORLY
CTG 31.1	BAMBOO
TG 31.1	ABLAZE
Com Fleet Air Wing 20	CUTIE
" " " 20	BOZO
Com Fleet Air Wing 30	DABBLE
" " " 30	DARWIN
Com Fleet Air Wing 40	ELMER
" " " 40	HAMMER
Com Fleet Air Wing 50	IODINE
" " " 50	TACK
Com Fleet Air Wing 60	GAZOLLO
" " " 60	KINGKONG
X - 1	LABEL
X - 13	FATAL
CTG 31.2	DINAH
TG 31.2	ADAM
CVL-2	DAGO
CVE-3	JASON
AV-1(F)	BANJO
AVP-2	EXTREME
AVP-3	HARRY
ComDesDiv-1	INFLATE
DesDiv-1	JACOB
DD-1	FANCY
DD-2	JAUNTY
DD-3	LADY
DD-4	HANG
DD-5(F)	MAILBOAT
DD-6	INSPIRE
Com Fleet Air Wing 10	EMBARK
" " " 10	LADDER
CTF 32	AWARD
TF 32	MAJESTY
AS-2(FF)	CAVIAR
ComSubRon-1	GERM
SubRon-1	JAGUAR
ComSubDiv-1	ENERGY
SubDiv-1	MANHOOD
X-2	KINGSLEY
X-3	FEARLESS
X-4(F)	MILD
ComSubDiv-2	EYEBALL
SubDiv-2	AMOS
X-5	FARMER
X-6	KITESTRING
X-7	HERO
X-8(F)	JOSHUA
ComSubRon-2	LAG
SubRon-2	BANJO
AS-3(FF)	FATSTUFF
ComSubDiv-3	IRONIC
SubDiv-3	AVAIL
X-9	CELLO
X-10	JOVIAL
X-11	MARINE
X-12(F)	GLADHAND
ComSubDiv-4	LAPEL
SubDiv-4	MARTHA
X-14	HEPCAT
X-15	MAPLE
X-16(F)	MAYO
CTF 33	KNICKNACK
TF 33	ENDOW
ComCarDiv-1	MINOW
CarDiv-1	IRISH
CVL-1	JONAS
CVL-2	LATE
ComCruDiv-1	BARNACLE
CruDiv-1	CATGUT
CB-1(F)	HAYRIDE
CB-2	JOIN
ComCruDiv-2	ANNOY
CruDiv-2	IRIS
CA-1(F),(FFF)	KISMET
CA-2	LAVISH
CA-3	MAIN
ComCruDiv-3	ERUPT
CruDiv-3	FEEBLE
CA-4(F)	KITKAT
CA-5	LAZY
CA-6	JOE
ComCruDiv-4	ELBOW
CruDiv-4	CASINO
CL-1(F)	LECTURE
CL-2	GENIAL
CL-3	KLEENEX
CL-4	ENRAGE
CL-5	JOY
CL-6	LEDGER
ComDesRon-1	ANGEL
DesRon-1	BUTCHER

These reports were to be broadcast to the 3rd Fleet on the Fox Schedules and the Operational Intelligence Circuits. The submarine Fox Schedules were to be repeated at least three times because of the necessity for the submarines to stay submerged most of the time in their operating areas except during short periods of darkness. There were also designated channels for early warning and independent fighter direction as well as the Air-Sea Rescue Channel. In addition to the radio call signs, the Officer in Tactical Command of each force was designated in control of the use of radar, radar countermeasures, and Identification Friend or Foe. OTCs were also responsible for tactical intercept and jamming plans, and the Senior Officer Present Afloat in each harbor would control communications. Radio Kashiwabara would be the main communications headquarters for Commander, 3rd Fleet, while Radio Etorofu was the relief station. Individual aircraft calls were to be composed of the

Appendix (1) to Annex (E)

RADIO CALLS

ComDesDiv-2	HASHMARK
DesDiv-2	KRAMER
DD-7	FAN
DD-8	IMPEDE
DD-9	JUSTICE
DD-10(F)	LABEL
ComDesDiv-3	ALPACA
DesDiv-3	BYWORD
DD-11	KOKOMO
DD-12	GEISHA
DD-13	LILY
DD-14	MAHATMA
DD-15(F)	JUSTIFY
ComDesRon-2	CASHBOX
DesRon-2	LOUD
ComDesDiv-4	FUDGE
DesDiv-4	IMPISH
DD-19	KNOCKLE
DD-20(F)	ELISHA
ComDesDiv-5	AMBER
DesDiv-5	BURRO
DD-21	CHARM
DD-22	KIMONO
DD-23	DIME
DD-24	IGLOO
DD-25(F)	HORN
ComDesDiv-6	BATTER
DesDiv-6	CHICO
DD-26	JELLY
DD-27	LOBSTER
DD-28	KEYHOLE
DD-29	MAGPIE
DD-30(F)	DISNEY
CTF 34	JUNE
TF 34	EDITH
CVE-4	FUTILE
DD-16	HOWL
DD-17	INCLINE
DD-18	GARLIC
ComSerRon-1	LOCAL
SerRon-1	MAGNIFY
ComSerDiv-2	ANDREW
SerDiv-2	DOLL
AO-4(FF)	LONELY
AO-5	KEEPSAKE
AO-6	JAKAL
ComSerDiv-3	CINEMA
SerDiv-3	HOLD
AO-7	GAUDY
AO-8	LOVE
AO-9	EBONY

CTF-35	ALE
TF-35	BUXOM
DM-1(FF)	MADCAP
ComMinDiv-1	FROSTY
MinDiv-1	INCITE
DMS-1	JUICE
DMS-2	MABEL
DMS-3	KAPOK
DMS-4(F)	ENDOW
ComMinDiv-2	ENABLE
MinDiv-2	ADAM
AM-1	CHOW
AM-2	DANDY
AM-3	HOTHEAD
AM-4	FOSTER
AM-5	ICECAP
AM-6(F)	EARDRUM
ComMinDiv-3	BARREL
MinDiv-3	ANGER
AM-7	CALEB
AM-8	DIGEST
AM-9	GAY
AM-10	EARN
AM-11	HORSEFLY
AM-12(FF)	DONUT

Appendix (2) to Annex (E)

CODES

1. SHACKLE CODE

0	C	E	J	V	P
5	I	S	R	G	
1	D	L	K		
2	T	M			
3	U	Q			
4	A	N			
6	H	Z			
7	X	Y			
8	F	W			
9	O	B			

Fig. 111
Radio call signs and shackle codes, Purple Staff Solution, Operations Problem 2

unit or base call followed by the type, letter, and number of the plane. For instance, a plane designated as 11-F-1 from light carrier CVL-1 had a call sign of "Jason Fox 1." "Jason" was the base or the ship, *F* (Fox) stood for the plane's type (in this case a fighter), and the number of the plane was 1. The collective call for an aircraft unit would be formed by adding "Gang" to the basic call. Therefore, all fighters from light carrier CVL-1 would be addressed as "Jason Fox Gang," while all fighters were known as "Fox Gang." The unit leader's call would be indicated by "Boss" preceding the unit call. (Pages 94–95, 99.)

XII Operations Problem 2
Even More Purple, September–October 1946

In late September 1946, Rear Admiral Smith issued an instruction, "Detail of Student Officers for the Solution of Task Force Commanders' Problems," with staff officers' roles and identities. The roles played by the students give us another indication of how the Operations Problems were played. For instance, the roles of Admiral PA's Chief of Staff and Operations Officer were brought into the exercise, as were four Assistants for Planning to Rear Admiral PJ, Commander, Task Force 33, the Submarine Force. Admiral PJ's Chief of Staff and Operations Officer were also detailed. In addition, eight Assistants for Planning for Rear Admiral PK, Commander, Task Force 35, the Fleet Air Force, were introduced, as were his Chief of Staff and Operations Officer; ten students played as Assistants for Planning to Rear Admiral PB, Commander, Task Force 37, the Striking Force, along with his Chief of Staff and Operations Officer; and Commodore PI, Commander, Task Force 39, the Service Force, had five students as Assistants for Planning in addition to his Chief of Staff (Commander Antrim—see chapter 1) and Operations Officer.[1]

For the Purple Chart Maneuver, Admiral Smith directed that these groups would, under the direction of students designated as student officer task force commanders, solve the Problem and draw up Operation Plans for the task forces to which they were assigned. These plans and annexes were to be in sufficient detail to carry out a Chart Maneuver according to Naval War College Maneuver Rules. (In a reminder that we are in 1946, he encouraged early submission in order that stencils could be cut and the necessary copies run off and distributed.) Final assignments of players and rooms were issued on 1 October, and the Chart Maneuver itself commenced at 0900 on 2 October.[2]

At this point, Admiral PA put forth his Operation Plan 2-46, again for review and with some additional details. "M-Day" would be the day that war was declared by the Purple homeland or that hostilities commenced. This event would be announced on the Fox Schedule; in any case, hostilities would be considered to exist if any Blue surface forces were discovered west of longitude 170 degrees east and north of latitude 44 degrees north. Should Purple territory be attacked by major surface forces, units were to operate in accordance with Battle Plan, Commander, Pacific Fleet Operation Plan 3-46. In addition, current tactical orders and doctrine

Chief of Staff, Commander, Purple Pacific Fleet (Capt. Frank Bruner, USN)

Operations Officer, Commander, Purple Pacific Fleet (Cdr. Thomas Briggs, Royal Navy)

Chief of Staff, Commander, Purple Submarine Force (Capt. James Lane, USN)

Operations Officer, Commander, Purple Submarine Force (Cdr. William Richards, USN)

Chief of Staff, Commander, Purple Fleet Air Force (Lt. Col. Stanley Sawicki, USA)

Operations Officer, Commander, Purple Fleet Air Force (Capt. Dwight Agnew, USN)

Chief of Staff, Commander, Purple Striking Force (Capt. Philip Snyder, USN)

Operations Officer, Commander, Purple Striking Force (Cdr. Albert Kaplan, USN)

Operations Officer, Commander, Purple Service Force (Cdr. Ronald Woodaman, USN)

publications were effective, modified to allow for greater dispersion of forces. Initial movements were to be in accordance with the Operation Plan; later movements would be directed by the task force commanders themselves. Search and rescue operations were as prescribed in current tactical publications. All units were to conserve fuel since the availability of fleet oilers was limited; conservation was to be taken into consideration in speeds maintained and engineering conditions (e.g., number of boilers on line, use of auxiliary steaming) employed.[3]

PURPLE PACIFIC Fleet
and THIRD Fleet
Operations Problem 2
Detail of Student Officers
for Solution of
Task Force Commanders Problems

Commander PURPLE PACIFIC Fleet and Commander THIRD Fleet, Adm. PA	Captain Smith
Chief of Staff	Captain Bruner
Operations Officer	Commander Briggs
T.F. 33 Submarine Force	
C.T.F. 33, R. Adm. PJ	Commander Roudebush
Assistants for Planning	Commander Richards
	Colonel Wood
	Commander Van Every
	Captain Van Deurs
T.F. 35 Fleet Air Force	
C.T.F. 35, R. Adm. PK	Captain May
Assistants for Planning	Colonel Sawicki
	Commander Nisewaner
	Captain Hardin
	Lt. Col. Lane
	Commander Munholland
	Commander Woodaman
	Captain Madeira
	Captain Agnew
T.F. 37 Striking Force	
C.T.F. 37, R. Adm. PB	Captain Hummer
Assistants for Planning	Captain Outerbridge
	Lt. Col. Shell
	Colonel Koonce
	Captain Hoffheins
	Captain Snyder
	Commander Andrews
	Colonel Privett
	Commander Kaplan
	Colonel Vaughan
	Colonel Erlenkotter
T.F. 39 Service Force	
C.T.F. 39, Commodore PI	Captain Sullivan
Assistants for Planning	Commander Antrim
	Colonel Shores
	Captain Heberton
	Captain Lane
	Colonel Grossetta

Fig. 112
Purple Pacific Fleet Detail of Student Officers, Operations Problem 2

Operations Problem 2

Detail of Student Officer Commanders

PURPLE PACIFIC Fleet

Command	Rank	Student Officer	Room
CinC PURPLE PACIFIC Fleet	Adm. PA	Capt. Smith	141
Chief of Staff	---	Capt. Bruner	141
Operations Officer	---	Cdr. Briggs	141
T.F. 33, ComSubFor	R.Adm. PJ	Cdr. Roudebush	164
Chief of Staff	---	Captain Lane	164
Operations Officer	---	Cdr. Richards	164
ComSubRon 1	Capt. P-12	Cdr. Van Every	163
C.O. AS-2	Capt. P-12	" " "	163
ComSubDiv 1	Capt. P-14	" " "	163
ComSubDiv 2	Capt. P-15	" " "	163
ComSubRon 2	Capt. P-13	Colonel Wood	161
C.O. AS-3	Capt. P-13	" "	161
ComSubDiv 3	Capt. P-16	" "	161
ComSubDiv 4	Capt. P-17	" "	161
T.F. 35, ComFltAirFor	R.Adm. PK	Capt. May	143
Chief of Staff	---	Col. Sawicki	143
Operations Officer	---	Capt. Agnew	143
ComFAW-10	Commo. PL	Col. Vaughan	149
C.O. AV-1	Capt. P-21	" "	149
C.O. AVP-2	Capt. P-22	" "	149
C.O. AVP-3	Capt. P-23	" "	149
ComFAW-20	Commo. PM	Capt. Hardin	162
ComFAW-30	" PN	Cdr. Munholland	151
ComFAW-40	" PO	Cdr. Nisewaner	153
ComFAW-50	" PP	Col. Erlenkotter	144
ComFAW-60	" PQ	Cdr. Andrews	150
T.F. 37, Com Striking For	R.Adm. PB	Capt. Hummer	155
Chief of Staff	---	Capt. Snyder	155
Operations Officer	---	Cdr. Kaplan	155
ComCruDiv 1	R.Adm. PB	Col. Grosetta	154
ComCruDiv 2	R.Adm. PC	" "	154
ComCruDiv 3	R.Adm. PD	Col. Koonce	154
ComCruDiv 4	R.Adm. PE	" "	154
ComCarDiv 1	R.Adm. PF	Col. Privett	156
C.O. CVL-1	Capt. P-18	" "	156
C.O. CVL-2	Capt. P-19	" "	156
ComDesRon 1	Commo. PG	Capt. Hoffheins	156
C.O. CL-5	Capt. P-7	" "	156
ComDesDiv 1	Capt. P-1	" "	156
ComDesDiv 2	Capt. P-2	" "	156
ComDesDiv 3	Capt. P-3	" "	156
ComDesRon 2	Commo. PH	Lt.Col. Lane	156
C.O. CL-6	Capt. P-8	" " "	156
ComDesDiv 4	Capt. P-4	" " "	156
ComDesDiv 5	Capt. P-5	" " "	156
ComMinDiv 1	Cdr. P-1	" " "	156
T.F. 39, ComSerFor	Commo. PI	Capt. Sullivan	149
Chief of Staff	---	Cdr. Antrim	149
Operations Officer	---	Cdr. Woodaman	149
C.O. CVE-3	Capt. P-20	Lt.Col. Shell	142
C.O. CVE-4	Capt. P-11	" " "	142
ComSerRon 1	Commo. P-I	Col. Von Shores	142
ComSerDiv 2	Capt. P-9	" " "	142
ComSerDiv 3	Capt. P-10	" " "	142
ComMinRon 1	Commo PR	Capt. Heberton	148
C.O. DM-1	Capt. P-24	" "	148
ComMinDiv 2	Cdr. P-2	" "	148
ComMinDiv 3	Cdr. P-3	" "	148
ComDesDiv 6	Capt. P-6	Lt.Col. Shell	142

Fig. 113
Purple Pacific Fleet Student Officer Commanders, Operations Problem 2

The General Concept to his Operation Plan reiterated that all the Purple bases except for Vladivostok were within reconnaissance range of Blue very-long-range aircraft. Blue possessed the atomic bomb and task force commanders were reminded to disperse their forces in order to prevent "serious" losses. Task force commanders were initially to operate under a "conservative" policy in light of the restraints imposed by the defense of a fixed area and the superiority of Blue naval forces. It was of the "utmost" importance to discover the enemy as soon as possible, to strike the Blue carriers at the maximum distance possible with dive-bombers, and for the submarines to inflict the maximum damage on the Blue carriers just before the latter arrived at their initial launch points: "Control of the air can best be accomplished by delivering blows at the CVs [carriers] while their planes are on board. This may be accomplished by VB [dive-]bombing and submarines attacks before the carriers arrive at a launching point." The key to destroying the enemy carriers was repeated attacks at the extreme range of his dive-bombing squadrons, that is, more than three hundred miles, and the key to that, in turn, was not allowing Blue to approach to within that range undetected. However, destruction of the Blue Striking Force was in itself only a prelude to the destruction of the Blue Expeditionary Force.[4]

Admiral PA ordered his subordinate commanders on 1 June to be prepared to get under way on signal at 2000 on the 3rd for extended operations in the Kurile Islands area and to report on board his flagship at 1500 on that day for a command

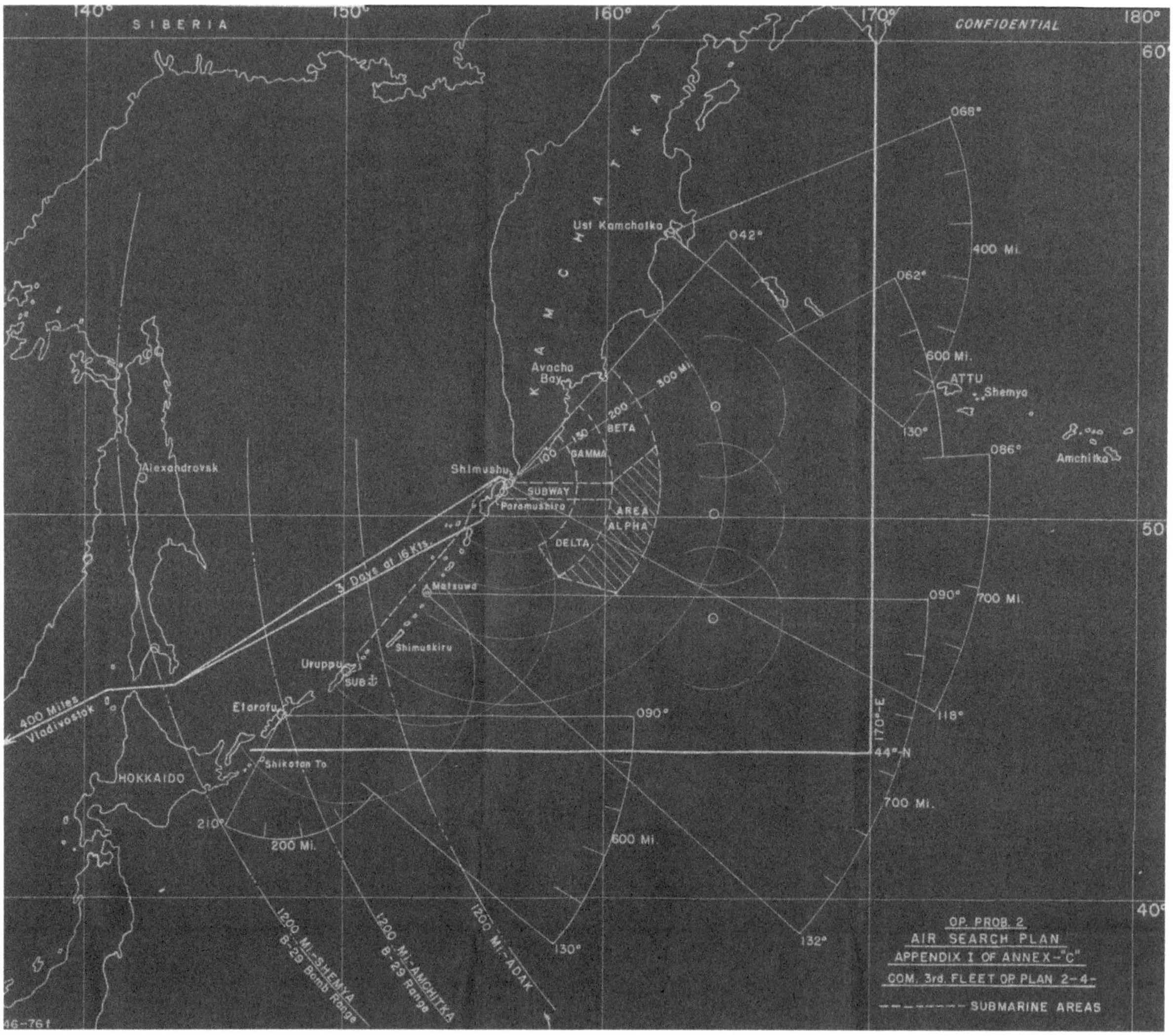

Fig. 114
Purple Air Search Plan, Operations Problem 2

conference. At the same time, he sent a message to Commander, Pacific Submarine Force to continue Submarine Division 2's patrol activity in the Aleutians until relieved; SUBDIV 3 was to depart on 4 June to take its station by the 10th, its four submarines being stationed off Attu, Adak, Dutch Harbor, and Unimak Pass. While making reports on types, numbers, composition, and course of Blue ships, they were to avoid being detected or making any "warlike" acts east of 170 degrees east. Commander, Pacific Submarine Force was to have SUBDIV 2 overhauled in the Kuriles when it returned to base and have SUBDIV 1 back to the Kuriles by 9 June for the same purpose. Admiral PA instructed Commander, North Pacific Air Command that the three seaplane tenders were to report to Commander, Cruiser Division 1 and be prepared to get under way at 2000 on 3 June for Paramushiru. Once there, the seaplane tenders would report to him for operational control.[5]

The Logistics Plan ordered immediate steps to ferry as much aviation gasoline and emergency supplies to the airfields and landing strips on southern Kamchatka as necessary for emergency use and dispersal of aircraft. At fields where no storage facilities were available, gasoline in fifty-gallon drums, to be hand pumped, was to be provided. Seaplane tender AV-1 had a thirty-day supply of aviation gasoline; operating from Ust Kamchatka, it would eliminate the necessity and undue risk of sending gasoline tankers north of Paramushiru more than once per month. Airfields were to keep their supplies of gasoline dispersed and be prepared to fuel planes in spite of damage to storage tanks. Maintenance crews were to be equipped and augmented such that any landing field that had been bombed could be put back into operation in thirty minutes. Along these lines, oilers, escort carriers, and tenders were to carry extra mines, depth charges, and fifty-gallon drums of gasoline from Vladivostok to the Paramushiru area. Finally, large ships were directed to refuel smaller ones and battle-damaged ships were to proceed—either on their own power or under tow—to the nearest ports where repair facilities were available.[6]

Admiral PA's Communication Plan ordered that there be no point-to-point traffic between Vladivostok, Paramushiru, and Alexandrovsk. All other bases were to preserve radio silence below thirty megacycles per second except for warnings or for defensive operations against the enemy. Purple bases were to use visual communications whenever possible. In addition, task force commanders and Officers in Tactical Command were to prescribe conditions of radio silence, and OTCs would control the use of radar, Identification Friend or Foe, and radar countermeasures. There were also Fox Schedules for the various commands as well as specific circuits for OTCs, task force commanders, and Commander, Purple 3rd Fleet.[7] The task force commanders had both primary and secondary radio circuits and there were specific circuits for task force, task group, and task unit commanders. There were also specific Air and Submarine Operational Intelligence Circuits for contact reports from aircraft and submarines. In addition, there were specific frequencies for submarine contact reports as well as authenticator and call lists.[8]

Purple had a Joint Army–Navy–Air Force Intelligence Center at Vladivostok, fully equipped to perform all the intelligence services for all forces in the North Pacific Ocean Area, including strategic, foreign, and operational intelligence. Information on Blue forces was being supplied in letter form in weekly estimates of enemy strength. In addition, there were daily intelligence bulletins, special ULTRA messages, and North Pacific Intelligence Bulletins Nos. 1–4.[9] Admiral PA's Joint Intelligence Center also put out bulletins of detailed information on enemy equipment and armament as well as air target maps, charts, and terrain maps. Fleet Weather Central supplied meteorological data and tables on tidal data, daylight periods, sunrise, sunset, moonrise, and moonset. Special Information from Chief of Naval Staff messages had been passed to all task force and administrative commanders. Additionally, there was information from enemy prisoners and captured documents; radio intelligence units attached to force commanders' staffs; and radar countermeasures equipment. "Cryptoanalysis" units were also assigned to the Joint Intelligence Center, Vladivostok; PA was confident they would furnish "pertinent" information to those required to know. Reconnaissance and observation reports were to be made to the nearest base or task force, whose commanders were responsible for relay to Admiral PA.[10]

```
SHACKLE CODE

O E G M S Y Z          Authenticator:
1 C K T                  Use first letter of message
2 O R                  as ROOT.
3 I U                    Enter shackle table and
4 N Q B                  use one of the adjoining
5 A J D W                letters on either side above
6 H X                    or below.
7 V P                    (Only 2 letters in the authenticator
8 L                      i.e. M F or M I or M D)
9 F
- - - - - - - - - - - - - - - - - - - - - - - - - - - - -

                         Call List.

CW       Command                VOICE
9XN      ComTHIRDFlt(CTF 30)    STORK                          TF Calls
7UC      THIRDFlt (TF30)        GERONIMO
7UM      CTF 33                 RUGBY
3DA       TF 33                 TOMAHAWK
5DP      CTF 35                 ANZAC
7TL       TF 35                 MOROCCO
9GS      CTF 37                 WAREEAGLE
6WA       TF 37                 SNOWFLAKE
2LU      CTF 39                 FROLIC
1GA       TF 39                 LOCUST
3AM      All T.F. Comdrs.       SYNDICATE - COLLECTIVE
9NT      Sub. Rescue vessels    LIFE UARD
7BR      Any spotting A/C       SPYGLASS
4RD      COMCrudiv 1,2,3,4      PADUCAH ONE, TWO, THREE, FOUR
7EW         Crudiv 1,2,3,4      CALICO  ONE, TWO, THREE, FOUR
1UT      Comcardiv 1,2,3,4      TOPEKA  ONE, TWO, THREE, FOUR
1LK         Cardiv 1,2,3,4      BLOWHOLE ONE, TWO, THREE, FOUR
2TG      Comdesron 1,2          CARNIVAL
1DX         Desron 1,2          POPCORN
2KH      Comdesdiv 1,2,3,4      SMOKEY
9CS         Desdiv 1,2,3,4      CHEERIO
5KR      Comminron 1            LIGHTFOOT
9DT         Minron 1            ROMANTIC
2HA      Commindivs 1,2,3       STEAMBOAT
3QD         Mindivs 1,2,3       LOVEJOY
9LB      ComSevron              TOMBOY
            Seron               CAIRO
1ED-1,2  CB 1,2                 WALDO ONE, TWO
4WJ      CA 1,2,3,4,5,6         TURLOCK 1,2,3,4,5,6
2RP      CL 1,2,3,4             WINNEPEG 1-4
5JD      CVL 1,2                LORENZO 1,2
2CY      CVE 3,4                PARADISE 3,4
5UM      DD 1-30                YUKON 1-30
3KG      AO 4-9                 POLKADOT 4-9
         Picket #1              SILVER                         FD calls
         Picket #2              GOLD
         CVL 1                  BLUE
         CVL 2                  GREEN
         CVE 1                  YELLOW
         CVE 2                  RED
         PBMs            M                                     AIRCRAFT
         PDYs            B
         PV-2            P
         PB4Y-2          F
Squadron Number, Type Letter, Plane No.
         9         PETER         101
```

Fig. 115
Purple 3rd Fleet authenticator and call lists

In a series of messages, Rear Admiral PJ, Commander, Purple Submarine Force, assigned his various squadrons and divisions, as well as his two submarine tenders, to their patrol stations, routes, and duties. He formed the twelve boats of Submarine Divisions 1, 2, and 4 into an Attack Group under Captain P-14, and the other four boats into a Patrol Group under Captain P-16 (Colonel Wood—see chapter 1); the two tenders constituted a Training and Logistics Group under Captain P-13 (also Colonel Wood). The Attack Group was to engage the Blue force as soon as its location was known. The submarines would cover the withdrawal of the Purple Striking Force if directed. The Patrol Group was also to strike Blue forces, but only after detecting and trailing them and reporting their location to the Attack Group. Rear Admiral PJ stressed accurate navigation to keep Purple submarine, air, and surface attacks on Blue coordinated. He was at Fleet Headquarters in Vladivostok,

Captain P-14 (Cdr. Salem Van Every, USN)

and Captain P-13, Commander, Training and Logistics Group, was second in command, embarked in AS-3.[11]

Admiral PJ's Communication Plan emphasized submarines' guarding their assigned radio frequencies, especially the Fox Schedule, times for reports, and voice frequencies for each division during "wolf-pack" operations. Very-high-frequency channels were to be used for coordination with surface ships and aircraft. Submarines were also to attach weather reports to their messages if possible. PJ's Training and Logistics Plan instructed submarines to fuel alongside tenders singly and clear the bay as soon as possible. It was the responsibility of Commander, Training and Logistics Group to schedule refueling, training, upkeep, and overhauls so that the minimum number of submarines was concentrated in the harbor.[12]

Submarines en route to or from patrol were to use "economical" speed, using discretion as to zigzagging. Blue submarines could be expected in practically any part of the Pacific Ocean. With "efficient" use of lookouts and radar, however, it was normally safe to proceed on the surface in daytime in enemy-patrolled waters. In low visibility when on the surface, however, commanding officers were to ensure an especially alert radar watch since Blue could bomb by radar. Radios were to be used minimally; submarines were to withdraw from their patrol stations along their return routes to transmit essential information, with short summaries of patrol results, position, and weather if time permitted. Admiral PJ ordered his boats to avoid contact with friendly forces, about which he would keep them informed.[13]

Weather (Report 180600 Z
(
(~~Forecast~~

Area K, L, M

1. Meteorological visibility, 1/4 miles.
2. Wind and sea: force 2 , direction S .
3. Assured sonar range, 2000 yards.
4. Percentage cloud cover 100 , ceiling 100 ft.
5. Rain, snow, ice conditions:
   Raining steadily since 1200 Z
6. Air temperature 45°

Remarks:

Forecast for next 24 hrs.
Rain stopping about 2400 Z
followed by heavy fog.
Clouds up to 10,000 ft.

Weather (Report 180600 Z
(
(~~Forecast~~

Area G, H and I

1. Meteorological visibility, 1/4 miles.
2. Wind and sea: force 1 , direction SW .
3. Assured sonar range, 2000 yards.
4. Percentage cloud cover 100 , ceiling 200 ft.
5. Rain, snow, ice conditions:
   Heavy fog
6. Air temperature

Remarks:

Forecast for next 24 hours
No change.
Clouds up to 10,000 ft.

Fig. 116
Meteorological data, Purple Forces, Operations Problem 2

Submarines leaving Paramushiru were to be escorted until dark on the day of departure. They were to leave the tender's commanding officer their estimated routes through Purple air-patrolled areas outside submarine operating zones; the tender was to forward this information to Commander, 3rd Fleet and Commander, North Pacific Air Command. Submarines were to keep accurate records of circumstances in which they found themselves and to take periscope photos of targets, harbors, facilities, and sinking ships—images that would get red frames in Moscow![14]

For dawn attacks on carriers, submarines were to attain position at night, spreading upwind at intervals of not less than three miles along the bearing from which the wind was blowing, since the carriers would turn into the wind for dawn or predawn launches. When carriers were launching aircraft, the screening ships' sound gear was likely to be ineffective; submarines were to take advantage of this factor. For night attacks on the carriers, submarines were to attain positions ahead of the enemy formation, in line normal to the enemy's course and about three miles apart. At the first attack, Blue would probably turn, likely heading directly toward one end of the Purple submarine line while permitting the other to advance and encircle him.[15]

Rear Admiral PK, Commander, Fleet Air Force and North Pacific Air Command, had Commodore PL commanding the Kamchatka Bay Group of two seaplane tenders with eighteen of the PBM Mariner naval patrol bombers in two squadrons and twelve Hellcat fighters organized into a fighter division. Commodore PO commanded the Shimushu Group, which consisted of the Avacha Bay Fighter Unit of twenty-four Hellcat fighters and the Kataoka Bay Group. The latter comprised one seaplane tender with eighteen PBM Mariners in two squadrons, thirty-six Harpoon naval patrol bombers in two more squadrons, and eighteen Harpoons in a squadron stationed at Kurilskoye Strip. This group was also composed of twenty-four Hellcat fighters in a squadron at the Cape Lopatka strip, thirty-six Hellcat fighters in an additional fighter squadron, and forty-eight Hellcat night fighters organized into two squadrons. Commodore PN commanded the Paramushiru Group, which consisted of a squadron of eighteen Liberator naval patrol bombers, a squadron of eighteen Harpoons, twelve Catalina naval patrol bombers organized into two squadrons, and two squadrons of Hellcat night fighters, for a total of sixty planes. Commodore PM's Matsuwa Group consisted of a squadron of eighteen Catalinas, a squadron of twelve Harpoons, a squadron of twenty-four Hellcat fighters, and another squadron of six Hellcat night fighters.

Commodore PQ's Etorofu Group had twelve Catalinas organized into two squadrons, a squadron of nine Liberators, a squadron of twelve Hellcat fighters, and a squadron of six Hellcat night fighters. There was also Commodore PP's Shikuka Unit, with twenty-four Liberators in two squadrons, as well as twenty-four Hellcat fighters and twelve Hellcat night fighters, the latter in two squadrons.

Captain P-20 (Lt. Col. George Shell, USMC)

Captain PPA's Konotoro Unit of fifteen Liberators constituted one squadron. Captain PPB's Moskalvo Unit of thirty Liberators was organized in two squadrons, while Captain P-20's Carrier Group of two escort carriers and four destroyers rounded out the organization.[16]

Admiral PK ordered the Kamchatka Bay Group to locate, report, and track enemy forces in the area assigned to it, the Shimushu Group the same in its area. The Paramushiru Group was to search a third area and fly photographic reconnaissance of Shemya and Attu every third day. Commencing on 5 June, fifteen copies of these photos were to be forwarded by air to the Joint Intelligence Center, Vladivostok without delay. The Paramushiru Group was also to maintain a twenty-four-hour antisubmarine patrol over Onnekotan Strait and destroy any Blue submarines detected. (Pages 2–3.)

The Matsuwa and Etorofu Groups were to locate, report, and track enemy forces in their respective areas, while the Sakhalin Group was to be prepared on short notice to reinforce the forward air groups. In addition, the Konotoro Unit was to maintain a twenty-four-hour antisubmarine patrol over La Perouse Strait and destroy any Blue submarines detected while the Moskalvo Unit transported drummed gasoline. However, upon contact with the enemy, the latter was to cease its logistical activity and report to the Attack Groups at the forward bases for operations. The Carrier Group was to proceed in the general direction of the enemy once it received the initial contact report and report to Commander, Striking Force. Until then, it was to continue training west of Paramushiru. (Pages 3–4.)

Each Attack Group was also to be composed of search units and attack units under a task group commander who would provide fighter protection for local bases and Purple shipping in the area. Task group commanders would also provide fighter cover for the Purple Striking Force within the limits of fighter range in their assigned areas and would be prepared to furnish antisubmarine patrols over fleet anchorages when requested. Admiral PK emphasized to his commanders that any task group making contact with a sizable Blue force within bombing range was to attack with all available planes and all other task groups were to be prepared to support it. The target priorities were Blue carriers, then battleships, attack transports, cruisers and destroyers, and finally targets of opportunity. (Pages 4–5.)

Commodore PL's Fleet Air Wing 10 would deploy from Vladivostok, staging Patrol Plane Squadrons VP-101 and VP-102 through Kataoka Bay with eighteen PBM Mariners. These squadrons would report to Commander, Task Group 35.1 for duty upon arrival. VP-103 and VP-104, totaling another eighteen Mariners, would report to the same location and the same commander, as would the six Catalinas of VP-105. The latter squadron was directed to Tennei Field to fly missions out to a hundred miles beyond the coast of Hokkaido. Another six Catalinas would be directed to Matsuwa and would report on arrival to CTG 35.4.

Commodore PM's Fleet Air Wing 20 would also fly from Matsuwa, sending fifteen Privateer naval patrol bombers of VP-201 to Shikuka for duty with CTU 35.6.1 and another fifteen of VP-202 to Konotoro to operate with CTG 35.6.2. At the same time, Commodore PN's Fleet Air Wing 30 would redistribute its aircraft and tenders from Paramushiru. He would send the 1st Division of Fighter Squadron VF-302—twelve Hellcats—to the Ust Kamchatka airstrip, reporting to CTG 35.1. Another twenty-four Hellcat fighters from VF-302—less its 1st Division—would be sent to Staraya Tarya for duty with Task Unit 35.2.2, and twenty-four Hellcats from VF-303 would fly to the Cape Lopatka airfield to dispatch to CTG 35.2. At the same time, eighteen Privateers would be sent to Moskalvo for duty with TU 35.6.3. (Page 6.)

Fleet Air Wing 30 would sail seaplane tenders AV-1 and AVP-2 to Kamchatka Bay, there to report to Commander, Task Group 35.1. In addition, AVP-3—also after arrival at Paramushiru—would sail to Kataoka Bay to report to CTG 35.2 for duty in TU 35.2.1. Fleet Air Wing 30 would provide fighter cover for AV-1 and AVP-2 as practicable en route to Kamchatka Bay. AV-1 was to obtain antisubmarine nets at Avacha Bay for use at Kamchatka Bay. At the same time, Commodore PO's Fleet Air Wing 40 would fly eighteen Harpoons from VP-403 to the Kurilskoye Lake airstrip, where they would be based for purposes of dispersal. This squadron would also be at the disposal of CTG 35.2. Nine Privateers from VP-501 were to be directed to Etorofu for duty with CTG 35.5, while six Catalinas from VP-503 would report to Matsuwa for operations with CTG 35.4. The last of these units was Commodore PQ's Fleet Air Wing 60, whose VP-601 of twelve Privateers would fly from Etorofu to Moskalvo and, upon arrival, report to CTG 35.6 for duty with TU 35.6.3. (Page 7.)

Summarizing by area, at Ust Kamchatka were seaplane tenders AV-1 and AVP-2 with twelve Hellcat fighters from VF-302, six Mariners from VP-101, and six more Mariners from VP-102. The tenders were expected to arrive there about 8 June; the first tanker would arrive about 18 June, others every fourteen days thereafter. Logistical estimates indicated that twelve naval patrol bombers per day could be supported in patrols out to four hundred miles for fifteen days and two fighters per day for twelve hours for fifteen days.

Petropavlovsk would have a capacity of eighteen flying boats, with another eighteen flying boats and sixty naval patrol bombers at Tarinski Bay. Staraya Tarya would have a capacity of forty-eight fighters. Commander, Fleet Air Wing 30 would provide the maintenance crews and thirty men for each airfield as well as fly in spare parts and other necessities. Commander, Service Force would pick up at Vladivostok and deliver by tanker in the first trip to Avacha Bay ten portable ten-thousand-gallon tanks for forwarding to the airfields in the area. Planes and stocks of fuel at all of these airfields were also to be dispersed and protected; tankers en route to Ust Kamchatka were to replenish every fourteen days. Every effort was also to be made to replenish the supply of fifty-gallon drums of aviation gasoline and lubricants. (Pages 8–9.)

Among the southern Kamchatka fields, Kurilskoye Lake had a capacity of eighteen naval patrol bombers, with Pauzhatka and Zheltovski thirty-six fighters each. Commander, Fleet Air Force was to fly in forty drums of lubricants and a thousand drums of aviation gasoline to Kurilskoye Lake, Pauzhatka, and Cape Lopatka, and twenty drums of lubricants and five hundred drums of aviation gasoline to Zheltovski. Six Catalinas from VP-404 would make four trips daily from Shimushu to bring supplies of aviation gasoline and lubricants as far above minimum requirements as possible. Six Catalinas from VP-204 would do the same from Paramushiru. Eighteen Privateers from VP-302 and another twelve from VP-601 would make two trips daily from the northern Sakhalin bases. These two squadrons would also assist in the early buildup of aviation gasoline and lubricants but would be released upon the first report of combat. At the start of the air transportation operation, it was estimated that approximately seventy-five thousand gallons of gasoline and lubricants would be flown into the area daily. Daily consumption was not estimated to exceed twenty-five thousand gallons. (Pages 9–10.)

In the Paramushiru–Shimushu area was AVP-3, along with 102 naval patrol bombers and 180 fighters. A tender was to arrive at Kashiwabara about 6 June. Tankers were expected to deliver aviation gas and lubricants directly to local facilities to the amount of 2.4 million gallons of aviation gasoline and 120,000 gallons of lubricants per month, which was the monthly expenditure estimate. Complete repair facilities were available at each field and fuel stowage was to be maintained to capacity by the Service Force. At Matsuwa, thirty-nine naval patrol bombers and thirty fighters were stationed, along with the same repair capacities; there was fuel stowage for a one-month expenditure of 1.2 million gallons of aviation gasoline and sixty thousand of lubricants. Etorofu similarly had twelve naval patrol bombers and eighteen fighters based; its monthly expenditure and capacity rates were 1.5 million gallons of aviation gasoline and seventy-five thousand of lubricants. (Page 10.)

The commanding officers of seaplane tenders AV-1, AVP-2, and AVP-3 were to obtain from Net Depot, Vladivostok an antitorpedo net each with the necessary installation equipment. The nets were to be used at all times when at anchor in the forward areas. Tenders, bases, and wings were to keep ammunition to authorized capacity, fuel and lubricants to 60 percent of capacity, and aviation gasoline to (in an apparent change) 20 percent. They were additionally to keep dry provisions and general stores for 120 days, fresh provisions to maximum capacity, and aircraft spares for ninety days. Units under Rear Admiral PK's command were required to submit reports on the availability of spares. Shortcomings in these would reflect unfavorably on "administrative efficiency"! (Pages 10–11.)

Air search areas were to be searched twice daily, except for one that was to be searched only once daily. Searches were to commence from Paramushiru, Shimushu, Matsuwa, and Etorofu on 5 June as well as from Kamchatka Bay as soon as possible. Contacts were to be maintained in order to provide accurate tracking information to all the task force commanders. Further, Admiral PK's forces were to be prepared to strike enemy forces in the area and to keep dispersed and protected to the maximum extent consistent with the mission. Units were to utilize airstrips at Kurilskoye Lake, Cape Lopatka, and Staraya Tarya to relieve the congestions at the fields on Shimushu. In the designated Submarine Area, attacks on submarine contacts were prohibited. In the designated Surface Area, no surface forces were to be attacked unless recognized beyond doubt as hostile. Day and night recognition signals were provided to prevent attack by friendly forces. (Pages 12–13.)

In the event of contact with major enemy forces, the unit making the contact report would trail the enemy; the task group commander would then arrange additional or relief search planes. Meanwhile, all other searches were to be continued unless by "express" orders of the force commander. A task group commander whose unit was reporting contact was to ready all available bombers for attack without further orders and take full advantage of twilight and low visibility in pressing an attack. The task group commander whose unit was carrying out the strike was to act personally as Target Coordinator, directing succeeding waves of Purple aircraft to the target; his plane was to be configured for maximum endurance, armed with machine guns only. When the Sakhalin task group commanders learned of contact with major Blue surface forces, they were to arm all naval patrol bomber squadrons and direct them to Matsuwa and Paramushiru for refueling as well as strike orders if the enemy force was located north of latitude 45 degrees north. If the enemy force had been located south of that latitude, two of the naval patrol bomber squadrons were to proceed to Etorofu and the two others to Matsuwa. All squadrons were to refuel at these locations and obtain strike orders. If the assigned fields were under attack, the planes were to proceed to the next adjacent field farther from enemy contact. (Page 14.)

The armament for the planes was eight armor-piercing bombs or two torpedoes for the Liberators, one torpedo for the Harpoons, and full gun ammunition for all planes. Casualties and enemy damage were to be reported "promptly" to the force commander; aircraft were to use any emergency field in case of damage. Identification Friend or Foe equipment was to be turned off. In the strike itself, units were to attack simultaneously from as many widely separated directions as possible and not to fly over the target again after dropping their ordnance but to use the speed from their dives to clear the area. Upon completion of an attack, planes were to return, refuel, and rearm for another strike; Admiral PK estimated that it would take nineteen minutes to rearm and refuel his heavy bombers. Task group commanders were ordered to launch additional strikes without further orders and then inform the task force commander. (Pages 14–14A.)

Rear Admiral PB's Purple Striking Force included the Main Body, which consisted of his own command of two battle cruisers, organized as Cruiser Division 1; the six heavy cruisers of CRUDIVs 2 and 3; and the four light cruisers of CRUDIV 4, with light cruisers CL-5 and CL-6 attached. In addition, the Striking Force included the two light carriers of Rear Admiral PF's Carrier Division 1, designated the Air Group, and Commodore PG's Screen Group, twenty-five destroyers from Destroyer Squadrons 1 and 2 (less Destroyer Division 6) and four destroyer-minesweepers from the Mine Division. (Pages 14–14A.)

Admiral PB described in detail Cruising Disposition 1-C, an anti-atomic-bomb disposition designed to afford antisubmarine and antiaircraft defense as well. The Formation Guide was designated as destroyer DD-25, occupying station 0000 (the center) and any rotation of the formation was to be performed about the Fleet Center. Ships were to turn simultaneously to new courses, but the destroyer-minesweepers would retain their relative screening positions on the carriers. The direction of the Fleet Axis was also normally to be in the direction from which the enemy surface attack was most likely; it would not change for flight operations. When the OTC required pickets, the Screen Commander was to designate the required number of destroyers and station them out to twenty-five miles from the Fleet Center at positions bearing 045, 135, 225, and 315 degrees true. Cruising Disposition 1-C had been designed for "expeditious" change into Approach Disposition No. 1. (Page 14.)

For emergency deployment, Cruiser Divisions 1 and 2 were to turn to a signaled deployment course and work gradually into column, with CRUDIV 1 in the Van, thereafter using course changes small enough to avoid blanketing fire (i.e., to keep ships out of each other's firing arcs). Oilers and carriers were to turn to the reciprocal of the enemy bearing, steam until well clear, and then parallel the battle line or proceed as otherwise signaled. The destroyer-minesweepers were to accompany the light carriers; destroyer DD-26, temporarily assigned to the Striking Force, would accompany the carriers as an additional screen ship. All sonar-equipped ships were to stand continuous echo-ranging watch unless directed otherwise. The destroyer-minesweepers screening the light carriers would also perform plane-guard duties, and the sonar-screen ships were to utilize sonics. (Page 14.)

When an air attack was considered imminent, Cruising Formation 1-CR, an antiaircraft defense formation, was to be assumed. This formation, which Admiral PB felt could be assumed "readily," was considered suitable for defense against both aircraft and submarines but was primarily for defense against carrier-based aircraft. DD-25 was again the Guide on station 0000, and the formation would rotate about the Center. Turns would again be simultaneous. Patrols, searches, strikes, and defensive measures were to be directed by the Officer in Tactical Command. (Pages 14–15.)

Two other formations might also be ordered. Cruising Disposition 2-C was another anti-atomic-bomb formation, with the task force in two groups, one the heavy combatant ships and the other the carriers, supported by the three light cruisers, destroyers, and destroyer-minesweepers. This formation was designed to give greater dispersion to the force as a whole and give greater flexibility tactically to each of the groups. These groups would be strategically concentrated yet tactically independent. The Guide for the combined formations was to be battle cruiser CB-1, those for each group CB-1 and light carrier CVL-1, respectively. Cruising Disposition 2-C permitted rapid deployment into either Approach or Battle Dispositions. (Page 16.)

Commodore PI, Commander, Service Force, had the Logistics Group of two escort carriers, the five destroyers of Destroyer Division 6, and the six oilers of Service Squadron 1. In addition, there was the Mine Group, under Commodore PR, which included the Minelaying Unit of one light minelayer and the four minesweepers from Mine Division 3, as well as the Minesweeping Unit of eight minesweepers from Mine Division 2. The Logistics Group was to furnish support for petroleum products to the fleet at sea as well as to advanced seaplane and land-based planes. The Mine Group was to sweep mines as directed; both groups would have their headquarters at Kashiwabara. The Minelaying Unit was to conduct strategic mining operations as directed and be prepared to engage in tactical mining. The Minesweeping Unit was to allocate its minesweepers as directed by the task group commanders. The Commanders, Logistics and Mine Groups were responsible for refueling their ships, while Commander, Fleet Air Force was responsible for replenishing aircraft, aviation material, and aviation personnel for the escort carriers. The primary source of supply for petroleum products was to be Moskalvo and Otomari in Sakhalin Island. Commodore PI designated Captain P-9—Commander, Logistics Group—as his second in command, at Kashiwabara.[17]

NOTES

1 Senior Class of June 1947, "Operations Problem 2: Purple—Section A," 25 September 1946, p. 1, folder 2603, box 138, RG 4, NHC.

2 Ibid., pp. 1–2.

3 Ibid., pp. 3–4.

4 Ibid., "Annex A: General Concept," p. A-1, and "Annex C: Air Plan," p. C-1.

5 Senior Class of June 1947, "Operations Problem 2: Purple—Section A, Dispatches All by Hand Officer Messenger," 25 September 1946, folder 2603-T, box 138, RG 4, NHC.

6 Senior Class of June 1947, "Operations Problem 2: Purple—Section A, Annex D, Logistic Plan," 25 September 1946, pp. D-1 to D-2, folder 2603, box 138, RG 4, NHC.

7 Senior Class of June 1947, "Operations Problem 2: Purple—Section A, Annex E: Communication Plan," 25 September 1946, p. E-1, folder 2603, box 138, RG 4, NHC.

8 Ibid., pp. E-1 to E-3.

9 This casual reference to ULTRA, then a highly classified cryptoanalytic program, may seem surprising. But the term had by this point become something of a shorthand, even outworn, among American veterans whose knowledge remained limited to operations rather than the details of intelligence sources. In this context, a classified war game, the fact that the term appears within documents relating to the scenario is not as provocative as might now seem. The author is indebted to David Kohnen of the Naval War College's College of Operational and Strategic Leadership for this insight.

10 Senior Class of June 1947, "Operations Problem 2: Purple—Section A, Annex F: Intelligence Plan," 26 September 1946, p. F-1, folder 2603, box 138, RG 4, NHC.

11 Senior Class of June 1947, "Operations Problem 2: Purple—Section A, Submarine Force Dispatches," 30 September 1946, pp. 1–3, folder 2603-K, box 138, RG 4, NHC.

12 Ibid., pp. 4–5.

13 Ibid., p. 6.

14 Ibid.

15 Ibid., p. 7.

16 Ibid., folder 2603-L, 1 October 1946, p. 1. Page references until the final endnote of the chapter are to this source.

17 Ibid., folder 2603-N, 1 October 1946, pp. 1–2.

EXIT

XIII *Operations Problem 2*

The Blue Statement and Staff Solution, September–October 1946

The Blue Statement outlined the same four purposes of the problem as had Purple and the same General Situation in terms of Purple–Blue relations since the defeat of Germany and Japan. Blue too did not see much hope for improving these relations: "Her aggressive foreign policy has resulted in economic and political penetration and control of bordering countries. She has consistently and effectively blocked all efforts by the [United Nations] to compose differences, in which her own interests were not uppermost."

The Blue Statement of the Problem

It appeared that Purple was willing to accept war to maintain its stand. Accordingly, the Joint Chiefs of Staff had "been occupied with the making of plans to meet any emergency in the event of breakdown of diplomatic negotiations." The General Situation spoke to Purple's base complex in Kamchatka, especially Petropavlovsk, but also its new bases at Bolsheretsk, Ust Kamchatka, and Korf Bay. It dealt in the same detail with Purple's development of former Orange bases in the Kuriles and Sakhalin, indicating that for purposes of the Chart Maneuver this information had been provided to both Blue and Purple and was not the result of intelligence collection.[1]

Intelligence, however, had told Blue that the Purple naval bases and airfields in the Kuriles and Sakhalin were well defended by coastal and antiaircraft batteries and by local defense craft, and were being equipped with "adequate" communications facilities. Purple bases in Kamchatka, however, were known not to be fully developed. Purple surface forces had been concentrating in Vladivostok since the first of the year and small forces of cruisers, destroyers, and "small" carriers had been operating in the Pacific north of 45 degrees north latitude. It was estimated that Purple had about four carriers, similar to Blue's escort or light carriers. In addition, there were thought to be two "large, heavily gunned, armored" cruisers; four or five heavy and six light cruisers; thirty destroyers; and twenty new, long-range, high-speed submarines. Purple was also estimated to have a large number of naval patrol planes—both flying boats and land planes—and of land-based aircraft of all types. Pointing out that Purple surface ships and submarines resembled Blue's types in terms of speed, radius, armament, and age, the General Situation surmised

Senior Class

Operations Problem 2

Detail of Student Officer Commanders

BLUE NORTH PACIFIC Force

Section B

Command	Rank	Student Officer	Room
ComNorPACFor	V.Adm. BF	Capt. Stryker	250
Chief of Staff	---	Capt. Fitzsimmons	250
Operations Officer	---	Capt. Higgins	250
T.F. 71, ComSubFor	Capt. B-13	Capt. Purmort	246
Chief of Staff and Operations Officer	---	Mr. Strong	246
ComSubRon 98	Capt. B-13	Capt. Purmort	246
C.O. AS-19	Cdr. B-13	" "	246
ComSubDiv 981	Cdr. B-6	" "	246
ComSubDiv 982	Capt. B-7	" "	246
ComSubRon 2	Capt. B-14	Col. Bidwell	246
C.O. AS-15	Capt. B-14	" "	246
ComSubDiv 11	Cdr. B-8	" "	246
ComSubDiv 12	Cdr. B-9	" "	246
T.F. 73, ComCover Force	R.Adm. BA	Capt. Greenacre	251
Chief of Staff and Operations Officer	---	Cdr. Hydeman	251
ComCarDiv 2	R.Adm. BA	Capt. Greenacre	251
C.O. CV-21	Capt. B-1	Cdr. Hay	251
C.O. CV-36	Capt. B-2	" "	251
C.O. CV-38	Capt. B-3	Capt. Heming	251
C.O. CV-39	Capt. B-4	" "	251
ComCarDiv 3	R.Adm. BM	Capt. Goulet	253
C.O. CV-37	Capt. B-33	" "	253
C.O. CV-33	Capt. B-34	" "	253
ComCruDiv 1	R.Adm. BD	Col. Cooley	253
ComCruDiv 3	R.Adm. BN	" "	253
ComCruDiv 15	R.Adm. BE	" "	253
ComDesRon 3	Capt. B-12	Col. Lyle	253
ComDesDiv 31	Capt. B-12	" "	253
ComDesDiv 32	Cdr. B-2	" "	253
ComDesRon 9	Capt. B-9	" "	253
ComDesDiv 91	Capt. B-9	" "	253
ComDesDiv 92	Cdr. B-4	" "	253
ComDesRon 15	Capt. B-29	" "	253
ComDesDiv 151	Capt. B-29	" "	253
ComDesDiv 152	Cdr. B-14	" "	253
C.O. CLAA-95	Capt. B-5	" "	253
C.O. CLAA-98	Capt. B-7	" "	253
T.F. 74, ComFltAirFor	R.Adm. BL	Capt. Price	250
Chief of Staff and Operations Officer	---	Col. Johnson	250
C.O. AV-13	Capt. B-32	" "	250
ComFAW-1	Capt. B-24	Capt. Tackney	252
C.O. AV-7	Capt. B-24	" "	252
C.O. AVP-48	Cdr. B-13	" "	252
ComFAW-3	Capt. B-25	Lt.Col. Negri	252
ComFAW-5	Capt. B-26	Cdr. Castree	252
ComFAW-7	Capt. B-27	Capt. Sweeney	252

Command	Rank	Student Officer	Room
T.F. 75, ComExpForce	R.Adm. BB	Capt. Greer	133
Chief of Staff and Operations Officer	---	Cdr. Stevens	133
C.O. CA-73	Capt. B-36	Capt. Ashford	137
ComBatDiv 5	R.Adm. BC	" "	137
ComCarDiv 19	R.Adm. BG	Cdr. Lyons	125
C.O. CVE-112	Capt. B-37	" "	125
C.O. CFE-115	Capt. B-28	" "	125
C.O. CVE-119	Capt. B-6	" "	125
ComCruDiv 13	R.Adm. BH	Capt. Ashford	137
ComDesRon 1	Capt. B-11	" "	137
ComDesDiv 11	Capt. B-11	" "	137
ComDesDiv 12	Cdr. B-1	" "	137
ComDesRon 7	Capt. B-8	Col. Luckey	137
ComDesDiv 71	Capt. B-8	" "	137
ComDesDiv 72	Cdr. B-3	" "	137
ComDesRon 13	Capt. B-10	Lt.Col. Bowen	137
ComDesDiv 131	Capt. B-10	" " "	137
ComDesDiv 132	Cdr. B-5	" " "	137
ComPhibGroup 7	R.Adm. BB	Capt. Greer	133
C.O. AGC-11	Capt. B-30	Cdr. Stevens	133
ComTransRon 1	Commo. B1	Cdr. Matter	125
ComTransDiv 11	Commo. B1	" "	125
ComTransDiv 12	Capt. B-15	" "	125
ComTransDiv 13	Capt. B-16	" "	125
ComTransDiv 14	Capt. B-17	" "	125
ComTransRon 2	Commo. BJ	" "	125
ComTransDiv 21	Commo. BJ	" "	125
ComTransDiv 22	Capt. B-18	" "	125
ComTransDiv 23	Capt. B-19	" "	125
ComTransDiv 24	Capt. B-20	" "	125
ComMinRon 1	Capt. B-23	Cdr. Woods	126
C.O. CM-5	Capt. B-23	" "	126
ComMinDiv 5	Cdr. B-10	" "	126
ComMinDiv 7	Cdr. B-11	" "	126
ComMinDiv 9	Cdr. B-12	" "	126
T.F. 78, ComSerFor	Commo. BK	Capt. Tucker	241
Chief of Staff and Operations Officer	---	Col. Simenson	241
ComSerRon 3	Commo. BK	Capt. Tucker	241
ComSerDiv 31	Capt. B-21	Col. Simenson	241
ComSerDiv 32	Capt. B-22	Col. Simenson	241
ComSerDiv 33	Capt. B-31	Capt. Ricketts	241
C.O. CVE-114	Capt. B-35	" "	241
ComCortDiv 1	Cdr. B-15	" "	241
ComCortDiv 5	Cdr. B-16	" "	241

Fig. 117
Student Officer Commanders, Blue North Pacific Force, Operations Problem 2

that Purple was operating a limited number of heavy bombers, though Blue did not know anything about their characteristics. According to reports, these heavy bombers resembled Blue Privateer naval patrol bombers and Purple's flying boats resembled Blue's Catalina and Mariner naval patrol bombers. As far as Blue knew, Purple had no very-long-range heavy bombers. Purple's land-based naval air fleet in the theater of operations was estimated to number about 150 medium and heavy bombers, 250 fighters, and fifty flying boats or seaplanes. "Purple is known to be sparing no expense to produce atomic bombs, but none has been produced so far." (Page 12.)

Blue's Special Situation as of 1 June listed the Blue North Pacific Force—including the 7th Fleet—as commanded by Vice Admiral BF and based in Puget Sound except for some units of the Submarine Force and the Fleet Air Wings. In addition, Blue had been assembling and training an expeditionary force of two Marine Corps divisions. Now at Seattle or its vicinity, these units were obviously in excess of the requirements to defend Alaska and the Aleutians from Purple attacks but they were also the only expeditionary forces that would be assigned to Vice Admiral BF.

Vice Admiral BF (Capt. Joe Stryker, USN)

Transport Squadrons were in Puget Sound, ready to sail to any designated port. No Marines or equipment were on board but the transports could be combat-loaded within five days of receiving the order. The Service Squadrons in Puget Sound were fully loaded and ready to sail on one day's notice. On 1 June, Admiral BF returned from a conference with the Commander-in-Chief of the Blue Pacific Fleet (not played by a student officer). The Joint Chiefs of Staff (no designated student officers) were planning an early invasion of Purple Asiatic territory in case of war with Purple. To make that invasion feasible, it was necessary first to seize a position in Kamchatka or the Kuriles as an advanced base. This seizure was to take place as soon as possible after the declaration of war; the selection of the position was to be made by Vice Admiral BF. It was thought that war with Purple was imminent and that the president could obtain a declaration of war anytime he desired. The declaration was to be put off, however, until Blue forces crossed the 180th meridian en route to their objective. (Pages 12–13.)

In the General Plan, the Blue North Pacific Force was officially assigned the task of seizing a position suitable for an advanced base west of longitude 165 degrees east and north of latitude 46 degrees north. Admiral BF was also given amplifying instructions to organize a Covering Force from units of the Blue 1st Fleet. The Covering Force would be under his command until his task was complete and it was to be ready in Pearl Harbor by 0000 local (Zebra) time on 5 June. The Blue Expeditionary Force was to cross the 180th meridian en route to its objective at 0000 Zebra time on 20 June, at which time "unrestricted" operations could commence if war had not already been declared. Vice Admiral BF's orders stated that the preliminary deployment of his forces could commence at any time after they were ready. After the seizure of the objective, a second force would follow for the development and garrisoning of the base but the responsibility for this phase of the operation was not his. He was warned that Purple agents in the "Blue Homeland had been ordered to spare no effort or expense to learn his intentions." (Pages 13–14.)

Not surprisingly, given that Blue was the invading power, its forces were even greater than Purple's. The Blue North Pacific Force, based at Puget Sound, included Admiral BF's heavy cruiser flagship, equipped with airborne-early-warning (presumably, in modern terms, electronic-support measures) gear. Rear Admiral BC (Capt. William Ashford) commanded Battleship Division 5, consisting of two prewar modernized battleships, one of which also was equipped with airborne-early-warning gear. Similarly equipped was one of the three escort carriers of Rear Admiral BG's CARDIV 19.

Rear Admiral BH (also Captain Ashford) commanded Cruiser Division 13's three light cruisers while Captain B-11 (also Captain Ashford) led Destroyer Squadron 1,

Rear Admiral BC (Capt. William Ashford, USN,)

Rear Admiral BG (Cdr. Charles Lyons, USN)

Captain B-8 (Col. Robert Luckey, USMC)

Captain B-10 (Lt. Col. Robert Bowen, USMC)

Captain B-13 (Capt. George Purmort, USN)

Captain B-14 (Col. Bruce Bidwell, USA)

Rear Admiral BB (Capt. Julian Greer, USN)

Commodore BI (Cdr. Alfred Matter, USN)

Commodore BK (Capt. Thayer Tucker, USN)

Captain B-23 (Cdr. William Woods, USN)

Rear Admiral BL (Capt. Walter Price, USN)

Captain B-24 (Capt. Stephen Tackney, USN)

Captain B-25 (Lt. Col. Peter Negri, USMC)

Captain B-26 (Cdr. John Castree, USN)

Captain B-27 (Capt. John Sweeney, USN)

Rear Admiral BA (Capt. Alvord Greenacre, USN)

Rear Admiral BM (Capt. Wildred Goulett, USN)

Rear Admiral BD (Col. Albert Cooley, USMC)

Captain B-31 (Capt. Claude Ricketts, USN)

composed of nine destroyers: the five ships of Destroyer Division 11 and the four radar-picket destroyers of Destroyer Division 12. Captain B-8 commanded DESRON 7, another nine ships similarly organized into DESDIVs 71 and 72; Captain B-10 led DESRON 13, ten ships, with five each assigned to DESDIVs 131 and 132. Blue's submarines consisted, first, of Submarine Squadron 98, commanded by Captain B-13, in a submarine tender as flagship, its ten boats organized into Submarine Divisions 981 and 982, each division having five boats. Second, Captain B-14, at Kodiak, commanded Submarine Squadron 1, which also consisted of a submarine tender flagship and ten boats, Submarine Divisions 11 and 12. Blue's submarines had operational characteristics that were identical to Purple's. Also at Puget Sound was Amphibious Group 7, led by Rear Admiral BB, in an amphibious command ship (AGC). Amphibious Group 7 consisted of Transport Squadron 1, commanded by Commodore BI and consisting of twelve attack transports, six attack cargo ships, and six high-speed transports organized as Transport Divisions 11, 12, 13, and 14. Commodore BJ commanded Transport Squadron 2 with the same number and types of ships and comprising Transport Divisions 21, 22, 23, and 24.

Service forces consisted of Commodore BK's Service Squadron 3. This in turn was broken down into Service Division 31, which consisted of eight attack cargo ships, two destroyer tenders, and two ammunition ships. Service Squadron 3 also consisted of Service Division 32 (six oilers) and Mine Squadron 1. Commanded by Captain B-23, the latter outfit had one minelayer as flagship, eight destroyer-minesweepers, and six minesweepers. (Pages 15–16.)

Blue's three escort carriers each had one fighter and one attack squadron. Accordingly, on each carrier there were eight Fireball fighters, eight Hellcat night fighters, and two Hellcat photographic reconnaissance fighters. In addition, each air group had twelve Avenger attack planes. Battleship Division 5 had four observation planes and CRUDIV 13 had an additional six observation planes. Fleet Air Command, Aleutians, under Rear Admiral BL, was based at Puget Sound. Admiral BL flew his flag in a seaplane tender; two additional seaplane tenders were assigned to Captain B-24's Fleet Air Wing 1, also at Puget Sound. Fleet Air Wing 1 additionally consisted of four squadrons of naval patrol bombers, with nine Mariners in each squadron. Fleet Air Wing 3, based in Amchitka and commanded by Captain B-25, consisted of another three squadrons of Privateers, with nine planes each; a squadron of six Catalinas; and six Superfortress heavy bombers from the 404th Bombardment Group of the Army Air Forces' Alaska Air Command and temporarily under the operational control of Fleet Air Command, Aleutians. Fleet Air Wing 3 also contained a squadron of sixteen Hellcat day fighters and another twenty Hellcat night fighters. (Pages 16–17.)

Fleet Air Wing 5, on Adak, was commanded by Captain B-26. This unit had the same organization and composition of Catalinas and three Mariner squadrons, an attached squadron of Superfortresses, and the two squadrons of, respectively, day and night Hellcat fighters. Fleet Air Wing 7, based on Shemya under Captain B-27, consisted of two squadrons of Mariners, with nine planes each, and a squadron of six Catalinas, plus twelve Hellcat day fighters and twelve Hellcat night fighters. Only the Superfortresses could carry atomic bombs. As the players already knew, land-based airplane and aircrew replacements could be made available in Adak in thirty-six hours, but the total number of planes in service at any one time could not be increased. (Page 17.)

The Blue Covering Force, at Pearl Harbor, was commanded by Rear Admiral BA, who was also Commander, Carrier Division 2 and flew his flag in one of the four fleet carriers in that unit. One of the four fleet carriers was to conduct night operations. CARDIV 2 was complemented by CARDIV 3, commanded by Rear Admiral BM and consisting of two fleet carriers, one of which was also dedicated to night operations. Like the flagships at Puget Sound, all the fleet carriers were equipped with airborne-early-warning gear, though not all embarked airborne-early-warning aircraft. In addition, the Blue Covering Force had one escort carrier for antisubmarine warfare. Two of CARDIV 2's fleet carriers each had two fighter squadrons and two attack squadrons, each carrier carrying forty-one Bearcat fighters, four Hellcat photographic reconnaissance fighters, and sixteen Hellcat night fighters. Each of the two carriers also had twenty-four Helldiver attack planes. One carrier was additionally equipped with eleven Avenger night-attack planes, five Avenger VA(Q) planes, and four Avenger carrier AEW planes. The second carrier had on board fifteen Avenger night-attack planes and five Avenger tactical electronic-warfare aircraft. One of the other fleet carriers of CARDIV 2 had fifty-five Bearcat fighters, twenty-eight Hellcat night fighters, four Hellcat photographic reconnaissance fighters, nine Avenger night-attack planes, four Avenger tactical electronic-warfare planes, and four Avenger AEW aircraft, all organized into three fighter and one attack squadrons. The fourth carrier of CARDIV 2, the night carrier, had two fighter and two attack squadrons, with forty-nine Hellcat night fighters and twenty-four Avenger attack planes between them. (Pages 17–19.)

Carrier Division 3's two fleet carriers consisted, as mentioned, of a night carrier, also with two fighter and two attack squadrons, altogether forty-nine Hellcat night fighters and twenty-four Avenger attack planes. The day carrier also had two fighter and two attack squadrons, in all forty-one Bearcat fighters, four Hellcat photographic reconnaissance fighters, twelve Hellcat night fighters, twenty-four Helldiver attack planes, fifteen Avenger night-attack planes, and five Avenger tactical electronic-warfare planes. The escort carrier had one fighter and one attack squadron on board, eighteen Hellcat night fighters and twelve Avenger attack planes. (Page 19.)

As for surface warships, the Blue Covering Force consisted of the four heavy cruisers of Cruiser Division 1, under Rear Admiral BD, three of the ships based in Puget Sound and a fourth in Seattle. Rear Admiral BN (Col. Albert Cooley) commanded the two heavy cruisers of CRUDIV 3 and Rear Admiral BE (also Colonel Cooley) the three light cruisers of CRUDIV 15. Captain B-12 (Colonel Lyle—see chapter 1) commanded DESRON 3, eight radar-picket destroyers organized as DESDIVs 31 and 32, with four ships in each division. Captain B-9 (also Colonel Lyle) commanded DESRON 9, nine ships, five of them in DESDIV 91 and the other four in DESDIV 92. Captain B-29 (also Colonel Lyle) commanded DESRON 15, another nine ships, with five destroyers in DESDIV 151 and four radar-picket destroyers in DESDIV 152. In addition, there were two antiaircraft cruisers; Escort Division 1, with six destroyer escorts, under Commander B-15; and Escort Division 5, an additional four destroyer escorts, under Commander B-16 (also Commander Matter). The surface forces were rounded out with the auxiliaries of Service Division 33, six oilers under Captain B-31. These surface units also had observation planes, eight in CRUDIV 1, four in CRUDIV 3, and six in CRUDIV 15. (Pages 18–19.)

Blue's bases were "adequately" equipped with the latest types of radars, antiaircraft batteries, communications gear, and ground-control-approach equipment. Blue's information on its own bases in Alaska and the Aleutians mirrored the information that Purple had on them. For purposes of Operations Problem 2, Blue's atomic bombs could be dropped only by Superfortresses equipped with four-thousand-pound-bomb racks. Bomb-damage-radius data were the same as was available to Purple, but it was revealed that the Blue Commander-in-Chief as of 1 June had four atomic bombs at the Adak base. He could not, however, expect resupply for several months. (Pages 19–28.)

The Blue Staff Solution to the Problem

On 10 September, the Blue Staff Solution was presented. Vice Admiral BF, Commander, North Pacific Force and Commander, Blue 7th Fleet, had conferred with the Blue Commander-in-Chief of the Pacific Fleet as well as the Joint Chiefs of Staff about an early invasion of Purple Asiatic territory in case of war, for which purpose he had to seize an advanced base either in the Kuriles or on the Kamchatka Peninsula as soon as possible after a declaration of war and when the invasion location was made known to him. He had the Blue Covering Force—organized from the 1st Fleet—under his command until these tasks were completed; the Blue Covering Force would be ready by 5 June.[2]

Admiral BF assessed the weather, climate, and hydrographic characteristics of the region. Hydrographic studies demonstrated that currents did not have any particular significance for the Operations Problems and that during the summer, the area was clear of drift ice except for some harbors in Kamchatka. (Pages 2–3.)

As to weather, snow should not be factor in July, or even much in June, but over half of the days in June, July, and August would be foggy. Matsuwa, for instance, was enveloped in fog every day during these three months, as were the southeastern shores on the lee sides of islands and landmasses. There could be two or three cold fronts per month aligned northeast–southwest and moving eastward and north of the islands. These fronts brought heavy rains. There were waves and swells from the southeast, a fifth of them between three and eight feet high. In August, aircraft operations were likely to be curtailed by fronts and fog. (Pages 3–4.)

In Kamchatka, summers were short, cool, damp, and cloudy. In addition, the mean position of the Pacific Polar Front was over Kamchatka and cyclonic storms there occurred at frequent intervals. At Petropavlovsk in July, for instance, visibility was less than six miles over a ten-day period. At Ust Bolsheretsk, the average number of foggy days was twenty-four in June and twenty-five in July. Therefore, air operations there would be hindered by clouds, rugged topography, low ceilings, poor visibility, and conditions favorable for icing. While conditions at Petropavlovsk were favorable most of the time, they were not representative of the entire peninsula. Low clouds were frequent over the Bering Sea and the Sea of Okhotsk, where air operations were again likely to be curtailed. Admiral BF, in fact, thought close air support might be "negligible," but that FIDO equipment permitted limited operations. He warned that weather predictions more than twenty-four hours in advance were no more than 60 percent reliable. (Pages 4–5.)

The Aleutians he placed in the semipermanent low pressure area, what he called the "breeding ground" of cyclone systems. Weather conditions would curtail flying from "time to time" but not prevent it for long periods, and fog disposal at these air bases would permit flying in foggy conditions. Carrier aircraft in the Aleutian Islands could be recovered provided that visibility was not less than six hundred yards and the ceiling was not under a hundred feet. Weather conditions, however, would be of major concern to the Officer in Tactical Command, who had to recover his carrier- and cruiser-based aircraft before fog set in. Adverse weather might also force a few days' postponement of offensive operations, but there would still be sufficient favorable weather to permit successful operations. Given how major a factor weather was going to be, he stressed the importance of collecting weather information. (Pages 5–6.)

The base to be chosen as an advanced staging site needed to be near sea routes, route intersections, narrow passages, mouths of large rivers, and terminal points. The amount of shipping passing by, the ability to accommodate a large force, and the ability to permit the safe departure of a force at any given time also had to be considered. In addition, this base would need to provide continual support to the Blue forces until the mission on the Asian mainland was complete, and it needed to be protected from ground, sea, and air attacks. Selection of the advanced base thus would be heavily dependent on terrain features and "artificial" defenses such as fixed fortifications, antiaircraft batteries, airfields, minefields, nets, and radar installations. The base should also have shallow approaches to force hostile submarines to or near the surface. (Pages 6–7.)

Additionally, the base needed more than one entrance, well separated. Anchorage depths needed to be about eight to twelve fathoms, and with deeper areas for floating dry docks. Anchorages also needed to be sheltered from the seas and strong winds, be free from excess currents, and have "reasonable" rises and falls of tide. The base also needed to be on ground that could be easily held from land attack, and its entrances and approaches had to be capable of being mined and netted at their narrowest points. Admiral BF also saw the need for an entrance that was not a straight line from the sea to the anchorage, and for surrounding topography that was high close to the harbor (so as to preclude aerial torpedo attacks) and that could support good land, sea, and air defensive installations. The base, however, needed enough level terrain to accommodate personnel, dry docks, foundries, machine shops, warehouses, piers, and protected fuel and ammunition stowage. There additionally had to be suitable terrain nearby for airfields and sea- and land-based planes as well as close proximity to a supply of both skilled and unskilled labor, raw materials, an inland water supply, a "healthful" climate year-round, and adequate communications by cable, telegraph, telephone, and radio. (Pages 7–8.)

What then were the possible advanced bases west of 165 degrees east and above 46 degrees north, his prescribed area of operations? According to Vice Admiral BF, there was no feasible base on the Sea of Okhotsk unless Blue established and maintained control of one or more passages between the Kurile Islands and controlled the air in that vicinity. Since this control would mean establishing an air base within easy range, no such site could be considered except possibly in the Kuriles. Harbors on the Kamchatka Peninsula and in the Kuriles, however, could be considered for an advanced base site. The harbors on Uruppu, Shimushiru, Matsuwa, Shasukotan, Onnekotan, and Araido Islands were either too small or too exposed. In addition, Purple sea and air bases on Paramushiru, Shimushu, and Kamchatka flanked the great-circle sea routes, so islands in this part of the Kuriles were not suitable sites. In Shimushu and Paramushiru Islands, Kataoka Bay and Kashiwabara were within the mile-wide Paramushiru Strait that separated the two islands. Combined, they constituted a major fleet anchorage. Shimushu was a low, rolling island eighteen miles long and twelve miles wide, separated from Kamchatka by the seven-mile-wide Shumusho Strait. The surface was a partially dissected plateau about four or five hundred feet high, its greatest elevation 624 feet. The coasts had steep cliffs about 150–500 feet high, and there was a great deal of well-drained land. Movement anywhere on the island was possible. It had a Purple naval-aviation field, a seaplane base, and two dry-weather landing fields. (Pages 8–9.)

Paramushiru was the second-largest island and owing to its location, Vice Admiral BF thought, the most important in the Kuriles. It had the largest population, extensive military installations, a fishing industry, and sulfur mines. Its military installations included four military airfields and a Purple Army and three Purple Navy fields, including the Purple Navy's headquarters for its Naval Air Arm. In addition to Kataoka and Kashiwabara, small anchorages were located at Kakumabetsu Bay and Musashi Bay; Suribachi Bay was an open anchorage, except for a small harbor protected from the east by a breakwater. Paramushiru was sixty miles long and twelve miles wide and was divided into three "well-defined" mountain units. To the northeast, the area was mostly rugged, with peaks over three thousand feet in altitude and valleys with gentle slopes permitting cross-island movement. The southwestern mountains, the most formidable barrier on the island, included two active volcanoes. The third section was a steep mass isolated from the southwestern mountains by a broad lowland belt running southward from the north coast. Coastal and cross-island benches and lowlands of the southwestern third of the island had low slopes with many beaches, affording easy movement. BF did not know how many Purple troops were on Paramushiru and Shimushu but because of their strategic importance he expected their bases to be of recent and extensive development and with sufficient garrisons—absent contrary information, he was going to assume so. He was also going to assume that bases were well defended by coastal batteries, antiaircraft batteries, and local air- and naval-defense craft, and that they had adequate communication facilities. (Pages 9–10.)

FROM/TO	ATTU	NIKOLSKOE	PARAMUSHIRO	PETROPAVLOVSK	UNIMAK ST	180th MERIDIAN	PUGET SOUND	ADAK
ADAK	400	660	970	900	470	120	2025	---
AMCHITKA	250	525	885	765	60	570		220
ATTU	---	270	630	520	600	270	2400	400
BOLSHERETSK	580	400	135	70	---	---	---	---
DUTCH HARBOR	800	---	1400	1260	---	500	1700	470
ETOROFU	1120	---	480	640	---	---	3300	---
KISKA	180	460	810	720	600	90	---	---
KODIAK	1300	---	---	---	570	---	1250	1050
KORF BAY	490	310	---	520	---	---	2800	800
MATSUWA	800	640	194	360	---	---	---	---
NIKOLAEUSK	---	840	895Sea 570Air	645	--	---	---	---
NIKOLSKOE	275	---	460	295	--	490	---	650
OTOMARI		1010	580	705	--	--	---	---
PARAMUSHIRO	630	460	---	160	1340	800	---	---
PEARL	2320	---	2800	2760	--	--	2350	2030
PETROPAVLOVSK	520	295	160	--	1303	765	---	---
SEATTLE	2438	---	3060	3000	1638	2176	---	---
SHEMYA	55	330	690	570	750	215	---	---
UNIMAK ST.	- --	---	---	---	---	538	1638	470
UST KAMCHATSK	400	140	400	235	---	580	---	800
VLADIVOSTOK	---	---	1080	1350Sea 1190 Air	--	---	---	---

Fig. 118
Table of approximate distances

The Kamchatka Peninsula was 750 miles long and between eighty and three hundred miles wide; it terminated only seven miles from Shimushu. Lying between the Bering and Okhotsk Seas, it was rugged country with greatly varied topography. It included marshy tundra on the west side of the peninsula; the vicinity of Petropavlovsk contained several airfields and afforded sites for several more. The Gulf of Kamchatka was never completely frozen, though it was often blocked by drift ice. It was packed with ice in February to March but usually clear by sometime in April. However, there was still heavy drift ice in the Bering Sea until early June. The coast between Osyplakhtak and Mys Kamchatski was often free from ice when the rest of the gulf was solidly packed. Ust Kamchatka and Kruger Bay were the only two anchorages in the Gulf of Kamchatka; the former was a small, unprotected harbor with three five-hundred-yard and six four-hundred-yard berths. Kruger Bay had the same facilities but offered the only sheltered anchorage in the gulf. Purple was understood to be developing air facilities at Ust Kamchatka; there was also an incomplete seaplane base there. The entire area was about three miles north of the great-circle routes from the Blue Mainland, four hundred air miles from Paramushiru, and the same distance from Attu. Vice Admiral BF thought it would be lightly defended because ice prevented the area from serving as a base during a considerable part of the year. (Pages 10–11.)

Avacha Bay had Petropavlovsk, "by far" the best port in Kamchatka. It was the only port in Kamchatka that could provide well-protected anchorage berths and landing facilities for oceangoing vessels, with a capacity of fifty thousand tons per month. Since it was also the capital of Kamchatka, it had good all-weather roads across the peninsula, though there were no other communications with Purple territory except by sea. Petropavlovsk was not used as an anchorage, because of limited room, but it had eight wharves in the harbor, and there were six commercial wharves in the other parts of the bay. There might even be one or two dry docks of five-thousand-ton capacity there. Rakovaya Harbor, in the southeast portion of Avacha Bay, was reported to have two docks for ships of fifteen thousand tons and one capable of taking a forty-thousand-ton ship. Tarinskaya Harbor, also in the southeast portion, was a seaplane base; it was landlocked and so barges were needed to land materials there. (Page 11.)

One great advantage of Avacha Bay, to Admiral BF, was that it was free from solid ice year-round except in narrows and coves, and much of that could be handled by icebreakers. Drift ice appeared by early June but should not interfere with the bay's use as a base. Wind conditions were similar to those in the Kuriles, and the weather was generally good from June to October. Only July and August were snow-free, however, and the snow in March was sometimes more than ten feet deep. The entrance channel to Avacha Bay was thirty-six feet deep at low tide and forty-one feet at high tide. Harbors and anchorages could be protected by antisubmarine nets, the bay approaches could be mined, and three beaches were suitable for landings. However, Purple had three seaplane bases, one fighter field, and one bomber field in the area; also, it would be necessary to inoculate Blue troops for typhus, plague, cholera, typhoid, paratyphoid, yellow fever, tetanus, and smallpox. Morzhovaya Bay, about sixty miles from Avacha Bay, was about four miles wide between Araali Cape and Morzhoui Island. It was surrounded by mountains more than two thousand feet high, and there were no settlements in the vicinity. The depth of the waters was between nine and twenty-nine fathoms; it was free from ice in the summer and suitable for operating seaplanes from a tender. (Page 12.)

Admiral BF next looked at the Komandorski Islands, lying about 125 miles off of Kamchatka. These two, narrow, mountainous islands had lengths of thirty and forty-three miles and coasts almost entirely marked by steep slopes and cliffs. June, July, and August saw fourteen, fifteen, and eight days of fog, respectively, and cloudiness prevailed 75 percent of the time. Prevailing winds from the south produced a gale every month, but the islands were free from ice in the summer. Purple had developed a base at Nikolski, at the northern end of Bering Island, for air, surface, and submarine operations. There was room there for twenty-five berths in ten to fourteen fathoms of water but no appreciable protection from westerly winds or seas. Heavy swells entered the anchorage during these westerly winds, which would make handling cargo difficult. Medni had no good anchorages or level stretches for airfields, and vessels could anchor only in small bays or coves. Still, Admiral BF wanted the Komandorskis reconnoitered for Purple activities and considered for use by Blue tender-based seaplanes.

No sites south of Paramushiru would be suitable, because any would be flanked by Purple sea and air bases in Paramushiru, Shimushu, and Kamchatka. The Paramushiru-Shimushu site was too strongly defended to seize with his limited force. Avacha was the only base west of 165 degrees east and north of 46 degrees north that could provide well-sheltered anchorage berths and possessed "sufficient other" necessary characteristics to qualify as a suitable advanced base. Also, it was lightly defended. It was, therefore, his choice for seizure. (Pages 12–13.)

Vice Admiral BF had information that Purple was developing at Petropavlovsk what was to be a major air and naval base but was now still lightly defended. It was additionally reported that Purple was developing other auxiliary land- and seaplane stations on both coasts of the Kamchatka Peninsula, but little was known of them except that "considerable activity" was being noted at Bolsheretsk, Ust Kamchatka, and Korf Bay. There was a small seaplane base at Korf Bay and another at Ust Kamchatka, each capable of handling eighteen flying boats. Further, he knew of three airfields and three seaplane stations in the vicinity of Avacha Bay with a combined capacity of sixty naval patrol bombers, forty-eight fighter planes, and fifty-four flying boats; another forty-eight fighters could be accommodated at Khayryuzovo and Bolsheretsk. The three fighters fields on Cape Lopatka had a total capacity of 120 planes, while the two airfields on Shimushu could handle ninety naval patrol bombers and eighty-five fighters. The four fields on Paramushiru had room for 175 fighters and 120 naval patrol bombers. When it had been under Japanese control, Shimushu had had four coastal-defense and about twenty dual-purpose guns. Additionally, it had antiaircraft batteries concentrated about its airfields. Paramushiru had eighteen coastal-defense and thirty-four dual-purpose guns, again concentrated about the airfields and the headquarters area. Matsuwa had one airfield that could accommodate forty naval patrol bombers and thirty fighters, while Etorofu could take twenty-four naval patrol bombers and thirty-six fighters. Sakhalin had seven airfields and stations, but with unknown capacities. (Page 15.)

Vice Admiral BF had no information on enemy troop concentrations. In 1945, Purple had had an estimated twenty-five thousand troops in Kamchatka, an estimated 10,500 of which were in fixed defenses and the remainder—including a mountain regiment—mobile forces. The former Orange garrison for the entire Kuriles had been about ninety-nine thousand troops, too greatly separated to be mutually supporting except for Paramushiru and Shimushu. The Orange garrisons on these two islands had numbered about thirty-two thousand, of which ten thousand were in the southern Paramushiru area. That island was "extremely" rugged, swampy in coastal areas, and had numerous streams that an advancing force would have to cross; movement of troops and equipment was difficult in many parts of the island and impossible in others. Given these conditions, he did not consider forces in northern and southern Paramushiru to be mutually supporting. His own two Marine Corps divisions in Seattle numbered about forty thousand men but would be sufficient only against a Purple force of no more than eleven to sixteen thousand strong. Cruisers, destroyers, and small carriers concentrated in Vladivostok were operating in the Pacific north of latitude 38 degrees but BF did not know the disposition of the Purple submarines or shore-based aircraft.

The Blue 1st Fleet, less his flagship, the heavy cruiser CA-73, was in Pearl Harbor, and part of this fleet would constitute the Blue Covering Force. Blue had four atomic bombs at Kodiak Island, and he did not think that Purple had any of these new weapons. (Pages 16–17.)

Vice Admiral BF saw the Blue Strength Factors as having the initiative; fleet carriers that were superior to Purple's light and escort carriers; and very-long-range aircraft that could carry the four atomic bombs. He also saw the Covering Force's aircraft as numerically superior to Purple's light-carrier airpower in a ratio of three to one, and he thought (as in fact Admiral PA assumed) that Blue air attacks could be concentrated to win temporary control of the air in small areas. Blue Weakness Factors, however, included land-based fighters that could not operate within 180 miles of the Purple bases and naval patrol bombers that had to sacrifice part of

Aircraft

BLUE	Shore Based	PURPLE
18	VLR B-29	0
72	VP (HL) PB4Y-2	150
0	VP (ML)	
36	VP (MS) PBM-5	50
18	VP (MS) PBY-6A	
44	VF	250
52	VF(N)	

BLUE Covering	BLUE Expeditionary	Ship Based	PURPLE
		CV	
356	0	VF	0
160	0	VA	0
19	0	VA(Q)	0
8	0	VA(W)	0
		CVE or CVL	
18	60	VF	96 or 160
12	36	VA	48 or 0
		BB, CA, CL	
18	10	VO	26
26	28	DD 32 knots, life 1.6, 6-5", 5 or 10 torpedoes	30
10	0	DE 21 knots, life 1.3, 3-3"	?
6	6	AO 18 knots	?
	1	AGC 16.2 knots	?
	24	APA 18.0 knots	?
	12	AKA 14.5 knots	?
	12	APD 18.0 knots	?
	1	CM 18.0 knots	?
	8	DMS 31.0 knots	?
	6	AM 15.5 knots	?
0	2	AE 15 knots	?
0	2	AD 18 knots	?
0	2	AV 18.3 knots	?
0	1	AVP 16.5 knots	?
0	2	AS 17.2 knots	?

BLUE Covering	BLUE Expeditionary		PURPLE
	2 Marine Divisions	Troops	Unknown
		Submarines	
	10	High Speed	20
	10	Fleet Type	0
6	0	CV 31 knots, life 10.6, 101 A/C, 12-5	0
0	0	CVL 31 knots, life 4.6, 42 A/C, 40 mm	4 (possible)
1	3	CVE 17 knots, life 4.6, 30 A/C, 2-5"	4 (if not CVL).
0	0	CC or CB Characteristics unknown.	2
0	2	OSB 18 knots, life 18.5, 8-16", 16-5"	0
6	0	CA 31 knots, life 5.5, 9-8", 12-5"	4 or 5
3	2	CL 30 knots, life 4.6, 12-6", 12-5"	6

Fig. 119
Relative Blue and Purple combat power

their payloads to reach some of the Purple bases. In addition, Blue fleet carriers had to withdraw from critical areas for replacement aircraft, fuel, and other logistical needs. The Blue forces also had no reinforcements, poor intelligence, and limitations on carrier- and cruiser-based flight operations because of the fog that might impact both scouting and air attacks on Purple surface forces at critical times. (Pages 20–21.)

To Admiral BF, Purple had only a couple of Weakness Factors: the prevalence of fog hindering effective searches in critical areas, and planes being mostly land based, negating the effective use of fighters more than three hundred miles out to sea. Purple had a number of Strength Factors, according to Admiral BF. These included land-based air forces in critical areas and the relative ease in replacing aircraft losses within twenty-four hours. Admiral BF thought that Purple's combined light-carrier and land-based fighters were about equal to the Blue Covering Force's fighters and the Purple forces could replace these planes much more quickly than could Blue. Purple also had airfields equipped with fog intensive disposal equipment to facilitate the operation of aircraft, it had better knowledge about local weather and conditions, and its forces could follow weather fronts moving eastward.

In addition, the high ground in the Avacha area facilitated low-level air attacks that could not be detected by shipboard radar. Vice Admiral BF thought he could partially nullify this advantage by establishing radar and fighter-direction stations in isolated places ashore. Still, there were other Purple advantages. For instance, Purple had numerous landing fields in critical areas (which permitted dispersing aircraft), alternative landing fields for use should carriers be damaged, and the option of operating carrier aircraft from land bases. All of this meant that there were ample fields within range of the Blue objective; Purple, he surmised, could deduce that objective with certainty and concentrate forces accordingly. Purple additionally had superior surface forces in terms of cruisers and destroyers. Moreover, its combined surface forces were superior to the Blue Expeditionary Force in terms of numbers and speed of fleet carriers, carrier aircraft, heavy and light cruisers, and destroyers. Finally, Purple had more high-speed submarines than did Blue. (Pages 20–21.)

Given that Purple air strength in the northern Kuriles–southern Kamchatka area was about equal to what Blue could concentrate, it was vital for Blue to seize the initiative. Admiral BF presumed that Blue holding this initiative could turn equality into superiority through early, coordinated carrier- and land-based air attacks—including the use of atomic bombs. "Our superiority in carriers and carrier aircraft gives our Covering Force a considerable advantage over Purple's mobile force during weather which permits carrier flight operations." His superiority in carrier aircraft was more than three to one. While his fleet carriers were at a speed disadvantage with respect to Purple's light carriers, he had superiority in heavy ships, light cruisers, and destroyers when his surface forces were collected as the Covering Force. While Purple had a superiority of combined surface forces over the Blue Expeditionary Force, by itself, he could change the destroyer balance by employing his eight destroyer-minesweepers in the destroyer role, though this would reduce his convoy screen. Still, the loss of two or more of his fleet carriers or severe flight-deck damage would greatly weaken his force, whereas similar damage to the Purple carriers would be of much less consequence for the enemy. Purple also had twenty submarines to his ten. These constituted a threat of "considerable magnitude" because of their long submerged-cruising range and their ability to make high-speed attacks. Blue did not yet have reliable antisubmarine detection devices or weapons developed but he could partially overcome this disadvantage by the use of a maximum number of screening vessels, "vigorous" offensive action, and improvised tactics, including the employment of submarines in his tactical organization. The Blue Covering Force, however, had to withdraw from the operational area to replenish its planes, refuel, and rearm because it was too small to rotate units for replenishment. Purple, on the other hand, had considerable airpower and Admiral BF thought that Purple could reinforce its ground troops at Avacha by parachute, though it would be difficult to supply such forces given the Blue Covering Force's power. (Pages 22–23.)

Vice Admiral BF could get no troop, ship, or aircraft reinforcements and did not know what reinforcements were available to Purple. He assumed that the Purple commander had no ship reinforcements but thought that Purple troops and planes that were in the theater but not in action could be shifted against Blue's forces. This capability would depend on terrain and Admiral BF's control of the air and seas. He also did not know about Purple's logistics situation but he was going to assume it was adequate. Reviewing his replenishment assets, he concluded that he would have to figure out his Course of Action before he determined the adequacy of his logistics. He did not, however, have any hospital ships. In addition, transportation from Avacha would have to be in the ships of his transport squadrons until airfields there had been made ready for an air transport command. Purple, however, could evacuate by air if no other means were available, since its main bases were close at hand. (Pages 23–24.)

Looking at Purple's potential capabilities, Admiral BF thought that Purple could concentrate its forces in the northern Kuriles–southern Kamchatka area to attempt to destroy the Blue surface forces as they came within three hundred miles. Purple surface forces could be supported by submarines and land-based air and coordinated action would allow the Purple commander to oppose Blue with superior forces, provided Purple detected Blue in time to concentrate. Alternatively, the Purple commander could use his land-based air and submarines first to reduce Blue strength by mining, bombing bases, and attrition attacks on surface forces within and outside Purple territory. Purple's third option was to reinforce Avacha with troops and defensive equipment to the point that the Blue Expeditionary Force would be too weak to seize it. This reinforcement, however, would have to take place by ship convoy as Blue could disrupt road communications between Bolsheretsk and Avacha. In addition, Purple could mass its submarines, land-based air, mine capability, and combined surface forces to destroy the Blue Expeditionary Force. This plan would mean having to evade the Blue Covering Force, which might be possible given the adverse weather in the area. Another capability, of course, was to destroy the Blue Covering Force with combined Purple surface forces or so damage it that it could not support the amphibious assault. If Purple immobilized enough Blue carriers, the amphibious assault would not succeed. Last, Purple could invade the Aleutians, so one of Admiral BF's tasks was to defend them. Such an attack might delay the Blue amphibious assault and allow Purple to reinforce Avacha but Admiral BF did not think he would have to respond to this possibility unless a Purple offensive toward the Aleutians really did materialize. (Pages 24–25.)

Avacha Bay was the most logical Blue objective—Admiral BF had the initiative, and he could strike without warning after 0000 Zebra time on 20 June. He saw the fleet carriers in his Covering Force striking the Purple bases, withdrawing, replenishing, and returning by the time the Expeditionary Force approached the critical area. He also, however, had to protect the latter en route to the objective area as well as provide direct air support during the landings. His General Plan of Action was to do all of this by keeping Purple constantly on the defensive through vigorous offensive action. More specifically, Admiral BF could neutralize the Purple air bases, destroy Purple shipping in the area, photograph the landing beaches for the Expeditionary Force, and then retire to replenish. Then he could return to keep the Purple air bases neutralized and destroy enemy aircraft, submarines, and surface ships before they could threaten the Expeditionary Force as it approached, captured, and then defended Avacha Bay. Lastly, he could attempt to capture, occupy, and defend Avacha in one phase. In this case, air strikes would be delayed until D-2 or D-1 (two days, or one, respectively, before the attack), and there would be no replenishment without the retirement of one group of carriers. (Pages 25–26.)

Blue Course of Action One, then, provided for attacks on Purple that he thought would so weaken the enemy that approach and landing operations could succeed. Better and more timely photographic reconnaissance of the landing areas would also be possible, as would replenishment. These operations would allow more adequate coverage of the Expeditionary Force by Blue land-based air in the Aleutians and enable the Covering Force to interpose itself between the Expeditionary Force and Purple surface forces as the former approached Avacha. Blue Course of Action Two would consist of about two days of carrier- and land-based air strikes. It had a number of disadvantages. First, there could be no replenishment of forces until after the landings had been made. Neutralization of Purple air installations would be more difficult because there would be fewer strikes with a smaller number of aircraft since the Covering Force would have to provide fighter protection to the Expeditionary Force. There would additionally be a delay in photographic reconnaissance of the landing areas. Initial operations would allow Purple to deduce where the landings were to occur. Still, he reasoned, the advantage of neutralizing Purple's air bases, timely reconnaissance of landing beaches, and early opportunity to destroy Purple surface forces outweighed the disadvantage of giving the objective away to Purple. He was confident that he had forces sufficient to accomplish this mission. (Pages 26–27.)

Under Purple's first assumed possible Course of Action, the Purple commander would deploy his land-based air forces to provide maximum striking power against Blue forces as they approached the southern Kamchatka or northern Kurile areas. He thought the Purple commander would keep his surface forces out of range of Blue carrier fighters until the carriers were within range of Purple's land-based fighters. Purple could then make repeated, heavy air strikes to reduce Blue carrier strength and then destroy the transports when they were present or it could continue focusing on destroying the Blue Covering Force.

Vice Admiral BF assumed that Purple's tactics would include coordinated carrier- and land-based strikes in maximum numbers in an effort to saturate Blue fighters and antiaircraft fire. Submarines would also be used to the extent possible, with target priorities as the carriers, transports, cruisers, screening ships, oilers, and then other auxiliaries. The Blue Covering Force, however, could approach Paramushiru from the east or southeast, thereby reducing the number of Kamchatka-based fighters that could reach Blue forces and forcing the remainder to operate at longer ranges. Strikes could then also be made on the southern Kuriles. If Blue approached southern Kamchatka from the east, some of the Purple airfields in the Kuriles would be out of fighter range. If Blue could succeed in a surprise approach, it should be able to gain control of the air by temporarily immobilizing Purple fighters on fields out of range, maintaining a superior number of fighters in the area, destroying the Purple light carriers if the opportunity arose, and destroying Purple air facilities at the same time. Blue land-based bombers could also be used in these strikes. By destroying Purple aviation-fuel supplies, hangars, and ammunition dumps in southern Kamchatka, Blue could make these airfields "almost useless" to Purple for an extended period. One or both groups of the Blue Covering Force could then withdraw to replenish and then return to cover the approach of the Blue Expeditionary Force to the objective. Blue had to accomplish these strikes, however, without incurring loss to its carrier strength, lest it not have strength left to cover the amphibious landings. (Pages 27–28.)

Blue's success against Purple Course of Action One would depend in large measure on its own ability to effect the surprise and failure of Purple to have its forces ready and disposed. Unless Blue could achieve surprise and use its atomic bombs to weaken Purple considerably, it might not have sufficient air strength to accomplish its mission. By aggressive offensive action, Admiral BF thought, Blue could destroy Purple shipping, aircraft, and installations so as to allow photographic reconnaissance, location and clearance of minefields, and replenishment of forces. Even if Purple succeeded in limiting Blue successes here, however, BF thought he had the power to capture, occupy, and defend Avacha Bay, keeping Purple bases neutralized and destroying their air, surface, and submarine forces. Purple focus on destroying Blue forces with submarines and land-based airpower would result, he thought, in weaker Purple submarine forces in the critical area but would allow greater chances of striking Blue forces as they approached the critical area of operations. This Purple Course of Action, however, would require Blue to allocate more land-based air for defensive tasks and force it to take greater risks in sending task units to operate independently outside the critical area. Admiral BF, however, could route his forces within supporting distance of his land-based air in the Aleutians and organize his task units to take the best advantage of escort carriers, destroyers, and destroyer-minesweepers. (Page 29.)

If Purple tried to reinforce Avacha Bay by the movement of troops by sea and accomplished this by 0000 Zebra on 20 June, Blue would not be able to prevent it. After the Blue Covering Force arrived in the area, however, it would be strong enough to prevent further reinforcement of Avacha by sea. Blue would be able to defeat Purple in this scenario by neutralizing or destroying its ships, aircraft, and base installations, and Blue land-based air could be used to augment attacks on Purple reinforcement groups. If Purple tried to attack the Blue Expeditionary Force with coordinated strikes by submarines, aircraft, and surface ships, Blue would have to maintain "efficient" reconnaissance to locate the Purple combined surface forces in time to permit the Covering Force to defeat the Purple surface forces before they threatened the Expeditionary Force. He thought it possible that Purple would operate its combined force to the north or the south of the Aleutian chain in a position to strike the Expeditionary Force. Doing so, however, would expose it to attacks by Blue land-based air and Covering Force attacks without the support of Purple land-based fighters. This Course of Action by Purple might divert the Blue Covering Force from its preinvasion air strikes but Purple's chance of destroying the Expeditionary Force would be small. It might delay the Blue landings, but at the loss of the Purple combined surface forces. (Page 30.)

If Purple attempted a surprise carrier air strike on the Blue Covering Force by its combined surface forces, followed by a surface action, Purple would have the advantage of weather fronts moving eastward behind which to approach and so achieve surprise. Purple in this scenario could also augment its attack with land-based bombers, if conditions were favorable. If Purple succeeded in immobilizing three or more Blue carriers, Blue's combined forces would be so weak that Purple could repel the invasion with its land-based air forces alone. In general, a carrier duel would be in Blue's favor because of its superiority in carrier forces and its ability to augment those carrier forces with land-based air. If, however, bad weather prevented carrier operations, Purple could strike with its surface forces. Vice Admiral BF knew that he had to take adequate precautions against surprise attacks by Purple but also that he had ample planes to safeguard against surprise if he employed them properly. If Purple attempted to invade the Aleutians, Blue would have to prevent it if it constituted a major threat to Blue communications. Obviously, Blue amphibious operations could be delayed, but Admiral BF did not think that Purple had sufficient surface forces to attempt an amphibious operation in Blue territory. If Purple did, Blue reconnaissance would have to locate its forces to prevent any occupation that could constitute a serious threat to Blue. (Pages 30–31.)

In the end, Admiral BF decided to pursue Blue Course of Action One. Accordingly, he would in Phase One neutralize the Purple base in the northern Kuriles, destroy Purple bases and shipping in the southern Kamchatka Peninsula, and exploit favorable opportunities to destroy Purple surface forces in the area. He would then obtain information on installations and landing beaches in Avacha Bay by aerial and submarine observation as well as from photographic reconnaissance. Also, minefields would be located and destroyed. Phase One would conclude with retirement and replenishment of the Covering Force, which would then cover the approach of the Expeditionary Force. Phase Two would consist of the actual capture, occupation, and defense of Avacha Bay by destroying Purple aircraft, submarines, and surface ships; covering and supporting the landing; and keeping neutralized Purple air bases that were within effective range of Avacha. Favorable weather conditions would be exploited to the greatest extent possible. The seizure of Avacha Bay would provide an advanced Blue base for subsequent operations against Purple "Asiatic" territory. Vice Admiral BF now had to determine the details: routes, searches, and the employment of submarines; Fleet Air Command, Aleutians; the Covering Force; the Service Force; the Expeditionary Force; and special groups. Finally, he had to figure out logistics, movements, and task organization. (Pages 31–32.)

NOTES

1 Senior Class of June 1947, "Operations Problem 2: The Blue Statement, Section B," 13 September 1946, pp. 1–11, folder 2603, box 138, RG 4, NHC. Subsequent page references until the next endnote are to this source.

2 Senior Class of June 1947, "Operations Problem 2: The Blue Staff Solution," 8 October 1946, pp. 1–2, folder 2603-C, box 138, RG 4, NHC. Remaining page references in this chapter are to this source.

XIV Operations Problem 2

The Blue Solution Continued, September–October 1946

In his Solution of the Complementary Problems, Vice Admiral BF first tackled the routes his forces would take. He decided first to move the Expeditionary Force and the Service Force to Avacha Bay via Adak while his flagship and the Mine Force went via Attu. He was going to have his submarines top off in Attu en route to their patrol stations and the submarine tenders shift to various islands in the Aleutians. The Covering Force was to pass close to Attu to allow the flagship to join up. The shortest routes to Attu and Adak from Puget Sound and Kodiak were the great circles via Unimak Pass. He therefore figured that Unimak Pass was the best place for Purple submarines to deploy. Since he needed vital information that only his own submarines could provide, his own submarines would have to move by this route, especially since he considered them the best protection from Purple's. By transiting Unimak Pass, the Mine Force might also give Purple submarines the impression that its task was to conduct an antisubmarine sweep in advance of a larger force. If the Main Body of the Expeditionary Force proceeded to Adak south of the Aleutians, it might escape detection prior to arrival at the staging point. This evasive route was 250 miles longer, but time was not pressing. In addition, the Covering Force did not need to penetrate the Aleutians should close support be necessary. After departing Adak, the Expeditionary Force was to proceed north of the chain to save eighty miles. The route for the Expeditionary Force also had the advantage that until ships passed Attu, attacking aircraft from Paramushiru had to pass over the Aleutians. He also determined that the Service Force at Pearl Harbor could proceed to Adak unless it was diverted by Commander, Covering Force.*

With respect to searches, the Purple–Blue international boundary ran from a point halfway between Saint Lawrence Island and Siberia to midway between Attu and Medni Islands. Purple might object to any search beyond this point prior to the initiation of hostilities but Admiral BF thought that security demanded that Blue locate any Purple surface force in a position to inflict serious damage. The

---

* Senior Class of June 1947, "Operations Problem 2: The Blue Staff Solution," 8 October 1946, p. 33, folder 2603-C, box 138, RG 4, NHC. Subsequent in-text page references, until the next footnote, are to this source.

Purple surface force could do real damage to Blue forces and the Aleutian installations if it succeeded in making a surprise attack. Consequently, searching to the Purple shoreline was required even prior to the commencement of hostilities; it could be accomplished by radar with flights to within seventy-five miles of the Purple coast. Any Purple protests, however, would have to be placated in order to avoid declaration of war prior to Blue's initial strikes. (Page 36.)

Upon the initiation of hostilities, searches were to be extended to Kamchatka in order to reconnoiter Korf Bay and Kruger Bay. Any Purple aircraft and installations on the Komandorski Islands would have to be destroyed, and sea areas would have to be searched in order to locate and destroy Purple forces as well as to guard against surprise attack by surface forces. BF wanted his initial searches to extend southwest to 38 degrees north as well as north and south of the Aleutians either to locate the Purple surface force or to provide reasonable assurance that those forces were not present. Thereafter, he would reduce the searches north and south of the Aleutians, maintaining them between Attu and Kamchatka west of Cape Olyutorski and counterclockwise to a point south of the central Aleutians. The radius of the searches to the southwest and south of the Aleutians could also be reduced.

In addition, as the Expeditionary Force moved to the west, the areas to be searched could be progressively reduced from the eastward up to the longitude of Amchitka. After the Purple surface force had been located, further reductions of the areas to be searched could also be made. He saw the conservation of search aircraft as very important, because their numbers were very limited and they were also needed for antisubmarine patrols, antishipping sweeps, weather flights, and bombing tasks. The present distribution of aircraft was neither efficient nor economical. He wanted to move Fleet Air Wing 1 and the three seaplane tenders to the Aleutians as early as practicable. Attu had a naval air station capable of supporting a large number of land planes of the type assigned to Fleet Air Command, Aleutians. This redeployment of aircraft was required before commencement of

ABLE	48 - 35 N 125 - 00 W	KANSAS	53 - 45 N 150 - 00 W
ANGER	24 - 15 N 160 - 00 W	KING	51 - 05 N 176 - 52 W
BAKER	49 - 45 N 130 - 00 W	LAMAR	54 - 00 N 155 - 00 W
BLUNT	32 - 00 N 165 - 00 W	LOVE	51 - 15 N 177 - 30 W
CHARLIE	50 - 55 N 135 - 00 W	MADRID	54 - 10 N 160 - 00 W
COTTON	38 - 25 N 170 - 00 W	MIKE	51 - 00 N 179 - 00 E
DOG	51 - 40 N 140 - 00 W	NORMAN	54 - 10 N 165 - 00 W
DISMAL	43 - 40 N 175 - 00 W	ORANGE	54 - 15 N 170 - 00 W
EASY	52 - 10 N 145 - 00 W	OSAGE	52 - 10 N 170 - 00 E
EDWARD	48 - 00 N 180 - 00	PARIS	54 - 10 N 175 - 00 W
FERRY	51 - 32 N 175 - 00 E	PREP	52 - 35 N 165 - 00 E
FOX	52 - 25 N 150 - 00 W	QUEEN	52 - 45 N 160 - 00 E
GAP	50 - 15 N 130 - 00 W	ROGER	52 - 10 N 177 - 30 W
GEORGE	52 - 40 N 155 - 00 W	ROSE	53 - 50 N 180 - 00
HARLEM	51 - 30 N 135 - 00 W	SUGAR	52 - 35 N 180 - 00
HYPO	52 - 30 N 160 - 00 W	TARE	53 - 15 N 175 - 00 E
INGOT	52 - 30 N 140 - 00 W	UNIT	53 - 20 N 170 - 00 E
ITEM	52 - 10 N 165 - 00 W	VICTOR	53 - 10 N 165 - 00 E
JASPER	53 - 15 N 145 - 00 W		
JIG	51 - 52 N 168 - 45 W		

CHURCH passes through the following points:
ANGER, BLUNT, COTTON, DISMAL, EDWARD, FERRY, to ATTU.

FRIGIDAIRE through ROGER, SUGAR, TARE, UNIT, VICTOR, QUEEN to AVACHA.

PONTIAC through ABLE, BAKER, CHARLIE, DOG, EASY, FOX, GEORGE, HYPO, ITEM, JIG, KING to ADAK.

REPUBLIC through ABLE, GAP, HARLEM, INGOT, JASPER, KANSAS, LAMAR, MADRID, NORMAN, ORANGE PARIS, ROSE, TARE to ATTU.

TABLE through LOVE, MIKE, FERRY, OSAGE, PREP, QUEEN to AVACHA.

Fig. 120
Route points, Blue Staff Solution, Operations Problem 2

search operations and would continue as conditions changed but he thought all this could be accomplished with ease. (Pages 36–37.)

Submarines were required to obtain vital information in areas that could not be subjected to aerial search. Purple forces in Kruger Bay and on the Komandorski Islands were the only shore-based forces north of Petropavlovsk that could offer an immediate threat. These areas, he thought, could be investigated to a limited extent before hostilities began by his submarines, which could also investigate minefields and check on air and sea activity in the vicinity of Avacha. His submarines could also investigate Purple air and sea activity around the Cape Lopatka–Shimushu–Paramushiru area as well as, prior to the start of hostilities, the Kurile and Amphitrite Straits since the latter were the most likely routes from the Asian mainland to Avacha. Blue submarines were to report any sightings of Purple surface or expeditionary forces. Single-ship movement, however, was not to be reported if it would jeopardize more-important missions. In addition, submarines could be used advantageously for such tasks as weather observation, lifeguard duty, radar picketing, and attacks on shipping. (Page 37.)

Some submarines would be required for weather observation as their primary mission. This was especially true for the Avacha–Paramushiru areas because the date for commencing operations depended on an accurate fair-weather forecast for a period of a day at Paramushiru and its vicinity, followed by about four days of fair weather in the area of Avacha Bay. Weather information in the western Aleutians was of lesser importance but had to be given consideration in order to plan and conduct land-based air operations. Also, since he was short of destroyers, his submarines would have to be employed as radar pickets off Attu; early warning of Purple planes approaching the Aleutian bases was "most essential." However, he surmised that he could get the best advantage out of his submarines by using them tactically. None of their roles were entirely tactical; the five submarines in the vicinity of Paramushiru and the two at Avacha could be assigned to the Commander, Blue Covering Force on D–3 and D–1 days, respectively, for tactical uses but even they could also be employed as lifeguards, for weather observation, and as radar pickets. Since their scouting function was so vital, however, he was going to put all of his submarines to sea and move the submarine tenders westward. (Page 38.)

The priority tasks for submarines, after weather observation and reconnaissance, were the location of Purple mines in the Avacha area and then radar picketing. Lifeguarding came next, then minelaying and the destruction of enemy shipping. Weather observations were needed in the Sea of Okhotsk, the area to the east of the Kuriles, and the vicinities of Paramushiru and Avacha, where aircraft could not fly prior to hostilities. BF was going to employ seven boats in the Okhotsk Sea and one off Bolsheretsk, all with a primary mission of weather observation and a secondary mission of reconnaissance; another boat about fifteen miles off Cape

Elizabeth and Sakhalin with the same primary missions would have the added task of planting two mines off Cape Mary after hostilities commenced. Two boats would be in La Perouse Strait with a primary mission of reconnaissance and then secondary missions of weather observation, minelaying, and antishipping. Three more boats would have primary missions of weather observation between 48 and 51 degrees north latitude and 149 and 150 degrees east longitude. His final boat in the Okhotsk Sea would be within fifty miles to the west of Shimushu with a primary mission of weather observation, except during air strikes, when lifeguarding would be primary. It would also plant two mines in the Kurile Strait. (Pages 38–39.)

In the Kuriles, Admiral BF was going to station one submarine about four miles off the eastern entrance to Paramushiru Strait and another about fifteen miles east of Cape Lopatka. After hostilities commenced, the submarine off the Paramushiru Strait was to plant two mines. Also as of D–3, these boats would operate at the direction of Commander, Covering Force. The submarine stationed in Amphitrite Strait would also come under Commander, Covering Force's operational control on D–3, as would the one eighty miles east of Cape Lopatka. This latter submarine would have a tertiary mission of reporting on enemy air formations en route to Blue bases or surface forces. East of the Kuriles would be two submarines for weather observation and then lifeguard duty and reconnaissance. (Pages 39–40.)

Admiral BF decided to station two boats off the Kamchatka Peninsula to reconnoiter the entrance of Avacha Bay, locate minefields, observe defenses, and photograph beaches. One of these submarines would relay this information to a naval patrol plane a hundred miles east of Avacha at 1900 Zebra time on 17 June. These boats then were to observe Purple air activity and shipping, furnish weather reports when requested, and operate at the discretion of Commander, Covering Force from D–3 on. There would be one more submarine stationed off the entrance to Kruger Bay and one off the Komandorski Islands. Finally, as noted above, the two submarines off Attu as radar pickets would be essential because of the need for early warning of Purple aircraft approaching the western Aleutians, in spite of the disadvantageous employment of submarines that this entailed in terms of destroying shipping. (Pages 40–41.)

Admiral BF thought that his use of such a large number of submarines was necessary to provide for adequate weather coverage as well as "reasonable" reconnaissance and the location of mines. He argued that the planting of the few mines mentioned in restricted areas that were being used by Purple should cause the enemy to devote much effort toward minesweeping, and so the mines should have a high "nuisance" value. The Submarine Force, after its missions of weather observation, reconnaissance, mining, and airborne early warning, would destroy Purple shipping that was encountered. Moving the submarine tenders westward would maximize the number of days the submarines could remain on station by reducing the transit time. Submarine tender AS-15 would be based at Amchitka, and Submarine Squadrons 1 and 98 could arrive at Attu by 0400 on 7 June. These boats could refuel and arrive on station by 18 June, in time to carry out their tasks. SUBRON 98, he thought, could best carry out its mission by being where enemy antisubmarine action was most probable. He planned to have Commander, Submarine Force establish his headquarters at Adak. (Pages 41–42.)

With respect to the employment of Fleet Air Command, Aleutians, Vice Admiral BF first looked at Fleet Air Wings 1, 3, 5, and 7, stationed at Puget Sound, Amchitka, Adak, and Shemya, respectively. His shore air facilities could support five PBM Mariner naval patrol bombers at Attu, another six of these planes at Adak, and a further twenty-four at Dutch Harbor. Prior to the opening of hostilities, he was going to have two primary tasks for his shore- and tender-based aircraft. One was to search for the Purple surface force and conduct antisubmarine patrols against Purple submarines. The second task was to destroy submarines reconnoitering the Aleutian area, commencing on D–3; initial searches, he thought, would not require more than thirty-five aircraft for a 95 percent probability of sighting. The remainder of his available aircraft would be employed for antisubmarine patrols, antisubmarine sweeps ahead of his own surface forces, weather flights, and utility. This initial search plan was designated "Search Plan A." After it was completed, a search of lesser extent would be maintained to provide "continuing assurance" against surprise attack on his surface forces or the Aleutian bases. This would be designated "Search Plan B" and the aircraft it freed from search missions would augment the planes assigned to antisubmarine patrol and utility, as well as bombing. As his surface forces moved westward, the number of search planes needed to search with 95 percent probability would progressively decrease. This search area would be limited in radius in the direction of the northern Kuriles and southern Kamchatka to avoid hostile acts prior to the beginning of war. But only one search flight per day was possible and additional security from surprise attack was required. Therefore, he needed a barrier patrol to augment the area search and cover the most likely Purple approach. Superfortresses could be used to transport the four atomic bombs from Kodiak to Adak and then on to Amchitka or Shemya, where they would be available for the patrols. (Pages 43–44.)

Vice Admiral BF knew that the highly efficient submarines available to Purple were difficult to locate because of the short range at which the *schnorchel* could be sighted visually or detected by radar. Purple had twenty of these boats, and their high submerged speed and ability to attack while deeply submerged made them "most potent" weapons. BF saw aircraft and coordinated air-surface attacks as the most effective methods of locating and destroying these submarines; aircraft were most needed because a shortage of destroyers and destroyer escorts prevented him from assigning these ships to antisubmarine units. He planned, when the weather permitted, on employing carrier aircraft for antisubmarine patrols to secure their ship formations; one escort carrier of the Expeditionary Force would conduct antisubmarine sweeps ahead of that force. It would be necessary to augment such patrols ahead of the large surface formations with special sweeps by tender- and shore-based aircraft. Each of these flights would consist of three or four naval patrol bombers, spaced closely enough to detect a *schnorchel.* The flights would sweep back and forth across the projected track of the formation, about a hundred miles ahead if the formation was maintaining its own antisubmarine patrol. Otherwise, the sweep flight would be deployed as the formation's antisubmarine patrol. He additionally planned to have replacement aircraft with crews so that mobile and shore-based units could be moved westward and made readily available for use. (Page 44.)

Admiral BF's Fleet Air Command had a total of eighteen Superfortress heavy bombers and seventy-two heavy, land-based naval patrol planes. As he noted above, an increasing number would be available for attack as Blue forces moved west. All ninety were capable of bombing Purple installations and forces in the Kurile and Kamchatka bases. He saw the Covering Force's air strikes against the Purple bases as greatly assisted by coordinated, land-based air attacks on the enemy. He specifically wanted the four atomic bombs to be used to destroy Purple surface forces if those forces were located at anchor. He did not think, however, that was likely, because of Purple's capability to scout for Blue forces and anticipate attacks, and he did not think that atomic bombs would be "most profitable" against surface forces at sea because of their ability to maneuver and disperse. Alternatively, should the Covering Force be threatened with a surface engagement, it probably would be advantageous to destroy one or more of Purple's heavy ships in an atomic-bomb strike.

The use of two or more of these bombs against Purple's most important airfields would result in the complete elimination of those targets, both planes and facilities. Because of the strength of Purple fighters at these airfields, fighter cover would be necessary for the Superfortresses unless they were sent in under instrument conditions to bomb by radar. Before he chose atomic-bomb targets, however, he wanted the air installations at Paramushiru and Shimushu reconnoitered, which could not be done until hostilities commenced. The initial strikes by his carrier aircraft would furnish this information, as well as on shipping in the Paramushiru Strait. Commander, Covering Force would be in a position to coordinate these land-based and carrier-based air strikes, and the Superfortresses would stand by in the area so they could be furnished targets by voice radio and proceed when ordered by Commander, Covering Force. (Pages 44–45.)

The use of the Komandorski Islands for air strikes against Blue, for search, and as a base for tenders had to be denied. Admiral BF thought that Fleet Air Command,

Aleutians could neutralize enemy strength in this area, as well as any Purple tenders in Kruger or Korf Bay, without assistance from Blue's carriers. He intended to have Fleet Air Command, Aleutians set up a weather unit on Attu and a standby unit on Adak. Provision also had to be made to use land-based planes for support and reconnaissance in the Avacha area; to furnish replacement aircraft and crews to mobile forces; and to collect photography and other intelligence from submarines and ensure it was distributed. Fleet Air Command, Aleutians was best suited for all these tasks. (Pages 45–46.)

Commencing at 0000 Zebra on 5 June, Fleet Air Command, Aleutians was to patrol the Aleutian area for Purple submarines and broadcast warnings to the North Pacific Force. At 1800 Zebra on 7 June, it was to conduct air searches in accordance with Search Plan A and then daily searches in accordance with Search Plan B. Further, it was to locate, track, and report Purple surface forces before hostilities commenced. After hostilities commenced, it was to assist other Blue forces in destroying Purple surface forces that were encountered and destroy Purple submarines that were threatening Blue forces. It was also to maintain an antisubmarine barrier patrol a hundred miles ahead of the Blue Expeditionary Force and Blue Service Force as well as provide an antisubmarine barrier patrol ahead of the Expeditionary Force while the latter was en route to Avacha Bay. While the Expeditionary Force was at Avacha Bay, Fleet Air Command, Aleutians would provide an antisubmarine patrol for that force; establish its headquarters and weather unit at Attu; establish the standby weather unit at Adak; and provide weather forecasts for the Kurile–Kamchatka–Aleutian area to the North Pacific Force. When hostilities commenced, Fleet Air Command, Aleutians would strike Purple tenders and air installations in the Komandorskis, the northern Kuriles, and southern Kamchatka as far north as Korf Bay. At Korf Bay itself and in the Komandorskis, strikes by Fleet Air Command, Aleutians would be coordinated with strikes from the Blue Covering Force's carriers. All of these air strikes were to be directed, controlled, and coordinated by Commander, Covering Force. (Pages 46–48.)

As noted above, Fleet Air Command, Aleutians would then provide replacement aircraft and crews to Blue's mobile forces, provide direct air support for the assault on Avacha Bay, and provide photographic reconnaissance of that area. It would additionally establish fighter-direction stations in Kamchatka as requested by Commander, Blue Expeditionary Force; provide air units for operation from captured Avacha air facilities as requested by the same commander; and obtain photographs and information on minefields from Blue submarines a hundred miles east of Petropavlovsk at 1900 Zebra on 17 June. Finally, Fleet Air Command, Aleutians would provide search and rescue service for the Covering Force commencing on D–3. Admiral BF was most concerned with ensuring that replacement aircraft be ferried from the Aleutians area early and that the aircraft available be maintained in as large numbers as practicable in order to keep complements of operating aircraft filled.

The Covering Force had to cover the other forces that were en route to their rendezvous and objective areas as well as provide its own scouting where searches by Fleet Air Command, Aleutians were not adequate. The Covering Force's major task, however, was to isolate the objective area from interference by Purple. The initial strikes by the Covering Force would be delayed until forecasts permitted successful carrier air operations that could be supported by Blue land-based air forces. He tentatively assigned six fleet carriers, five heavy and three light cruisers, two antiaircraft cruisers, eighteen destroyers, and eight radar-picket destroyers to the Covering Force. He would join the Covering Force in his flagship at the fueling area off Attu before the initial carrier strikes, thereby adding a sixth heavy cruiser to the force. By coordination of his carrier-based and land-based air strikes, including the use of the atomic bombs, he thought that the Purple air strength in the Kurile–southern Kamchatka area—which equaled his own air strength—could be so reduced that Blue would have air superiority. Commander, North Pacific Force would decide when the initial attacks were to be launched, primarily waiting until the weather gave Blue the greatest advantages. Commander, Covering Force would coordinate the strikes. (Page 49.)

The logistics for the Covering Force would be covered by what he called the "Special Fueling Group," which would be under the control of Commander, Covering Force and would replenish fuel oil, aviation gasoline, bombs, and ammunition. The units of the Special Fueling Group—Service Division 33, Escort Divisions 1 and 5, and escort carrier CVE-114—were at Pearl Harbor. He saw a great need for "expeditious" reservicing of the Covering Force after the initial air strikes; six attack cargo ships and an ammunition ship would be included in the Service Group so as to permit simultaneous ammunition transfer. The Service Force itself, based at Attu, would also send a unit to the replenishment area to supply the Covering Force. (Pages 49–50.)

Since the two escort carriers assigned to the Expeditionary Force could not alone furnish adequate close air support and protect the Expeditionary Force from Purple attacks, the Covering Force had to take on the latter task also. In addition, the Covering Force had to provide photographic reconnaissance and surveillance of the Paramushiru–Shimushu and Avacha areas so as to augment the efforts by Fleet Air Command, Aleutians. Minesweeping operations at Avacha also had to be supported and protected from Purple air attack and the Covering Force was the only unit that could carry out this mission, though it might be opposed by Purple surface forces during its strikes. (Page 50.)

In sequence, the Covering Force was first to cover the other Blue units en route to and at the objective. When hostilities commenced, it was to destroy Purple air bases in the northern Kuriles and southern Kamchatka and then surface forces in a position to threaten Blue forces. Commander, Covering Force would coordinate his strikes on enemy bases, defenses, and surface forces with those of Fleet Air Command, Aleutians, except for any operations controlled by the Air Support Control Unit (explained below). The Blue Covering Force would then provide direct air support to the landings at Avacha Bay as requested by Commander, Expeditionary Force; direct submarine operations in the Paramushiru and Avacha areas; and mine the eastern entrances to Paramushiru Strait, if Purple surface forces were present in that harbor at the commencement of hostilities. Admiral BF thought that the units that he had available were adequate to all these tasks, except for the Service Force units. (Pages 50–51.)

The problem with the Service Force was that his Special Fueling Group could not replenish bombs or large-caliber ammunition to the Covering Force. It could be made part of the Service Force at Attu, and the Service Force could itself meet the need. Also, if based at Attu, the Special Fueling Group would have the protection of the land-based air and harbor defenses while being sufficiently near the operating area for flexibility. The Service Force additionally had to support the Expeditionary Force after the latter crossed the 180th meridian. The Service Force would then proceed to Adak with the Expeditionary Force, replenish at Great Sitkin, and continue on to Attu. It was then to remain at Attu until it departed for the replenishment area to service the Covering Force. (Page 52.)

All eight attack cargo ships and all eight oilers would be employed to expedite replenishment. Two destroyer tenders, four oilers, and two ammunition ships would be left at Attu and eight attack cargo ships, eight oilers, eight destroyers, ten destroyer escorts, and two escort carriers would be sent to the fueling area. Increased protection from Purple air and submarine attacks would result from the extra escort ships and escort carriers. The Service Force would continue to support the Expeditionary Force and the Covering Force after the Special Fueling Group was released by Commander, Covering Force. Admiral BF was confident that with these arrangements, the number of ships was adequate to the task. (Page 52.)

Once the Expeditionary Force crossed the 180th meridian, the Expeditionary Force would proceed at its best speed, twelve knots, to the objective unless D-Day was delayed by weather. Any mines in the approaches had to be cleared before the initial bombardment and landings and the minesweeping units needed fighter cover. The Covering Force was going to provide this fighter protection on D–2. His assignment of an escort carrier and eight destroyers to the Service Force left two escort carriers and twenty destroyers for the Expeditionary Force, which he thought would suffice along with the destroyer-minesweepers. The operation would also entail the tactical control of aircraft, fire support, transportation, and the landing of the troops. An Air Support Control Unit would be embarked in the amphibious flagship to exercise tactical control of aircraft in the target area. A Transport Group would have all the attack transports, attack cargo ships, and high-speed transports under its control. (Pages 53–54.)

(Unless otherwise noted, figures are for Sustained Speed 14.5 knots).

	Capacity	Gal/hr.	Gal/mi.	Range
Batdiv 5				
2 BB-46	1,454,891	1734	119.5	12,190
Cardiv 19				
3 CVE-112	1,274,200	800 at 14.5 1130 at 17.0	55.2 66.5	23,100 19,300
Crudiv 13				
3 CL-65	581,700	1000	68.9	8,450
Desron 1 & 13				
19 DD-884	139,608	390 at 14.5 4480 at 32	26.9 140.0	5,180 995
Desron 7				
9 DD-722	139,608	345 at 14.5 4480 at 32	23.8 140.0	5,850 995
Amphibious Group				
1 AGC 11	947,611	455	31.4	30,000
24 APA-128	314,644	440	30.3	10,600
12 AKA-1	336,198	325	22.4	15,000
12 APD-87	95,148	198 at 12 490 at 18	16.5 27.2	5,770 3,500
Mine Group				
1 CM-5	54,000			No data
8 DMS-19	129,359	430	29.6	4,370
6 AM-100	64,000	143 at 15.5	9.2	7,550
Service Group				
2 AE-3	345,000	270 at 15	18.0	15,000
8 AKA-1	336,198	325	22.4	15,000
2 AD-14	631,384	800	55.2	11,450
6 AO-22	644,000	765	52.8	12,200
Subfor				
2 AS-11	650,390	627 at 16.5	38.0	17,200
	Capacity	Gal/hr.	Gal/mi.	Range
Fleet Air				
2 AV-7	610,850	759	50.6	10,900
1 AVP-10	74,590	143 at 13		7,900

Fig. 121
Blue Expeditionary Force fuel-oil logistics

The Expeditionary Troops would be embarked in these ships, with escort ships provided by the Support Group. The amphibious force flagship, twenty-four attack transports, twelve attack cargo ships, and twelve high-speed transports would actually land the troops—the two Marine Corps divisions that would seize, occupy, and defend Avacha Bay. The Support Group was to drive off any enemy forces attempting to interfere with the movement to the objective area or the landings. It was to provide fire support, close air support, and antisubmarine patrols as well as to clear minefields. He had two prewar modernized battleships, two escort carriers, three light cruisers, twenty destroyers, one minelayer, eight destroyer-minesweepers, six minesweepers, and one seaplane tender for these tasks. As soon as the seaplane tender could operate PBM Mariners around Avacha, a squadron of these planes would be made available to Commander, Air Force, Aleutians. Admiral BF also ordered that if Morzhovaya Bay was undefended or could be cleared of mines, it was to be used for tender-based seaplane antisubmarine patrols. The Expeditionary Force was to route and provide escorts for all shipping moving from the objective area. (Pages 54–55.)

BF's flagship would operate with the Covering Force. He would move to Adak for conferences, then by air to conferences with his commanders at various bases in the Aleutians, rejoin his ship at Adak, and then with his flagship join the Covering Force at Attu. Leaving Puget Sound on 3 June and arriving at Adak on 9 June, he would refuel at the Covering Force's rendezvous near Attu when that unit had been released by Commander, Blue North Pacific Force. Meanwhile, the Special Fueling Group would refuel the Covering Force en route to the Aleutians from Pearl Harbor, again upon arrival off Adak, and then again just prior to the initial strike. After the refueling off Adak, the Special Fueling Group would replenish itself at Great

	Capacity	Gal/hr. at (15 (20 (25 (30	Gal/mi.	Range Hours	Range Miles
Cardiv 2 & 3					
6 CV-9	1,753,410	1740 2800 4625 8350	116 140 185 276	1000 625 380 210	15,000 12,500 9,500 6,400
Crudiv 1 & 3					
6 CA-68	623,250	1247 2240 3764 6740	83 112 150 225	500 278 165 92	7,500 5,500 4,100 2,700
Crudiv 15					
3 CL-55	581,700	1055 1778 3395 5580	71 89 136 189	530 320 168 102	8,200 6,500 4,200 3,100
CL (AA)					
2 CL-95	410,237	832 1472 2640 5210	55.5 73.6 105.6 173.7	490 278 155 78	7,400 5,500 3,900 2,400
Desron 3 & 15					
17 DD-710					No data use figures for DD-692 Class.
Desron 9					
9 DD-692	139,608	492 848 1604 3260	32.8 42.4 64.2 108.7	284 164 87 43	4,200 3,300 2,100 1,200
1 CVE-105	1,274,200	835 at 15 kts. 1130 at 17 "	55.6 66.5		22,900 19,200
Cortdiv 1 & 5					
10 DE-51	99,364	198 at 12 kts. 600 at 21 "	16.5 28.6		6,000 3,400
Servdiv 33					
6 AO-22	644,000	765 at 14.5	52.8		12,200

At 15 knots the Covering Force will use per hour:

CV	6 x 1740	10,440
CA	6 x 1247	7,460
CL	3 x 1055	3,170
CL(AA)	2 x 832	1,660
DD	26 x 492	12,800
		35,530 gallons fuel.

At 20 knots:

CV	6 x 2800	16,800
CA	6 x 2240	13,440
CL	3 x 1778	5,330
CL(AA)	2 x 1472	2,940
DD	26 x 848	22,000
		60,610 gallons fuel.

At 25 knots:

CV	6 x 4625	28,800
CA	6 x 3764	22,600
CL	3 x 3395	10,200
CL(AA)	2 x 2640	5,300
DD	26 x 1604	41,800
		108,700 gallons fuel.

At 30 knots:

CV	6 x 8350	50,000
CA	6 x 6740	40,400
CL	3 x 5580	16,800
CL(AA)	2 x 5210	10,400
DD	26 x 3260	84,700
		202,300 gallons fuel.

Fig. 122
Blue Covering Force fuel-oil logistics

Sitkin. After again refueling the Covering Force off Attu, it would join the Service Force at Attu and become part of it. (Page 56.)

All fueling from tankers by the Expeditionary Force was to be completed prior to crossing the 180th meridian. The amphibious force flagship, ammunition ships, destroyer tenders, attack transports, and attack cargo ships all had enough fuel for the trip. At the two-thousand-mile mark, however, the two battleships would each need 239,000 gallons of fuel oil, the three escort carriers would each need 133,000 gallons, the three light cruisers would each need 137,800 gallons, and the twenty-eight destroyers would each need 93,000 gallons, for a total, according to the Staff Solution, of 1,650,400 (actually 3,894,400) gallons of fuel. Each tanker carried almost three million gallons, so the supply was sufficient. After being refueled, the battleships and escort carriers would refuel the destroyers as necessary until the first replenishment after the landings. By fueling at Adak, the Service Force would also be close to the fueling station at Great Sitkin. Each escort carrier had 159,000 gallons of aviation gasoline on board, and each tanker 823,000 gallons. Given the assistance of land-based air cover, Admiral BF did not foresee the escort carriers having any gasoline problem until after the landings. In addition, he thought he could shift his airplane and submarine tenders as far west as he wanted without any demands on his mobile fuel supply. (Page 58.)

The Covering Force could make Attu at thirty knots on less than sixteen million gallons and his tankers carried seventeen million. Since he was not pressed for time, the problem was where and when to refuel; Attu was the prime location. The Covering Force Escort Group would not have any fuel or gasoline concerns since it would be deployed with the tankers. Moreover, the Covering Force carriers had 1,753,410 gallons of aviation gasoline, enough to outlast their fuel oil. In addition, the aircraft tenders could replenish their supplies of aviation gasoline at Great Sitkin and Attu as well as receive a first replenishment from the Service Force as that unit passed Kiska. Ammunition would be the real limiting factor for this operation. The air strikes by land-based air units would help the Covering Force conserve bombs; that would permit five days of continuous bombing by the fleet carriers. Fuel at this point, however, would be "dangerously" low, so he had to provide for flexibility to allow the refueling of the Covering Force. (Pages 60–61.)

Aircraft:

Type	Capacity	20% of fixed tank	Usable
F6F-5	600	50	550
F8F-1	533	36	497
SB2C-5	653	70	583
TBM-3	810	67	753

Assuming three flights per plane to the maximum range per day as the worst condition, CV 21 will expend:

3 x 41	F8F-1	61,000
3 x 16	F6F-5	26,000
3 x 24	SB2C-5	42,000
3 x 20	TBM-3	45,000
		174,000 gallons

CV-39 will expend:

		52,700
3 x 32	F6F-5	85,000
3 x 55	F8F-1	38,300
3 x 17	TBM-3	176,000 gallons.

Fig. 123
Blue Covering Force aircraft fuel-oil logistics

Since speed was essential in this operation, his Movement Plan called for the Expeditionary Force to have enough fuel to complete the run to the objective area without delay. The sustained speed of his attack cargo ships was fourteen and a half knots, but he could not expect them really to make more than twelve. By 1600 Zebra on 22 June, sixty-four hours after crossing the 180th meridian, the Expeditionary Force was to be off the landing beaches leading to Petropavlovsk. This would give his forces a fast approach but too fast to allow more than a day and a half of strikes if he was to refuel and rearm the Covering Force so it could give maximum support to the landing. In order to maximize his freedom of action, he was going to reprovision the Expeditionary Force in the Aleutians. This would reduce the number of aircraft required for search by reducing the area to be covered and give his troops a chance to "stretch their legs." The Expeditionary Force could be in Adak eight and a half days after departing Puget Sound but after crossing the 180th meridian it would have to regulate its speed in order to coordinate its movements with the Covering Force's. (Page 62.)

Admiral BF's eight destroyer-minesweepers could make a sustained speed of thirty-one knots, the six minesweepers fifteen and a half knots, the minelayer eighteen knots, and the twelve high-speed transports also eighteen knots. In formation, these ships would all be limited to about thirteen knots. To arrive off the beaches forty-eight hours before the Expeditionary Force, the underwater demolition teams and the Mine Force had to cross the 180th meridian forty-three hours earlier. He could shift them to Attu in seven days and four hours in order to sortie from there forty hours before they were needed at Avacha. The longer he could have them at sea without being detected, the better. While this force was strong enough to beat off minor air attacks, aid from the Covering Force would be required to carry out the mission fully. If he shifted the underwater demolition teams and the mine craft to Attu, he would not need to divert the Covering Force. Moreover, by delaying the commitment of the minesweepers to the task, he could give his submarines more time to explore the minefields. Unfortunately, he had a communication problem with this aspect of his Movement Plan in that the nearest high-powered radio station was in Adak. (Pages 62–63.)

Allowing twenty-four hours for refueling en route, the Covering Force could proceed to reach the vicinity of Attu at sixteen knots seven days after departing Pearl Harbor. At twenty knots, it could strike Petropavlovsk in twenty-one hours and Paramushiru in twenty-three and a half hours. At twenty-five knots, the Covering Force could cut these times to seventeen and twenty-one hours, respectively. Also, by being near Attu at the commencement of hostilities, the Covering Force could be in a position to defend the Aleutians should Purple attack before 20 June. It could also interpose itself between Purple and the Blue Expeditionary Force if weather prohibited reconnaissance. (Page 63.)

The Blue high-speed submarines in Seattle could be off Avacha Bay in ten days and ten hours at twelve knots. At twenty-five knots, they could be at Attu in a hundred hours. Allowing twelve hours for fueling and assuming a six-knot speed of advance for the last 360 miles, the submarines could be off Kamchatka with a good fuel supply seven and a half days after leaving Seattle. At fourteen knots, his slower submarines could arrive at Attu a hundred hours after departing Kodiak. Constantine Harbor in Amchitka, which was 765 miles from Petropavlovsk and 885 miles from Paramushiru, would provide the submarine tenders with an advanced base that BF thought would be secure from Purple land-based-air attack. The tender at Kodiak could also be escorted by the Blue Alaskan Sea Frontier Force (organized for coastal defense), and his other tender could move with the mine craft. In order to keep the tenders separated, he was going to put one in Adak and shift it to Attu after Avacha Bay had been seized. (Page 63.)

Fleet Air Wing 7—which consisted of twenty-four fighters, six very-long-range aircraft, eighteen land-based naval patrol bombers, and six seaplane-based naval patrol bombers—was on Shemya, which could also accommodate twelve more very-long-range aircraft and three additional fighters. Attu could handle forty-eight fighters and seventy-two land-based naval patrol bombers and had an ample seaplane base but could not take anything larger than Mariners. Amchitka had Fleet Air Wing 3 with thirty-six fighters, six very-long-range aircraft, twenty-seven land-based naval patrol bombers, and six seaplane-based naval patrol bombers, and it could take another fourteen fighters, seventeen naval patrol bombers, and ninety-four very-long-range aircraft. Adak's Fleet Air Wing 5 had the same numbers and types of aircraft as at Amchitka and could accommodate up to 210

Force	From	To	Can be RFS	Limiting Speed	Distance	Minimum Time for Transit	Then Required at Destination	ETD	SOA	ETA	(Note: All times ZEBRA Dates June).
AV-13	PUGET	ATTU	0000/3	18	2438	135	--	2000/3	16/17	0100/10	16 kts. to UNIMAK 17 thereafter.
AV-7	PUGET	ADAK	0000/3	18	2350	130	--	2000/3	16	2300/9	Leave formation at UNIMAK PASS. Report to Support Grp. 1200/19.
AVP-48	PUGET	KISKA	0000/3	16	2400	150	--	2000/3	16	1300/10	do
AS-15	KODIAK	ATTU		13	1300	100	--	0000/3	13	0400/7	Replenish
AS-15	ATTU	AMCHITKA		17	250	15	--	1900/11	15.5	1200/12	
AS-19	PUGET	AKAK	0000/3	17	2350	138	--	2000/3	16/17	0100/10	
Covering Force	PEARL HARBOR	ADAK Area	0000/5	CL 28	1900	68	1800/16	0800/8	15	2000/14	Fuel en route, also fuel off ADAK.
	ADAK Area	Fuel near ATTU		CL 28	420	15	1800/18	0600/15	5	1000/18	Fuel off ATTU.
	Near ATTU	PARAMUSHIRU		28	400 530	19	0000/20	0800/19	25	0000/20	
	PARA	AVACHA	0600/20	28	160	6	1900/20	6600/20	--	1900/20	
	AVACHA	Replenish	0100/21	28	270	10	1100/21	0100/21	28	1100/21	Replenish
	Replenish	AVACHA	0400/22	28	300	11	1900/22	0400/22	20	1900/22	
Fueling Group	PEARL	42°-15 N 173°-35 W	0000/5	CVE 15	1475	98	1800/12	1000/8	15	1200/12	
	42°-15 N 173°-35 W	Off ADAK		CVE 15	600	40	2000/14	0400/13	15	2000/14	
	Off ADAK	GREAT SITKIN		CVE 15	140	14		0600/15	10	2000/15	Replenish
	GREAT SITKIN	Off ATTU	1000/16	CVE 15	420	28	1800/18	0600/17	15	1000/18	
	Off ATTU	ATTU		CVE 15	150	10	0800/20	0800/19	15	1800/19	Replenish
Flagship	PUGET	ADAK		30	2438	82	--	2000/3	16	2300/9	16 kts. to UNIMAK
Amphibious Force	PUGET	AKAK	0000/6	AKA 12	2350	196	1500/17	1400/7	12	1800/15	Fuel ADAK Area
	ADAK	180 Meridian	1800/18	AKA 12	100	8.3	0000/20	1500/19	12	0000/20	
	180 Meridian	AVACHA	0000/20	AKA 12	765	67	1900/22	0000/20	11.5	1900/22	
Service Force	PUGET	ADAK	0000/2	AKA 12	2350	196	1500/17	1400/8 1400/7	12	1800/15	
	ADAK	GREAT SITKIN	1800/16	AKA 12	40	4	0000/17	1800/16	12	2200/16	Fuel ADAK Area
	GREAT SITKIN	ATTU	2000/17	AKA 12	400	33	1800/18	0900/18	12	1800/19	
	ATTU	Replenish-ment Area	1800/20	AKA 12	300	25	1100/21	1800/20	12	1100/21	
Subron 98 Less AS-19	PUGET	ATTU	0000/3	25	2440	98	--	0000/3	25	0200/7	
Subron 1	KODIAK	ATTU	0000/3	13	1300	100	--	0000/3	13	0400/7	
Mine Force	PUGET	ATTU	0000/2	AM 13	2438	188	1800/17	2000/3	13	1300/11	
DMS only	PUGET	ATTU	0000/2	DMS 31	2438	61	1800/17	2000/3	16/17	0100/10	16 kts. to UNIMAK 17 thereafter.
Mine Force	ATTU	AVACHA	1200/12	AM 13	520	40	1900/20	0300/19	13	1900/20	

Fig. 124
Blue Force movement characteristics

fighters, 130 naval patrol bombers, and thirty-six very-long-range aircraft. Finally, Fleet Air Wing 1 was at Puget Sound with thirty-six seaplane-based naval patrol bombers. (Page 63.)

Seaplane tender AVP-48, which had a capacity for twelve naval patrol bombers, was in Puget Sound, as were AV-7 and AV-13, each with a capacity for thirty naval patrol bombers. These ships could be shifted to Attu, Sand Bay, Adak, or Dutch Harbor. Admiral BF would have AV-7 accompany the Blue Expeditionary Force from Adak to Avacha. However, he wanted his initial air strikes against Purple to be as strong as possible and he still needed his land-based and tender-based air to search for and cover the Expeditionary Force. He compared the advantage of having all the planes based within striking distance of the enemy versus the disadvantage of a concentration providing a target worthy of great sacrifice by Purple. He accordingly sent the three seaplane tenders in Puget Sound forward immediately since a tender was required to operate seaplanes from Attu, Kiska, and Adak. Since Dutch Harbor had facilities for twenty-four Mariners, Adak could take six, and Attu could handle five more, twenty-four planes of Fleet Air Wing 1 would depart for Dutch Harbor right away. The atomic bombs were all at Kodiak and could be lifted from there by the Superfortresses. (Page 64.)

Addressing his Task Organization for Operation Plan No. 17-46, Vice Admiral BF thought the number of escort vessels barely adequate given their various antisubmarine and antiaircraft capabilities. He therefore organized two Special Groups. One was referred to as the Fleet Flagship, under the command of Captain B-36 (Captain Ashford—see chapter 13), and consisted of his heavy cruiser flagship and two escorting destroyers. The other was the Fueling Group of one escort carrier, ten destroyer escorts, and six oilers, all under Captain B-31 (Captain Ricketts—see chapter 13). Next was Rear Admiral BA's Covering Force of six fleet carriers in Carrier Divisions 1 and 2; six heavy cruisers in Cruiser Divisions 1 and 3; three light cruisers in CRUDIV 15; two attached antiaircraft cruisers; and the eighteen destroyers and eight radar-picket destroyers of Destroyer Squadrons 3, 9, and 15. Rear Admiral BB was to command the Expeditionary Force, which consisted of Captain B-38's Air Support Control Unit, as well as Rear Admiral BC's Support Group, which comprised two prewar modernized battleships; Carrier Division 19's two escort carriers; CRUDIV 13's three light cruisers; DESRONs 1 and 13, with sixteen destroyers and four radar-picket destroyers; and Mine Squadron 1, with the minelayer, the six minesweepers, and the eight destroyer-minesweepers. The Expeditionary Force also was to include Rear Admiral BB's Transport Group, comprising the amphibious force flagship and Transport Squadrons 1 and 2, with their twenty-four attack transports, twelve attack cargo ships, and twelve high-speed transports. The Expeditionary Force was rounded out by the Expeditionary Troops, two Marine Corps divisions commanded by Major General BO (not played by a designated student officer). There was, of course, Fleet Air Command, Aleutians, commanded by Rear Admiral BL and consisting of Fleet Air Wings 1, 3, 5, and 7, with ninety-six fighter planes, 126 naval patrol bombers, eighteen very-long-range aircraft, and three seaplane tenders. Captain B-13's Submarine Force comprised two submarine tenders and twenty submarines of SUBRONs 1 and 98. Rounding out the entire organization was Commodore BK's Service Force of two destroyer tenders, two ammunition ships, eight attack cargo ships, and six oilers, all from Service Squadron 3; one escort carrier; and eight destroyers from DESRON 7. (Pages 67–69.)

The Blue Alaskan Sea Frontier would defend the Aleutian bases and maintain logistical services at those bases. The Operation Plan reiterated that BF would receive no assistance from Blue occupation forces in Japan or Korea. It repeated also that his force was to capture, occupy, and defend Avacha Bay immediately following the outbreak of hostilities; destroy enemy air bases within effective supporting distance of Avacha; support amphibious assault operations on favorable landing beaches; and then provide an advanced base for support of subsequent operations against Purple's Asian territory. (Pages 70–72.)

The Operation Plan gave Commander, Covering Force discretion to employ the four atomic bombs as he saw fit. If it became necessary to postpone D-Day because of weather, all groups of forces were to delay their departure from port or retire along their previous tracks, refuel if advisable and possible, and resume forward movement on the basis of the new D-Day. Also, if the original D-Day was postponed, the initial strikes by the Blue Covering Force would be made on D–3, provided that weather conditions were favorable then; Commander, North Pacific Force would make the final decision. All commands were further ordered to avoid not only detection but any hostile acts, unless attacked; the commencement of hostilities was planned to coincide with the initial strike by the Covering Force and Commander, North Pacific Force would broadcast to all commands the date and time of the commencement of hostilities. All commands were to inform Commander, North Pacific Force by operational priority message of any hostile acts by Purple forces. Meanwhile, all units were to enforce strict security measures and conserve fuel. As Commander, North Pacific Force and Commander, 7th Fleet, Vice Admiral BF would fly his flag in the heavy cruiser CA-73, while Rear Admiral BA, Commander, Covering Force, designated second in command, would fly his flag in fleet carrier CV-21. (Pages 72–74.)

Under Air Search Plan A, Fleet Air Command, Aleutians would carry out searches to execute Operation Plan No. 17-46. On 7 June, it was to conduct air searches as soon after 1800 as conditions permitted. The searches were thought to have a 90 percent probability of detecting a large Purple formation by radar; the planes could also approach Purple territory to within seventy-five miles of the coast. Fleet Air Command, Aleutians was to avoid visual contact by Purple forces and avoid hostile acts prior to the commencement of hostilities, unless attacked. The command's planes were not to fly over Purple territory; if hostilities opened before the search was made, the search was to be made as soon as possible. The search was to be extended to locate any large Purple forces off the east coast of the Kuriles and Kamchatka west of the 175th east meridian as well as at Korf Bay, Kruger Bay, Avacha Bay, Paramushiru, Buroton Bay, and Hitokappu Bay. Prior to the opening of hostilities, aircraft were to report Purple submarines that were sighted after their return to base; if hostilities had commenced, the search planes were to report on submarines and then destroy them. These attacks, however, were to be limited in time to permit sufficient search for the large Purple surface force, which was the main target. (Pages 75–76.)

Weather (Report 180600 Z
(~~Forecast~~

Area K, L, and M

1. Meteorological visibility, 1/4 miles.
2. Wind and sea: force 2, direction S.
3. Assured sonar range, 2000 yards.
4. Percentage cloud cover 100, ceiling 100 ft.
5. Rain, snow, ice conditions:

Raining steadily since 1200 Z

6. Air temperature 45°

Remarks:

Forecast for next 24 hours:
Rain stopping about 2400 Z
followed by heavy fog.
Clouds up to 10,000 ft.

Weather (Report 180600 Z
(~~Forecast~~

Area G, H and I

1. Meteorological visibility, 1/4 miles.
2. Wind and sea: force 1, direction SW.
3. Assured sonar range, 2000 yards.
4. Percentage cloud cover 100, ceiling 200 ft.
5. Rain, snow, ice conditions:

heavy fog

6. Air temperature 45° F

Remarks:

Forecast for next 24 hours:
No change.
Clouds up to 10,000 ft.

Fig. 125
Meteorological data, Blue Forces, Operations Problem 2

In Air Search Plan B, an area search and barrier search were required in addition to the antisubmarine patrols called for by Operation Plan No. 17-46. The primary task was still to locate, report, and trail the large Purple surface force (while making weather observations), with a secondary task of detecting and reporting on Purple submarines and, after the commencement of hostilities, destroying them. Commencing 10 June, Fleet Air Command, Aleutians was to provide a barrier patrol until the Covering Force carried out its initial strike. The barrier search area was between 53°20′ N, 161°30′ E and 48°00′ N, 162°20′ E. Planes were to be loaded for maximum endurance. It was thought that the barrier patrol would have a 95 percent probability of detecting the Purple surface force. These patrols would be discontinued after the initial Blue strike had been made. Searches were also to be flown once daily between 245 and 295 degrees from Attu and, commencing on D–6, that sector was to be continuously searched by radar. There was thought to be a 95 percent probability of detecting the Purple forces.

If contact was lost, the search was to resume until the Purple force was located again. For reasons previously explained, the searches would be decreased from the east as Blue forces moved westward and would be moved to areas necessary to protect against surprise attack. Prior to D–3, Purple submarines outside three hundred miles from the Blue Aleutian bases were also to be reported, and contact reports were to be made on Purple submarines within three hundred miles of those bases (the distinction between the two cases being left unclear). (Pages 77–78.) When requested, submarines were to be furnished with air escort when they were within a hundred miles of a Blue base, and damaged submarines would be provided with surface escorts.

The Logistics Plan gave significant emphasis to keeping ships as fully fueled as possible because of the demands not only of operations but of the expected weather as well. (Pages 79–83.) There was also a focus on force and Special Group commanders' ensuring they were fully supplied with antisubmarine weapons as well as, especially, anything needed for flight operations in low visibility. (Pages 84–89.)

A fairly detailed Movement Plan indicated how Vice Admiral BF was going to maneuver his forces for the coming battle. The Fleet Flagship would depart Puget Sound at 2000 on 3 June via Republic and Unimak Pass and proceed to Adak at sixteen knots, with an estimated time of arrival of 2300 on 9 June. It would depart Adak at 1200 on 14 June, rendezvous with the Covering Force at the fueling area off Adak at 2000 on 14 June, and return to Adak the next day after the command conference. It would again depart Adak on 17 June at 1000 for the rendezvous with the Covering Force at the fueling area off Attu, proceeding there at twenty-five knots and arriving at 0200 on 18 June. The Fleet Flagship would join the Covering Force when ordered to by Commander, North Pacific Force. The Fueling Group would depart Pearl Harbor at 0800 on 8 June and proceed to Great Sitkin as directed by Commander, Covering Force. It would fuel the Covering Force en route and arrive at Great Sitkin at 2000 on 15 June. The Fueling Group was to depart Great Sitkin at 0600 on 17 June and proceed to the fueling area off Attu at fifteen knots. It would leave the fueling area when released by Commander, Covering Force and proceed to Attu to join the Service Force. (Page 99.)

The Covering Force was to leave Pearl Harbor at 0800 on 8 June and proceed to the fueling area off Adak, arriving by 2000 on 14 June. It was to depart the fueling area at 0600 on 15 June for its position in the western Aleutians, then proceed to the fueling area off Attu designated by its commander, there to refuel by 1000 on the 18th, which would be on time to launch its initial strikes on the northern Kuriles–southern Kamchatka bases at 1100 three days before the landing day (referred to in abbreviated form as "Love Dog," for *LD*, in the phonetic alphabet of the time). If D-Day was postponed, the Covering Force would depart in time to strike at 0340 on Love Dog–3 (i.e., minus three) in accordance with the current Operation Plan. The exception was the flagship, heavy cruiser CA-73, which would join the Covering Force at the fueling area off Attu prior to the initial strikes. The Expeditionary Force would also depart from Puget Sound at 1400 on 7 June at twelve knots with an estimated arrival at Adak at 1800 on 15 June. It would leave Adak at 0200 on the 20th and cross the 180th meridian the next midnight. It would slow to eleven and a half knots to arrive at Avacha Bay at 0600 on the landing day. (Pages 99–100.)

There were numerous exceptions to these movements. In particular, Mine Division 9 was to depart Puget Sound at 2000 on 3 June. It was to proceed at thirteen knots and arrive at Attu at 1800 on 11 June. Mine Squadron 1—less Mine Division 9—would leave Puget Sound at the same time and day and proceed at sixteen knots to Unimak Pass by the same route, then increase speed to seventeen knots to arrive at Attu at 0100 on the 10th. Destroyer Division 132—less destroyer DD-872—would pass through Unimak Pass and escort seaplane tender AV-7, submarine tender AS-19, and CA-73 to Adak, while DD-763 and DD-764 would escort seaplane tender AVP-48 to Kiska, then return to Adak. All of these movements would be made at sixteen knots and between 9 and 11 June. The latter two destroyers would screen the entrance off Adak and rejoin the Expeditionary Force when it arrived at Adak on 15 June. Destroyers DD-765 and DD-871 would escort the flagship, CA-73, until released off Attu to rejoin the Expeditionary Force. Seaplane tender AV-7 would also depart Adak for Avacha Bay in company with the Expeditionary Force. (Page 100.)

Fleet Air Command, Aleutians would shift its air wings and squadrons to various air bases in the Aleutians as required. Seaplane tenders AV-7, AV-13, and AVP-48 would leave Puget Sound at 2000 on 3 June and proceed at sixteen knots via Unimak Pass, with AV-13 going to Attu by 0100 on 10 June, AV-7 entering Adak at 2300 on the 9th, and AVP-48 proceeding to Kiska by 1300 on 10 June. These tenders would shift to other Aleutian bases as the changing conditions dictated, except for AV-7, which would accompany the Expeditionary Force from Adak to Avacha. The Submarine Force's movements, now repeated in detail, were to be as recounted above. The Service Force was to leave Puget Sound at 1400 on 7 June, proceed at twelve knots to Adak by 1800 on 15 June, and depart from there at 1800 on the 16th for Great Sitkin. It was to arrive there by 2200 on 16 June, leave again by 0900 on the 18th, and reach Attu at 1800 on 19 June, where it was to base itself to provide logistical support to the North Pacific Force as required. (Pages 100–101.)

Vice Admiral BF again warned his units that Purple forces could open hostilities without warning and that all units had to be ready for any eventuality. They were to avoid detection, if possible, as well as hostile acts prior to the commencement of hostilities. After the commencement of hostilities, any Purple forces encountered were to be destroyed. All specified routes were to be adhered to unless circumstances dictated deviation, and units scheduled to leave from the same base at the same time and speed were to steam in company. Exceptions to this were that Commander, Covering Force would direct the movements of the Fueling Group until he released it. Also, submarines were to proceed independently to their patrol stations after their departure from Attu.*

* Ibid., p. 101. See also Senior Class of June 1947, "Operations Problem 2: The Blue Staff Solution," 1 October 1946, pp. 1–2, folder 2603-I, box 138, RG 4, NHC.

XV Operations Problem 2

More of the Blue Solution, September–October 1946

Admiral Smith issued the Detail of Student Officers for Blue. Vice Admiral BF, Commander, North Pacific Force and Commander, 7th Fleet, had a Chief of Staff and an Operations Officer. There were also three Assistants for Planning for Captain B-13, Commander, Submarine Force; six Assistants for Planning for Rear Admiral BA, Commander, Covering Force; six Assistants for Planning for Rear Admiral BL, Commander, Fleet Air Force; six Assistants for Planning for Rear Admiral BB, Commander, Expeditionary Force; and two Assistants for Planning for Commodore BK, Commander, Service Force.[1]

On 25 September, Commander, North Pacific Force sent Preliminary Movement Orders to Commander, Amphibious Group 7 and Commander, Submarines, Pacific. Commander, Amphibious Group 7 was ordered to embark two Marine Corps divisions on 5 June for extended operations and to make preliminary plans to have all ships in his area depart about 10 June for the Aleutians area. Commander, Submarines, Pacific was ordered to direct Submarine Squadron 1 to Adak when that unit was ready for deployment so that it could maintain an offensive patrol east of the main channels through the Kuriles. Commander, North Pacific Force wanted these submarines on their patrol stations as soon as possible; upon arrival, they would be provided mission details. Submarine Squadron 98 was also to head to Adak and from there to maintain an offensive patrol of three boats east of La Perouse Strait. These patrols too were to start as soon as possible, with details to follow once on station.[2]

Vice Admiral BF wanted pilots and crewmen—while in no way deviating from their primary mission of strikes—to endeavor by careful observation to supplant the present inadequate information concerning targets and defenses in areas over which they flew: "This information may be invaluable in the later stages of the war and especially for amphibious operations." In particular, any and all flak information would be of great value, particularly in reference to heavy guns and automatic weapons on Avacha Bay.[3]

Chief of Staff, Commander, Blue North Pacific Force and 7th Fleet (Capt. John Fitzsimmons, USN)

Operations Officer, Blue North Pacific Force and 7th Fleet (Capt. Ronald Higgins, USNR)

Admiral BF now issued his formal Communication Plan, an exact mirror of Admiral PA's for the Purple Pacific Fleet in terms of categories about security, schedules, circuits, and frequencies.[4] BF also gave more details about the Movement Plan. The Covering Force was to depart Pearl Harbor on 6 June, arrive at Adak on the 11th, depart Adak on 18 June, arrive at a rendezvous on 20 June to refuel its destroyers, and then proceed to its aircraft-launching area. Arriving the same day (D–3), it would carry out air strikes on Paramushiru, also flying reconnaissance on that base as well as Shimushu and Avacha Bay. On 21 June (D–2), the Covering Force would strike Avacha Bay and carry out reconnaissance of the Purple airfields in the lower Kamchatka Peninsula. The next day, 22 June, D–1, would see the Covering Force bombard Avacha Bay again. On D-Day itself, 23 June, the Covering Force would again strike Paramushiru and Shimushu and also provide cover to the Expeditionary Force. The Expeditionary Force itself would have departed Seattle on 10 June, arrived at Dutch Harbor on the 16th, left Dutch Harbor two days later, and crossed the 180th meridian by the 20th, to arrive off Avacha Bay and conduct its amphibious landings at H-Hour on D-Day.[5]

Rear Admiral BA, as Commander, Blue Covering Force and Carrier Division 2, outlined his Task Organization and Operation Plan. He divided his forces into several units, including the 1st and 2nd Fast Carrier Groups; the Service and Bombardment Groups; and the 1st and 2nd Battle Groups (what would more recently have been called "surface action groups"). The 1st CARGRU, which he commanded, included three fleet carriers based around CARDIV 2, three heavy cruisers based around Cruiser Division 1, one light cruiser from CRUDIV 15, one antiaircraft cruiser, and thirteen destroyers from DESRON 9 and DESDIV 32. The 2nd CARGRU, which was commanded by Rear Admiral BM, also had three fleet carriers from CARDIV 3, two heavy cruisers from CRUDIV 3, two light cruisers, one antiaircraft cruiser, and another thirteen destroyers from DESRONs 3 and 15.[6]

The missions of the Blue Covering Force were to gain and maintain control of the air in the target areas, destroy Purple land-based aircraft and base installations, and exploit the destruction of Purple submarine and surface forces. In addition, it would provide search and rescue in support of the capture and retention of Avacha Bay. Admiral BA's Operation Plan was to be effective at 1300 on 6 June, and unrestricted operations were to commence at 0000 hours on 20 June or upon the outbreak of hostilities, whichever was earlier. Rear Admiral BM was designated as the second in command; Admiral BA himself would sail in fleet carrier CV-21 with the 1st CARGRU. (Pages 1–3.)

Pilots were to be briefed and then interrogated after action about Purple air installations, runways, and planes; antiaircraft, coastal, and automatic weapons; shipping, harbors, and shore facilities; troops, encampments, roads, traffic, and new construction; and damage. Photographic intelligence was to be collected on landing beach areas, bombardment areas, air installations, harbor facilities, strategic gun emplacements, known and suspected flak batteries, and damage. There was also to be damage assessment on the atomic-bomb targets at Paramushiru and Shimushu. Finally, he noted that the effects of the atomic bomb strikes might oblige the Covering Force to shift a portion of its strikes to adjacent areas. (Pages 4–5.)

Rear Admiral BA then outlined his Launching Areas, numbered One, Two, Three, and Four, between 50 and 53 degrees north latitude and 159 and 163 degrees east longitude. His Movement Plan was elaborate. At 1800 on 6 June, his units would depart Pearl Harbor and take a great-circle course to Adak at a speed of 14.4 knots. The Covering Force was scheduled to arrive at Adak at 1600 on 12 June, passing through the Adak Strait. Task Groups 73.1, 73.3, and 73.2 would enter the strait in that order; units would be released south of Adak for independent passage through the strait to their assigned anchorages. Unit commanders would assign the order of passage of their ships and provide for effective destroyer protection of their own and adjacent units. At 2100 on 18 June, the Covering Force would depart Adak for Point X-RAY, 52°10′ N, 176°25′ W. The course was to be 275 true from Point X-RAY to Point YOKE, the latter ten miles north of the northern tip of Buldir Island. The Covering Force would steam on course 254 from Point YOKE to Point PACKARD (50°38′ N, 165°00′ E) at twenty-five and a half knots. Destroyers would be refueled in the vicinity of Point PACKARD from 0200 to 0600 on the 20th, and the Covering Force would proceed on course 287 to the center of Launching Area One, latitudes 50°45′ to 51°30′ N and longitudes 160° to 160°45′ E. (Page 6.)

At 1600 on 20 June, the Covering Force would arrive at Launching Area One and commence its air strike on Paramushiru as well as its reconnaissance and photography of Paramushiru, Shimushu, and Avacha Bay. The launching and recovery of these strikes were to be conducted at twenty-five knots, with other speeds and courses as necessary to remain in the area. At 1600 on 21 June, the Covering Force would arrive at Launching Area Two, latitudes 51°45′ to 52°30′ N and longitudes 160°40′ to 161°20′ E. There, it would commence its air strike on Avacha Bay and conduct its reconnaissance and photography of the bay and of the lower Kamchatka airfields. If practicable, it would detach the Bombardment Group, which would start bombarding its targets in Avacha Bay by 1700 on 22 June. (Rear Admiral BD, Commander, Bombardment Group, noted that the minesweepers of the Expeditionary Force might still be working at that time in or near the bombardment area.) At 1700 on the 22nd, the Covering Force would arrive at Launching Area Four, latitudes 51°00′ to 51°45′ N and longitudes 159°15′ to 160° E. From here, an air strike would be launched against Avacha Bay to support the bombardment and cover the Expeditionary Force. The Bombardment Group would rejoin the Covering Force when released and the entire Covering Force would retire by 1600 on 23 June. It would draw eastward during the night and then return to Launching Area Four for strikes on Paramushiru and Shimushu. By 0600 on 24 June at the latest, the Covering Force would depart Launching Area Four at a speed of eighteen and a half knots to Point CADILLAC (52° N, 172° E) to rendezvous with Service Group 2 for refueling and loading ammunition. Other rendezvous points were FORD, DODGE, CHEVROLET, and PLYMOUTH, all between 49 and 53 north latitude and between 158 and 172 east longitude. (Pages 6–7.)

En route from Pearl Harbor to Adak, the Covering Force would use Cruising Disposition 5-X from the *Fast Carrier Task Force Instructions*. TG 73.3 would occupy Station One and assume the Guide. TG 73.1 would take Station Two, while TG 73.2 would take Station Three. Picket Stations 1, 2, 3, 4, and 6 would be occupied by destroyers from the nearest task group. TG 73.3 would station pickets bearing 120, 180, and 240 relative from the Guide at distances of twelve miles. Task Groups 73.1 and 73.2 would use a modification of Cruising Disposition 5-LS, with destroyers on Screening Circle 6 and with one additional destroyer on each quarter. Disposition 5-W would be used in restricted waters. For refueling, all units were to use Cruising Disposition 5-F at a speed of ten knots, with all the fleet carriers in the forward fueling line, then the ships of CRUDIV 15 in the second. CRUDIV 15's ships would each fuel one destroyer and the fleet carriers would refuel the remainder, four each in most cases. After leaving Adak, the Covering Force was to employ Cruising Disposition 5-W. On strike days, however, two destroyers would be stationed by the northern task group bearing 300 true at a distance of thirty miles from the center of that task group. A similar unit would be stationed on strike days by other task groups bearing 180 true at the same distance. Formations within the Covering Force's task groups would be ordered by their commanders, normally Cruising Disposition 5-LS for low visibility and 5-R or 6-R for good visibility. These dispositions would be modified to suit the composition of the task groups and to place the antiaircraft cruiser near the task group's center and near the fleet carriers. (Page 7.)

By the Air Strike Plan, battle carrier CVB-39 would furnish all combat air patrols for Task Force 73, all antisubmarine patrols, photographic reconnaissance over Avacha Bay, and the Search and Rescue Cover.[7] At 0400 on 20 June, CVB-38 would launch thirty-two fighters as a sweep, sixteen additional fighters as an escort, and twenty-four attack planes, along with the Target Coordinator in a fighter plane. The targets for this strike would be Kataoka, Miyoshino, Bettobu, Imalsami, and Cape Lopatka. With the same targets, CVB-33 would launch the same number and mix of aircraft. So too would CVB-37, with the difference that this third carrier would launch thirty-six fighters as its sweep, sixteen fighter escorts, thirty-nine attack planes, three photographic reconnaissance fighters, and airborne early warning and tactical electronic warfare planes as required. Its targets were the same as the other two carriers. But CVB-36 would strike at Kakumabetsu, Suribachi, and Kurabu. This fourth carrier's number and mix of strike aircraft would be the same as CVB-37's and the 0900 and 1400 strikes would be the same as the 0400 strike from the other groups. (Pages 8–9.)

The next day, CVB-39 would again furnish the antisubmarine and combat air patrols as well as Search and Rescue Cover and photographic reconnaissance. CVB-33, -37, and -38 would have the same loads and planes but the targets would now be Bolsheretsk, Petropavlovsk, Staraya, Kalakhtyrka Lake, and Tarinski Bay. CVB-36 (duplicated by CVB-21) would also have the same loads and mix of planes as on the 20th, but now with targets at Kataoka, Miyoshino, Bettobu, Imalsami, Cape Lopatka, Kakumabetsu, Suribachi, and Kurabu. Target Coordinators were again designated. Subsequent strikes at 0900 and 1400 would again be conducted in the same manner. These mixes of planes, loads, and targets would continue on 22 June at 0400, 0900, and 1400 as well as on 23 June, with the exception that on 23 June there would be no fighter sweeps during the 0400 strike. Part of the Air Strike Plan entailed fighter cover over Avacha Bay. At 0200, fighters would launch, returning with the 0400 strike after being relieved on station. CVB-33 and CVB-38 would each contribute thirty-two fighters to this mission, with CVB-37 adding another thirty-six. At 0400, CVB-21 and CVB-36 would each launch another thirty-six fighters, which would also return upon relief. At 0830, the carriers would launch another flight similar to the 0200 group and at 1230 a flight similar to the 0400 group. A final flight for the day would be launched at 1630, the same as the 0830 group. All flights would return to their carriers after being relieved on station. (Pages 10–11.)

Rear Admiral BA, Commander, Covering Force and Carrier Division 2, now issued his Bombardment Plan. Admiral BD's Bombardment Group was divided into the 1st Bombardment Unit, three heavy cruisers from CRUDIV 1, and the 2nd Bombardment Unit, the latter unit commanded by Captain B-9 and consisting of six destroyers from DESDIV 91. Units were to have special target maps of enemy positions and firing points. Minesweeping was to commence off Blue and Green

- 13 -

7 JUNE, 1946

BLUE FLEET

FORCE COVER

AIRCRAFT FLIGHT FORM

WEATHER AREA

STUDENT

MOVE NO

MANEUVER TIMES OF MOVE

ROOM NO

List of Task Groups: Base Ship or Station	Squadron	Plane Numbers	Time of Start of Flight	Time of End of Flight	Duration of Flight	Air Speed, Knots	Altitude, Feet	Fuel in Planes at Start of Flight or Move	Bombs, Torpedoes or Special Equipment	Orders to Aircraft Task Group Commanders and Radio Frequency.
CV-21 (Requalification landings to be completed by 1000)	CVG-21	ALL	0700	1230	5-5	130 (Patrol)	500 (Patrol)	Aux. as necessary	None	A/S PATROL REMAINING PLANES REQUALIFICATION LANDINGS
CV-33 (Requalification landings to be completed by 1000)	CVG-33	ALL	0700	1000	3.0	130 (Patrol)	500 (Patrol)	Full No Aux.	None	SECTOR SEARCH. REMAINING PLANES REQUALIFICATION LANDINGS
CV-36 CV-37 CV-38 CV-39	CVG-36 CVG-37 CVG-38 CVG-39	ALL ALL ALL ALL	0700 0700 0700 0700	1000 1000 1000 1000	3.0 3.0 3.0 3.0			Full No Aux.	None	REQUALIFICATION LANDINGS
CV-36	CVG-36	Ship Designate	1200	1700	5.0	130 (Patrol)	500 (Patrol)	Aux. as Necesary	None	A/S PATROL
CV-38	CVG-38	Ship Designate	1200	1500	3.0	130 (Patrol)	500 (Patrol)	Full No Aux.	None	SECTOR SEARCH

46-42

- 14 -

8 JUNE, 1946

BLUE FLEET

FORCE COVER

AIRCRAFT FLIGHT FORM

WEATHER AREA

STUDENT

MOVE NO.

MANEUVER TIMES OF MOVE

ROOM NO.

List of Task Groups: Base Ship or Station	Squadron	Plane Numbers	Time of Start of Flight	Time of End of Flight	Duration of Flight	Air Speed, Knots	Altitude, Feet	Fuel in Planes at Start of Flight or Move	Bombs, Torpedoes or Special Equipment	Orders to Aircraft Task Group Commanders and Radio Frequency.
CV-39 (Requalification landings to be completed by 1000)	CVG-39	ALL	0700	1230	5.5	135 (patrol	500 (patrol)	Aux.as Necessary	None	A/S PATROL. REMAINING PLANES REQUALIFICATION LANDINGS FOR REMAINING PILOTS.
CV-37 (Requalification landings to be completed by 1000)	CVG-37	ALL	0700	1000	3.0	130 (patrol)	500 (patrol)	Full No Aux.	None	SECTOR SEARCH. REMAINING PLANES REQUALIFICATION LANDINGS FOR REMAINING PILOTS
CV-21 CV-33 CV-36 CV-38	CVG-21 CVG-33 CVG-36 CVG-38	ALL ALL ALL ALL	0700 0700 0700 0700	1000 1000 1000 1000	3.0 3.0 3.0 3.0			Full. No Aux.	None	REQUALIFICATION LANDINGS FOR REMAINING PILOTS.
CV-33	CVG-33	Ship Designate	1200	1700	5.0	130 (patrol	500 (patrol)	Aux.as necessary	None	A/S PATROL.
CV-21	CVG-21	Ship Designate	1200	1500	3.0	130 (patrol	500 (patrol)	Full. No Aux.	None	SECTOR SEARCH.

46-42

- 15 -

9 JUNE, 1946

BLUE FLEET

FORCE COVER

AIRCRAFT FLIGHT FORM

WEATHER AREA

STUDENT

MOVE NO.

MANEUVER TIMES OF MOVE

ROOM. NO.

List of Task Groups: Base Ship or Station	Squadron	Plane Numbers	Time of Start of Flight	Time of End of Flight.	Duration of Flight	Air Speed, Knots	Altitude, Feet	Fuel in Planes at Start of Flight or Move	Bombs, Torpedoes or Special Equipment	Orders to Aircraft Task Group Commanders and Radio Frequency.
CV-38 (Planes on squadron tactics land at 1100)	CVG-38	ALL	0700	1325	6.4	130 (patrol	500 (patrol)	Aux.as Necessary	None	A/S PATROL REMAINING PLANES SQUADRON TACTICS.
CV-36 (Planes returning from search rendezvous and land with own squadron).	CVG-36	ALL	0700	1100	4.0	130 (Sector	500 Search)	Aux.as Necessary	None	SECTOR SEARCH. REMAINING PLANES SQUADRON TACTICS.
CV-21 CV-39 CV-33 CV-37	CVG-21 CVG-39 CVG-33 CVG-37	ALL ALL ALL ALL	0700 0700 0700 0700	1100 1100 1100 1100	4.0 4.0 4.0 4.0			Aux.as Necessary	None	SQUADRON TACTICS.
CV-37 (Planes scheduled for squadron tactics will be launched at 1300)	CVG-37	ALL	1230	1700	4.5	130 A/S	500 Patrol	Aux.as Necessary	None	A/S PATROL REMAINING PLANES SQUADRON TACTICS.
CV-39 (Planes returning from search rendezvous and land with own squadron)	CVG-39	ALL	1300	1700	4.0	135 Sector	500 Search	Aux. as Necessary	None	SECTOR SEARCH. REMAINING PLANES SQUADRON TACTICS.
CV-21 CV-36 CV-33 CV-38	CVG-21 CVG-36 CVG-33 CVG-38	ALL ALL ALL ALL	1300 1300 1300 1300	1700 1700 1700 1700	4.0 4.0 4.0 4.0			Aux. as Necessary	None	SQUADRON TACTICS.

46-42

- 16 -

10 JUNE, 1946.

BLUE FLEET
FORCE COVER
STUDENT
ROOM NO.

AIRCRAFT FLIGHT FORM

MOVE NO.
MANEUVER TIMES OF MOVE

WEATHER AREA

List of Task Groups: Base Ship or Station	Squadron	Plane Numbers	Time of Start of Flight	Time of End of Flight	Duration of Flight	Air Speed, Knots	Altitude, Feet	Fuel in Planes at Start of Flight or Move	Bombs, Torpedoes or Special Equipment	Orders to Aircraft Task Group Commanders and Radio Frequency.
CV-21 (Planes on group tactics land at 1100)	CVG-21	ALL	0700	1325	6.4	130 A/S Patrol	500	Aux. as Necessary	None	A/S PATROL. REMAINING PLANES GROUP TACTICS.
CV-33 (Planes returning from search rendezvous and land with own squadron)	CVG-33	ALL	0700	1100	4.0	130 Sector Search	500	Aux. as Necessary	None	SECTOR SEARCH. REMAINING PLANES GROUP TACTICS.
CV-36 CV-37 CV-38 CV-39	CVG-36 CVG-37 CVG-38 CVG-39	ALL ALL ALL ALL	0700 0700 0700 0700	1100 1100 1100 1100	4.0 4.0 4.0 4.0			Aux. as Necessary	None	GROUP TACTICS.
CV-36 (Planes scheduled for group tactics will be launched at 1300)	CVG-36	ALL	1230	1700	4.5	130 A/S Patrol	500	Aux. as Necessary	None	A/S PATROL. REMAINING PLANES GROUP TACTICS.
CV-38 (Planes returning from search rendezvous and land with own squadron)	CVG-38	ALL	1300	1700	4.0	130 Sector Search	500	Aux. as Necessary	None	SECTOR SEARCH. REMAINING PLANES GROUP TACTICS.
CV-21 CV-33 CV-37 CV-39	CVG-21 CVG-33 CVG-37 CVG-39	ALL ALL ALL ALL	1300 1300 1300 1300	1700 1700 1700 1700	4.0 4.0 4.0 4.0			Aux. as Necessary	None	GROUP TACTICS.

46-42

- 17 -

11 JUNE, 1946.

BLUE FLEET
FORCE COVER
STUDENT
ROOM NO.

AIRCRAFT FLIGHT FORM

MOVE NO.
MANEUVER TIMES OF MOVE

WEATHER AREA

List of Task Groups: Base Ship or Station	Squadron	Plane Numbers	Time of Start of Flight	Time of End of Flight	Duration of Flight	Air Speed, Knots	Altitude, Feet	Fuel in Planes at Start of Flight or Move	Bombs, Torpedoes or Special Equipment	Orders to Aircraft Task Group Commanders and Radio Frequency.
CV-39 (Planes on wing tactics land at 1100)	CVG-39	ALL	0700	1325	6.4	135 A/S Patrol	500	Aux. as Necessary	None	A/S PATROL. REMAINING PLANES WING TACTICS.
CV-37 (Planes returning from search rendezvous and land with own squadron)	CVG-37	ALL	0700	1100	4.0	130 Sector Search	500	Aux. as Necessary	None	SECTOR SEARCH. REMAINING PLANES WING TACTICS.
CV-21 CV-33 CV-36 CV-38	CVG-21 CVG-33 CVG-36 CVG-38	ALL ALL ALL ALL	0700 0700 0700 0700	1100 1100 1100 1100	4.0 4.0 4.0 4.0			Aux. as Necessary	None	WING TACTICS.
CV-33 (Planes scheduled for wing tactics will be launched at 1300)	CVG-33	ALL	1230	1700	4.5	130 A/S Patrol	500	Aux. as Necessary	None	A/S PATROL. REMAINING PLANES WING TACTICS.
CV-21 (Planes returning from search rendezvous and land with own squadron)	CVG-21	ALL	1300	1700	4.0	130 Sector Search	500	Aux. as necessary	None	SECTOR SEARCH. REMAINING PLANES WING TACTICS.
CV-39 CV-37 CV-36 CV-38	CVG-39 CVG-37 CVG-36 CVG-38	ALL ALL ALL ALL	1300 1300 1300 1300	1700 1700 1700 1700	4.0 4.0 4.0 4.0			Aux. as Necessary	None	WING TACTICS.

46-42

- 18 -

12 JUNE, 1946

BLUE FLEET
FORCE COVER
STUDENT
ROOM NO.

AIRCRAFT FLIGHT FORM

MOVE NO.
MANEUVER TIMES OF MOVE

WEATHER AREA

List of Task Groups: Base Ship or Station	Squadron	Plane Numbers	Time of Start of Flight	Time of End of Flight	Duration of Flight	Air Speed, Knots	Altitude, Feet	Fuel in Planes at Start of Flight or Move	Bombs, Torpedoes or Special Equipment	Orders to Aircraft Task Group Commanders and Radio Frequency.
CV-38 (Planes on wing tactics land at 1100)	CVG-38	ALL	0700	1325	6.4	130 A/S Patrol	500	Aux. as Necessary	None	A/S PATROL. REMAINING PLANES WING TACTICS.
CV-36 (Planes returning from search rendezvous and land with own squadron)	CVG-36	ALL	0700	1100	4.0	130 Sector Search	500	Aux. as Necessary	None	SECTOR SEARCH. REMAINING PLANES WING TACTICS.
CV-21 CV-39 CV-33 CV-37	CVG-21 CVG-39 CVG-33 CVG-37	ALL ALL ALL ALL	0700 0700 0700 0700	1100 1100 1100 1100	4.0 4.0 4.0 4.0			Aux. as Necessary	None	WING TACTICS.
CV-37 (Planes scheduled for wing tactics will be launched at 1300)	CVG-37	ALL	1230	1700	4.5	130 A/S Patrol	500	Aux. as Necessary	None	A/S PATROL. REMAINING PLANES WING TACTICS.
CV-39 (Planes returning from search rendezvous and land with own squadron)	CVG-39	ALL	1300	1700	4.0	135 Sector Search	500	Aux. as Necessary	None	SECTOR SEARCH, REMAINING PLANES WING TACTICS.
CV-21 CV-36 CV-33 CV-38	CVG-21 CVG-36 CVG-33 CVG-38	ALL ALL ALL ALL	1300 1300 1300 1300	1700 1700 1700 1700	4.0 4.0 4.0 4.0			Aux. as Necessary	None	WING TACTICS.

46-42

Fig. 126
Air Patrol Plan, Blue Covering Force, Operations Problem 2

Beaches at 0400 on 23 June and be covered by the Bombardment Group while the Avacha Bay area itself would be bombarded by the Expeditionary Force on D-Day. On D–1, the 1st Bombardment Unit would destroy beach defenses according to schedule, destroy all radio and possible radar installations on inland high ground, and deliver counterbattery fire at enemy guns. (Page 19.)

The 2nd Bombardment Unit, on the same day, would deliver close-range covering fire for minesweeping, according to schedule. It would also deliver counterbattery fire against Purple's guns, especially those in the rear of the beaches, and be prepared to "smoke" high ground inland of the beaches using white-phosphorus projectiles on request by the minesweeping commander. Ships were to use armor-piercing and common projectiles only for direct fire against enemy emplacements and pillboxes. Individual ships assigned to fire at specific areas were also to be responsible for prompt counterbattery fire when minesweepers were taken under shore fire. Assigned targets were to be covered even if interruptions necessitated a reduction in volume. High-capacity projectiles were to be used for specified area and other targets. All ships were to use every opportunity to close the range and employ direct fire on pillboxes, dugouts, concrete shelters, and blockhouses. Harbor facilities, dock cranes, fuel tanks, and other assets that were to be converted to Blue use later on were to be left standing. Commander, Bombardment Group was to be embarked in heavy cruiser CA-74; Captain B-9 was second in command. (Pages 19–20.)

Rear Admiral BA's Bombardment Plan was assigned for execution to Rear Admiral BD's TG 73.4, the Bombardment Group. TG 73.4 consisted of Rear Admiral BD's own 1st Bombardment Unit, three heavy cruisers from Cruiser Division 1, as well as Captain B-9's 2nd Bombardment Group, six destroyers from DESDIV 91 plus destroyer DD-747. The Bombardment Group's mission was to bombard Avacha Bay on D-Day; accordingly, special target maps had been issued to indicate enemy positions and firing points. The Bombardment Group was also to cover minesweeping that would commence off Blue and Green Beaches at 0400 on 23 June. The bombardment mission itself would begin on D–1 in support of both the capture and retention of Avacha Bay. The 1st Bombardment Group, specifically, would destroy Purple beach defenses, including Points Daini and Venus, and all radio or possible radar installations on the high ground inland of Point Daini, and deliver counterbattery fire at enemy guns on Points Daini and Venus. The 2nd Bombardment Unit would maintain close-range covering fire for the minesweepers; deliver counterbattery fire at the Purple guns—especially those to the rear of the beaches; and be prepared to smoke the high ground inland of the beaches using white phosphorus, on request. (Pages 19–20.)

Admiral BA's Communication Plan stated that Commander, Covering Force could break radio silence if the circumstances dictated, but this was not to be

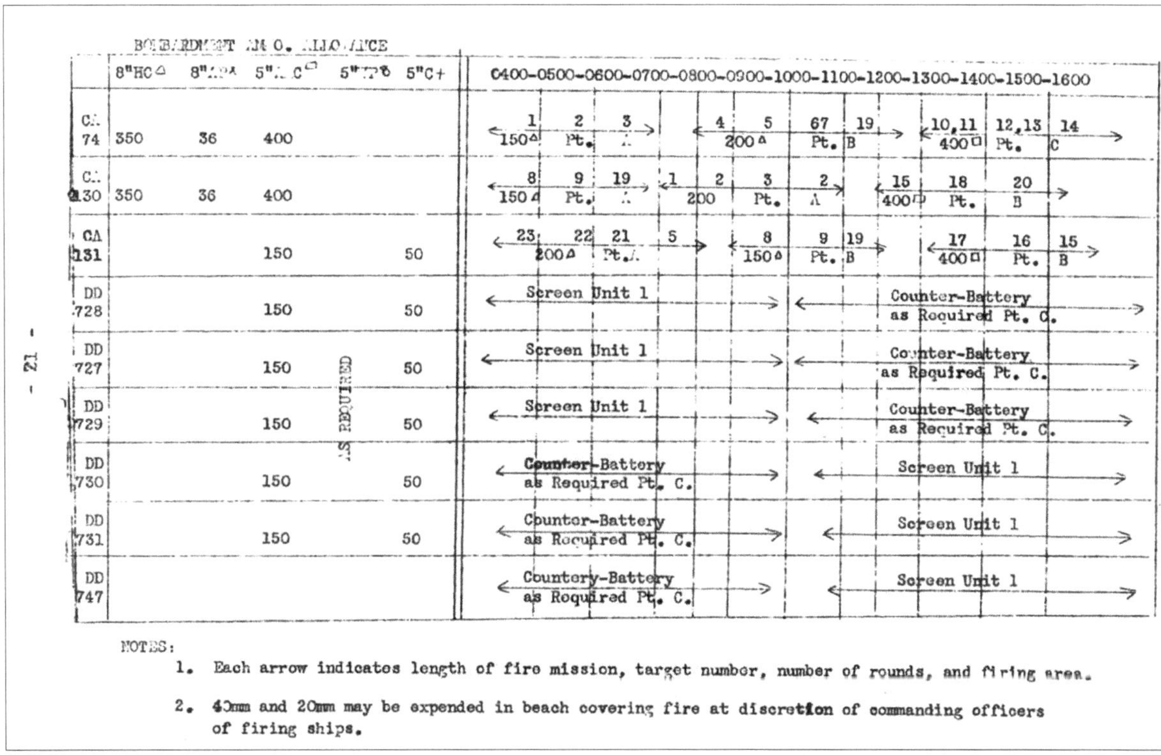

Fig. 127
Blue Bombardment Group schedule of fires, D–1, Operations Problem 2

considered as lifting radio silence generally. Transmissions were still to be kept to an absolute minimum. The 1st and 2nd Carrier Task Groups, if operating together, would be governed by the same conditions; if operating singly, they would be governed by their respective OTCs. (Page 22.)

Certain ships were designated as Combat Information and Combat Information Relief Ships. On odd-numbered days, fleet carriers CV-21 and CV-37 would be designated as the Combat Information Center Ships, while CV-36 and CV-33 would be the reliefs. On even-numbered days, CV-21 and CV-37 would again be the Combat Information Center Ships, the reliefs being CV-39 and CV-38. All heavy cruisers, light cruisers, and destroyer-division flagships were to be ready to assume Combat Information Center guard ship duties if so ordered by Commander, Task Group 73.1 or 73.2. When these two task groups were operating together, as they were expected to be the majority of the time, CV-21 would maintain continuous Combat Information Center Guard and CV-37 would be the 1st Relief Combat Information Center guard ship. Fighter-direction ships were the same as the Combat Information Center guard ships and with the same instructions. (Page 23.)

Concerning the radar guard ships, odd-numbered days would find CA-130 and CA-136 as Guard Able, CA-131 and CA-87 as Guard Baker, and CV-36 and CV-33 as Guard Easy. On even-numbered days, Guard Able was to be heavy cruisers CA-74 and CA-75, Guard Baker was to be CA-67 and light cruiser CL-104, and

Guard Easy was to be CV-21 and CV-37. Radar Guards Fox and George would both be assigned by Commander, Destroyer Squadron 9 and COMDESRON 3 in order to provide 360 degrees of coverage for surface searches. Radar Guards Charlie, Hypo, and Item would be assigned by message from Commander, Task Force 73 to fit changing circumstances. When two Fast Carrier Task Groups were operating singly, each Officer in Tactical Command would assign radar guards whenever required. Similarly, when it came to radar intercept guard ships, antiaircraft cruiser CLAA-95 and destroyer DD-742 would be the radar intercept guard ships on odd-numbered days, while on even-numbered days it was to be CLAA-98 and destroyer DD-728. Radar-jamming guard ships were to be the light cruisers CL-67 and CL-104 on odd-numbered days and the heavy cruisers CA-131 and CA-87 on even-numbered days. COMDESRON 9, COMDESRON 3, and COMDESRON 15 would each detail two destroyers to be ready to act as radar pickets and would identify these ships to CTF 73, who would order them to designated positions whenever circumstances so required. (Pages 23–24.)

Aircraft Early Warning guard ships were designated as CV-21, -36, -38, -39, -37, and -33 for the days of 5, 6, 7, 8, 9, and 10 June, respectively; the Relief Aircraft Early Warning guard ships for those same days would be CV-39, -37, -33, -21, -36, and -38. After 10 June, these watches would repeat, starting with CV-21. If Fast Carrier Task Groups were separated by more than a hundred miles, each would arrange its own Aircraft Early Warning guard ships. (Page 24.)

All Task Force 73 ships were to guard the Fleet Fox Schedule beginning at 0000 on D–10, except for unit commanders, who could make arrangements with their respective units. All ships were to guard the Force Talk Between Ships frequency at all times, except that the Service Group (Task Group 73.3) could use the guard ships arranged for it by CTG 73.3. The primary and secondary frequencies for the Task Force Commanders Circuit was to be guarded by the Force Flagship and monitored by the flagship of CTG 73.2. Whenever the Fast Carrier Task Groups were operating separately, this monitoring watch would be particularly important. Fleet carriers were to guard the very-high-frequency nets, while heavy and light cruisers guarded their own Spotting frequencies. Commanders, Task Groups 73.1 and 73.2 had Primary Air Search and Reconnaissance Plane frequencies to guard. The OTC would retransmit to task force and task group commanders any important information received, employing either visual or VHF methods. (Pages 24–25.)

Commanders, Task Groups 73.1 and 73.2 were additionally to guard frequencies, for receipt only, for Air and Submarine Operational Intelligence. CTF 73 would guard the Submarine Ship–Shore frequency or assign it locally, again as a receiver watch only. Special frequencies would also be set up and operated upon order of the OTC, specifically, Radar Reporting and Inter-Fighter-Direction Channels and

Radar Telling and Warning Nets. There were also Fire Support Group and Fire Support Unit frequencies established, one each for Fire Support Unit 1 and Fire Support Unit 2, as well as primary and secondary frequencies. All ships were to use "utmost" care in obtaining weather information given its importance to the campaign and the fact that weather conditions were expected to be poor and fluctuating. Admiral BA also repeated "for emphasis" that the basic communications doctrine provided for utmost flexibility in communications; if a task group or task unit commander found that a new channel was needed, he could assign the frequency without delay. If time allowed, however, that commander was to state his needs and make "suitable" recommendations.

Indicating how destroyers were to refuel at sea on 20 June from the carriers and cruisers as well as from Service Group 2 on 25 June, BA indicated that ammunition was also to be replenished at these times. Daily reports of fuel were to be made, but only by visual communications unless otherwise directed. Fuel and aviation gasoline were available from shore tanks and were to be delivered by barge at Attu; fueling bases were at Akutan and Great Sitkin. (Pages 25–27.)

Admiral BA's Task Organization was built around the 1st and 2nd Battle Groups as well as a Carrier Group. The 1st Battle Group was commanded by Rear Admiral BD and consisted of three heavy cruisers from CRUDIV 1, a light cruiser from CRUDIV 15, and thirteen destroyers from Destroyer Squadron 9 and Destroyer Division 32. Rear Admiral BN, COMCRUDIV 3, commanded the 2nd Battle Group, which comprised two heavy cruisers from CRUDIV 3, the antiaircraft cruiser CLAA-51, and CL-87 from CRUDIV 15, as well as the thirteen destroyers of DESRONs 3 and 15, less DESDIV 32. Rear Admiral BA's own Carrier Group consisted of the six fleet carriers from CARDIVs 2 and 3 and the antiaircraft cruisers CLAA-95 and CLAA-98. His assumptions comprised three possible scenarios —Purple surface forces would be superior to his, equal or inferior, and tactically concentrated. He assumed that the Blue Expeditionary Force's heavy ships might not be within supporting distance in case of an engagement but that no engagement would be a tactical surprise to either side. He further assumed that Purple would close for a decisive engagement regardless of air superiority and would try to force a decision before Admiral BA could concentrate the Expeditionary Force's heavy ships. He planned for two other scenarios: Case One, in which his air striking power was available; and Case Two, in which it was not because of weather, damage, or lack of ammunition, or for any other reason. In Case One, he would destroy the Purple forces by early and sustained air attack supported by surface gun and torpedo attacks. In Case Two, he would destroy the Purple forces by gun and torpedo attacks, preferably after concentrating with the heavy ships of the Expeditionary Forces but alone if necessary to exploit an opportunity that presented itself. (Page 28.)

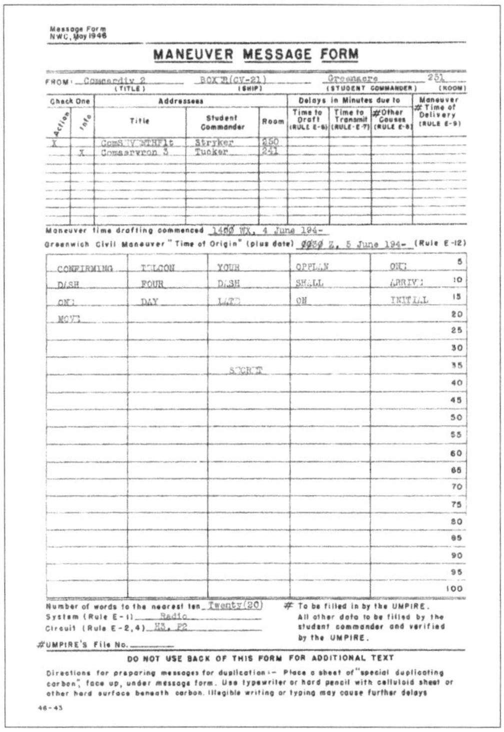

Message Form
NWC, May 1946

MANEUVER MESSAGE FORM

FROM: Comcardiv 2 (TITLE) BOXER(CV-21) (SHIP) Greenacre (STUDENT COMMANDER) 251 (ROOM)

Action	Info	Title	Student Commander	Room	Time to Draft (RULE E-6)	Time to Transmit (RULE-E-7)	#Other Causes (RULE E-8)	Maneuver #Time of Delivery (RULE E-9)
X		ComSEVENTHFlt	Stryker	250				
	X	Comservron 3	Tucker	241				

Maneuver time drafting commenced 1400 WX, 4 June 194-
Greenwich Civil Maneuver "Time of Origin" (plus date) 0030 Z, 5 June 194- (Rule E-12)

CONFIRMING	TELCON	YOUR	OPPLAN	ONE	5
DASH	FOUR	DASH	SHALL	ARRIVE	10
ONE	DAY	LATE	ON	INITIAL	15
MOVE					20

SECRET

Number of words to the nearest ten Twenty(20)
System (Rule E-1) Radio
Circuit (Rule E-2,4) H3, P2
#UMPIRE'S File No.

To be filled in by the UMPIRE. All other data to be filled by the student commander and verified by the UMPIRE.

DO NOT USE BACK OF THIS FORM FOR ADDITIONAL TEXT

Directions for preparing messages for duplication:— Place a sheet of "special duplicating carbon", face up, under message form. Use typewriter or hard pencil with celluloid sheet or other hard surface beneath carbon. Illegible writing or typing may cause further delays

46-43

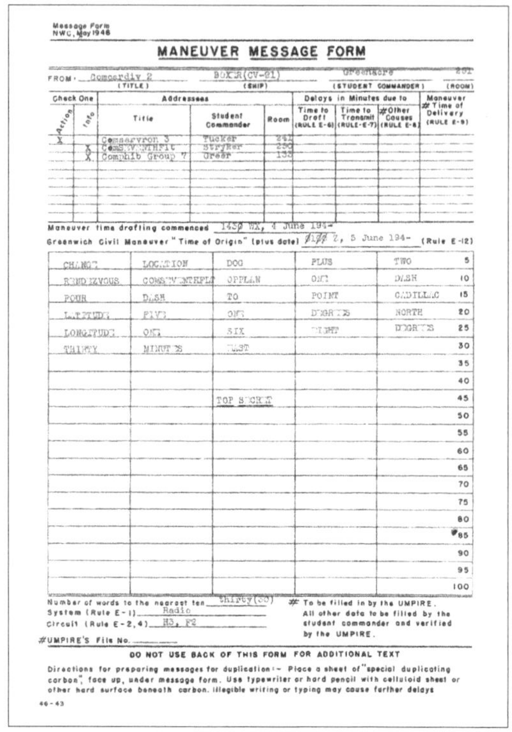

Message Form
NWC, May 1946

MANEUVER MESSAGE FORM

FROM: Comcardiv 2 (TITLE) BOXER(CV-21) (SHIP) Greenacre (STUDENT COMMANDER) 251 (ROOM)

Action	Info	Title	Student Commander	Room	Time to Draft (RULE E-6)	Time to Transmit (RULE-E-7)	#Other Causes (RULE E-8)	Maneuver #Time of Delivery (RULE E-9)
X		Comservron 3	Tucker	241				
	X	ComSEVENTHFlt	Stryker	250				
	X	Comphib Group 7	Greer	133				

Maneuver time drafting commenced 1430 WX, 4 June 194-
Greenwich Civil Maneuver "Time of Origin" (plus date) 0100 Z, 5 June 194- (Rule E-12)

CHANGE	LOCATION	DOG	PLUS	TWO	5
RENDEZVOUS	COMSEVENTHFLT	OPPLAN	ONE	DASH	10
FOUR	DASH	TO	POINT	CADILLAC	15
LATITUDE	FIVE	ONE	DEGREES	NORTH	20
LONGITUDE	ONE	SIX	EIGHT	DEGREES	25
THIRTY	MINUTES	EAST			30

TOP SECRET

Number of words to the nearest ten Thirty(30)
System (Rule E-1) Radio
Circuit (Rule E-2,4) H3, P2
#UMPIRE'S File No.

To be filled in by the UMPIRE. All other data to be filled by the student commander and verified by the UMPIRE.

DO NOT USE BACK OF THIS FORM FOR ADDITIONAL TEXT

Directions for preparing messages for duplication:— Place a sheet of "special duplicating carbon", face up, under message form. Use typewriter or hard pencil with celluloid sheet or other hard surface beneath carbon. Illegible writing or typing may cause further delays

46-43

In either case, the 1st and 2nd Battle Groups would interpose themselves between the enemy and the Carrier Group on lines of bearing about ninety degrees apart and at about equal ranges from the Purple forces. In Case One, the two battle groups would repel any Purple surface threat to the Carrier Group and attack with guns and torpedoes if a favorable opportunity developed. In Case Two, the 1st and 2nd Battle Groups would repel any Purple surface threat to the Carrier Group but would engage defensively only if the heavy ships of the Blue Expeditionary Force were within supporting distance. At the same time, the 1st and 2nd Battle Groups would maneuver to close with the heavy ships of the Expeditionary Force. If, however, these ships were not within supporting distance, the 1st and 2nd Battle Groups would "aggressively" engage with guns and torpedoes, pressing home their attacks regardless of the consequences. (Pages 28–29.)

In Case One, the Carrier Group would commence maximum air strikes on the enemy at the earliest possible moment and continue as long as it was out of Purple gunfire range. It would gain and hold its launching station on the unengaged side of the 1st and 2nd Battle Groups but within easy striking range of the Purple surface force. In Case Two, the Carrier Group would remain outside Purple gun and torpedo range "at all costs" but within voice-radio distance of the 1st and 2nd Battle Groups if possible. It would also take station on the unengaged side of the 1st and 2nd Battle Groups and, in either Case One or Two, maneuver at high speed for antisubmarine protection. Also in either Case One or Two, the three task groups were to maneuver independently, though Admiral BA emphasized the "alert"

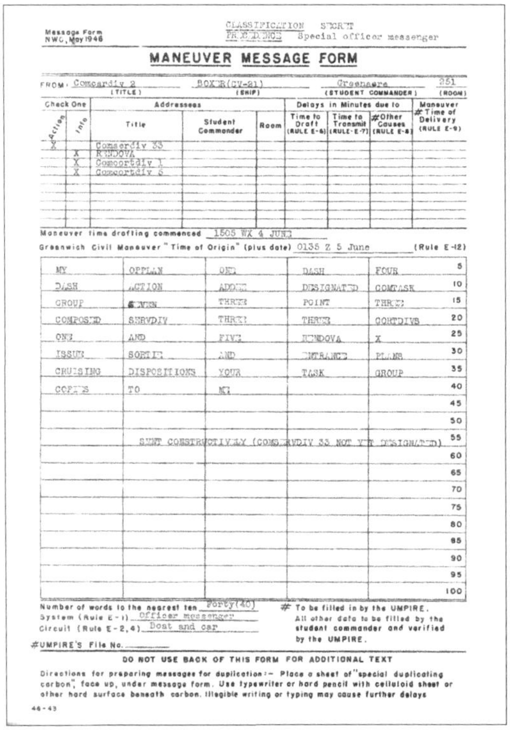

Message Form NWC, May 1946

CLASSIFICATION SECRET
PRECEDENCE Special officer messenger

MANEUVER MESSAGE FORM

FROM: Comcardiv 2 (TITLE) BOXER(CV-21) (SHIP) Greenacre (STUDENT COMMANDER) 251 (ROOM)

Action	Info	Title
X		Comservdiv 33
	X	RENDOVA
	X	Comcortdiv 1
	X	Comcortdiv 5

Maneuver time drafting commenced 1505 WX 4 JUNE
Greenwich Civil Maneuver "Time of Origin" (plus date) 0135 Z 5 June (Rule E-12)

MY	OPPLAN	ONE	DASH	FOUR	5
DASH	ACTION	ADDEE	DESIGNATED	COMTASK	10
GROUP	SEVEN	THREE	POINT	THREE	15
COMPOSED	SERVDIV	THREE	THREE	CORTDIVS	20
ONE	AND	FIVE	RENDOVA	X	25
ISSUE	SORTIE	AND	ENTRANCE	PLANS	30
CRUISING	DISPOSITIONS	YOUR	TASK	GROUP	35
COPIES	TO	ME			40

SENT CONSTRUCTIVELY (COMSERVDIV 33 NOT YET DESIGNATED)

Number of words to the nearest ten Forty(40)
System (Rule E-1) Officer messenger
Circuit (Rule E-2,4) Boat and car
#UMPIRE'S File No.

To be filled in by the UMPIRE. All other data to be filled by the student commander and verified by the UMPIRE.

DO NOT USE BACK OF THIS FORM FOR ADDITIONAL TEXT

Directions for preparing messages for duplication:— Place a sheet of "special duplicating carbon", face up, under message form. Use typewriter or hard pencil with celluloid sheet or other hard surface beneath carbon. Illegible writing or typing may cause further delays

46-43

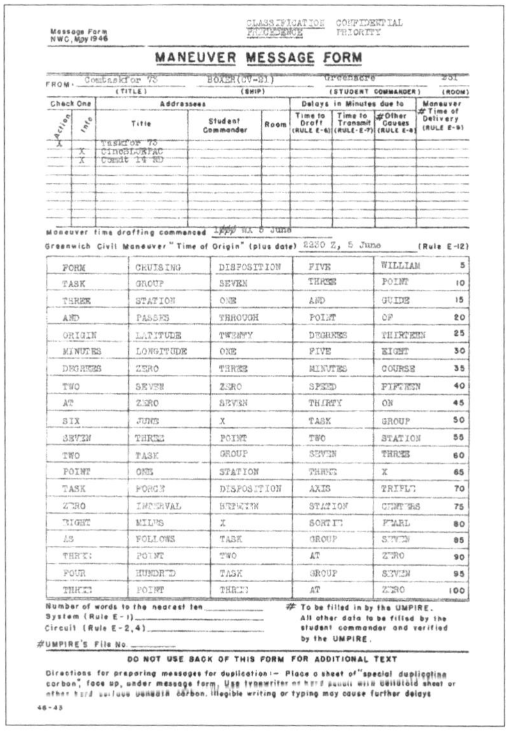

Message Form NWC, May 1946

CLASSIFICATION CONFIDENTIAL
PRECEDENCE PRIORITY

MANEUVER MESSAGE FORM

FROM: Comtaskfor 73 (TITLE) BOXER(CV-21) (SHIP) Greenacre (STUDENT COMMANDER) 251 (ROOM)

Action	Info	Title
X		Taskfor 73
	X	CincBLUEPAC
	X	Comdt 14 ND

Maneuver time drafting commenced 1ØØØ WX 5 June
Greenwich Civil Maneuver "Time of Origin" (plus date) 2230 Z, 5 June (Rule E-12)

FORM	CRUISING	DISPOSITION	FIVE	WILLIAM	5
TASK	GROUP	SEVEN	THREE	POINT	10
THREE	STATION	ONE	AND	GUIDE	15
AND	PASSES	THROUGH	POINT	OF	20
ORIGIN	LATITUDE	TWENTY	DEGREES	THIRTEEN	25
MINUTES	LONGITUDE	ONE	FIVE	EIGHT	30
DEGREES	ZERO	THREE	MINUTES	COURSE	35
TWO	SEVEN	ZERO	SPEED	FIFTEEN	40
AT	ZERO	SEVEN	THIRTY	ON	45
SIX	JUNE	X	TASK	GROUP	50
SEVEN	THREE	POINT	TWO	STATION	55
TWO	TASK	GROUP	SEVEN	THREE	60
POINT	ONE	STATION	THREE	X	65
TASK	FORCE	DISPOSITION	AXIS	TRIPLE	70
ZERO	INTERVAL	BETWEEN	STATION	CENTERS	75
EIGHT	MILES	X	SORTIE	PEARL	80
AS	FOLLOWS	TASK	GROUP	SEVEN	85
THREE	POINT	TWO	AT	ZERO	90
FOUR	HUNDRED	TASK	GROUP	SEVEN	95
THREE	POINT	THREE	AT	ZERO	100

Number of words to the nearest ten
System (Rule E-1)
Circuit (Rule E-2,4)
#UMPIRE'S File No.

To be filled in by the UMPIRE. All other data to be filled by the student commander and verified by the UMPIRE.

DO NOT USE BACK OF THIS FORM FOR ADDITIONAL TEXT

Directions for preparing messages for duplication:— Place a sheet of "special duplicating carbon", face up, under message form. Use typewriter or hard pencil with celluloid sheet or other hard surface beneath carbon. Illegible writing or typing may cause further delays

46-43

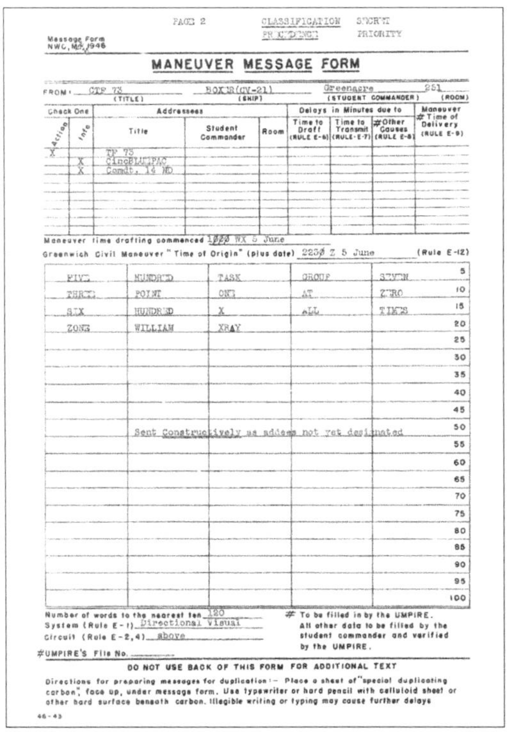

PAGE 2

Message Form NWC, May 1946

CLASSIFICATION SECRET
PRECEDENCE PRIORITY

MANEUVER MESSAGE FORM

FROM: CTF 73 (TITLE) BOXER(CV-21) (SHIP) Greenacre (STUDENT COMMANDER) 251 (ROOM)

Action	Info	Title
X		TF 73
	X	CincBLUEPAC
	X	Comdt. 14 ND

Maneuver time drafting commenced 1ØØØ WX 5 June
Greenwich Civil Maneuver "Time of Origin" (plus date) 223Ø Z 5 June (Rule E-12)

FIVE	HUNDRED	TASK	GROUP	SEVEN	5
THREE	POINT	ONE	AT	ZERO	10
SIX	HUNDRED	X	ALL	TIMES	15
ZONE	WILLIAM	XRAY			20

Sent Constructively as addee not yet designated

Number of words to the nearest ten 120
System (Rule E-1) Directional Visual
Circuit (Rule E-2,4) above
#UMPIRE'S File No.

To be filled in by the UMPIRE. All other data to be filled by the student commander and verified by the UMPIRE.

DO NOT USE BACK OF THIS FORM FOR ADDITIONAL TEXT

Directions for preparing messages for duplication:— Place a sheet of "special duplicating carbon", face up, under message form. Use typewriter or hard pencil with celluloid sheet or other hard surface beneath carbon. Illegible writing or typing may cause further delays

46-43

Fig. 128
Maneuver messages, Blue Covering Force, Operations Problem 2

cooperation and coordination that was required of all three task group commanders. Commanders, 2nd Battle Group and Carrier Group were to base their movements on those of the 1st Battle Group. Upon junction with the Expeditionary Force's heavy ships, Commander, Blue North Pacific Force's Battle Plan would become effective. At that time, if the Purple air threat was negligible, CLAA-95 and CLAA-98 would be directed to join the 1st and 2nd Battle Groups. Commander, Covering Force was embarked in CV-21 with the Carrier Group, and Rear Admiral BM, Commander, Carrier Division 3, was designated second in command.

At this point in the presentation, a telephone conversation was simulated between Rear Admiral BA and Vice Admiral BF, Commander, Blue North Pacific Force, in which BA informed BF, Rear Admiral BB (Commander, Expeditionary Force and Amphibious Group 7), and Rear Admiral BC (Commander, Battleship Division 5) about these dispositions and his Battle Plan. (Page 29.)

NOTES

1 Senior Class of June 1947, "Operations Problem 2: Detail of Student Officers for Solution of Task Force Commanders Problems, Section B," 25 September 1946, pp. 1–2, folder 2603-N, box 138, RG 4, NHC.

2 Senior Class of June 1947, "Operations Problem 2: Blue—Section B, Preliminary Movement Orders," 25 September 1946, p. 1, folder 2603-F, box 138, RG 4, NHC.

3 Ibid., Annex A: Intelligence Plan, p. 1.

4 Ibid., Annex B: Communication Plan, pp. 1–2.

5 Ibid., Annex C: Movement Plan, p. 1.

6 Senior Class of June 1947, "Operations Problem 2: Blue—Section B, Preliminary Movement Orders," 30 September 1946, p. 1, folder 2603-P, box 138, RG 4, NHC. Subsequent in-text page references in this chapter are to this source.

7 "Battle carriers"—as the real-world ships of the USS *Midway* class, which began to enter service immediately after the end of hostilities, were then known—incorporated the results of wartime combat experience in terms of, especially, armor and damage control. They were considered more survivable than the "fleet carriers" (equivalent to the *Essex* class) that were operated by Purple.

NORTH

XVI *Operations Problem 2*

The Blue Solution Concluded, September–October 1946

In these final phases of planning for the Blue Solution to Operations Problem 2, Rear Admiral BB, Commander, Expeditionary Force, came up with a number of scenarios for his Battle Plan. In Case One, in which the Blue Covering Force was absent, he would have under his command—from his amphibious force flagship—Commodore BI's Transport Group, comprising three transport units that were further subdivided into eight transport divisions and altogether included the twelve high-speed transports, twenty-four attack transports, and six attack cargo ships. Also under his command was Rear Admiral BG's Carrier Group, three escort carriers, with Captain B-50's (not played by a student officer) Carrier Screen Unit of the five destroyers from Destroyer Division 132 and three destroyer-minesweepers from Mine Division 7. In addition, there was Captain B-23's Minesweeping Group of five destroyer-minesweepers, six minesweepers, and one minelayer, further subdivided into Minesweeping Unit 1 and Minesweeping Unit 2, with ships from Mine Divisions 5 and 9. The Expeditionary Troops would consist of two Marine divisions plus attached units, while Rear Admiral BC's Expeditionary Striking Group was composed of his Battleline, two prewar modernized battleships from Battleship Division 5, and two destroyers. The Expeditionary Striking Group was composed of Captain CA's Right Flank Force, one light cruiser and seven destroyers from Destroyer Squadron 7, along with Rear Admiral BH's Center Force of one heavy cruiser, one light cruiser, and another seven destroyers, the latter from DESRON 1. The Expeditionary Striking Group was rounded out by Captain CB's Left Flank Force, another light cruiser and seven destroyers from DESRON 13.[1]

In Case Two, in which the Blue Covering Force was present, the Transport Group, the Carrier Group, the Minesweeping Group, and the Expeditionary Troops would assume the same dispositions as in Case One. Rear Admiral BC, however, would form a Combined Striking Group. Here, his battle line would consist again of the two prewar modernized battleships but would now additionally have the four destroyers of DESDIV 11. The Right Flank Force would now be commanded by Rear Admiral BD and consist of two heavy cruisers, two light cruisers from Cruiser Division 15, the nine destroyers from Destroyer Squadron 7, and three additional destroyers from DESDIV 91. Rear Admiral BD would now also command the Center Force, composed of two heavy cruisers from CRUDIV 1, two light cruisers from CRUDIV 13, five destroyers from DESRON 1, and seven destroyers from DESRON 15. The Left Flank Force was to be commanded by Rear Admiral BN, with two heavy cruisers from CRUDIV 3, two attached light cruisers, and seven destroyers from DESDIV 92. (Page 17.)

Case One assumed Purple forces to be equal or inferior to the Blue Expeditionary Striking Group, which would engage the Purple forces "decisively" and destroy them. In Case Two, in which Purple forces were superior to the Combined Striking Group, the latter was to interpose itself between Purple and the remainder of the Expeditionary Force. The Expeditionary Striking Group would fight a delaying action on a course to concentrate with the Covering Force while staying out of decisive gun range. Every opportunity that arose or could be created to destroy Purple units was to be exploited and, in case of contact with the Purple Main Body, the destruction of the Purple fleet would become the primary mission. In either Case One or Two, the Expeditionary or Combined Striking Group was to be prepared to fight a surface engagement using full radar control. Rear Admiral BB would be embarked in amphibious force flagship AGC-11, while Commander, Expeditionary Striking Group and Commander, Combined Striking Group would be embarked in prewar modernized battleship BB-48. (Page 18.)

Bombardment of the beach and the surrounding areas would be conducted by the Gunfire Support Group commencing at H-Hour minus 4 ("H minus four") on D-Day. The preliminary bombardment of possible enemy strongpoints in the vicinities of Mayachnny, Sredni, and Stanitskogo areas were to be conducted on D–1 by units of the Covering Force. The Expeditionary Force's bombardment was to be conducted in three phases. Phases One and Two would be controlled by Commander, Gunfire Support Group, while Phase Three would be controlled by Commander, Assault Group. These phases were to commence at H–4, H–2, and H-Hour, respectively, with specific geographic reference points for each phase. For example, Phase One would consist of gunnery runs between designated points, with lookout, fire-control, and gun positions on high ground as the priority targets. The battleships in particular were to destroy these gun positions with counterbattery fire when located. Phase Two would consist of separate runs by Gunfire Support Units 1 and 2. Gunfire Support Unit 1 would make runs from a designated point for twelve thousand yards on 072 degrees true and the reverse course. It was also to cover the beach areas and probable gun positions in direct support of the landings on Red Beach. Gunfire Support Unit 2 would make runs from a different point for twelve thousand yards on course 008 degrees and reverse, covering the beach areas and probable gun positions in direct support of Green Beach. Finally, Phase Three would consist of area bombardments, counterbattery fire, and "call fires" (no-notice requested missions) as directed by Commander, Assault Group. Commander, Assault Group would direct the shift of fire control from air or surface ships to the shore fire-control party. (Page 19.)

In his Cruising Plans, Rear Admiral BB outlined two plans, which he designated 6C2 and 6C3. The two plans were primarily for defense against submarines, aircraft, or both. Unless otherwise prescribed by the Officer in Tactical Command, the stations of the various units were as indicated in attached diagrams, showing circles with spacing of one mile and numbered one to seven. The Fleet Guide was to be the flagship of the Task Force Commander. When the battle line was not present, the Guide of the formation was the flagship of the Transport Squadrons. When the train (oilers, etc.) was not present, the aircraft carriers were to occupy the area that it normally occupied, except when launching or recovering aircraft. The destroyer screen was under the direct command of the senior destroyer squadron commander, who was to occupy Station No. 1. Because of the extent of the formation, it would not be possible for the Screen Commander to exercise adequate control of the entire screen as rapidly as required. Therefore, Admiral BB designated a Deputy Screen Commander, who would occupy Station No. 30, or the highest even-numbered station being occupied at any given time. Port and Starboard Screen Assistants would occupy stations on the beams of the formation. (Page 20.)

The object of BB's Sortie Plan, known as 1V, was the security of the heavy ships when their movements were restricted (e.g., while leaving port and taking station after clearing restricted waters). His plan assumed that channels to deep water had been swept and buoyed, that antisubmarine patrols and air patrols had been established, and that Blue submarines were well clear. In addition, it was assumed that Purple submarines—some deployed as minelayers—might be present off a port but that Purple surface forces would not be. (Page 21.)

His General Plan was to cover the sortie from port of the heavy ships by offensive measures by the light forces. His destroyer squadrons would sortie from their fleet anchorage through swept channels first. Commander, Destroyer Squadron 1 would pass through Point ABLE at minus 7.5 hours—that is, with reference to Zero Hour, 2330 Zebra—and proceed to Point BAKER, through which he would pass at minus 10 minutes. There, DESRON 1 was to establish offensive antisubmarine patrols seaward out to twelve miles with DESRONs 7 and 13 on either side, north and south. The mine squadrons would pass through Point ABLE at minus 7.3 hours, pass through Point BAKER at minus 2 minutes, and establish an Inner Anti-Submarine Patrol on bearing 310 degrees true from BAKER. (Page 21.)

Admiral BB's heavy cruisers would pass through ABLE at minus 7 hours and BAKER at plus 10 minutes, and then turn to 310. His light cruisers would pass through ABLE at minus 6.9 hours, pass through BAKER at plus 16 minutes, and take station on the heavy ships ahead. The battleships would pass through Point ABLE at minus 6.8 hours, pass through Point BAKER at plus 25 minutes, and then also set course to take station on the heavy ships ahead. The Force Flagship would proceed through Point ABLE at minus 6.6 hours and through Point BAKER at plus 35 minutes before it set course for the same heavy ships. Amphibious Group 7 would pass through Point ABLE at minus 6.5 hours and Point BAKER at plus 45 minutes, and set course for its station with the heavy ships while the carriers would pass through Point ABLE at minus 5 hours, Point BAKER at plus 2.4 hours, and then take up their stations with these same ships. Sortie speed was ordered as twelve knots. (Pages 21–22.)

The fundamental difference between Cruising Dispositions 6C2 and 6C3 was what he called an "Eccentric Screen" employed in one of them to economize in the sonar screen and thereby free ships for use as radar pickets. Since the purpose of the Cruising Dispositions was to afford maximum protection from submarine attack, and because BB's force was so large, no changes were to be made to effect greater mutual protection from aerial attack and the screen would not be reoriented for better concentration of antiaircraft fire. The Screening Plan was laid out to protect against a forty-five-knot torpedo fired at six thousand yards while the Expeditionary Force steamed at twelve knots. Screen Station No. 1 would lie on the Fleet Axis and in the Van, even-numbered stations to the left and odd to the right. In the Plan, red circles indicated stations to be left unoccupied if the number of screening ships that were available was reduced. (Page 23.)

The Sonar Screen was to be made up of destroyers and a limited number of destroyer-minesweepers, with the Inner Anti-Submarine Screen exclusively composed of destroyer-minesweepers. All of these ships were to be under the command of the Screen Commander. The Flank Anti-Submarine Screens were to be made up of high-speed transports and minesweepers under the command of Commander, Train. The Screen Commander was to maintain the strength of the Sonar Screen at the maximum level consistent with the requirements for radar pickets. When it was necessary to reduce the number of screening ships, the entire screen would be drawn in. (Page 24.)

When ships of the screen departed or otherwise separated themselves from the screen, all others were to close up automatically. Ships sighting mines were to report these sightings to the nearest heavy ship likely to be affected by it. Destroyers were to keep boilers on line for twenty-six knots at all times. (Pages 24–25.)

Rear Admiral BB also organized a Screening Group, under Captain B-11, consisting of Screening Unit 1 (the nine destroyers from Destroyer Divisions 11 and 12), Screening Unit 2 (the nine destroyers from DESDIVs 71 and 72), and Screening Unit 3 (five destroyers of DESRON 13). Rear Admiral BC's Gunfire Support Group was divided into Gunfire Support Units 1 (the two prewar modernized battleships from Battle Division 5) and 2 (a heavy cruiser and the three light cruisers from

Cruiser Division 13). BC also commanded the Expeditionary Striking Group when it was formed—two battleships, a heavy cruiser, three light cruisers, and twenty-three destroyers. Additionally, he commanded the Combined Striking Group when formed, consisting of the ships of the Expeditionary Striking Group plus three heavy cruisers from CRUDIV 1, the two heavy cruisers from CRUDIV 3, the three light cruisers from CRUDIV 15, and seventeen destroyers from DESRON 9.[2]

While the Expeditionary Force was still in port, its commander was the SOPA who therefore would coordinate all naval activities afloat, the debarkation of Major General BY's (also referred to as "BO" and "XX" for other aspects of the planning) two Marine divisions, and the equipment and supplies for the assault troops. On D-Day, he would coordinate all tactical aircraft at the objective area commencing at H–4. The Transport Group would embark the troops, equipment, and supplies in landing boats, with Transport Unit 1 landing on Red Beach and Transport Unit 2 on Green Beach at H-Hour. Transport Unit 3 was to carry out a demonstration off the entrance to Avacha Bay at H–1, reembark its troops, and then be held in reserve. During the landing, the Carrier Group was to destroy Purple air and naval forces that might attack the Expeditionary Force, furnish aircraft for close air support, coordinate and control aircraft for antisubmarine and combat air patrols, and coordinate screening and antisubmarine functions with Commander, Screening Group. The Minesweeping Group was to remove mines from the Red and Green Beach approaches, conduct salvage and towing, and clear the entrance to Avacha Bay. The Screening Group was to destroy Purple air, surface, and submarine forces that threatened the Expeditionary Force; send one Screening Unit to Commander, Gunfire Support Group for bombardment and call fires; and oppose threats from enemy naval forces. (Pages 2–3.)

The Gunfire Support Group was to carry out a preinvasion bombardment, destroy Purple forces and defensive positions at the objective, destroy enemy air and naval forces, and otherwise assist Expeditionary Troops to the maximum practicable extent. The Expeditionary Troops were to capture and hold Avacha Bay—including the harbor entrances—at the earliest possible date. The Expeditionary Striking Group was to destroy Purple surface forces if those forces were equal or inferior but conduct delaying actions if Purple was superior. The Combined Striking Group was similarly to oppose or delay and coordinate its air units with other Blue forces in the area. In addition, Admiral BB ordered that preliminary movements could be carried out but that no offensive operations would be carried out until 20 June. Rear Admiral BB, Commander, Expeditionary Force, would be embarked in AGC-11, while Rear Admiral BC, Commander, Gunfire Support Group, was to be embarked in BB-48. (Pages 4–5.)

Admiral BB's Air Plan consisted of three phases of air support for the Expeditionary Force. Phase One was to be the air support en route to the objective,

Phase Two was the Pre-Landing Aerial Bombardment, and Phase Three was the Air Support of Landing Troops. All three phases called for the air defense of the Expeditionary Force by antisubmarine and combat air patrols, air searches embedded in the Air Plan and as directed, and such air services as transportation, air reconnaissance, and message communication. Phase One was divided into two legs. In the first leg, from Seattle to Dutch Harbor, antisubmarine patrols would be flown according to schedules provided in the Air Plan and according to Fleet Tactical Publication 223A, *Anti-Submarine and Escort of Convoy Instructions.* Carriers were to be assigned this duty by days; off-duty carriers were to keep eight fighters and four attack planes in Condition Twelve from sunrise to sunset and two night fighters in the same operational readiness from sunset to sunrise. On the second leg, from Dutch Harbor to Avacha Bay, antisubmarine and combat air patrols as well as searches would be flown in accordance with Air Schedule Baker, which was attached. The search, unless otherwise ordered, would cover a sector described as 110 degree true to a distance of two hundred miles from the formation center.

Antisubmarine patrols would consist of six fighters and six attack planes in fighter and attack-plane teams, with each team covering a sixty-degree sector to a distance of twenty miles from the formation center and covering only the area outside the formation. Antisubmarine patrols were to be oriented to true north, with Sector No. 1 being 000 degrees true to 060 degrees true and the remaining sectors numbered consecutively clockwise. At 2100, each carrier was to put two night fighters in Condition One and maintain them until 0300 the following morning upon the completion of launching flights. Also, the carrier that launched the combat air patrol was to keep its deck prepared for landing cripples while the other carriers prepared their decks for the rapid launching of additional fighter planes. (Page 7.)

In Phase Two, the prelanding aerial bombardment was to consist of two carriers furnishing full groups—less antisubmarine patrols—for the saturation bombardment of the landing beaches, which was to commence at H–1 and last fifteen minutes. Combat air patrols were to be furnished by the Covering Force. Antisubmarine patrols would consist of six attack planes flying singly and covering the Expeditionary Force on its exposed sea flanks. Phase Three was to consist of air support, each carrier furnishing four attack planes and eight fighter-bombers on a two-hour schedule to the Tactical Air Commander as ordered during daylight periods. Combat air patrols would be provided by the Covering Force, and antisubmarine patrols would consist of six attack planes flying singly, with two planes from each carrier on a four-hour schedule. Air searches were to be conducted as ordered and each carrier would keep two night fighter planes in Condition Eleven from 2100 to 0300 each night. The Tactical Air Commander, embarked in the flagship of the Commander, Expeditionary Force, would control all aircraft tactically at the objective area from the time that aircraft reported into him after takeoff—or when arriving from other forces at the rendezvous points—until he released them to return to their parent carriers or other bases. Carrier captains would control their own planes prior to their reporting in and being relieved of this duty. The daily air schedules for the following day were also to be forwarded by the Tactical Air Commander to the carriers by message each evening. (Pages 7–8.)

Antisubmarine patrols were assigned to one carrier each day: for 10, 11, 12, 13, 14, and 15 June, escort carriers CVE-112, -115, -119, -112, -115, and -119, respectively. The antisubmarine patrols were to take off each day at 0315, 0700, 1045, 1430, and 1815 and land at 0715, 1100, 1445, 1830, and 2030. There were to be one attack plane assigned per patrol leg and three attack planes per patrol. Under Air Schedule Baker, there would be six flights throughout the day for combat air patrols, antisubmarine patrols, and searches. The first planes for these flights would be launched at 0300, 0630, 1000, 1200, 1330, and 1700, while the first planes to be recovered from these flights would land at 0645, 1015, 1345, 1500, 1715, and 2045. Duty 1 would consist of two flights of twelve fighters for the combat air patrol, along with two flights of six fighters and six attack planes each for the antisubmarine patrols. Duty 2 was to consist of one flight of twelve fighters for the combat air patrol, one flight of six fighters and six attack planes for the antisubmarine patrol, and two flights of eight fighters and eight attack planes for searches. Duty 3 comprised two flights of twelve fighters for the combat air patrols, with the two flights of six fighters and six attack planes repeated for the antisubmarine patrols and the one flight of eight fighters and eight attack planes for searches also repeated. Carriers would be assigned these duties the previous night by message. (Page 9.)

Admiral BB's Communication Plan called for radio silence down to thirty megacycles per second at staging and assembly areas except for warnings or defensive operations. Visual communications were to be used whenever possible and radio silence was to be observed while at sea en route from Puget Sound. All units were also to observe radar, Identification Friend or Foe, and radar countermeasures silence. Ships designated as radar and Identification Friend or Foe guards were to maintain continuous twenty-four-hour watches following the departure from Puget Sound, and ships designated as radar countermeasures ships would emit only on the order of the OTC. In addition, radar intercept ships were to monitor and log continuously Purple transmissions; at the objective, they were also to guard Talk Between Ships and Task Force Common (apparently to detect intrusion by Purple). Numerous circuits and frequencies were established and guards assigned. (Pages 10–12.) Radio Call Signs for the Amphibious Landing were provided. For landing craft and other assault waves, Boat Commanders were to call by wave number, beach color, and beach number, while call signs for tenders would be formed by using the call "Holstein" followed by individual identification.

While en route, flag hoist and semaphore communications were to be used whenever possible but flashing lights were only to be employed for half an hour after sunrise until a half-hour before sunset. Filters and low-power lights were to be used as much as possible and visible light was not to be used for signaling during deep twilight. While at the objective, reports were to be made by visual communications or by messenger boat. Radio was not to be used except for "important" operational traffic or when visual or messenger communications could not be carried out. Infrared signaling equipment could also be used when visual light was prohibited. (Pages 13–14.)

Rear Admiral BB's Communications Plan included procedures for reporting reef and beach obstacles. The results of the minesweeping and underwater demolition prior to 1000 on D-Day would be transmitted in full by Commander, Task Force 75 by radio cipher on Task Force Common. If later information of value was obtained after this, it would be broadcast by Commander, Task Force 75 on the Continuous Wave or Task Force Common frequencies as well as on Landing Craft, Infantry Common frequency. Plain-language reports to all ships present would also be sent, reporting items in sequence for each beach; if there was no information on that beach, it would be omitted. These procedures also outlined ways to report entrances to paths cleared by the demolition teams. Specifically, these procedures entailed grid coordinates on 1/20,000 scale air and gunnery target maps of designated points and the distance in feet between remaining mines, the number of rows of mines, and the interval in feet between rows. The procedures also indicated whether there were buried, exposed, or remaining trip-wire mines. (Pages 14–15.)

Student officers were next provided with Advance Information for the Chart Maneuver. This information included the time and day at which the maneuver would begin, the time and duration of the first move, the way that students were to express Maneuver Time, and the fact that previous reconnaissance had not yielded any information owing to bad weather. Admiral Smith instructed the student officers to complete the first move for the forces they were commanding prior to 0900 on 2 October 1946. The tracks of all forces under way at this time were to be drawn on appropriate charts and flimsies made in accordance with the Maneuver Rules. The locations of forces not under way during Move 1 were also to be indicated. The state of the weather was to be given to each force, group, or unit commander, but only for the area in which his units were moving. Information on weather in the various areas was to be collected and disseminated by each side's Commander-in-Chief, but only as provided for in his plans. Forecasts were to be furnished to each Commander-in-Chief by the Director of the Maneuver from time to time. These forecasts were to vary in completeness and coverage, again in accordance with each Commander-in-Chief's plans for collecting information on which to base a forecast, in terms of Meteorological Visibility; Assured Sonar Range; Percentage of Cloud Cover and Ceiling; Rain, Snow, and Ice Conditions; and Air Temperature.[3]

Student officers were especially cautioned to do all work in connection with the maneuver in the rooms assigned. Physically covering all their material when securing during the noon hour and at the end of the afternoon session was also suggested, as was avoiding receiving or giving out unauthorized information. They were also only to use designated Plotting Charts, ensure the completeness of data on the flimsies and forms, and ensure that the information entered on the Record of Move Form provided a brief and accurate history of all important events during the move. The Record of a Move was turned in when the flimsy for the following move was sent to the Master Plot. Accurate fuel records were to be kept in the prescribed forms and available for inspection by the Maneuver Staff. Forms requested by the staff were to be sent by pneumatic tube. Finally, student officers were to use downtime between moves to plan for future moves and to progress their reading in preparation for their theses.[4]

NOTES

1 Senior Class of June 1947, "Operations Problem 2: Blue—Section B, Annex D: Battle Plan," 1 October 1946, p. 16, folders 2603-P–U, box 138, RG 4, NHC. Subsequent pages references until the next endnote are to this source.

2 Senior Class of June 1947, "Operations Problem 2: Blue—Section B, Annex D: Battle Plan," 1 October 1946, p. 1, folder 2603-R, box 138, RG 4, NHC. Subsequent pages references until the next endnote are to this source.

3 Senior Class of June 1947, "Operations Problem 2: Advance Information for Chart Maneuver," 1 October 1946, pp. 1–2, folder 2603-H, box 138, RG 4, NHC.

4 Ibid., p. 2.

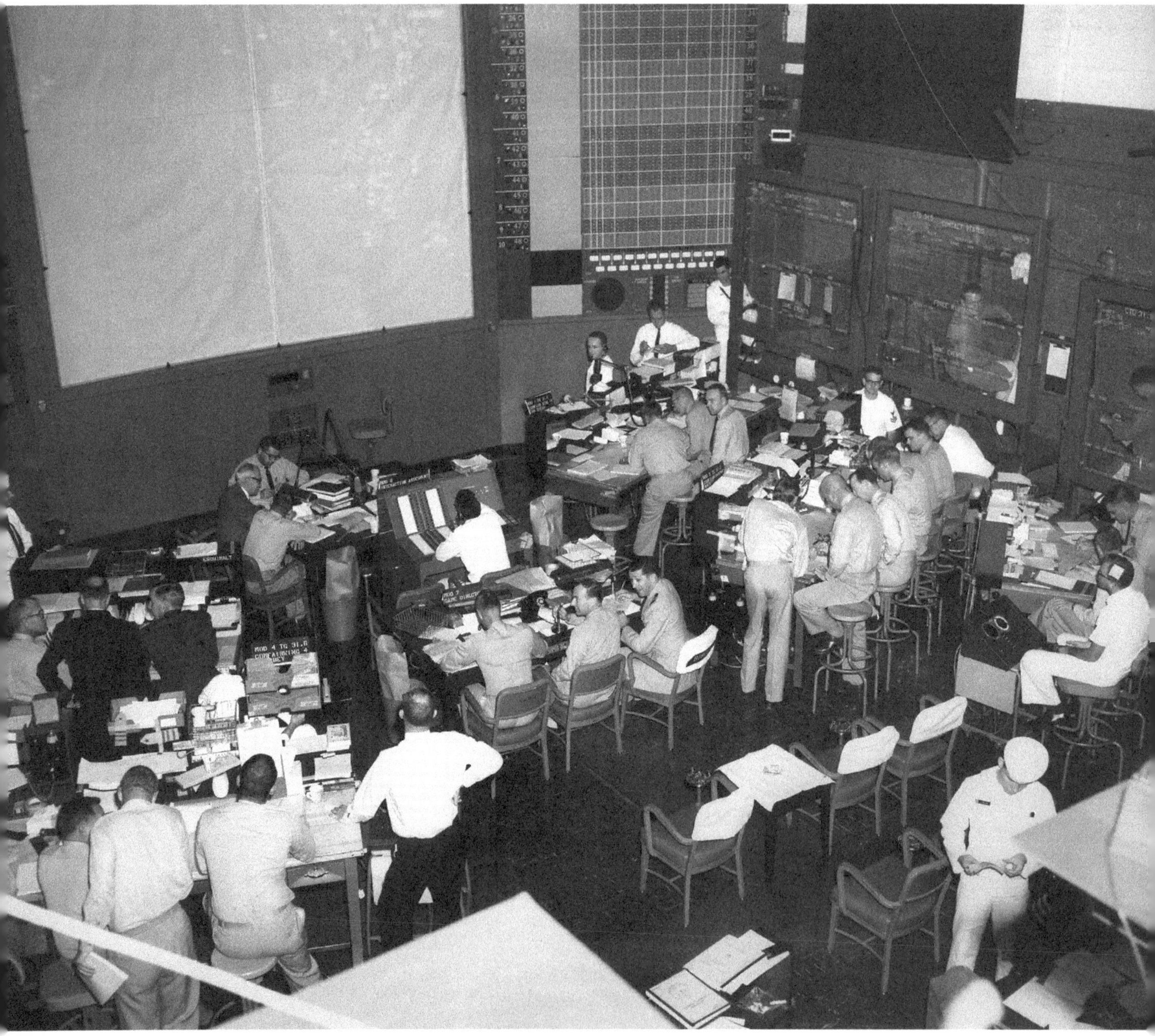

XVII *Operations Problem 2J*

Blue, October–November 1946

In late October 1946, the junior class of June 1947 began preparing for Operations Problem 2J. Between 8 and 25 November, Section A of the class drew up the Blue Statement of the Problem and Section B the Purple Statement. Over the next several days, each section created its Solution to the Problem while also attending lectures and presentations in the auditorium in Pringle Hall. The maneuver itself commenced on 16 November and continued through the 22nd, along with several additional presentations. Students and staff prepared for and conducted the Critique from 23 to 25 November.[1]

In Operations Problem 2J, both Blue and Purple played out both a Chart Maneuver (Phase One) and a Board Maneuver (Phase Two). The Maneuver Staff included a number of instructors from the junior class, including the Director of the Maneuver, who was again Captain Evans (see chapter 1); Commander Mayberry (see chapter 1), who was the Assistant Director for Blue and the Chief Damage Computer; Commander Reed (see chapter 1), who was the Assistant Director for Purple and the Submarine Umpire; Commander Kirkpatrick (see chapter 9), who was the Air Umpire; and Lieutenant Commander Holzapfel (see chapter 1), who was the Assistant Air Umpire. Lieutenant Commander Curran (see chapter 1), from the Department of Intelligence, was the Communication Umpire, while Commander Miller (see chapter 9), one of the junior class students, was the Historian. Blue and Purple each had Move Umpires and Force Damage Recorders—Commander Dalton (see chapter 8), Colonel Edson (see chapter 8), Lieutenant Commander Howell (see chapter 9), and Lieutenant Colonel Thompson (see chapter 9), all junior class students. During Phase One, there were two Air-Submarine Contact Evaluators, Lieutenant Commander Holzapfel again and Lieutenant Commander Taeusch (see chapter 8), from the junior class. There were also two Submarine-Surface Contact Evaluators, again Lieutenant Commander Taeusch and his classmate Lieutenant Commander Aymond (see chapter 8).[2]

Phase One required several Assistants to the Air Umpire, roles played by additional junior-class members: an Aerial Combat Loss Computer, who was Lieutenant Colonel Leary (see chapter 8); a Hit Computer, Commander Richardson (see chapter 9); an Anti-Aircraft Loss Computer, Lieutenant Commander Bonner (see chapter 9); Blue and Purple Status Board Recorders, Commander White (see chapter 8) and Colonel Mason (see chapter 9); an Operational Loss Computer, Commander Rynd (see chapter 9); Blue and Purple Radar Contact Recorders, Commanders Coleman (see chapter 8) and Foley (see chapter 9); and two civilian Plotters, draftsmen John Lawton (see chapter 9) and Frederick Wagner (see chapter 9) from the Department of Tactics' War Gaming Section. For Phase Two, Blue and Purple each had Damage Computers for their battle lines, this time Lieutenant Commanders Aymond and Bonner again. The Damage Computers for Center and Van Cruisers for Blue and Purple were again Commanders Coleman and Richardson, respectively, while Commanders Kirkpatrick and Rynd this time were the Blue and Purple Damage Computers for the Center and Van Destroyers. The Blue and Purple Damage Computers for the Rear Cruisers were Lieutenant Commander Taeusch again and Commander Foley again, while those for the Rear Destroyers were Lieutenant Colonel Downey (see chapter 8) and again Colonel Mason. Mr. Lawton was again the Plotter, while the two Assistant Plotters were again Mr. Wagner and Miss DeMarco (see chapter 9), all draftsmen.[3]

The Blue Statement of the Problem

The objectives of the maneuver were to provide student officers opportunities to exercise in preparing an Estimate of the Situation, formulating appropriate directives and dispositions, and executing planned actions in a Chart or Board Maneuver. In addition, Operations Problem 2J was to demonstrate to them the capabilities, limitations, and tactical employment of ships and aircraft; demonstrate the capabilities, limitations, and tactical employment of new submarine types and antisubmarine measures; and highlight the restrictions on a task force's freedom of movement when assigned to protect a convoy.[4]

In the General Situation, the Purple–Blue War had commenced on 15 August 1950 when the Purple Far Eastern Air Force flew large-scale, surprise raids on Blue forces and installations in Japan and the Aleutians, accompanied by the advance of Purple ground forces into Manchuria. Purple was now occupying southern Korea, Manchuria, and northern China. All Purple attempts to invade Japan had been "temporarily" forestalled by the Blue atomic bombing of Purple amphibious forces and transport aircraft units assembled at Vladivostok. At the commencement of hostilities, Blue had a small occupation force in Japan, augmented by a "moderately strong" air force of long-range and medium-range bombers as well as long-range fighters. Owing to "excellent" dispersion, most of the Blue air forces had escaped destruction in the initial Purple air attacks. Blue air strength had been dissipated in subsequent action but it still had an "operationally effective" force in the Japanese home islands. (Page 1.)

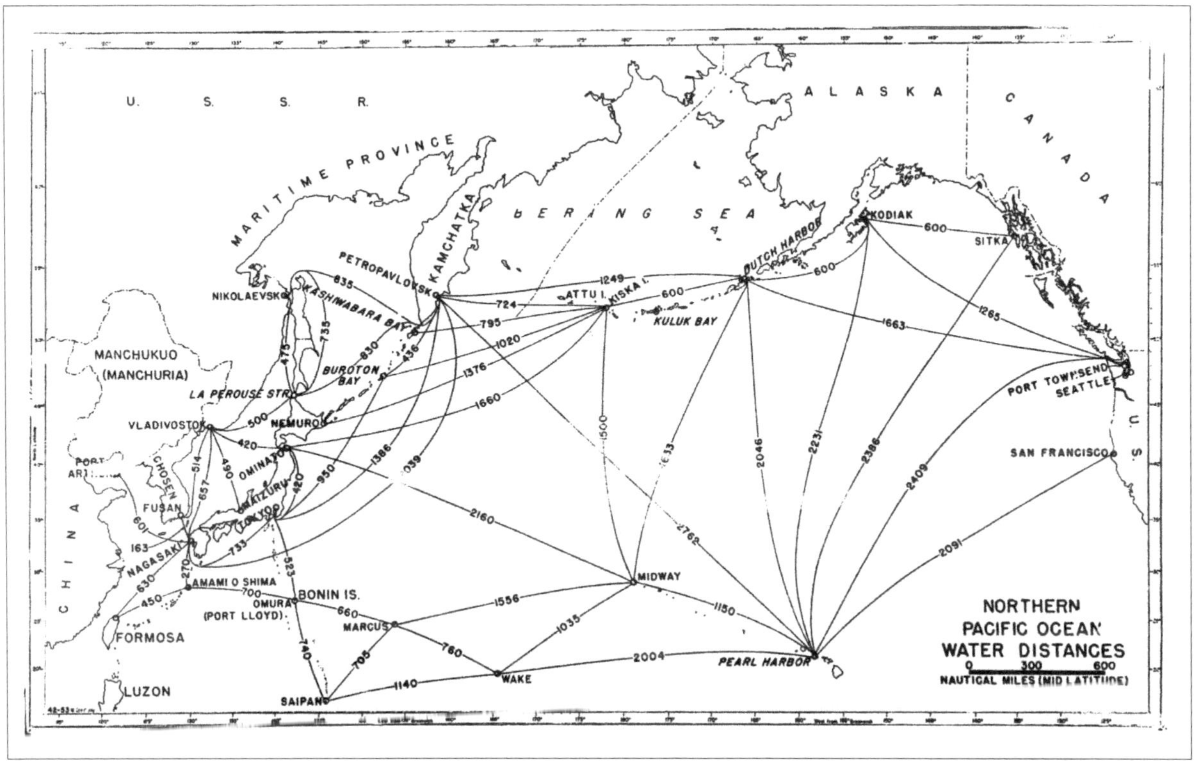

Map 13
Northern Pacific Ocean Water Distances

At the outbreak of war, the Purple Pacific Fleet, with headquarters at Vladivostok, was based in Siberia, on Sakhalin, and in the Kuriles. The Blue Pacific Fleet, headquartered at Pearl Harbor, was based in Hawaii. Other than submarines, the only Blue fleet combatant forces immediately available were a small Fast Carrier Task Force, a small amphibious force, a two-division force of Marines, and the fleet air wings. The small Blue fleet forces operating in Far Eastern waters had been withdrawn to Guam during the period of strained relations that had preceded the war. The mobilization of the Blue Army, however, was expected to permit the staging of four divisions monthly from the San Francisco Port of Embarkation. The Special Situation laid down that Blue plans to reinforce its occupation forces in Japan had been approaching fruition at the outbreak of hostilities. All the planning for an invasion of Avacha Bay was now set aside; the mission of Blue forces now was to get these reinforcements through to Japan, where they would resist the threatened Purple invasion.

The 2nd and 3rd Infantry Divisions and major elements of the 15th Army Air Force—personnel, fighter aircraft, and ground equipment—had arrived in Pearl Harbor on 13 August and had been scheduled to depart on 20 August in an escorted convoy designated HTF-1. Given the critical situation that had arisen because of the Purple surprise attack, it was decided to add the 1st and 3rd Marine Divisions to the reinforcement program even though they were assault trained. These units would be embarked in the available transports of the Blue Amphibious Force and would sail with Convoy HTF-1. The sailing date for HTF-1 was advanced to 19 August, and Fleet Admiral BA, Commander-in-Chief of the Blue Pacific Fleet (no designated student officer), accordingly issued his Operation Order No. 8-50. (Page 2.)

Fleet Admiral BA's orders were specifically issued to Vice Admiral BC, Commander, Blue Covering Force (Commander Baldridge—see chapter 9), designated Task Force 11. The Blue Covering Force consisted of three battle carriers organized into Carrier Division 2, three fast battleships organized into Battleship Division 1, five heavy cruisers organized into Cruiser Division 1, and one antiaircraft cruiser, along with twenty-three destroyers organized into Destroyer Flotilla (DESFLOT) 1. Vice Admiral BC was apprised of the weakness of Blue occupation forces for repelling a Purple invasion of the Japanese home islands, of the fact that Purple amphibious forces and transport aircraft had been damaged by Blue atomic attacks, and that Blue Army Air Forces in Japan had been weakened but were still operationally effective. Also, not only had all of Blue's atomic bombs that had been available in Japan at the commencement of hostilities now been expended but it was believed that Purple had produced a small number of its own. Blue thinking on the matter, however, was that Purple was reserving those weapons for use in more-critical areas; none had yet been employed in the Far East. (Page 3.)

Purple naval forces known to be in Far Eastern waters as of 10 August included three fleet carriers similar to the *Essex* class, three fast battleships like those of the *South Dakota* class, and six heavy cruisers equivalent to the *Baltimore* class. Also known were four squadrons of "modern" destroyers, an amphibious force assembling at Vladivostok that consisted of transport squadrons and other vessels capable of a major landing, a submarine force of about sixty high-speed submarines based at Vladivostok and Kashiwabara, and a fleet air force based at Shimushu, Paramushiru, Matsuwa, and Etorofu. Purple's submarines were believed to be on patrol in the Hawaiian, Central Pacific, and Aleutian areas. The Purple Fleet Air Force operated naval patrol planes whose characteristics were similar to those of Blue PBM Mariner naval patrol bombers. At least two fleet carriers and three fast battleships had been sighted on 14 August in Kashiwabara Harbor, among "many" other combatant ships. (Page 3.)

Blue submarines were engaged in war patrol and reconnaissance in the Yellow Sea, the Japan Sea, and the waters adjacent to the Kurile Islands. Blue submarines proceeding to and from stations would be routed clear of the Blue forces involved in this operation. Convoy HTF-1 was to depart Pearl Harbor on 19 August, and BC's Covering Force—along with an Escort Force, Task Group 51.7, commanded by Commodore BJ (Commander Morton—see chapter 9)—had been assigned as escort. Commander, Blue 5th Air Wing (not played by a designated student officer) was directed to provide air cover to the Blue Covering Force and Convoy HTF-1 within seven hundred miles of Oahu and Midway. Commander, 20th Blue Army Air Force (no designated student officer) was requested to provide air cover within seven hundred miles of Honshu. Blue Covering Force was to ensure the safe arrival of HTF-1 at Tokyo at the earliest possible date; Commander, Blue Covering Force was authorized to issue orders to Convoy HTF-1 when the occasion required. Prior to departure from Pearl Harbor, he was authorized to load two atomic bombs on each Covering Force battleship for delivery to the 20th Army Air Force in Japan. Rear Admiral BE (Commander Simonsen—see chapter 9), embarked in battle carrier CVB-41, was designated as the second in command of this operation. Finally, Vice Admiral BC was additionally provided information on the assignment of forces to TG 51.7, the composition and prescribed route of HTF-1, the assignment of forces to the overall Blue Covering Force, and the estimated characteristics of the Purple high-speed submarines (see below). (Pages 3–5.)

Task Force 11 and Convoy HTF-1, escorted by TG 51.7, departed Pearl Harbor at 0630 on 19 August. TF 11 maintained a position within air support distance of the convoy, which proceeded along its prescribed route at a speed of fifteen knots. At 2100 on 22 August, ships of TF 11 reported intermittent radar contact on a possible surface target but attempts to establish visual or sonar contact were unsuccessful. At the time of this incident, Vice Admiral BC's flagship was at 30° N, 179° W, on a course of 275 true at a speed of fifteen knots. At 2330, he received a message from Fleet Admiral BA that Blue submarine SS-127, patrolling off Kashiwabara, had reported a Purple force of about ten large and twenty small ships departing that base at 1000 on 23 August on a course of 160 at twenty knots. This force was evaluated as the Purple Striking Force; there was no evidence of another Purple surface force at sea in the Pacific. (Page 5.)

At 0800 on 24 August, following a submarine contact, attack cargo ship AKA-99 of HTF-1 was sunk by a torpedo. Destroyers participating in depth-charge attacks on the Purple submarine reported "extreme difficulty" in tracking the target and could not determine the results of their attack. During the afternoon of 24 August, destroyers of TG 51.7 were refueled from convoy tankers. The operation was completed by 1630 and all ships of TF 11 were topped off from the convoy oilers during the morning of 25 August. At 1700 on 25 August, with his flagship at 31° N, 166° E, Vice Admiral BC launched an X-RAY search between bearings 272 and 056 degrees true out to four hundred miles. One night fighter and one attack plane were assigned to each sector, except that airborne-early-warning aircraft were substituted for attack planes in Sectors 38, 42, 1, and 5. At 1927, a "bogey" (an unidentified air contact) was reported bearing 335 true at an altitude of ten thousand feet and a range of 105 miles. At 1930, the AEW plane in Sector 42 reported radar contact on a surface force at 38° N, 162° E. This force was estimated to be composed of nine large and sixteen small ships proceeding on 170 degrees at about twenty knots. BC's flagship was then at 31°10′ N, 165°55′ E. The center of HTF-1 was thirty-five miles to the southward; Blue was steering 280 at fifteen knots. Weather in the vicinity of TF 11 at 1930 on 25 August was wind from the southeast at Force 3, smooth seas, unlimited visibility, and two-tenths cloud cover. There was to be no appreciable change in weather for the next twenty-four hours. (Pages 5–6.)

Each student officer was to submit Vice Admiral BC's Estimate of the Tactical Situation as of 1930 on 25 August as well as his own Solution to the Problem. (It was the Staff Solution, summarized below, that would actually be played.) Students were to use data previously compiled and carried in the Running Estimate. Each student officer was also to submit a Commander, Task Force 11 Action Plan as of 1930 on 25 August and a Cruising Disposition for TF 11. The student officer ultimately selected to play CTF 11, assisted by members of his section, was also to submit any additional dispositions considered necessary for TF 11 as well as a Cruising Disposition for the convoy and the Escort. Modified or new Battle and Approach Dispositions for Phase Two were to be submitted at the conclusion of Phase One, the Chart Maneuver. There were also to be submitted a Communication Plan—to be devised in accordance with the Maneuver Rules—any additional annexes desired, flimsies and Aircraft Flight Forms covering all airborne aircraft at 1930 hours, and flimsies and Aircraft Flight Forms covering the movement of surface forces and aircraft during Move 1. Move 1 was to occur at 1930 on 25 August, game time. Students were to employ specific Plotting Sheets for strategic and tactical plotting. (Page 7.)

Forces assigned to TF 11 included the three battle carriers of Rear Admiral BE's Carrier Division 2, the three fast battleships of Rear Admiral BF's (Commander Baldridge) BATDIV 1, and five heavy cruisers from Rear Admiral BG's (Commander Baldridge) CRUDIV 1. Commodore BD's (Commander Baldridge) DESFLOT 1 would consist of antiaircraft cruiser CLAA-95 as the flagship; four radar-picket destroyers from Captain B-5's (Commander Morton) Destroyer Squadron 5; nine destroyers from Captain B-9's (Commander Bridewell—see chapter 9) Destroyer Squadron 9; and ten destroyers from Captain B-13's (Commander White) DESRON 13. All of Blue's destroyers were equipped with Type G-3 torpedoes. The three battle carriers each had air wings of two fighter squadrons. Two of the three carriers also had two attack squadrons, the third only one. Nevertheless, each carrier had the same number of planes. These consisted of fifty-three Corsair fighters, four Corsair photographic reconnaissance fighters, eight Corsair night fighters, thirty-two Helldiver attack planes, twenty-four Avenger attack planes, four Avenger tactical electronic-warfare aircraft, and four Avenger AEW aircraft. In addition, BATDIV 1 had six observation scout planes, while CRUDIV 1 carried ten additional planes of this type. Blue aircraft were armed with Type E-1 aircraft torpedoes. (Pages 8–9.)

Convoy HTF-1's route was from Pearl Harbor to Points TIN (22°35′ N, 159°28′ W), STEEL (30°00′ N, 178°00′ W), COPPER (31°00′ N, 168°30′ W), LEAD (33°10′ N, 144°00′ E), and thence to Tokyo. Its composition consisted of the Convoy Commodore—Rear Admiral BH (Commander Morton)—in amphibious force flagship AGC-7, the Convoy Vice Commodore—Commodore BV (Commander Webber—see chapter 8)—in attack transport APA-27, twenty attack transports in total, twelve attack cargo ships, seven Navy transports, four oilers, one ammunition ship, one provision ship, and eleven merchant ships. Also part of this force were two escort carriers loaded with Army Air Forces fighters; these planes, loaded by crane, could neither launch nor recover from the escort carriers. Forces assigned specifically to Commodore BJ's Task Group 51.7 included the antiaircraft cruiser CLAA-121, from which he flew his flag, five destroyers and four radar-picket destroyers of DESRON 1, nine destroyers from DESRON 7, and six destroyer escorts from Escort Division 1. (Page 9.)

	Type A.	Type B.
Endurance Surface (Miles-kts) on Diesel Engines.	15,000-10 kts 5,100-16 kts	7,300-10 kts
Endurance Submerged on Schnorchel (Miles-kts)	16,000-7 kts	7,500-7 kts
Submerged on Batteries (Miles-kts)	365-5 kts* 170-8 kts 110-10 kts 22-18 kts	100-4 kts* 55-6 kts
Submerged on Walther Turbine (Miles-kts)	None	224-15 kts 158-24 kts
Maximum Speed (Surface/Submerged)	16/18	11/24
Diving Time to Periscope Depth	20 Seconds	20 Seconds
Maximum Depth with 2½ Safety Factor	395 Feet	415 Feet
Torpedo Tubes	6 Bow.	4 Bow 6 Side
Torpedo Capacity	20	10

(Torpedoes believed similar to type F-3)

General:

All submarines are equipped with mast and periscope surface search radars and with mast air search radar. Air search and periscope surface search radars can be used both on surface and at periscope depth.

Radar interception is practicable on surface or at periscope depth.

Radio reception and transmission is practicable both on surface and at periscope depth.

Percent of battery depletion, while running at any given speed, will be computed as a direct relation between number of minutes at that speed and the endurance at that speed.

While charging batteries these submarines cannot make more than 7 knots.

Battery charging rates are as shown in "Maneuver Rules."

*Silent running speeds, comparable in noise emitted to 2-3 knots silent running speeds of conventional Fleet Type Submarines.

Fig. 129
Estimated Characteristics of Purple High-Speed Submarines, Operations Problem 2J

Blue knew that all the Purple high-speed submarines were equipped with mast- and periscope-mounted surface-search radars as well as mast-mounted air-search radar. All air-search and periscope surface-search radars could be used both on the surface and at periscope depth. Radar interception was considered "practicable" on the surface or at periscope depths, as was radio reception and transmission. The percentage of battery depletion while running at any given speed was to be computed as a direct relation between the number of minutes at that speed and endurance at that speed. While charging batteries, Purple submarines could not make more than seven knots; battery-charging rates were shown in the Maneuver Rules. Noise at silent-running speeds was comparable to that emitted at two or three knots, silent running, by conventional fleet-type submarines. (Page 10.)

The Blue Chart Maneuver

For the Blue Chart Maneuver, Task Force 11's Covering Force had student officers playing CTF 11 (Vice Admiral BC) and CTF 11's Chief of Staff (Lieutenant Colonel Wallace—see chapter 9); Commander, Battleship Division 1 (Rear Admiral BF); Commander, Cruiser Division 1 (Rear Admiral BG); and Commander, Destroyer Flotilla 1 (Commodore BD). In addition, student officers played Commander, Carrier Division 2 (Rear Admiral BE), his Chief of Staff (Colonel Manzo—see chapter

9), and his Operations Officer (Commander Rogers—see chapter 9). Players were assigned as commanders of each of the three carrier battle groups, built around CVB-41, -42, and -43 (again Lieutenant Colonel Downey, Commander Bridewell, and Commander Kirkpatrick). Finally, student officers played HTF-1's Convoy Commodore (Rear Admiral BH) and Convoy Vice Commodore (Commodore BV), as well as Commander, Task Group 51.7, the Escort Force.[5]

Vice Admiral BC's Task Organization laid down that Rear Admiral BE would command the Carrier Group, the three battle carriers of CARDIV 2. Rear Admiral BF would lead the Heavy Striking Group, three fast battleships from BATDIV 1 and nine destroyers from DESRON 9. Rear Admiral BG would command the Light Striking Group—the three heavy cruisers of CRUDIV 1 and nine additional destroyers from DESRON 9. Finally, Commodore BD would lead the Screen, which consisted of the antiaircraft cruiser and the twenty-three destroyers from DESFLOT 1. In this scenario, on the evening of 25 August, there would be bright moonlight after 2200. Since Blue forces had already been attacked by Purple submarines and were probably being shadowed by them, Admiral BC thought there was good probability that a detached Purple surface striking group was in the area but had not yet been detected. An unidentified plane had recently been detected near his force.[6]

BC knew that the Purple Main Body of ten large and twenty small ships had sortied from Kashiwabara Bay during the night of 23–24 August. He also knew that at 1930 on this date, a Blue search plane had reported the Purple Main Body at 38° N, 162° E, on course 170 at a speed of twenty knots, by this time consisting of nine large and sixteen small ships. Purple was known to have in Far Eastern waters on 10 August an even greater force, consisting of three fleet carriers, three fast battleships, six heavy cruisers, and four squadrons of destroyers. A sighting at Kashiwabara Bay on 14 August revealed two fleet carriers and three fast battleships. As for his own forces, there were no Blue submarines in the area, no reinforcements immediately available, and no land-based air forces to assist. Admiral BC assumed that Purple would press home an attack that night or the next morning on the Blue Covering Force and that these surface attacks would be supported by wolf-pack submarine attacks in the evening. He also assumed that Purple had a detached, high-speed surface striking group in the area that was still undetected and that Purple's ultimate objective was, of course, to prevent HTF-1 from getting to Japan.[7]

Therefore, the Blue Covering Force would first divert the convoy to the southwest and then deliver early, repeated, and overwhelming air attacks on the Purple forces. These strikes would be followed by a rapid closure of the range; Admiral BC would accept a day battle if it was necessary to annihilate the Purple surface forces. He would also conduct continuous air searches to detect other Purple surface units in the immediate area. These air searches would commence at 2000 on 25 August; meanwhile, the Blue Covering Force would remain in a position to protect HTF-1.

The convoy's withdrawal to the southwest would take place at night, and the Covering Force would gain a position downwind within thirty-five miles of the convoy and between it and the reported Purple striking force. The Covering Force would be prepared to launch major air strikes and conduct searches to the south and west during the night. Additionally, a fast battleship striking force could be formed at any time during day or night to repel any surprise Purple surface encounter. This plan was to be effective at 1930 on this day. Commander, Task Force 11 was embarked in fast battleship BB-57 and Rear Admiral BE, embarked in CVB-41, was designated as the second in command.[8]

BC's Task Organization had Rear Admiral BF commanding the Battleline, the three fast battleships from BATDIV 1, as well as the Center Force, one heavy cruiser. Rear Admiral BG commanded two heavy cruisers from CRUDIV 1 and ten destroyers from DESRON 13, together constituting the Right Flank Force. The Left Flank Force, commanded by Commodore BD, consisted of the two remaining heavy cruisers from CRUDIV 1, the antiaircraft cruiser, and nine destroyers from DESFLOT 1. Rear Admiral BE commanded the Carrier Group, the three battle carriers from CARDIV 2 and four radar-picket destroyers from Destroyer Squadron 5.[9]

Blue carrier aircraft were to attack the reported Purple surface forces and scout for any additional Purple surface forces to the west and south at dawn on 26 August. HTF-1 was to remain outside the reach of Purple carrier-based air. BC assumed that Purple surface forces would be reduced in strength by his carrier air strikes and that a detached surface force would have been detected and dealt with. He also assumed that no undetected Purple surface forces were within threatening distance of HTF-1 since searches had come up negative. Additionally, he envisioned a daylight battle, good weather (aircraft could fly and smoke would lie), and encounters with high-speed Purple submarines. He planned to have his battle line initially engage the Purple battle line at extreme range and close as rapidly as possible to between eighteen and twenty-two thousand yards, and then again to moderate ranges in conjunction with destroyer torpedo attacks. The battle line would provide its own Air Spot and Inner Anti-Submarine Patrol. The Center Force was to scout the enemy and report its dispositions, then retire and join the Van as reinforcement. The Right Flank Force, if in the Van, would defend the battle line against Purple light-force attacks. The Right Flank cruisers were also to support the destroyer torpedo attacks against the Purple battle line when the range closed to twenty thousand yards. Cruiser planes were to furnish the Air Spot and Intermediate Anti-Submarine Patrol. The Right Flank Force, if in the Rear, would defend the formation from Purple light forces.[10]

If the Left Flank Force found itself in the Van, it would also defend the Blue battle line against Purple light-force attacks. Otherwise, it was to defend against those light-force attacks from the Rear. The Carrier Group would furnish its own antisubmarine and combat air patrols as well as the combat air patrol for the battle line. It was additionally to coordinate air strikes against the Purple battle line. After deployment, it would operate on the disengaged side of the Blue battle line—outside the main-battery gun range but within supporting distance of HTF-1—so as to give it air cover. If that was not feasible, the Carrier Group was to detach one battle carrier with an escort to join and operate with the Blue Escort Force. The direction of the deployment would be signaled and all planes were to report and attack all submarines that had been sighted and could be struck. Rear Admiral BF, Commander, Battleline, was designated as second in command in BB-58. CTF 11 was embarked in BB-57.[11]

1. Word Code

North	Winter
South	Summer
East	Spring
West	Fall

(May be used combined for additional directions)

Carriers	Frenchman
Battleships	Italian
Cruisers	German
Destroyers	Dutchman
Deploy	Pope
Right	Cardinal
Left	Priest
Attack!	Panther
With Torpedoes	Leopard
With Gunfire	Dill Pickle
With Aircraft	Ambitions
Illuminate	Amphatheatre
With Searchlights	Hill Billy
With Starshells	Loganberries

2. TRUK Code

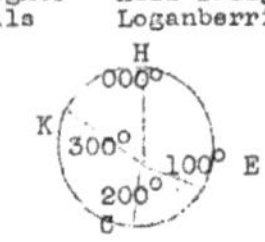

3. Shackle Code

∅	ZVWQJ
5	KHFG
1	XUC
2	YB
3	RE
4	MO
6	AL
7	TN
8	DP
9	IS

4. Authentication -- Use either of first two letters from last word of message. Enter the Shackle Table above with either letter. Any three adjoining letters become the authenticator.

Fig. 130
Blue Covering Force communication codes, Operations Problem 2J

All units were to maintain radio silence but not after contact. Radio frequencies consisted of numerous functional circuits, including a Talk Between Ships Emergency Tactical Voice Circuit, a Task Group Commanders Voice Command Channel, and a Task Force Common Circuit. There were also the Inter-Fighter Director and Tactical Voice Circuits for the battle line, the Center Force, the Right and Left Flank Forces, and the Carrier Group. Additionally, there were General Warning Voice, Air Spot, Air Search and Airborne Early Warning, Fighter Director, Strike Control, and Task Force Combat Air Patrol Common Circuits. There was no radar silence, and all ships and aircraft were to have their Identification Friend or Foe transponders on. Guard ships were also to have their interrogator-responder units on and ready for actuation at all times. BB-58 and BB-59, as well as heavy cruisers CA-130 and CA-131, were designated as radar intercept guard ships. The two fast battleships, along with CA-73, -74, -75, and -95, were designated as radio intercept guard ships. As was standard procedure, the Plan put the control of radar jamming and radio deception under the Officer in Tactical Command.[12]

OTC	AVENGER
Combatdiv 1	GLORIOUS
Comcrudiv 1	MASSACRE
Comcardiv 2	ACE HIGH
Comdesflot 1	ARGUS
Comdesron 5	SPEEDY
Comdesron 9	ACTIVE
Comdesdiv 92	ALERT
Comdesron 13	WHITE HORSE
Comdesdiv 132	BARKEEP
Battleline	HARDY
Center Force	BANDIT
Right Flank Force	RANDOM
Left Flank Force	HERBERT
Carrier Group	EVADER
Convoy Commodore	ANXIOUS
CTF 51.7	AGGRESSIVE
Comdesron 1	BASEBALL
Comdesdiv 12	BELL HOP
Comdesron 7	BAT BOY
Comdesdiv 72	BRANCH BANK
ComCortdiv 1	AESOP
AGC	LEARNED
APA	GRIFFITH
AKA	RENTAL
AP	ROWDY
AO	ANTHRACITE
AE 16	ARBUTUS
M/V	ANTHONY
CVE's	ANDIRON

For individual ships preface hull numbers by respective type or organizational call.

Fig. 131
Blue Covering Force voice calls, Operations Problem 2J

The Blue Staff Solution to the Problem

The Blue Staff Solution to the Problem reiterated that by 1930 on 25 August, radar contact had been made on a large Purple force. As the contact had been made by a carrier-based AEW aircraft from a distance of about eighty-five miles, it was inferred that the information might be considerably in error. An attack cargo ship from HTF-1 had been sunk the day before by submarine attack. Purple had about sixty modern, high-speed submarines in Far Eastern waters when hostilities commenced, some on patrol in the Central Pacific. As of 1930, the Purple force was 455 miles distant and bearing 335 degrees true from his flagship, which was by this time at 31°10′ N, 165°55′ E. Admiral BC assumed that his force was known to Purple, a bogey having been reported at 335, 105 miles. Since Convoy HTF-1 and the Escort Force were only thirty-five miles to the south, their positions were presumably known as well. His current position was outside the effective range of both Blue and Purple land-based airpower.[13]

He interpreted his orders to ensure the safe arrival of Convoy HTF-1 in Tokyo at the earliest possible date as meaning without either serious damage or undue delay. The Commander-in-Chief of the Pacific Fleet had assigned nearly all combatant forces available to this mission. In light of his orders and the strength of his forces, Admiral BC discarded the option of withdrawing his force to the protection of air cover that was available to the southwest. Instead, he had to take positive action against the Purple forces that were coming into range. If he had to divert HTF-1, that diversion could only be temporary. He could, however, consider his mission successful if he merely drove off the Purple forces rather than destroying them, as long as they did not continue to threaten the convoy. If Purple forces remained within threatening range, however, Admiral BC would be forced to destroy them. The Purple force just reported did not yet pose a potential threat to the convoy but it could within a few hours, so its destruction would be his immediate objective. (Pages 2–4.)

Formal factors of the Situation that could possibly affect his own Course of Action were Relative Combat Power, Reinforcements, Time and Space, Hydrography, Weather, Moonlight and Daylight Periods, and Logistics. As concerned Relative Combat Power, he had several sources of information. First, intelligence reports listed Purple combatants in Far Eastern waters on 10 August as the three fleet carriers, three fast battleships, six heavy cruisers, and four squadrons of destroyers reported above. Second, air reconnaissance of Kashiwabara on 14 August revealed "many combatant" types, definitely including two fleet carriers and three fast battleships. Third, a Blue submarine patrolling off Kashiwabara had reported the departure at 1000 on 25 August of a Purple force of ten large and twenty small ships. Commander-in-Chief, Pacific Fleet surmised that this was the only Purple force at sea in the Pacific and his AEW plane report of nine large and twenty small ships was close to the mark. Considering all these reports, Vice Admiral BC deduced that Purple had at sea at least two or three fast battleships, two or three fleet carriers, four to six heavy cruisers, and sixteen to twenty destroyers. He limited the estimated number of Purple destroyers to twenty because of these reports, because Purple's preparations for additional amphibious operations would require destroyers, and because he knew that Purple had only about thirty-six destroyers in the Far East. Admiral BC surmised that Purple might have an additional force of one to three large ships and four destroyers. He took all these numbers into account in case his aircraft report was inaccurate. If a second force was present, it had probably sailed with the first force that had been sighted and could not now be more than a few hundred miles south of the initial contact. Still, because Purple was limited in its number of destroyers, Admiral BC thought it doubtful that a second force existed. Nevertheless, he could not entirely discount the possibility until more concrete information was received. (Pages 5–6.)

Purple's carriers were known to carry about a hundred aircraft, of which about fifty-five were fighters. The fighters included night fighters, and the carrier complements included attack, AEW, and tactical electronic-warfare planes. If Purple had three fleet carriers, Blue had about twenty more fighters and forty-five more attack planes. If Purple only had two fleet carriers, Admiral BC's superiority would be about seventy-five fighters and ninety-five attack planes, respectively. He also estimated that the Purple fleet carriers carried about the same number of night fighter, AEW, and tactical electronic-warfare planes as did his battle carriers. In total, if he had three carriers to Purple's two, he would have 381 planes compared with two hundred. If numbers of carriers were even, he would still have 381 aircraft to Purple's three hundred. (Pages 6–7.)

	BLUE	PURPLE	
TYPE	3 CVB's	3 CV's	2 CV's
VF	165	140 to 150	95 to 100
VF(N)	24	24	16
VA(DB)	96	70 to 75	45 to 50
VA(TB)	72	36	24
VA(AEW)	12	12	8

Fig. 132
Blue versus Purple aircraft, Operations Problem 2J

Since Blue submarines were on combat and reconnaissance patrols in the Yellow Sea, the Sea of Japan, and the Kuriles but were going to be rerouted away from his forces, Vice Admiral BC assumed that any submarines encountered would be enemy ones. He understood that on about 10 August, Purple's sixty high-speed submarines were on patrol in the Hawaiian, Central Pacific, and Aleutian areas but his information about the location and employment of Purple submarines was limited to that and the fact that his forces had been attacked the previous day. He had no means of estimating the number of Purple submarines deployed against him but had to assume that it was substantial. He also surmised that these submarines were deployed at or near focal points on his route and that Purple would have had "ample" time to concentrate them. If Purple's surface and submarine forces were operating in tactical concentration, he thought it probable that the submarines would be deployed in the Van of Purple surface ships, so as to avoid mutual interference. If

Purple submarines were operating independently, he assumed that Purple would follow the World War II German pattern of concentrating wolf packs along the convoy route, especially in midocean, where Blue land-based airpower would be minimal. (Pages 7–8.)

Certain characteristics of the Purple submarines were worthy of special consideration. In particular, the maximum speed on the surface of Purple's Type A submarine was only sixteen knots and the submerged speed eighteen knots. The Type B submarine was capable of only eleven knots on the surface but twenty-four knots submerged. The submerged endurance of a Type A was 365 miles at five knots, 170 miles at eight knots, 110 miles at ten knots, and twenty-two miles at eighteen knots. The submerged endurance of a Type B was 224 miles at fifteen knots and 158 miles at twenty-four knots. Both types had very extensive ranges when submerged on a *schnorchel,* but their maximum speed when operating in this manner was only seven knots. He concluded that the Type A submarines could not make an end run around his convoy and that Type B could do so only when submerged. In view of the limited endurance of the Type B at high submerged speed, he regarded this possibility as remote; he deduced that his losses in submarine waters would be proportional to the time he spent in those waters: "My losses may be held to a minimum by 'bulling through' submarine infested waters." (Pages 8–9.)

Rear Admiral BC's midocean position, outside the effective range of friendly land-based air, was also outside the range of Purple land-based aircraft operating out of the Kurile Islands. An advance of about six hundred miles to the westward would bring his forces within range of Blue land-based air on Honshu, but after advancing only about five hundred miles his force would be within range of Purple land-based air on Etorofu, in the Kuriles. These latter aircraft were believed to be similar to Blue PBM Mariners so they would not constitute a serious threat at a distance of more than 750 miles. Still, they were capable of patrol operations out to a thousand miles. The naval patrol land-based air support available to him was questionable. The 20th Army Air Force had suffered serious losses and its combat tasks might preclude any assistance to Admiral BC's forces. Also, unless additional atomic bombs were flown separately into Japan, the 20th Army Air Force would not have any until TF 11 arrived. Accordingly, he did not consider land-based airpower to be a decisive factor in the upcoming operations. (Pages 9–10.)

CARRIERS	BLUE		PURPLE
CLASS	CVB-41	Similar	CV-9
NUMBER	3		2 to 3
MAX. SPEED	33		33
SUST. SPEED	33		31
AIRCRAFT COMPLEMENT	127	Approx.	100
LIFE	13.6		10.6
A.A. BATTERY & RANGE	18 - 5"/54 23,000		12 - 5"/38 16,000
AUTO WEAPONS	84 - 40MM 56 - 20MM		72 - 40MM 70 - 20MM

Fig. 133
Combat efficiency of Blue and Purple materiel, Operations Problem 2J

BC considered his battle carriers to be capable of higher sustained speed than Purple's fleet carriers and superior in their ability to withstand damage and put up antiaircraft fire. He also considered the two forces equal in operating efficiency. He additionally considered his force generally equal to Purple in numbers of battleships; that was true also of heavy cruisers, with a possible difference of one ship either way. His antiaircraft cruiser merely augmented his destroyers in air and torpedo defense and in antisubmarine work; having no torpedoes, it was of small value for offensive purposes in a fleet engagement. Finally, in the absence of more information, he assumed that all of Purple's destroyers were equipped similarly to his of the 692 class, with ten torpedoes each. (Pages 10–12.)

BATTLESHIPS	BLUE		PURPLE
CLASS	BB-57	Similar	BB-57
NUMBER	3		2 or 3
MAX. SPEED	27		27
SUST. SPEED	25		25
ARMOR	16"/6".25		16"/6".25
LIFE	22.6		22.6
MAIN BATTERY & RANGE	9 - 16"/45 34,000		9 - 16"/45 34,000
SEC. BATTERY & RANGE	*20 - 5"/38 16,000		20 - 5"/38 16,000
AUTO. WEAPONS	68 - 40MM 76 - 20MM		68 - 40MM 76 - 20MM
PLANES	2 VOS		2 VOS

*South Dakota 16 - 5"/38

Fig. 134
Blue and Purple battleship characteristics, Operations Problem 2J

Admiral BC also took his destroyers to be quite similar to Purple's in general performance but calculated that the Purple destroyers had about two hundred torpedoes to his 160, while he had at least three and possibly seven more destroyers than Purple. Further, he had more five-inch guns in his light forces for defense against destroyer torpedo attacks than did his adversary. In addition, Blue Destroyer Division 51, four *Gearing*-class destroyers, was equipped with the latest radar, radar countermeasures, and Combat Information Centers, factors that he thought made these ships particularly well suited as radar pickets. (Page 13.) In terms of antiaircraft weapons, Admiral BC had 320 five-inch guns in TF 11; Purple might have as many as 298 similar guns or as few as 208. His 40 mm guns numbered 996 to Purple's estimated 664–948.

HEAVY CRUISERS	BLUE		PURPLE
CLASS	CA-68	Similar	CA-68
NUMBER	5		4 to 6
MAX. SPEED	32		32
SUST. SPEED	31		31
LIFE	5.5		5.5
MAIN BATTERY & RANGE	9 - 8"/55 28,000		9 - 8"/55 28,000
SEC. BATTERY & RANGE	12 - 5"/38 16,000		12 - 5"/38 16,000
AUTO. WEAPONS	48 - 40MM 22 - 20MM		48 - 40MM 22 - 20MM
PLANES	2 VOS		2 VOS

Fig. 135
Blue and Purple heavy cruiser characteristics, Operations Problem 2J

No reinforcements would be available to either side. Still, there was a possibility that "unaccounted for" Purple forces could be within supporting distance of the reported Purple force, but the Purple commander would gain nothing by dividing his limited surface forces. Admiral BC planned on guarding against that possibility, but he was confident that the reported Purple force was the only one out there. (Page 13.)

The factors of time and space were important. The need for the reinforcement in Japan was urgent and delay would allow Purple to deploy submarines more effectively. In addition, he wanted to come within Blue land-based air range as soon as possible. The Purple Striking Force was now about five hundred miles north-northwest of HTF-1. At twenty-five knots, it could be about 210 miles from the convoy at 0430, if the convoy and its escorts continued on 280 degrees at fifteen knots. Purple torpedo planes were thought to have a radius of 425 miles at 130 knots true airspeed, assuming equal distances flown both before and after dropping the torpedo. To simplify calculations, he assumed the Purple torpedo planes could fly a total of 850 miles at 130 knots—that is, for 6.54 hours. Assuming a wind component of twelve knots along the planes' track and a

AA CRUISER	BLUE	PURPLE
CLASS	CL-95	None
NUMBER	1	
MAX. SPEED	32.5	
SUST. SPEED	30.0	
LIFE	3.3	
A.A. BATTERY & RANGE	12 - 5"/38 16,000	
AUTO. WEAPONS	24 - 40MM 16 - 20MM	
TORPEDOES	None	
PLANES	None	

Fig. 136
Blue antiaircraft cruiser characteristics, Operations Problem 2J

DESTROYERS	BLUE		PURPLE
CLASS	DD-692	DD-710	DD-692*
NUMBER	9	14	16 to 20
MAX. SPEED	34		34
SUST. SPEED	32		32
LIFE	1.6		1.6
GUNS (D.P.) & RANGE	6 - 5"/38 16,000		6 - 5"/38 16,000
AUTO. WEAPONS	12-40MM 20-20MM		12-40MM 20-20MM
TORPEDO TUBES	10 C.L.	5 C.L.	10 C.L.*
TORPEDOES	BLUE torpedoes are 24" wakeless type with a range of 26,000 yards at 50 knots and 40,600 yards at 40 knots. PURPLE torpedoes are believed to be similar.		

Fig. 137
Blue and Purple destroyer characteristics, Operations Problem 2J

(Purple) carrier speed of twenty-five knots, he thought that these planes had to fly 4.17 hours out at ninety-three knots relative speed and 2.37 hours back at 167 knots relative speed. In the same way, with their carrier closing at thirty knots, the times would be 4.3 hours out and 2.24 hours back. With Blue forces on course 225 degrees, Purple planes under present wind conditions could close at 109 knots in a 4.17-hour run if they were launched at 456 miles from Blue and a 4.3-hour run if launched 470 miles from Blue. With the Purple carriers closing at twenty-five knots and Blue forces still on course 225 degrees, Purple could close to 456 miles from the Blue forces by about 2155; an attack launched at that time could not reach Admiral BC's forces before 0205. Under the same conditions, an attack launched against TF 11 at 1945 would arrive at about 2355. (Page 14.)

With the enemy carriers closing at thirty knots and with Blue forces still on course 225 degrees, Purple could launch an attack as early as 2040, with the strike arriving over TF 11 at about 0058. An attack against TF 11, in other words, could be launched almost immediately and arrive shortly before midnight. He stressed, however, that Purple would have to continue its course and speed after launching and until its planes were back on board. If Blue forces turned south, Purple would have to close to 430 miles if it steamed at twenty-five knots or 443 miles if at thirty knots to launch a torpedo attack. At twenty-five knots, the Purple commander could close TF 11 to 430 miles by 0200 but an attack launched at that time could not reach the Blue forces during darkness. At thirty knots, the Purple commander could close to 443 miles by 2300 and attack at that time so as to see the strike arrive over TF 11 at about 0320. (Page 15.)

Admiral BC acknowledged that his analysis deliberately stretched the Purple torpedo planes' capabilities slightly to allow for maximum possible launching ranges. Therefore, it appeared that by turning Task Force 11 south during the night, he could preclude the possibility of a night torpedo attack unless Purple detached one or more carriers to close him at thirty knots. Even in that case, he surmised that an attack could not be made prior to 0320 and that the Purple force would have to continue to approach TF 11 at thirty knots until about 0530 to recover its planes. By turning TF 11 slightly toward the Purple Striking Force shortly before dawn, he thought that he could be well within a 335-mile striking range by 0500 and launch a full strike against the Purple carriers. He admitted, however, that his analysis did not include one-way or suicide missions by Purple, which were possible at Purple's present distance from TF 11. (Pages 15–16.)

To facilitate antisubmarine and combat air patrols for TF 11, Admiral BC wanted his carriers to remain as close as they could to Convoy HTF-1. At distances between the convoy and the Blue Covering Force up to thirty miles, it was practicable for a combat air patrol to protect effectively both forces. At distances up to 150 miles, it was considered practicable to maintain a combat air patrol and antisubmarine patrol over the convoy, with reliefs on station every two to two and a half hours. (Page 16.)

His air search to the northeast would cover the area from which a second but smaller Purple force would have to approach in order to close to within air strike range before midday on the next day. He was again assuming here that Blue forces could advance west and south at fifteen knots. He did not think that the air search entirely eliminated the possibility of a second Purple force to the west; a second force might in fact remain undetected and be within striking range of HTF-1 at dawn in the sector between southwest and northwest. Given these considerations of time and space, he concluded that he had to eliminate the possibility of a second force to the west by a search launched prior to 0100. He also determined that a change of course to the south was necessary to avoid aerial torpedo attacks at night. Admiral BC additionally thought that he could safely resume his westerly course at dawn and that, given his air superiority, he could interpose the Blue Covering Force at such a distance from Convoy HTF-1 that effectively precluded Purple strikes on it. If tracking the enemy force at night disclosed that it was not attempting to close his force, he could resume his westerly course without danger of a night torpedo attack. (Pages 16–17.)

NOTES

1 Junior Class of June 1947, "Operations Problem 2J: The Schedule of Events," 29 October 1946, pp. 1–2, folder 2604, box 139, RG 4, NHC.

2 Junior Class of June 1947, "Operations Problem 2J: Maneuver Staff," 18 November 1946, p. 1, folder 2604, box 139, RG 4, NHC.

3 Ibid.

4 Junior Class of June 1947, "Operations Problem 2J: The Blue Statement," 29 October 1946, p. 1, folder 2604, box 139, RG 4, NHC. Subsequent in-text page references until the next endnote are to this source.

5 Junior Class of June 1947, "Operations Problem 2J: Chart Maneuver," 18 November 1946, p. 1, folder 2604, box 139, RG 4, NHC.

6 Junior Class of June 1947, "Operations Problem 2J: Blue—Section A," 18 November 1946, p. 1, folder 2604-E, box 139, RG 4, NHC.

7 Ibid.

8 Ibid., p. 2.

9 Junior Class of June 1947, "Operations Problem 2J: Annex (A) Battle Plan," 18 November 1946, p. 1, folder 2604-G, box 139, RG 4, NHC.

10 Ibid., pp. 1–2.

11 Ibid., p. 2.

12 Junior Class of June 1947, "Operations Problem 2J: Annex (F) Communication Plan," 18 November 1946, pp. 1–2, folder 2604-F, box 139, RG 4, NHC.

13 Junior Class of June 1947, "Operations Problem 2J: The Blue Staff Solution," 21 November 1946, pp. 1–17, folder 2604-O, box 139, RG 4, NHC. Remaining page references in this chapter are to this source.

CARGO UNLOADING
AREA
AND PRIORITY SUPPLY
SALVAGE
65
70
60

XVIII *Operations Problem 2J*

The Blue Solution Continued and Concluded, October–November 1946

Admiral BC expected effective sonar ranges of 1,500 to two thousand yards in the area of operations. With a "fairly calm" sea—which existed at present—1,500 yards was a "fairly safe" prediction. Surface sea temperatures would be about 79 degrees Fahrenheit, and no change was expected in these conditions for the next twenty-four hours. He therefore concluded that the conditions were favorable for Purple submarine operations, in that the time available for his surface escorts to conduct an attack after making sonar contact on a high-speed submerged submarine would be "extremely short."[1]

The direction of the wind was favorable to Purple, in that its commander could close Task Force 11 while both launching and recovering aircraft, though the radius of action of his planes was reduced by the necessity of carrying heavy bomb loads against a headwind. By the same token, Blue would have difficulty closing Purple but would have the advantage of a tailwind when planes were loaded. Clear visibility and unlimited ceiling would be an advantage for Purple if the Purple commander contemplated night torpedo attacks. Clear visibility, however, would assist fighter interception for both sides. He also thought that the advantages of a smooth sea for submarines and for the antisubmarine units were about equal. Smooth seas, for instance, introduced temperature gradients that enhanced a submarine's chances for an undetected approach at periscope depth or deep. The same sea conditions, however, enhanced visual or radar detection of a submarine near the surface. (Pages 17–18.)

As concerned daylight and dark periods, sunset on the 25th was 1700, while sunrise on the 26th was 0445. Moonrise on the 25th was 2200 and moonset on the 26th was 1200. The moon was in its last quarter with a nineteen-degree north declination. Admiral BC had been concerned over Purple night air attacks and now grew even more so given the clear visibility during daylight and the illumination provided by the moon at night. The overall advantage was still with the attacker. Night aerial torpedo attacks against targets silhouetted against a low moon were far more likely to be successful then were interceptions of the attacking aircraft by night fighters. Considering the advantages that accrued to the defense from forcing a submarine to remain submerged, however, illumination from the moon was a decided disadvantage for the Purple submarines. He assumed that Purple submarine radar was sufficiently developed to provide all the target information required but that conditions permitting visual observation would aid the defense more.

If the action was not prolonged, he did not think that logistics would become an important factor. All of TF 11's ships had been topped off that morning, and Convoy HTF-1's escorts had been fueled the day before. However, since the Purple Striking Force had departed Kashiwabara only two days ago, Admiral BC assumed that the enemy's logistics situation was favorable too. (Pages 18–19.)

Vice Admiral BC saw his own forces as superior in total numbers of aircraft. In addition, he had aircraft carriers that were capable of higher sustained speed and better able to resist damage. Because of his greater number of planes, he also saw his force as superior in terms of antiaircraft defense and ability to retain fighters for defense in an exchange of air strikes. He additionally had the superiority of having an antiaircraft cruiser, having at least three more destroyers than Purple, and perhaps having as many as seven more destroyers. Blue's bombers could carry bomb loads farther owing to tailwinds en route to the target. Weaknesses included a lack of freedom of action due to the necessity of protecting the convoy and the need to get it to Japan as soon as possible. He also had to use some of his strike aircraft in a search role, had to maintain antisubmarine and combat air patrols over both his own force and HTF-1, and faced poor sonar conditions in which to try to detect Purple submarines. Purple's strengths included an unknown number of high-speed submarines and unlimited freedom of action and initiative. Purple also might have forces yet unlocated by Blue, whereas Purple knew the locations of all of Blue's forces. A predominant southeast wind permitted Purple to close with Blue while conducting flight operations, and weather and visibility conditions favored submarine and night air attacks. Purple surface forces also had sustained speeds ten knots faster than the Blue convoy, and Purple had a possible superiority of forty torpedoes in its destroyers. However, Purple was inferior in terms of its number of aircraft and in light forces, and it had to accomplish its mission in the next two days or be subjected to land-based air attack. (Page 20.)

Admiral BC concluded that the Purple strengths most likely to be decisive were the offensive power of its submarines and the fact that the Purple commander had such freedom of action and initiative. He expected these factors to be most "skillfully" exploited by Purple. Blue factors most likely to be decisive were its air superiority and the opportunity to exploit that superiority under favorable conditions. The distance separating the Blue and Purple forces should, he thought, enable him to avoid combat during the night, and once his force came under the protection of Blue land-based air forces, six hundred miles to the west, Purple would be forced to act under unfavorable conditions. Although he considered himself greatly handicapped by the need to protect HTF-1, he thought that his forces were adequate and that conditions were favorable to the accomplishment of his mission. (Page 21.)

Looking at things from Purple's viewpoint, and given its past success in the employment of secret agents in Blue territory, Admiral BC assumed that Purple's intelligence on the movement of Blue forces was excellent. He also assumed that the importance of the prompt and safe arrival of his convoy in Japan was known to Purple. The commander of the Purple forces to the north was in a position to force an engagement that might conceivably result in the destruction of HTF-1, which would "greatly enhance" Purple's prosecution of the war. The Purple commander had also to be aware of the superiority of Admiral BC's forces, with the exception of submarines. The Purple commander, he thought, would therefore be counting heavily on Blue's lack of knowledge concerning the location and employment of the Purple submarines. The Purple commander's actions would be influenced by time and space considerations. In the Purple Estimate, it would be vital to prevent BC's forces from getting into range of Blue land-based air. Moreover, Purple could not be certain that Blue land-based air in Japan had expended all its atomic bombs. BC accordingly thought it logical that Purple's inferior force would attempt to gain equality or superiority by attrition attacks, especially by the tactical concentration of its high-speed submarines. In any case, he surmised that Purple had to realize that its force would be more and more at risk the closer that Admiral BC's forces got to Japan and that this reality would force Purple into early offensive efforts. (Pages 22–23.)

Admiral BC saw Purple's first task to be to attack the Blue Covering Force so as to get at Convoy HTF-1. BC listed Purple's inferable capabilities in terms of destroying Blue forces by either submarine, air, or surface attack, or in combination. Admiral BC's primary and only mission being protection of HTF-1, he had no speed advantage over Purple and could only close with Purple surface forces if the latter desired an engagement. Also, given his inferior speed, he could not evade Purple forces; he had to destroy them or drive them off. Under these circumstances, Admiral BC had to let Purple take the initiative. Purple submarine attacks seemed the primary threat to the Blue forces. He did not see Purple surface and air forces, in theory at least, as sufficiently strong to overwhelm the Covering Force and destroy HTF-1. In contrast, a large number of Purple's high-speed submarines could inflict serious damage on TF 11, and lacking positive intelligence, he had to assume that these submarines would be employed to the maximum of their capabilities and that there had been ample time to concentrate them along the convoy route. (Pages 23–26.)

The best defense against submarine attack would be, in view of the superior effectiveness of the Purple submarines, to prevent them from gaining favorable attack positions. It followed that Purple's ability to exploit the offensive strength of its submarines was largely dependent on its present position relative to the Blue forces. Admiral BC thought it advisable to change course south to 180 degrees true; he could hold a southerly course until dawn before resuming his westerly heading. He did not see this course change delaying his arrival in Japan for more than thirteen hours. He now revisited his antisubmarine air searches. Search Plan Love was now being flown, entailing 360-degree coverage by twenty-four Helldiver attack planes to a distance of 150 miles in the Van and on the flanks, forty miles astern. These planes were each loaded with two 350-pound depth charges and had orders to conduct hold-down tactics in the event of submarine contacts. The overall purpose was to prevent Purple submarines from approaching Blue forces on the surface. The difficulty of detecting submarines employing the *schnorchel,* especially at night, was admitted, but submarines so deployed could not exceed seven knots and would not be able to overtake the Blue convoy at that speed. Moreover, the Purple submarines, with their short submerged, high-speed endurance, could attack his convoy only from favorable initial positions. By changing his course to the south, Vice Admiral BC also prevented Purple from launching an effective air strike against Blue forces at night. He was similarly limited against Purple surface forces but this turn gave him the advantage of continuing to employ his attack planes on antisubmarine patrols. He thought his diversion to the south, therefore, would keep his losses to Purple submarines to a minimum. (Pages 26–28.)

After the Purple submarine strikes, the threat of Purple air attacks was considered the most dangerous. Because the two forces were still 460 miles apart, nothing but one-way attacks were currently possible. Since Blue forces had a superiority of at least eighty planes and possibly 180 over Purple carrier air forces, Admiral BC thought he should be able to inflict more damage on Purple than Purple did on him. He had to maintain combat air patrols over both the Covering Force and Convoy HTF-1 as well as extensive air antisubmarine patrols but he was going to abandon the latter if he had the opportunity to launch an "effective" strike against Purple. If he was to launch such a strike and prevent Purple from doing the same against him, continuous accurate information on the enemy had to be obtained and a condition of readiness for rapid launching of strike aircraft had to be maintained. He thought that he could obtain the necessary intelligence on Purple forces by keeping three airborne-early-warning aircraft in position as snoopers to track the Purple force. These aircraft were to be provided with a fighter escort during the day. (Pages 28–29.)

Admiral BC assumed that Purple's substantial air attacks would wait until it had eroded Blue air superiority by attrition attacks in the form of submarine and one-way air strikes. Purple might have submarines pick up aircrews that ditched after the attacks. To maximize their possibilities of these strikes, Purple would, Admiral

BC thought, coordinate them from a direction somewhat north of west within two or three hours of moonrise. Therefore, Blue forces would have to maintain a strong nighttime combat air patrol and adequate AEW picket coverage. Another possibility for employing Purple air strength was to use that strength for purposes of deception in an effort to draw Blue forces through submarine waters. Since Purple submarines could not conduct an end run around the Blue forces if the latter maintained course and speed, Purple might try in this way to get Blue to change course and speed so as to enable the Purple submarines to take up attack positions. Therefore, Admiral BC planned instead on maintaining course and speed and traversing submarine waters as quickly as possible. His air superiority would more than offset the Purple air threat and inflict "disproportionate" damage on Purple. (Page 29.)

The possibility of a surface engagement between Blue and Purple forces appeared remote to Admiral BC but he had to consider it. If Purple submarine and air attacks reduced Blue air superiority through the destruction or damage of BC's carriers, further submarine, air, and surface action could bring further destruction on the Blue Covering Force and Convoy HTF-1. If, however, he could prevent the initial damage, Admiral BC was certain that initial Purple failure would preclude a second, surface phase. Given the possible scenarios for a surface engagement, Admiral BC did not foresee any surface action until such time as carrier-based air on both sides had been eliminated or reduced to "impotency." He also assumed that ample time was available for modification or preparation of new surface-action plans in that case. Blue forces would not be diverted, and the necessity of interposing his surface forces between HTF-1 and the enemy would result in a normal action. He therefore devised three possible Plans of Action. Plan No. 1 assumed that combatant surface strength had not been reduced, except in carriers. Given his lack of definitive information on Purple forces, he could not make detailed evaluation of Purple strengths and weaknesses under this plan but he felt he had to consider Purple using part of its surface forces as a decoy to draw off the major Blue strength and then engage the convoy with the remaining majority of its surface forces. Here, battleship and cruiser scout planes would be most useful in limited search missions. (Pages 29–31.)

Under Plan No. 2, Blue surface strength was still superior. Under this scenario, Admiral BC could expect Purple to take advantage of all conditions or circumstances to neutralize Blue superiority, even seeking engagement at night or in reduced visibility. Efforts to evade the Covering Force and attack HTF-1 directly would be intensified, and the offensive capabilities of Purple submarines that could be brought into position would be exploited to the fullest. Under Plan No. 3—in which the remaining Blue surface forces were inferior—Admiral BC would be forced to fight a retirement action to protect HTF-1 to the limit of the Blue Covering Force's remaining capabilities. Admiral BC would try to hold a westerly course and would request a strike mission from Commander, 20th Army Air Force. He did not think that even in this scenario he would be inferior in light forces and he would entertain the thought of employing his destroyers in night attacks. He also had to consider that Purple might be able to coordinate its submarine and air attacks; he thought, however, that Purple could not coordinate all weapons simultaneously without first negating some portion of the Blue force and, ultimately, that Purple coordination of its forces could not prevent him from carrying out his mission. Still, he had to make provision for damaged ships. Any damaged ship capable of fifteen knots could join the convoy; others were to be sent to Guam or Eniwetok with a suitable escort, under tow if necessary. Air cover and tugs were to be requested from Guam or Eniwetok. (Page 32.)

In approaching his Complementary Problems, Admiral BC noted the tasks that had to be performed by his aircraft: antisubmarine operations, the search for and tracking of enemy surface forces, AEW patrols, defense against air attack, and the destruction of enemy surface forces by air attack. He understood that his air antisubmarine and air strike tasks were in conflict in that extensive employment of aircraft for antisubmarine use would reduce the number available for strikes. However, in view of the serious threat posed by Purple submarines and because of the unlikelihood of Purple surface forces coming within air striking range until Blue air superiority was reduced by Purple submarine and one-way air attacks, Vice Admiral BC was going to subordinate his air strike requirements, at least for the time being. He did, however, have to maintain an accurate track of Purple surface forces and be ready to launch strong air strikes when the opportunity presented itself. (Page 33.)

In terms of air antisubmarine operations, Admiral BC considered the chief value of his aircraft to be one of interference with attempts by Purple submarines to approach his forces. If he could force the Purple submarines to remain submerged, he might destroy them by air attack. More importantly, he thought, hold-down tactics would interfere to some extent with submarine communications.

Accordingly, he planned to use two types of antisubmarine patrols. Plan A entailed Search Plan Love, already described. These planes were to be launched at two-and-a-half-hour intervals. Plan B was a close-in antisubmarine patrol during daylight hours to a distance of thirty to forty miles outside the screens of both Task Force 11 and HTF-1. Twelve to eighteen Avenger attack planes would be airborne for this purpose during daylight hours. He planned to cancel this patrol if a Purple air attack was probable. In case of enemy air attacks while antisubmarine planes were airborne, these planes would be informed of the attack and directed to proceed immediately at minimum altitude to a distance of at least seventy-five miles from either Blue surface force. (Pages 33–34.)

Admiral BC planned to keep three AEW planes aloft during the night, launching them at three-hour intervals. These planes would employ snooper tactics, especially low-altitude flying to avoid interception. He realized that these tactics were inconsistent with the task of early warning but he thought that his planes could still detect Purple carrier flight operations. At dawn, these planes would retire to a safe distance, and four additional AEW planes—each with four fighters as escorts—would be employed during daylight hours. One of these planes would remain generally southeast of the present contact and at maximum effective tracking range while a second would remain southwest of the contact at the same range. The remaining two would be stationed 150 miles from TF 11, twenty degrees to either side of the general bearing. One of these two planes would be at twenty thousand feet and the other at ten thousand, alternating altitude assignments at two-hour intervals to relieve the crews on oxygen. To eliminate the possibility of a second Purple surface force going undetected, Admiral BC would also launch a reinforced X-RAY search out to a distance of four hundred miles between bearings 240 degrees and 312 degrees at 0115. For this search, he would employ eighteen fighters and nine attack planes, each attack plane protected by two fighters in each sector. This search would ensure the discovery of a second Purple force—if one existed—in time to launch air strikes against it at dawn. He also thought that his antisubmarine search would provide "ample" protection during the night against any attempted surprise surface attack. (Pages 34–35.)

BC did not think a Purple night air attack was very probable before about 0200. Still, he would launch four night fighters at 1945 for use against enemy search planes and snoopers, and at 0030 he would launch eight night fighters as a night combat air patrol, landing them at 0445. At dawn, he could then launch twenty-four fighters for a day combat air patrol, which could be increased to forty-eight fighters if Purple air forces were within striking range. About seventy fighters—including those airborne—plus all the night fighters would be retained for defense when and if Blue day air strikes were launched.

Related to the other uses of his airpower was the destruction of enemy surface forces themselves. He knew that if Purple approached within night striking range, Purple surface forces would be exposed at dawn to a strike by his carriers. Therefore, he did not plan on launching any night air attacks but would retain all possible aircraft for day strikes. If Purple surface forces did approach to within air striking range, at dawn he would immediately launch the strongest possible air strike for the destruction of the Purple carriers. Antisubmarine patrols would be reduced or discontinued as necessary to utilize all "suitable" aircraft for strike operations. If the strike range was greater than 335 miles, Helldivers would be continued on their antisubmarine patrols. All carrier electronic-warfare aircraft would also be used; the primary targets would be the Purple carriers. These ships would be struck until sunk or so damaged that further Purple air operations were impossible. Because Purple cruisers and destroyers might raid the convoy, these ships would be the next targets. Admiral BC's ability to launch repeated strikes, he knew, would largely depend on the success of the initial strike and the effectiveness of his defense against Purple air strikes. If he was able to eliminate Purple air strength, he could give antisubmarine operations first priority. (Page 35.)

The employment of his surface forces would first involve the detection and destruction of the Purple submarines, followed by the destruction of Purple aircraft. The latter would involve early warning of air attacks by picket destroyers and assigned radar guard ships, fighter direction, and antiaircraft gunfire. Finally, his surface forces would need to be ready to destroy Purple surface ships by gunfire. TF 11 was disposed in Cruising Disposition 11-R, which employed all available destroyers in screening stations. While the twenty-three destroyers assigned to his command were not adequate, 11-R employed them to the maximum of their capabilities. The eccentric screen, centered two thousand yards ahead of the Disposition Center, was in the direction of the course and was designed to provide maximum protection against a forty-six-knot torpedo fired from a range of five thousand yards. Considering the high speed of Purple's submarines and the low speed of his convoy, he did not think it was possible to achieve immunity from submarine attack from astern nor did he think that the frequent course reversals necessary to conduct air operations would improve matters. Antisubmarine tactics for his screening ships were based on the necessity of having one destroyer track a contact while it coached a second destroyer on to the target for the attack. Sonar contacts were to be plotted on a moving grid centered on the ship in contact, which would keep the attacking ship constantly informed of the target's location by grid-square coordinates. Communications were to be by VHF voice radio; aircraft of the Inner Anti-Submarine Patrol would guard this circuit and join in the attack. (Pages 36–37.)

The two-destroyer unit would remain with the contact only until TF 11 had reached a position such that the submarine was no longer in a position to attack. Harassing attacks and random depth charges were to be delivered to hamper the submarine's fire control. In case of multiple sonar contacts, all of the destroyers in the vicinity would drop depth charges and take all measures to complicate the submarines' attack problem. This would result, he argued, in TF 11's passing through submarine-infested waters as quickly as possible. These tactics had been rehearsed by TF 11's screening ships and aircraft and a "high degree" of proficiency had been achieved. While these tactics were admittedly not adequate to preclude an effective submarine strike, BC thought they and his aircraft that were engaged in antisubmarine search and patrol would enable him to keep his losses from Purple submarines within "acceptable" limits. He therefore considered his antisubmarine arrangements to be as complete as possible with the forces under his command. (Page 37.)

BC had deployed Destroyer Division 51—his radar-picket ships—in the antisubmarine screen around his heavy ships rather than on the picket station. However, loss of his AEW planes would necessitate the employment of DESDIV 51 on picket stations, and provisions were made for that in Cruising Disposition 11-R. Additional radar guards had been assigned to Task Force 11 so as to afford complete radar coverage. Radar intercept and jamming assignments had also been made. All of TF 11's heavy ships were capable of acting as standby fighter-direction ships, as were DESDIV 51's. Additionally, air-sea rescue services by battleship and cruiser aircraft would be available upon request. (Pages 37–38.)

BC did not think it possible to formulate Surface Action Plans with any degree of assurance, since not all possible circumstances could be foreseen. Nevertheless, he promulgated three Surface Action Plans in "more or less" skeletal form based on the assumptions that Blue and Purple forces would be of about equal strength, that Blue forces would be superior to those of Purple, and that Blue forces would be inferior, respectively. It would be "simple" to make any necessary changes by signal or message. In view of the submarine threat, the Surface Action Plans would contemplate deployment directly from Cruising Disposition 11-R. He was prepared for a night action with Purple but preferred a day action. (Page 38.)

The Blue Board Maneuver now began. Rear Admiral BE (Commander Simonsen) continued in command of the Carrier Group, the three battle carriers from Carrier Division 2. Rear Admiral BF (Colonel Manzo) was now commanding the Heavy Support Group, which consisted of the three fast battleships from Battleship Division 1 and the five heavy cruisers from Rear Admiral BG's (Commander Rogers) Cruiser Division 1.[2]

Vice Admiral BC now surmised that since Purple carrier-based aircraft could not carry atomic bombs, Purple would employ one-way missions to deliver torpedo attacks on the Blue carriers this night, strikes that would be aided by present weather conditions. He also surmised that Purple would offensively employ its surface force sometime in the next thirty-six hours after Blue forces had been whittled down by air and submarine attrition attacks. (Page 40.)

The Heavy Support Group and Commodore BD's (Commander Morton) Screen Group were to destroy enemy air and submarine forces as well as provide radar pickets, radar guard intercept, jamming, fighter direction, and air-sea rescue per current instructions and orders. When directed, the Heavy Support and Screen Groups were also to destroy Purple surface forces; all units were to employ Cruising Disposition 11-R, from which the Blue Covering Force would deploy. CTF 11 was embarked in fast battleship BB-57 with the Heavy Support Group, and Rear Admiral BF, in fast battleship BB-58, was the second in command. (Page 41.)

In the Air Plan, Rear Admiral BE's Carrier Group was again ordered to assist in destroying Purple submarine, air, and surface forces so as to ensure Convoy HTF-1's safe arrival in Japan. His three battle carriers specifically were to contribute by providing the early warning and air defense. The Carrier Group was to prioritize

Purple carriers, and Target Coordinators, who were assigned approach altitudes and were to devise the strike tactics, were to direct all air attacks to the enemy carriers until those ships were sunk or further Purple air operations were "unmistakably" impossible. Targets for "Charlie Strikes" (see below) would be assigned by message. (Pages 42–45.)

Plan Able was laid down for strikes 335 to 370 miles distant. Strike A-1 was to consist of ninety-six fighter planes, each with two drop tanks and "window" (i.e., metal strips, "chaff," to obscure the enemy radar image); forty-nine to seventy-two Avenger attack planes, each with two drop tanks and window, as well as one torpedo; and twelve tactical electronic-warfare aircraft with radar-countermeasures equipment. All available Avenger attack planes were to be employed, even to the point of canceling antisubmarine patrols that involved these planes if they could be landed and rearmed in time to be used. Plan Baker was for strikes closer than 335 miles. This strike, Strike B-1, was to employ fifty-two fighter planes, each with window and one or two drop tanks.

In addition, there would be forty-eight to ninety-six Helldiver attack planes, each equipped with window and carrying one thousand-pound general-purpose bomb and a drop tank or, if the range permitted, one two-thousand-pound general-purpose bomb. There would again be twelve tactical electronic-warfare aircraft equipped with radar countermeasures. In Strike B-2, forty-eight fighters would have window, a drop tank, and a thousand-pound general-purpose bomb if the range permitted. Forty-nine to seventy-two Avenger attack planes would again be employed, this time with window, one or two drop tanks, and a torpedo each. All attack planes were again to be taken from antisubmarine search and patrol if they could be readied for the strike. Strike B-2 was to coordinate with Strike B-1 so as to begin just as the latter was finishing. Finally, Plan Charlie—for use after Purple air strength had been eliminated—was also divided into two types. Strike C-1 would see seventy-two fighters with maximum bomb and rocket loads used inside their target range, along with window. In addition, there would be thirty to sixty attack planes with window and either torpedoes or bombs, as ordered. Additionally, there would be six tactical electronic-warfare aircraft, again equipped with radar countermeasures equipment. Strike C-2 had a similar composition. Charlie Strikes were seen as mopping-up attacks that employed all available aircraft to destroy the remaining Purple surface forces. Antisubmarine operations were to take precedence over Charlie Strikes unless otherwise ordered by the Officer in Tactical Command. (Pages 45–46.)

In Surface Action Plan No. 1, Rear Admiral BF's Battleline consisted of the three fast battleships of Battleship Division 1 along with the four radar-picket destroyers from DESDIV 51. Commodore BD's Right Flank Group consisted of heavy cruisers CA-130 and CA-131, antiaircraft cruiser CLAA-95, and five destroyers from

Time	Land or Launch	Flight No.	Planes No.	Planes Type	Loading	Orders or Remarks
1945	Launch	NCAP-1	4	F4U-4N	1 Drop Tank	Standard NCAP 4 hrs. 45 min.
1945	Launch	SS-1	24	SB2C-5	2x350#D.B., 1 Drop Tank; Radar	5 hr. anti-submarine Search Plan LOVE. Report all contacts. Attack any submarine contact, then conduct hold down tactics until relieved or forced to return to base by fuel shortage.
2000	Launch	AEW-1	3	TBM-3W	Full load fuel	8 hr. flight. Track enemy force using snooper tactics. Avoid interception. Make frequent reports using MHF radio.
2200	Land	S-1A	18 14	F4U-4N TBM-3	---------	[illegible] search less VA(W) planes.
2215	Launch	SS-2	24	SB2C-5	Same as SS-1	Same as SS-1
2300	Launch	AEW-2	3	TBM-3W	Same as AEW-1	Same as AEW-1. Retire at dawn
0015	Launch	NCAP-2	8	F4U-4N	Same as NCAP-1	NCAP and DADCAP.
0030	Land	NCAP-1	4	F4U-4N	---------	----------
0045	Launch	SS-3	24	SB2C-5	Same as SS-1	Same as SS-1.
0045	Land	SS-1	24	SB2C-5	---------	----------
0100	Land	S-1B	4	TBM-3W	---------	4 AEW planes from 1700 search
0115	Launch	S-2	9 18	TBM-3 F4U-4	4x350#D.B., 2 Drop Tanks; Radar 2 Drop Tanks	Reinforced X-ray search between 240° and 312° T to 400 miles. 1 VA, 2 VF in each sector. Report all contacts via MHF radio. Attack any submarine contact. Avoid combat. Conserve fuel.
0315	Launch	SS-4	24	SB2C-5	Same as SS-1	Same as SS-1.
0315	Land	SS-2	24	SB2C-5	---------	----------
0400	Land	AEW-1	3	TBM-3W	---------	----------
0415	Launch	AEW-3	4 16	TBM-3W F4U-4	Same as AEW-1 2 Drop Tanks	4 VF escort for each AEW. Stations: 1. SE of enemy force at maximum effective tracking range. 2. SW of enemy force at maximum effective tracking range. 3. 20° to left of enemy bearing line, 150 miles from own force at 10 and 20,000 ft. (alternate at oxygen altitude with #4 every 2 hours.) 4. 20° to right of enemy bearing line 150 miles from own force at 10 and 20,000 ft. (alternate at oxygen altitude with #3 every 2 hours.) 1 & 2 use snooper tactics.
0415	Launch	CAP-1	24	F4U-4	1 Drop Tank	DCAP 4½ hours. Note: Be prepared to increase DCAP to 48 VF if ordered.
0415	Launch strikes as and if ordered. See Appendix 2.					
0415	Launch	ASP-1	14	TBM-3W	4x350#D.B. 2 Drop Tanks	Close in, low altitude ASP 30° sectors to a distance of 30 miles outside screens. 6 planes in outboard sectors of TF-11, 6 in opposite sectors from HTF-1 and 2 planes in between convoy and Task Force. 6 hour flights. Clear area at minimum altitude if air attack is imminent.

NOTE: Continue routine flights during daylight hours except as modified by strike requirements:
DCAP - relieve at 4½ hour intervals
AEW - relieve at 6 hour intervals (Relieve VF escort earlier, if necessary)
SS - search (24 VA) at 2½ hour intervals (5 hr. flights)
ASP - relieve at 6 hr. intervals

Air Search Plan LOVE is centered on T.F. 11
Planes flying Air Search Plan LOVE remain outside destroyer screens

Fig. 138
Blue Covering Force Flight Plan, Operations Problem 2J

DESDIV 131. Rear Admiral BG's Left Flank Group was composed of CA-73, -74, and -75 as well as nine destroyers from DESRON 9. Finishing off this Task Organization were the three battle carriers of Rear Admiral BE's CARDIV 2 and five destroyers from DESDIV 132, these units together now comprising the Air Support Group. In this scenario, Blue and Purple carrier-based air strength had both been substantially neutralized and Purple surface strength was about equal to Blue's. In addition, Purple was attempting to evade the Blue Covering Force in order to destroy HTF-1. When the Blue Covering Force interposed itself under these conditions, it was assumed that a normal action would result. Therefore, the Blue Covering Force was to destroy the Purple force by a daylight surface engagement, initially at maximum gun range and then closing to decisive range when the Purple battle line had been reduced by gunfire and torpedo attacks. (Page 47.)

Admiral BC's battle line was to destroy the Purple battle line but keep itself between the Purple ships and HTF-1. It was to provide its own Air Spot and Anti-Submarine Patrol as well as be prepared to destroy any detached Purple surface forces. The Right Flank Group was to hold position while the Left Flank Group defended the Blue battle line against Purple light forces opposite its own flank. The

Left Flank Group was also to attack the Purple battle line with torpedoes when favorable opportunities presented or when ordered to do so by the OTC. The Left Flank cruisers were to provide their own Air Spot and Anti-Submarine Patrol as well as be prepared to conduct limited air search missions when ordered by the OTC. The Carrier Group was to provide such air support as its material condition allowed, coordinating its air efforts with the attacks by the Left Flank Group. If incapable of further offensive action, the Carrier Group was to join HTF-1 and release DESDIV 132 for reassignment by the convoy OTC. (Pages 47–48.)

Surface Action Plan No. 2, an identical Task Organization, reflected the same assumption about Blue and Purple carrier airpower being mutually neutralized. Here, however, Purple was seeking a night action; its efforts to evade the Blue Covering Force would therefore involve the division of its forces. Thus, to ensure the safe arrival of Convoy HTF-1, Vice Admiral BC had to plan for a night surface engagement. Specifically, his battle line would interpose itself between the Purple surface striking group and HTF-1 and endeavor to destroy the Purple battle line. The Right Flank Group would again hold position while the Left Flank Group defended the Blue battle line from the Purple light forces opposite its own flank. As soon as the Blue and Purple battle lines were engaged, the Left Flank Group would also attack the Purple battle line with torpedoes. The Carrier Group would again provide air support as its condition permitted and again coordinate that air support with the flank-group attacks. If incapable of air operations, the Carrier Group was again to join HTF-1, releasing DESDIV 132 for assignment by the OTC. (Pages 49–50.)

In Surface Action Plan No. 3, Commander, 20th Army Air Force was requested to strike the Purple forces with all available bombing aircraft. Again assuming the mutual neutralization of both Blue and Purple carrier air forces, in this case Purple surface forces would be superior to his and would close Blue as rapidly as possible in an attempt to destroy the Blue convoy by gun and torpedo fire. Admiral BC's plan now would be to protect HTF-1 by a retirement action, presumably as TF 11 came under the air umbrella of the 20th Army Air Force. The Blue Covering Force again was to keep itself interposed between the Purple forces and HTF-1. All weapons capable of inflicting damage on Purple were to be exploited and preparations to launch nighttime light-force attacks were to be made. (Page 51.)

Student officers played CTF 11 as well as his Chief of Staff. Other roles to be filled were Commander, Battleline, who was also Commander, Battleship Division 1 (Rear Admiral BF). Commander, Center Force wore a second hat as Commander, Destroyer Squadron 5 (Captain B-5—Commander Morton), and there was a separate officer in the Center Force to command CA-75 (Captain B-6—now Lieutenant Colonel Leary). The student officer who played Commander, Right Flank Force was also Commander, Cruiser Division 1 (Rear Admiral BG—Commander Rogers); a second officer in this unit played Commander, Destroyer Squadron 13

(Captain B-13—now Commander White). Commander, Left Flank Force was also Commander, Destroyer Flotilla 1 (Commodore BD—Commander Morton); another student officer commanded CA-130 and CA-131 (Captain B-4—also Lieutenant Colonel Leary). Additionally, there was a Commander, Observation Aircraft (also Commander Simonsen); the Convoy Commodore (Rear Admiral BH—now Commander Webber); and Commander, Task Group 51.7, the Escort Force (Commodore BJ—also Commander Webber).[3]

NOTES

1 Junior Class of June 1947, "Operations Problem 2J: The Blue Staff Solution," 21 November 1946, p. 17, folder 2604-O, box 139, RG 4, NHC. Subsequent in-text page references until the next endnote are to this source.

2 Junior Class of June 1947, "Operations Problem 2J: Requirement B," 20 November 1946, p. 39, folder 2604-O, box 139, RG 4, NHC. Subsequent in-text page references until the next endnote are to this source.

3 Junior Class of June 1947, "Operations Problem 2J: Board Maneuver," 22 November 1946, p. 1, folder 2604-L, box 139, RG 4, NHC.

XIX *Operations Problem 2J*

Purple, October–November 1946

Purple's Statement of the Problem contained the same Schedule of Events, Objectives, and General Situation information, with the exception that Purple did not know how many Blue Army divisions were assembling near San Francisco for embarkation to East Asia. Admiral PA, Commander-in-Chief of the Purple Pacific Fleet (no designated student officer), listed his forces as fast combatant surface ships based at Kashiwabara; a submarine force based at Vladivostok, with a partially deployed advanced base at Kashiwabara; an amphibious force; a service force; a mine force; and his fleet air wings, based in the Kuriles. As a consequence of the enforced delay in launching Purple's invasion of Japan, Admiral PA issued Operation Plan No. 7-50 to Vice Admiral PB, Commander, Purple Striking Force (Commander Loveland—see chapter 9). Operation Plan No. 7-50 laid out the forces available to PB, namely, two fleet carriers from Carrier Division 1, three fast battleships from Battleship Division 1, four heavy cruisers from Cruiser Division 3, and eighteen destroyers from Destroyer Squadrons 2 and 4. There were also six submarine tenders and sixty submarines from Submarine Squadrons 1, 2, 3, 4, 5, and 7, all under the command of Rear Admiral PF (Commander O'Connell—see chapter 9) and constituting the Submarine Force. Fleet units for the Attack Force, the Mine Force, the Service Force, and the fleet air wings had not yet been designated.[1]

The Purple Statement of the Problem

Blue atomic bomb strikes on Purple amphibious forces had severely damaged those forces, which were being reorganized for another invasion attempt. Blue reinforcement of Japan might prevent an early Purple occupation of the Japanese home islands. The Blue 20th Army Air Force in Japan still had one wing of Superfortress heavy bombers and six squadrons of fighters in an operational status. Admiral PA thought that all available atomic bombs had been expended on Purple forces. Nevertheless, he knew that the Blue Joint Chiefs of Staff (no designated student officers) had ordered a reinforcement of the Blue occupation forces and that Blue air bases at Pearl Harbor, Johnston Island, Eniwetok, Guam, and Saipan were now fully operational, though the air bases at Wake and Marcus would not be activated for a period of at least thirty days. He did not know about the status of the Blue air base at Iwo Jima but he did know that Guam was fully operational as an advanced fleet base and capable of supporting Blue surface and submarine forces to the extent of the tender facilities that were available there.[2]

Additionally, Admiral PA knew that Blue surface forces had been concentrated in the Hawaiian area and that on 15 August they were estimated to be three battle carriers, three *South Dakota*–class fast battleships, eight *Baltimore*-class heavy cruisers, "about" five squadrons of destroyers, and "many" escort and transport ships. Minor surface forces of "negligible" combat strength were now at Guam, where two auxiliaries with submarines moored alongside had been spotted on 14 August. Unidentified submarines that were probably Blue had been sighted in the vicinity of Vladivostok, Port Arthur, Tsingtao, the La Perouse Strait, and the Kuriles during July and early August. Admiral PA additionally informed Vice Admiral PB that Blue possessed a limited number of high-speed submarines whose characteristics were similar to Purple's. The Purple Far Eastern Air Force was to continue its neutralization of Blue air strength in Japan and the Aleutians in anticipation of Blue attempts to reinforce its occupation forces in Japan. This reinforcement effort was assumed to entail the transportation by sea of ground, service, and air forces; supplies; and munitions, the latter including an additional supply of atomic bombs. It was further assumed that Blue reinforcements would be staged through Pearl Harbor and would travel in "strongly protected" convoys. It was also assumed that considerations of time prevented Blue from sending its convoys on routes that were entirely under shore-based air cover. Additionally, Blue commitments in other theaters precluded substantial reinforcement of the Blue Pacific Fleet during the early phases of operations in the western Pacific.[3]

Accordingly, the Purple Pacific Fleet was to support the Purple occupation of Japan by preventing reinforcement by sea of Blue forces in Japan and by establishing Purple Army forces ashore on the island of Honshu in order to extend Purple control over the northwestern Pacific area. When directed by Admiral PA, the Attack Force would establish Purple Army forces "firmly" ashore on Honshu, and the Striking Force would prevent Blue reinforcement by sea. The Purple Striking Force was to coordinate its operations with units of the Submarine Force, which was deployed across the Hawaii–Japan sea routes. The fleet air wings were to locate, report, and track enemy sea forces within the search radius of the Kurile Islands bases and destroy any Blue submarines encountered. The Submarine Force was also to locate, observe, and report on the operations of Blue sea forces in the Hawaiian, Central Pacific, and northwestern Pacific areas. It was to destroy enemy sea forces when the opportunity permitted in accordance with specific orders that were to be issued to the submarine units. The Submarine Force was additionally to assist the Striking Force in preventing Blue reinforcement of its occupation forces in Japan by sea. Further, its submarines deployed across the Hawaii–Japan sea routes were to coordinate their operations with Commander, Striking Force when ordered by that officer.[4]

The Mine Force was to plant defensive minefields in accordance with the Mine Plan and provide minesweeping service in accordance with the Minesweeping Plan. It was also to provide antisubmarine patrols at the Purple fleet bases, while the Service Force was to provide logistical services to Purple Pacific Fleet forces in accordance with Purple Pacific Fleet Logistic Plan No. 2-50. Operation Plan No. 7-50 was to become effective upon receipt and included charts that indicated submarine operating zones as well as zones with bombing and attack restrictions. Commander, Submarine Force was also to promulgate Submarine Notices to all commands concerned with submarines departing from their assigned patrol areas. At the same time, task force commanders were to ensure that all surface ships so equipped made the required challenge on their sound gear in accordance with the *Purple Recognition Manual* when operating in Purple submarine patrol zones. Finally, Admiral PA would be at Fleet Headquarters, Vladivostok, during these operations.[5]

The Purple Chart Maneuver

The Chart Maneuver for Purple Section B began on 18 November 1946, starting with the student officer assignments for the maneuver. The Chart Maneuver had student officers playing the roles of Commander, Task Force 32, the Purple Striking Force (Vice Admiral PB—Commander Loveland), and of his Chief of Staff (Lieutenant Colonel Street—see chapter 9). In addition, there were student officers playing the roles of Commander, Carrier Group (Rear Admiral PD—Commander Owers [see chapter 9]); his Chief of Staff (Colonel Hawthorne—see chapter 9); and his Operations Officer (Commander Lundgren—see chapter 9). Student officers would also play COM CVG 1 (Commander Marvin-Smith—see chapter 9) and COM CVG 2 (Commander Church—see chapter 9). Additionally, there was a Commander, Support Group (Rear Admiral PE—also Commander Loveland); a Commander, Screen (Captain P-2—also Commander Loveland); and four student officers for Task Force 34, the Submarine Force. These student officers played the Commander, Submarine Force (Rear Admiral PF—Commander O'Connell); Commander, Submarine Squadron 3 (also Commander O'Connell); Commander, Submarine Division 21 (Captain P-6—Commander Gillette [see chapter 9]); and Commander, Submarine Division 22 (Captain P-7—Lieutenant Commander Ramage [see chapter 9]).[6] The Chart Maneuver itself entailed Air Action Plan No. 1-50, which in turn revolved around a Task Organization of Rear Admiral PD's Carrier Group, the two fleet carriers from Carrier Division 1; Rear Admiral PE's Support Group, the three fast battleships from Battleship Division 1 and the four heavy cruisers from Cruiser Division 3; and Captain P-2's Screen, the eighteen destroyers from Destroyer Squadrons 2 and 4.[7]

The General Situation revealed that the Purple invasion of the Japanese home islands was planned for 29 August, making Blue's reinforcement of its occupation forces imperative before that date. The Blue forces were at 31′10° N, 165′55° E at 1930 on 25 August, on an estimated course of 273 degrees and at a speed of eighteen knots. This force was now identified as three *Midway*-class battle carriers, three *South Dakota*–class fast battleships, four or five *Baltimore*-class heavy cruisers, and twenty-three *Sumner*-class destroyers. Another Blue convoy of thirty to forty fast auxiliaries, three to four heavy cruisers, eighteen to twenty-three destroyers, and possibly some destroyer escorts was thirty-five miles to the south. By 1928, Vice Admiral PB knew from a reported bogey that Blue had sighted the Purple Striking Force.[8]

At the same time, Vice Admiral PB had SUBRONs 2 and 4 maintaining a patrol to the south of his present position with orders to coordinate their operations with his. He assumed that the Blue forces would maintain a speed of eighteen knots as well as a nighttime course between 225 and 270. By dawn, he expected Blue to be on a course between 270 and 280, by which time he also expected Blue to attempt an "all-out" daylight air strike on the Purple Striking Force if it was in range. He planned to have his force reduce Blue carrier strength during the night by air and submarine attacks to gain air parity or superiority as a preliminary to thwarting Blue's reinforcement attempt. As such, the Carrier Group was to launch night air attacks, track the Blue force, and destroy enemy snoopers. The Support Group was to provide antiaircraft protection to the Carrier Group during this operation while the Screen provided antiaircraft and antisubmarine protection to both the Carrier and Support Groups as well as plane guard services to the Carrier Group. The Submarine Group was to concentrate and deliver its attacks against the Blue Covering Force. Admiral PB would be embarked in fast battleship BB-2, while his second in command, Rear Admiral PD, Commander, Carrier Group, would be embarked in fleet carrier CV-1.[9]

Under Annex A to the Air Plan, airborne early warning planes were to be maintained in Sectors 19 and 21 to track the Blue forces. Reliefs for these planes were to be provided in order to maintain continuous contact and tracking. Between 0030 and 0130 on 26 August, Vice Admiral PB would launch his nighttime air strike so that the Blue forces were hit at 0330. This strike was to be a torpedo attack carried out with twenty-four Avenger attack planes and twelve night fighters. This was to be a decisive attack on the Blue battle carriers, with each Avenger armed with a torpedo and the night fighters each carrying five-inch rockets. The fighters were to create a diversion by destroying Blue aircraft on their flight decks by rocket and strafing attack. The fighters were also to carry out strafing attacks on the Blue screening ships; Admiral PB stressed early and accurate reporting about damage to the Blue carriers. He additionally established 30°00′ N, 164°00′ E as a ditching point for planes that could not return to their carriers. Aircrews at this ditching rendezvous would be picked up by Purple submarines. Further, during the night, he would maintain night fighters in readiness on deck to be catapulted to intercept and destroy any Blue bogeys. A total of four night fighters would be retained for this purpose after the night strike had been launched.[10]

At dawn, if sufficient success had been attained by night air and submarine attacks, an all-out air strike would be made. This dawn air strike would consist of forty-eight attack planes, each armed with a thousand-pound armor-piercing bomb, and forty-five fighters, twenty of which would be armed with thousand-pound armor-piercing bombs. The primary targets would again be the Blue battle carriers, which were to be attacked until they were rendered "definitely harmless." Secondary targets were the fast battleships and heavy cruisers. The fighters armed with bombs, however, were to jettison their ordnance if it was necessary to protect the attack planes or themselves. Vice Admiral PB expected to launch the planes at 0500 on 26 August but the strike would be initiated only on his order.[11]

In Annex B, the Submarine Plan, the location of the Blue forces was given again, as were their probable nighttime moves. SUBRON 2 was given the specific mission of inflicting maximum damage on the Blue carriers by concentrated attack prior to dawn. Submarines SS-11, SS-12, and SS-13 were to proceed at maximum submerged speed on course 185 while SS-14 and SS-15 proceeded at their maximum submerged speeds on 165. SS-16, -17, -18, -19, and -20 would proceed toward 31°10′ N, 165°55′ E at a speed of eleven knots. All boats were to guard schedules for five minutes every hour on the half hour. SUBRON 3 was to inflict maximum damage on either the Blue Covering Force or the convoy, as directed, by concentrating in two groups for the nighttime attacks. SS-21, -22, -23, -24, and -25 were to proceed to Point LOVE—30°35′ N, 155°48′ E—and arrive by 0400 on 27 August. Fleet submarines SS-26, -27, -28, -29, and -30 were to proceed to Point MIKE—30°30′ N, 158°00′ E—to arrive by 1600 on 26 August. They were to establish patrol lines north of these points at five-mile intervals and guard schedules for five minutes every hour on the hour.[12]

Visual communications were to be employed in lieu of the radio whenever possible. The use of radar was unrestricted and the radar guard ships had been assigned in Cruising Disposition 32-N. All ships and aircraft were to keep Identification Friend or Foe transponders energized; interrogators were to be restricted to emergency use except for recognition guard ships.[13]

All ships having two Talk Between Ships sets were to have both the Task Force Common freqency and their respective Task Group frequencies set up and task group commanders were to relay information received on the Task Force Common Circuit to ships of their groups that were only guarding the Task Group Circuit. There was also a visual chain of command, a Task Force Commanders Circuit, and directions for the Aircraft-Submarine Common Circuit. Concerning Communications Intercept and Countermeasures Plan, radio intercept guard ships and assigned frequencies were assigned to BB-1 and BB-2 as well as heavy cruisers CA-2, -3, and -4. When ordered by the Officer in Tactical Command, Blue radio frequencies were to be jammed.[14]

Vice Admiral PB put forth a Surface Action Plan, No. 2-50, with a Task Organization consisting of three fast battleships from Battleship Division 1, under Rear Admiral PE, as the Battleline, as well as Rear Admiral PF's Center Forces, four heavy cruisers from Cruiser Division 3 and eighteen destroyers from Destroyer Squadrons 2 and 4. With the same General Situation and the same composition of Blue forces, Surface Action Plan No. 2-50 noted that Purple SUBRON 3 was trailing the Blue Covering Force to sink ships damaged in the Air Phase. The plan assumed by this time that the carrier air forces on each side would have been neutralized, though spotting planes were available to both sides. It further assumed that Blue had passed through Purple submarine waters. Last, it assumed that the Blue Covering Force was still pressed for time and limited tactically in having to keep itself interposed between the Blue convoy and the Purple Striking Force. Within a General Plan of destroying the Blue Covering Force by daylight surface engagement as a prelude to destroying the convoy, Admiral PB set out the details for his component commands. The Purple battle line was to engage the Blue battle line by commencing action at "extreme range," closing to long range, and keeping Blue "well occupied" while a surface torpedo attack was pressed home. If the enemy "countermarched" to avoid Purple torpedoes, the Purple battle line was not to follow but should bear toward the convoy. The Purple battle line was to remain concentrated and be alert to any tactical confusion in the Blue Covering Force that might present an opportunity to concentrate fire on unsupported

1. SHACKLE CODE

00	N	F	B
0	M	Y	T
1	P	D	R
2	A	I	H
3	V	J	
4	Z	C	
5	E	P	
6	L	S	
7	G	U	
8	O	K	
9	X	W	

2. F•r Authentication enter shackle code with the first letter of the second word or numeral group in the text of the message. Any two letters, numerals, or letter numeral comprises authentication: For example - text "Rotate fleet axis to 180 degree" authenticators are NY or BY.

3. Messages intended solely for deception purposes shall be identified by the authenticators JP, RL, or UV.

Fig. 139
Purple shackle codes, Operations Problem 2J

1. Damage Code.

0-9%	10-19%	20-29%	30-39%	40-49%	50-70%
Apple	Peach	Orange	Lemon	Grape	**Cherry**
Pear	Persimmon	Plum	Tomato	Cranberry	Cantaloup
Lime	Tangerine	Crabapple	Loganberry	Raisin	
Pineapple	Mango	Papaya			
Kumquat	Blackberry	Razzberry			
Blueberry	Elderberry				
Over 70%					
Apricot					
Strawberry					

2. Voice Radio Calls.

CTF 32		Helen, Goldie, Prunella, Lena
CTG 32.1	Carrier	Mary, Josephine
32.2	Support	Jane, Patricia
32.3	Screen	Georgia
34.2	Submarines	Edith, Barbara
	ComSubron 2	June, Mabel
	" Subron 3	Susie, Agatha
	" Cardiv 1	Beatrice
	" Batdiv 2	Constance
	" Crudiv 3	Faith
	" Desron 2	Dorothy
	" Desdiv 21	Grace
	" Desdiv 22	Hazel
	" Desron 4	Katie
	" Desdiv 41	Lorraine
	" Desdiv 42	Maisie
	" Subdiv 21	Abigail
	" Subdiv 22	Virginia
	" Subdiv 31	Carolyn
	" Subdiv 32	Christine
	Battleline	Patsy, Dinah
	Center	Rachel, Rosie
	Right Flank	Clementine, Florence
	Left Flank	Tillie, Louise
	CAG-1	Wilma
	CAG-2	Olivia
	Battleships	Airdales, puppies, pointers
	Carriers	Cockers, Dalmations, Chows
	Cruisers	Greyhounds, retrievers, poodles
	Destroyers	Chesapeakes, setters, springers
	Merchant Ships	Scotties, Terriers
	Enemy (BLUE)	Greek or Trojan
	Own (PURPLE)	Roman or Etruscan

Individual Ship calls is division called followed by hull number. Thus BB-1 is "Constance One". DD-17 is "Maisie One Seven".

To call Combatdiv 2, call "Constance".
To call Batdiv 2, call "Constance's boys".

3. Special Code Words.

Attack -- Seduce
With Torpedoes - Blonds
" Aircraft - Brunettes
" Gunfire - Red Heads
Retire -- Kentucky
Deploy -- Colorado
Affirmative -- Football
Negative-- Baseball
Illuminate -- Blanket
With searchlights - requisition
" starshells - invoice
Lay smoke screen - Prostitute
With destroyers - robbers
" aircraft - burglars
Open Fire - Exhibit

Fig. 140
Purple voice calls and codes, Operations Problem 2J

units. If Blue deployed a battleship Detached Wing, the Purple battle line was to make every effort to isolate it from mutual support of the Blue battle line and destroy it in detail.

The Center Forces were to gain station "expeditiously" and push home a torpedo attack against the Blue battle line as soon as the opposing battleships became engaged. The heavy cruisers were to support the destroyer torpedo attack, repel any Blue torpedo attack, and—depending on the tactical situation—shift one or two cruisers to the inner cruiser area if that became necessary. The Center Forces were not to come within gun range of any Blue battleship until that ship was engaged by the Purple battle line. The Center Forces were to assign three planes to locate, track, and report on the convoy. Commander, Purple Striking Force was embarked in BB-2 with the battle line, and Rear Admiral PE, the second in command, was in BB-1.[15]

In Annex C to the Surface Action Plan, PB's Deployment Plan Able envisioned that the Blue Center Force—if there was one—would be weaker than his. Admiral PB's Center Forces would deploy to the right at maximum formation speed when forty-five thousand yards from the Blue battle line on a course so as to reach an advanced position just outside of Blue gun range when the two battle lines became engaged. Reiterating that the Center Forces were not to come within Blue battle line gun range during this deployment, he also ordered his battle line to deploy to the right so as to complete the movement by the time it reached maximum gun range on the Blue battle line. Under Deployment Plan Baker, all units would deploy in the same manner as in Deployment Plan Able except to the left instead of the right. Under Deployment Plan Charlie, the assumption was that the Blue Center Force—with or without a fast battleship—was stronger than his Center Forces. Accordingly, his Center Forces would commence deployment to the right at the best formation speed when seven thousand yards outside of Blue gun range, if there was no Blue battleship, and twelve thousand yards if there was one. This deployment was to be a retiring deployment about 115 degrees to the right of the enemy bearing and was to be an effort to draw the Blue Center Force within range of the Purple battle line. Center Force units were to remain outside of the gun range of any superior Blue force and take advantage of any opportunity to turn and concentrate on an inferior Blue force. When the situation permitted, the Purple Center Forces were to gain station for a torpedo attack. Also under Deployment Plan Charlie, the Purple battle line would close at its best speed in an attempt to engage the Blue Center Force before it could be supported by the Blue battle line. Upon reaching the extreme gun range of the Blue battle line, the Purple battle line was to deploy to the right or left, whichever best suited the tactical situation at the time. Finally, Deployment Plan Dog was the same as Charlie, except the Purple Center Forces would deploy to the left.[16]

Annex D of the Surface Action Plan, the Torpedo Attack Plan, had as its objective maximum damage on the Blue battle line by launching a destroyer torpedo attack as soon as the opposing battle lines were engaged. Admiral PB assumed in this plan that his own light forces were strong enough to break through the Blue defenses and reach an attack point. He also assumed that the Blue battle line would turn away from a strong torpedo attack and that the torpedoes, to be effective, needed to be limited to a twenty-minute run to the target if possible. Under Attack Plan X-ray, Purple destroyers would conduct their approach in line of divisions with an axis approximately normal to expected torpedo tracks and with standard distances and intervals. The destroyers were to use maximum formation speed and, upon deployment, effect a quick deployment ahead by divisions, commencing at about twenty-two thousand yards from the Blue battle line. At a range of eighteen thousand yards, the Purple destroyers would make a simultaneous turn into column on a firing course, fire their torpedoes when steady, and retire under a smoke screen. Smoke screens, in fact, were to begin as soon as the torpedoes were fired.[17]

All ships were to fire simultaneously by division, and each division was to fire by the Base Torpedo Course Method. The Division Leader would—on the basis of a normal gunfire distribution—fire at the Blue Van, with the Division Leader selecting the leading Blue ship as his point of aim. He would signal his base torpedo course to his formation while the next adjacent division would apply a three-degree correction to the base torpedo course in the direction of the Blue Rear. Each succeeding division would apply an additional three-degree correction to the Blue Rear, the wing divisions thus having base torpedo courses that were nine degrees apart. If the approach was made within thirty degrees of the enemy's bow or stern, the Control Division Leader would displace his own base torpedo course three degrees toward the Blue Van. All ships were to use high-speed settings and half-degree spread angles on their torpedoes and were to fire full salvos.[18]

Under Attack Plan Yoke, the assumption was that there would be considerable difficulty in reaching the torpedo firing point because of Blue gunfire and that the Purple destroyers might therefore have to fire at long range. Thus, Attack Plan X-ray would be followed insofar as practicable except that the standard Base Torpedo Course Method would be used. Finally, under Attack Plan Zebra, the assumption was that Blue opposition was exceptionally strong, preventing the employment of either Attack Plan X-ray or Yoke. In addition, Admiral PB was assuming in this case that his destroyers had suffered heavy damage, had uncertain communications, and had ragged or broken formations. Therefore, the attack procedure was to conduct attacks in large groups when practical from the best position that could be attained using an individual-target plan. Every effort here was to be made to launch torpedoes at a Blue target before the Purple destroyers in question were sunk.[19]

The Purple Staff Solution to the Problem

Vice Admiral PB, having prevented reinforcement by sea of Blue forces in Japan, was to assist in establishing Purple Army forces ashore on Honshu to extend Purple control over the northwestern Pacific area. He did not see anything that was not feasible in this.[20]

PB knew the Blue 20th Army Air Force was based in Japan and had one wing of long-range bombers and six squadrons of fighters that were operational. Pearl Harbor, Eniwetok, Johnston Island, Guam, and Saipan were all fully operational Blue bases. He did not know the status of Iwo Jima and he also assumed that Wake and Marcus would not be operational for at least thirty days. Guam was capable of supporting Blue surface and submarine forces to the extent of its available tender facilities. PB estimated that Blue would have three battle carriers, three *South Dakota*–class fast battleships, eight *Baltimore*-class heavy cruisers, five squadrons of destroyers, and many escort and transport ships in the Hawaiian area. Blue submarines were operating in the vicinity of Vladivostok, Port Arthur, Tsingtao, La Perouse Strait, and the Kuriles. He understood Blue ground forces in Japan to be small occupation forces but that a two-division Marine force was in Hawaii, as well as "several divisions" of Blue Army troops in San Francisco. (Pages 2–3.)

Vice Admiral PB's own air bases and air forces consisted of the fleet air wings of naval patrol planes and bombers, which had the task of locating and tracking the Blue surface forces and destroying the deployed Blue submarines. His sea forces consisted of the Purple Attack Force, which included the Amphibious Force, whose mission was to land on Honshu about 29 August. He also had the Mine Force for defensive mining and minesweeping, the Service Force for the logistical needs of the Purple Pacific Fleet, and the Submarine Force, sixty boats that were to locate, observe, report on, and then destroy the Blue surface forces in the Hawaiian, Central Pacific, and northwestern Pacific areas in coordination with his Striking Force. Purple ground forces were those of the Purple Army, which was in the process of occupying Manchuria, southern Korea, and North China. He endorsed the assumptions of Commander, Pacific Purple Fleet regarding the coming action. (Pages 3–4.)

At this point, the scenario began to develop, and planning adjusted to accommodate it, effectively in real time. On the afternoon and evening of 23 August, Purple submarine SS-35 sent three contact reports of Blue naval forces at sea, in two groups of ships. The first group was an estimated ten large and twenty small ships, in a circular disposition, on course 270 at speed eighteen knots. The second group had a larger number of ships; it was east of the first group, also on 270, at fifteen knots. At 1700 on 25 August, a Purple X-ray search was launched, and an AEW aircraft reported contact with the two Blue forces. The first group was now estimated to be eleven large and twenty small ships. The second group, now reported thirty-five miles south, was again reported to be simply a large force. Admiral PB assumed that these various contact reports referred to the same two groups of Blue ships; he surmised that the smaller group was the Blue Covering Force, the larger one the troop convoy protected by escort, and possibly large combatant, ships. These contact reports reinforced his understanding of the Blue objective as being the reinforcement of occupation troops in Japan. (Pages 4–5.)

Given that the Blue Pacific Fleet could not be reinforced, his submarine and aircraft contact reports meant, he concluded, that three to four heavy ships and some twenty smaller ships were absent from the Covering Force. Either not all the ships initially available had sailed, or the unaccounted-for ships were with the convoy to strengthen its defenses, or there was another Blue force at sea. "Reliable" Purple intelligence had indicated that at the outbreak of war Blue had had available a small carrier task force. He did not think that Blue would split into two separate groups a unit that had trained and operated together as an effective Fast Carrier Task Force. In addition, such a division would violate the principle of concentration. So he discarded the possibility of another Blue force in the area and assumed that Blue combatant strength was concentrated in one carrier task force that constituted the Blue Covering Force. Knowing that Blue would make every effort to sail with maximum aircraft and battleship strength—even delaying departure to do this—Admiral PB concluded that the Blue Covering Force was composed of eleven large ships and about twenty small warships. He now estimated that his two fleet carriers, three fast battleships, four heavy cruisers, and eighteen destroyers were facing three battle carriers, three fast battleships, five heavy cruisers, and eighteen to twenty-three destroyers. He therefore assumed that the Blue convoy, with its large number of transport ships and an unknown number of escort ships, had also three heavy cruisers and twenty to twenty-five destroyers. (Pages 5–7.)

Admiral PB had ninety Corsair fighters, sixteen Corsair night fighters, and eight Corsair photographic reconnaissance fighters. He also had forty-eight Helldiver and twenty-four Avenger attack planes, eight Avenger tactical electronic-warfare planes, and eight Avenger AEW aircraft. There were also fourteen observation planes and the naval patrol bombers from the fleet air wings, as well as an undetermined number of army fighter and rescue planes. Blue had an estimated 153 fighters, twenty-four night fighters, twelve day fighters, 168 attack planes, twelve tactical electronic-warfare planes, twelve AEW planes, sixteen observation planes, an unknown number of naval patrol bombers, 150 army fighters, and 120 Superfortress heavy bombers. In the absence of more information, Vice Admiral PB assumed that Blue planes were comparable to Purple's. (Page 8.)

Admiral PB took the sides to be equal in the combat efficiency of their personnel. As to the relative combat efficiency of materiel, only the comparison of the

Purple Striking Force to the Blue Covering Force mattered, because Blue would attempt to keep the maximum distance between the Purple Striking Force and its convoy, whereas Blue could not weaken the convoy's antisubmarine and antiaircraft protection by detaching units from the Covering Force. In any case, the Blue convoy's escort strength would be negligible in comparison to the Purple Striking Force. The maximum combatant strength opposing him at any one time would be represented by the Blue Covering Force. (Pages 8–9.)

Carriers	PURPLE	BLUE
Class	ESSEX	MIDWAY
Number	2	3
Max. Speed	33	33
Sust. Speed	31	33
Life	10.6	13.6
Armor	Hanger Deck - 2.5"	Flight Deck - 3.5"
	4th Deck - 1.5"	Hangar Deck - 1.5"
Main Battery & Range	12 - 5"/38 16,000	18 - 5"/54 23,000
Auto. Weapons	72 - 40MM 70 - 20MM	84 - 40MM 56 - 20MM

Fig. 141
Purple and Blue aircraft carrier characteristics, Operations Problem 2J

Admiral PB took Blue battle carriers to be "much superior" to Purple's fleet carriers in terms of age, armored flight decks, antiaircraft protection, and sustained speed. They were, therefore, much more difficult to put out of action. In addition, Blue had one more carrier than Purple and larger air wings for each carrier, resulting in an almost two-to-one superiority in fighter planes and better than two to one in attack planes. Blue and Purple were equal in battleships on paper, and Admiral PB considered the two forces equal ship for ship in cruisers, but Blue had five of these to Purple's four. All of Purple's destroyers were equivalent to the *Sumner* class, with ten torpedo tubes each, except for five that only had five tubes each. He thought it probable that in all Blue possessed a superiority of two to five destroyers and a probable superiority in numbers of torpedoes. In the absence of more information, he also took Blue and Purple torpedo performance to be "at least" equal. Finally, he assumed Purple to have a "great superiority" in the numbers of submarines available for tactical employment, a factor he took to be of critical importance. (Pages 9–12.)

Battleships	PURPLE	BLUE
Class	SOUTH DAKOTA	SIMILAR TO SOUTH DAKOTA
Number	3	3
Max. Speed	27	
Sust. Speed	25	
Life	22.6	
Armor (Side-Deck-Forward)	16" - 6¼" - 11.3"	SAME
Main Battery & Range	9 - 16"/45 34,000	
Sec. Battery & Range	20 - 5"/38 16,000	
Auto. Weapons	68 - 40MM 76 - 20MM	
Planes	2 VO	

Fig. 142
Purple and Blue battleship characteristics, Operations Problem 2J

The destructive effect of the atomic bomb was of such magnitude that he considered it necessary to evaluate separately the probability of its use by the Blue 20th Army Air Force against the Purple Striking Force. He was convinced that Blue still had atomic bombs in Japan, since they had been used on Purple amphibious forces in the Far East, although Commander, Purple Pacific Fleet believed that Blue had expended all of these weapons. If Blue did still have atomic weapons, they must be in short supply, since it had been briefed that the Blue convoy was carrying additional atomic bombs. Considering that the Purple Striking Force was not the only or even the main Purple force attempting to occupy the Japanese home islands, Vice Admiral PB did not think that Blue would use any remaining atomic bombs on his force; there were much more "profitable" targets closer to Blue

Cruisers	PURPLE	BLUE
Class	BALTIMORE	SIMILAR TO BALTIMORE
Number	4	5
Max. Speed	32	
Sust. Speed	31	
Life	5.5	
Armor (Side-Deck-Forward)	6" - 2.5"-5.5"	SAME
Main Battery & Range	9 - 8"/55 28,000	
Sec. Battery & Range	12 - 5"/38 16,000	
Auto. Weapons	48 - 40MM 22 - 20MM	
Planes	2 VO	

Fig. 143
Purple and Blue heavy cruiser characteristics, Operations Problem 2J

land-based air forces. He had no factual data on which to predict results if, nevertheless, an atomic bomb were used, but he estimated that the maximum damage Blue could expect from one bomb would be the loss of one ship and damage to one or more others. Even this presupposed Blue's being able to locate the Purple Striking Force, make a successful attack against his defense, and obtain a hit. (Pages 12–13.)

Destroyers	PURPLE	BLUE
Class	SUMNER	SIMILAR TO SUMNER-GEARING
Number	18	18 - 23
Max. Speed	34	
Sust. Speed	32	
Life	1.6	
Armor	None	
Main Battery & Range	6 - 5"/38 16,000	SAME
Auto. Weapons	12 - 40MM 20 - 20MM	
#Torp. tubes	10 CL (5 CL)	
Type torp.	G-3	

Fig. 144
Purple and Blue destroyer characteristics, Operations Problem 2J

Admiral PB reviewed hydrographic factors in detail, concluding that sonar conditions would be poor, that the lowest probability of sonar detection would occur at periscope depth and at deep depth, and that these conditions should assist Purple submarine operations but also complicate the detection of Blue submarines. (Pages 14–15.)

From an operational standpoint, therefore, the weather was favorable for Purple air, except that the Purple Striking Force would have to close the enemy during flight operations if the present wind conditions prevailed. High visibility, he surmised, would favor scouting operations and daylight defense against enemy air attack. With respect to submarine operations, if present sea conditions continued, the visual detection of submarines during daylight was possible because of periscopes "feathering," wake indications at the surface, and impulse bubbles. (Pages 15–16.)

Submarines	PURPLE		BLUE
Class	A	B	
Number	10	10	Limited numbers
Surface Speed/Endurance	10/15,000	10/7,300	
Submerged Speed/Endurance			
Schnorchel	7/16,000	7/7,500	Characteristics
Batteries	5/365	4/100	similar to
	18/22		PURPLE
Walther	---	15/224	Submarines
		24/158	
Tubes	6 Bow	4 Bow 6 Side	
Torp. Capacity	20(Type F-3)	10 (Type F-3)	

Fig. 145
Purple and Blue submarine characteristics, Operations Problem 2J

Visibility conditions at night should aid in the detection of surface submarines but also assist aircraft in night attacks. He expected, moreover, Blue submarine attacks from "down moon" (that is, from the direction of the moon, which would illuminate the target for the submarine). At this time the two groups of Blue ships were 450 miles southeast of the Purple Striking Force. At 1928, a bogey was detected by the Purple Striking Force, indicating that Blue air search planes had probably located the Purple Striking Force. Given the comparative speeds of the two forces, however, PB did not think that Blue could close his force unless he so desired; in fact, he could force Blue to approach by threatening the convoy, which was limited to only fifteen knots. The closest Purple submarine was about 120 miles to the south; if PB continued southward, he had either to shift the submarines or to keep them fully informed of his movements and trust their recognition procedures; he would try to stay clear of any Purple submarines. (Pages 16–17.)

In light of the distances to the various Blue air bases, it appeared that a day's run by the Purple Striking Force at twenty knots on any course would not materially increase the danger from conventional, land-based air attack except from Japan and Iwo Jima. (This assumed that Blue bombers were not sent on one-way missions.) In any case, he recalled from World War II that Iwo Jima could be used for emergency landings but that its usefulness for large-scale bombing attacks was limited. Any significant land-based bombing, therefore, had to originate in Japan itself. (Page 17.)

Purple submarines were located along two patrol lines. The center of the closest line to Blue was about 200 miles from the Blue position. If Blue forces continued on their present course and speed, it would be possible for some of Purple's submarines to reach attack positions shortly before dawn. (Page 17.)

The Purple Striking Force had fueled to capacity and loaded necessary supplies prior to its sortie from Kashiwabara at 2000 on 23 August. On the morning of 25 August, the destroyers had topped off. Submarine Squadrons 2 and 3 had ample diesel fuel and all of their torpedoes. As for Blue's logistics, PB knew that Blue had recently departed Hawaii, so he assumed that Blue fuel and supplies were also ample. He further assumed that the Blue commander could top off his destroyers as necessary. He did not see logistics as a critical factor for either side. However, both Purple and Blue were approximately a thousand miles from the nearest base to which damaged ships could retire. Admiral PB therefore planned to station submarines that were not able to reach primary attack positions along the probable retirement courses of damaged Blue ships. (Page 18.)

NOTES

1 Junior Class of June 1947, "Operations Problem 2J: The Schedule of Events," 29 October 1946, pp. 1–2; Junior Class of June 1947, "Operations Problem 2J: The Purple Statement," 29 October 1946, pp. 1–3; both folder 2604, box 139, RG 4, NHC.

2 Junior Class of June 1947, "Operations Problem 2J: The Purple Statement," p. 3.

3 Ibid., p. 4.

4 Ibid.

5 Ibid., p. 5.

6 Junior Class of June 1947, "Operations Problem 2J: Purple—Section B, The Chart Maneuver," 18 November 1946, p. 1, folder 2604-D, box 139, RG 4, NHC.

7 Junior Class of June 1947, "Operations Problem 2J: Purple—Section B, Air Action Plan," 18 November 1946, p. 1, folder 2604-K, box 139, RG 4, NHC.

8 Ibid.

9 Ibid., pp. 1–2.

10 Ibid., p. 3.

11 Ibid.

12 Ibid., p. 4.

13 Junior Class of June 1947, "Operations Problem 2J: Purple—Section B, Annex (C), Communication Plan," 18 November 1946, p. 1, folder 2604-H, box 139, RG 4, NHC.

14 Ibid., pp. 1–2.

15 Junior Class of June 1947, "Operations Problem 2J: Purple Section B, Surface Action Plan, CTF 32 No. 2-50," 18 November 1946, pp. 1–2, folder 2604-J, box 139, RG 4, NHC.

16 Ibid., p. 3.

17 Ibid., p. 4.

18 Ibid.

19 Ibid., p. 5.

20 Junior Class of June 1947, "Operations Problem 2J: The Purple Staff Solution," 21 November 1946, pp. 1–2, folder 2604-P, box 139, RG 4, NHC. Subsequent in-text page references are to this source.

XX Operations Problem 2J

More Purple, October–November 1946

Formally summarizing his Strength Factors, Vice Admiral PB saw one strength as Purple's two squadrons of submarines that were within interception distance of the Blue forces. In addition, his forces had freedom of action and the submarines could expect poor sonar conditions. In contrast, he considered himself "vastly inferior" to Blue in carrier aircraft, though Admiral PB thought that if aircraft strength could be eliminated the relative strength of the Blue Covering Force and the Purple Striking Force in a surface engagement would be "nearly equal." Meanwhile, any advantage he had was with his submarines. They would be a potent weapon if used "properly," though he had to remain out of attack aircraft range as long as Blue held its present carrier superiority while he exploited his own submarine advantage.[1]

Admiral PB understood that the Blue convoy headed to Japan might be only one of several, but since it was the only one on which he had information, he would make it his target. He had to destroy this convoy or at least reduce its shipping and defensive strength to the point where it was no longer feasible for it to continue to Japan. It was possible for Purple submarines to get directly at the convoy, but for his surface and air forces to do so he had first to destroy the Blue carrier forces or destroy, drive off, or evade the Blue Covering Force. Otherwise, PB would have to attack the convoy and the Covering Force at the same time. On this basis, he considered the following actions. First, he could attempt to destroy the Blue carrier air forces and then the Blue convoy by air attack. He might also attempt to destroy or drive off the Blue Covering Force and then, by air and surface attacks, destroy the Blue convoy. His third alternative was to drive off the Blue convoy by submarine attrition attacks. Lastly, if he could evade the Blue Covering Force, he would be able to employ air, surface, and submarine attacks to destroy the convoy. (Pages 20–21.)

Admiral PB knew that Blue's movements would be highly restricted. Blue could still evade Purple forces to the south, however. In addition, the Blue Covering Force could interpose itself between the convoy and the Purple Striking Force. It could also drive off Purple forces that approached to within striking range. Finally, the Blue Covering Force could destroy the Purple forces by "relentless" attacks, disregarding close support of the convoy. Evasion to the south and toward the cover of land-based aircraft could occur only if the Blue commander was able to delay the arrival of the reinforcements in Japan. In any case, given Purple superiority in speed, the Blue commander could only delay, not prevent, action with the Purple Striking Force. If the Blue Covering Force interposed itself between the Purple Striking Force and the convoy, its mere presence would discourage offensive action by Purple. Admiral PB surmised that if he persisted in attempting to reach the convoy, Blue would have to be prepared to destroy him since PB had no intention of being driven off. (Pages 21–23.)

Admiral PB considered specifically a Course of Action in which he destroyed the Blue carrier air forces with his own airpower, then used that airpower to destroy the Blue convoy. If he could destroy the Blue battle carriers by air attack, submarine attack, or a combination, he would be able to send his remaining air strength over the Blue Covering Force to strike directly at the convoy without having to engage the Covering Force with his surface forces. This was the desirable option since it would expose the least number of his units to Blue. However, Blue carrier air superiority was such that Admiral PB did not think he had much chance of success in it. Submarines could attack the Blue carriers but he did not think that they could completely destroy them given the carriers' "excellent" underwater protection. He estimated, in fact, that it would take six to eight torpedo hits per carrier. Still, submarine attacks might reduce Blue carrier air superiority to nearly that of Purple. If the Blue carriers were completely destroyed, Purple carrier planes could strike at the convoy without further reducing the Blue surface forces by air attack, but Admiral PB thought it more likely that he would have to make further air strikes on them. The probable result would be the effective elimination of air strength on both sides, in which case the Purple Striking Force would still have to deal with the remainder of the Blue Covering Force. Since he did not think that this Course of Action had great prospect for success, he would retain it for future consideration as possible but not probable. (Pages 23–24.)

He was still faced, however, with the need to eliminate Blue's superior carrier strength. He did not think that could be done by day air attack alone. He therefore contemplated a night aircraft strike. If he steamed directly toward Blue at maximum speed, he might be able to close for a night air strike. This strike, however, would depend on Blue movements. He expected Blue to divert the convoy to the south and interpose the Blue Covering Force upon detection of the Purple Striking Force. Admiral PB thought it might be possible to launch a night attack a few hours after 1930. To hold off for moonlight might be desirable but in either case he did not think that surprise was likely. Given this reality, he assumed he would be open to counterattack and he did not think the damage he would take justified the damage he might inflict on Blue. In addition, with the wind from the southeast, launches and recoveries would require Purple to steam toward Blue. If any delays occurred, he might find himself in daylight within Blue aircraft range and not able to withdraw. (Page 25.)

Therefore, Admiral PB returned again to the idea of eliminating Blue air superiority by submarine attacks on Blue's carriers. To do that, Blue's submarines had to take up new attack positions. Once there, he thought, the submarines should be able to pick out targets and inflict serious damage. As of 1930, Purple Submarine Squadron 2 was so disposed that at least some of its boats could attain a predawn attack position provided that the Blue Covering Force made good any course between 235 and 345. He saw it as "most probable" that this would be the case, so SUBRON 3 was also to be in an attack position by the following morning. He proposed to regulate the movements of the Purple Striking Force so as not to interfere with these boats and maybe even draw the Blue Covering Force into Purple submarine waters. He thought it essential that he maintain constant scouting of Blue forces and relay updates to SUBRONs 2 and 3. The submarines' movement to station would require considerable night surface steaming. Blue antisubmarine patrols might force the submarines to submerge temporarily but even limited use of high submerged speeds should allow four to six submarines to achieve an attack position. These boats, in turn, should enable Admiral PB to engage the Blue Covering Force on more favorable terms. (Pages 25–26.)

As long as Blue possessed its carrier superiority, PB did not think it advisable to remain very long within aircraft striking range. As of 1930, Purple and Blue were approximately 450 miles apart, out of range. Although Blue might decide to close with him, Vice Admiral PB thought that he could maintain this distance unless Blue detached one or more carriers minus their heavy-ship antiaircraft support. In the latter case, Admiral PB thought he could strike the Blue carriers. He expected action to start on the evening of 25 August and the early morning of 26 August. While he remained out of range, the Purple submarines would attack the Blue Covering Force. After the reduction of Blue carrier strength, Admiral PB could strike with his carrier air forces, then engage and drive off the Blue Covering Force. Following this engagement, he could expect to overtake the convoy late the same afternoon or early that evening. By late on the afternoon of 26 August, he would be approximately eight hundred miles from Blue's land-based air cover in Japan. Since this land-based air cover was a threat of limited proportion, he thought that closing the distance to Japan was a justified, calculated risk. (Pages 26–27.)

For the submarine attrition attacks, he had twenty submarines with a total of three hundred torpedoes. He thought it doubtful that anywhere near all the submarines could get into the attack. Sooner or later, Blue would detect his submarines and by a combination of evasive courses and antisubmarine tactics be able to "bully" through submarine waters to Japan, even with losses. Once the Blue convoy managed to pass one of Purple's submarines, that submarine was eliminated as a threat—as long as the convoy continued on the same course and speed—given the submarines' slow surface speed and restricted range at high underwater speeds. At most, any one submarine could expect to make only one attack and many would be unable even to reach attack positions. Therefore, PB did not think he could destroy or drive off the convoy by submarines alone. (Page 27.)

The Blue Covering Force had the advantage of interior lines, meaning that the Purple Striking Force could not evade it. He might be able to detach a small combatant group to get at the convoy but the success of this action would depend on Blue failing to detect the division of Purple forces. In view of Blue air superiority and the probability that Blue had already detected the Purple Striking Force, PB did not think it suitable to risk piecemeal destruction. He might get within aircraft striking range of the convoy and send some planes to attack the Blue Covering Force while the majority of the Purple Striking Force engaged the Blue convoy. This action too did not appear to be acceptable, however, in view of Blue's "overwhelming" air superiority. So not even submarine, air, and surface strength combined could defeat both Blue forces, at least by day. Therefore, he decided, he would destroy or drive off the Blue Covering Force and then destroy the convoy, first reducing Blue air and surface strength by night and early-morning submarine attacks. These would be followed by daylight submarine and carrier air strikes to destroy the carriers and reduce the surface forces. The remaining Blue surface forces would then be destroyed or driven off by daylight surface engagement, after which all available Purple forces would attack the convoy. These attacks were to take place out of range of Blue land-based air forces in Japan. (Pages 27–28.)

Having decided on his Course of Action, Vice Admiral PB divided his tasks into four phases of execution. Phase One entailed cruising and tactical scouting by the Purple Striking Force and then the reduction of Blue air and surface strength by night and early-morning submarine attack. Phase Two would be the destruction of Blue carrier air forces and the reduction of the Blue Covering Force's surface units by daylight submarine and air attack. Phase Three was the destruction or driving off of the remaining surface units of the Blue Covering Force by daylight surface engagement. Phase Four was the destruction of the convoy. (Page 29.)

In Phase One, the mission was to track the Blue force, remaining outside aircraft attack range while submarines reduced Blue carrier air strength by attacking the carriers themselves. In this phase, it seemed unlikely that Blue had a considerable number of submarines in the present area of operations since there was no reason to expect Blue to have had advance information on the proposed movements of the Purple Striking Force. It seemed more likely that Blue submarines were off port entrances and at focal traffic points, which is exactly where they had been sighted.

Only the normal antisubmarine precautions and patrols were required, therefore. (Pages 29–30.)

Admiral PB saw air attack from Blue land-based air forces in Japan—including atomic-bomb strikes—as a more likely and constant threat. He also had to take into account the possibility of a surprise attack by carrier-based air forces if he lost contact with the Blue Covering Force. Therefore, he chose Cruising Disposition 32-R, which he thought would provide good protection against "normal" air attacks and "fair" protection against submarines. In the event of early warning of an air attack, Vice Admiral PB planned to shift to a tighter disposition, Air Defense Disposition 32-V. Should an atomic-bomb attack become likely, he would open up Cruising Disposition 32-R to a new disposition. (Page 30.)

Movements by the Purple Striking Force during Phase One would be as required to stay out of the Blue air striking range, to remain clear of Purple submarines, to draw Blue through Purple submarine waters, and to approach the Blue convoy's eventual track to Japan. But what was the Blue air striking range? By comparison to his own planes' characteristics, fighter planes had a maximum combat radius of 370 miles with two auxiliary fuel tanks. Attack planes carrying a torpedo would have a maximum combat radius of 425 miles, while attack bombers carrying a thousand-pound bomb would have a maximum combat radius of 335 miles. Any strike beyond the combat radius of fighter aircraft would have little prospect of success, so if he remained farther than about four hundred miles from the Blue Covering Force until his submarines had completed their attacks, he could expect to be safe from an effective Blue air strike. On this basis, he planned to steam initially to the southwest; later changes in his movements would depend on information he received from his search planes. In tracking the Blue forces, he noted that one of his airborne-early-warning planes had made contact at 1930. He had directed this plane to maintain and develop (i.e., collect more information about) the contact. The plane had a remaining endurance of approximately 5.7 hours, so he would launch two additional AEW planes at once to take over. From this point on, he would launch an additional two AEW planes every three hours to permit continuous tracking of Blue movements. (Pages 31–32.)

He again saw the effective employment of his submarines as the key to his "difficult" situation. He concluded that if only a few submarines were able to make their attacks, the results would be more decisive if these strikes were delivered against the Blue Covering Force rather than the convoy's transports. If he placed all of his submarines in one line that ran roughly north–south, he could not expect more than one or two of his submarines to get in attacks and once Blue had passed this line, his submarines would no longer be effective. The optimum, from the submarine viewpoint, was to station his boats in depth along an east–west line and along the projected track of the Blue Covering Force, thus permitting each boat to make an attack. He could not count on the route Blue would take, however, so he had to compromise and attempt to obtain submarine coverage in both depth from east to west and from north to south across Blue's most probable track. (Pages 32–33.)

If PB ordered his submarines to close immediately on Blue's last reported position, they would probably all be flanked later by Blue movements. He expected Blue to divert its convoy southward during the night but eventually to come back to a course that would take it to southern Japanese ports. He would have submarines attempt to get ahead of the Blue Covering Force and continually interpose themselves between the Blue forces and Japan. The Blue forces could not divert greatly from their route because of the urgent need for reinforcements in Japan. To cover any probable diversions, he ordered Commander, Submarine Force to divide his submarines into four wolf packs, each of five boats organized as a submarine division. Submarines were to be directed to adjust the present positions so as to establish patrol lines by about sunrise on 26 August. The submarine with the lowest hull number of each division would occupy reference points; the remaining boats of a division would take station at ten-mile intervals bearing 205 degrees from the reference point. Submarine Division 21's reference point would be 31°00′ N, 163°12′ E; SUBDIV 22's reference point would be 30°10′ N, 163°24′ E; SUBDIV 31's was 32°25′ N, 161°06′ E; and SUBDIV 32's boats would take up stations on 31°38′ N, 162°54′ E. Commander, Submarine Force was to reroute his submarines if such action was indicated by information from the search planes. To permit individual submarines to receive early intelligence on Blue movements, an Air Search frequency was established; AEW planes were directed to report on it and submarines were to maintain a listening watch. (Pages 33–34.)

Admiral PB expected the new patrol lines to allow his submarines to attain attack positions even if Blue steamed directly west or north of west from its 1930 position. The submarine patrol lines would also cover any Blue diversion to the south for up to ten hours. Additionally, he planned to deploy his submarines in depth along any other probable Blue route. After these initial deployments, however, he would be forced to rely on Commander, Submarine Force to assist the submarines in gaining their final attack positions. He would keep Commander, Submarine Force fully informed of enemy movements and keep the Purple Striking Force west of 160 degrees east. If he was forced to go east of this line, he would so inform the submarine commander so that the latter's boats could be notified. Purple air and surface forces would be directed not to attack submarines encountered east of 160 east and to make appropriate sonar challenges when forced to operate there. Submarines that could not close the Blue Covering Force were to attack the convoy. Since he could not forecast when and where the air and surface engagements would take place, Commander, Submarine Force would be informed of his intentions when events developed "sufficiently," at which point he would be directed to deploy his submarines in the action or take up positions along the most probable course along which Blue cripples would retire. (Page 34.)

Phase Two operations—the destruction of Blue carrier air forces and the reduction of the Blue Covering Force's surface units by daylight submarine and air attack—were to commence upon the completion of Phase One. Admiral PB knew a follow-on air strike by the Purple Striking Force was necessary. If his Corsair fighters were equipped with two auxiliary fuel tanks, they had a radius of 375 miles. His Helldiver attack planes had a radius of 335 miles if carrying an auxiliary fuel tank and a thousand-pound bomb; a radius of 230 miles if armed with a torpedo; and a 370-mile radius if carrying a 1,600-pound bomb and two wing tanks of fuel. His Avenger attack planes had a radius of 425 miles if carrying a torpedo and two auxiliary fuel tanks, or a radius of 435 miles with the same auxiliary fuel tanks but a thousand-pound bomb replacing the torpedo. From these figures, his maximum striking distance for a coordinated attack employing Helldiver attack planes was 370 miles; for an attack employing the Avenger attack planes with bombs, torpedoes, or some of each, plus fighter escorts, it was 375 miles. Primary targets for the initial air strikes were the Blue battle carrier flight decks and their onboard aircraft. By using "toss bombing" techniques and thousand-pound armor-piercing bombs, he hoped to achieve penetrations of the battle carriers' armored flight decks. With the 1,600-pound armor-piercing bombs, he hoped for penetration of both flight and hangar decks. A coordinated attack approaching from different directions and preceded by deceptive measures and fire-control radar jamming would give the best chance for success. If he was able to destroy the Blue carriers without the loss of all of his aircraft, the rest could attack the surface ships in the Blue Covering Force. (Pages 35–36.)

With forty fighters, sixteen night fighters, and eight photographic reconnaissance fighters, the Carrier Group would augment the combat air patrol and defend the Purple Striking Force from Blue air attack. With seventy-two attack planes and fifty fighters, it would complete the destruction of the Blue carriers and their air forces and then inflict maximum damage on the enemy surface forces. Priority of targets after the carriers was the fast battleships and then the heavy cruisers. Three AEW planes would provide a basic early warning watch while five others scouted for contacts. Eight tactical electronic-warfare planes would jam enemy radars and employ radar deception when ordered by the Officer in Tactical Command. PB did not contemplate further coordination of submarine operations after Phase One. However, submarines of Submarine Squadron 3 that had been unable to close the Blue Covering Force during Phase One should be able to attack that unit during Phase Two. (Page 36.)

Phase Three operations were to consist of the destruction or driving off of the Blue Covering Force's surface ships by a daylight surface engagement. It was impossible to predict the situation this far in advance but as the basis for Phase Three planning he assumed that by this time Blue carrier air strength had been either destroyed or greatly reduced and that Purple carrier air strength was in the same state. According to his estimate, the opposing forces would still be two fleet carriers, three fast battleships, four heavy cruisers, and eighteen destroyers for Purple facing three battle carriers, three fast battleships, five heavy cruisers, and eighteen to twenty-three destroyers for Blue. He saw the two forces as essentially equal in fast battleships but with a Purple inferiority in both heavy cruisers and destroyers. This imbalance gave Blue a "probable" superiority in light forces and the number of surface torpedoes. If, however, he had any remaining air strength, Admiral PB was prepared to exploit it, combining aircraft attacks with submarine torpedo and surface gun and torpedo attacks. (Page 37.)

Present indications were that the Blue convoy would be to the southwest and a southwesterly deployment course would not cause much difficulty in aircraft operations with a southeasterly wind. Admiral PB expected the Blue Covering Force to attempt to remain between the Purple Striking Force and the convoy. He also expected that moving the Purple Striking Force toward the convoy would cause the Blue Covering Force to seek a surface engagement. He did not think he could gain the windward position with its attendant advantages for his light forces and the employment of smoke. He further assumed that whichever way he deployed, Blue would deploy in the same direction so as not to be outflanked. The best distribution of his light forces would be two-thirds of them in the Van and one-third in the Rear. Blue would attempt to flank his position and probably have a similar distribution of light forces.

To ensure the flexibility, he was going to organize a Center Force that he could use to strengthen either flank. Since the two battle lines were equal in numbers and characteristics, there were no favorable ranges. Fire, therefore, would be opened at extreme range to obtain early hits. As soon as the Blue battle line was under Effective Fire from Purple's main batteries, the light forces in the Van would be ordered to attack. Any remaining air strength would be coordinated by the OTC to support this attack. After air and light-force attacks had reduced the Blue battle line's speed, the enemy would be closed to decisive range. His submarines would be directed by message as the situation developed. Once the Purple battle line engaged at long range and the air and light-force attacks had reduced the Blue battle line so it could be engaged at short range by Admiral PB's battleships, the Purple Striking Force would have access to the convoy. (Pages 38–39.)

The Purple Striking Force's fourteen observation planes would scout tactically, provide Air Spot for their parent ships, and detect and destroy Blue submarines.

The Purple battle line of three fast battleships would engage at long and then short ranges in coordination with the air and light forces, while the Center Group (two heavy cruisers and five destroyers) would defend the area ahead, defend the area between the Left and Right Flank Groups, and join the forces in the Van upon deployment. The Right Flank Group of one heavy cruiser and five destroyers along with the Left Flank Group of one heavy cruiser and four destroyers would defend the battle line from Blue light-force torpedo attacks and attack the Blue battle line with torpedoes when ordered by the OTC. Finally, the Carrier Screen's four radar-picket destroyers would escort the two fleet carriers, furnishing them with antisubmarine protection and plane-guard services. (Pages 39–40.)

Phase Four would entail the destruction of the convoy. The convoy was estimated to consist of a large group of transport ships protected by an unknown number of escort vessels and possibly a few larger combatant ships. Admiral PB had concluded earlier that the combatant strength protecting the convoy would be negligible as compared to the strength of the Purple Striking Force. However, he could not yet make "intelligent" plans for Phase Four; it depended on the outcomes of Phases One, Two, and Three, the remaining strength of the Purple Striking Force, the location of the Blue convoy, and the position of Purple's submarines. All of his efforts, however, were directed toward the successful accomplishment of Phase Four. He thought that his running estimate would provide him with the necessary information for issuing required instructions, the latter to be delivered to his ships by Talk Between Ships. (Page 40.)

The Purple Board Maneuver

On 21 November (real time), Vice Admiral PB (again Commander Loveland for the Board Maneuver) outlined for the Purple Board Maneuver his operation plan—Commander, Battleship Squadron 1 Plan No. 1-50. The Task Organization consisted of Rear Admiral PD's (again Commander Owers) Carrier Group, the two fleet carriers of Carrier Division 1; Rear Admiral PE's (now Colonel Hawthorne) Heavy Support Group, the three fast battleships from Battleship Division 1 and four heavy cruisers from Cruiser Division 3; and Captain P-2's (now Commander Lundgren) Screen of eighteen destroyers from Destroyer Squadrons 2 and 4. The General Situation, recent events, broad assessments and assumptions, objectives, and tasks for the projected four phases were the same as for the Chart Maneuver. (Pages 41–43.)

Vice Admiral PB's Surface Action Plan added new details. Its Task Organization had Rear Admiral PD's two fleet carriers from Carrier Division 1 and four radar-picket destroyers from Destroyer Division 41 collectively constituting the Air Group. The three fast battleships from Battleship Division 1 were now organized as Rear Admiral PE's Battleline, while Rear Admiral PF's (now Commander Gillette) Center Group was organized from heavy cruisers CA-1 and CA-2, four destroyers

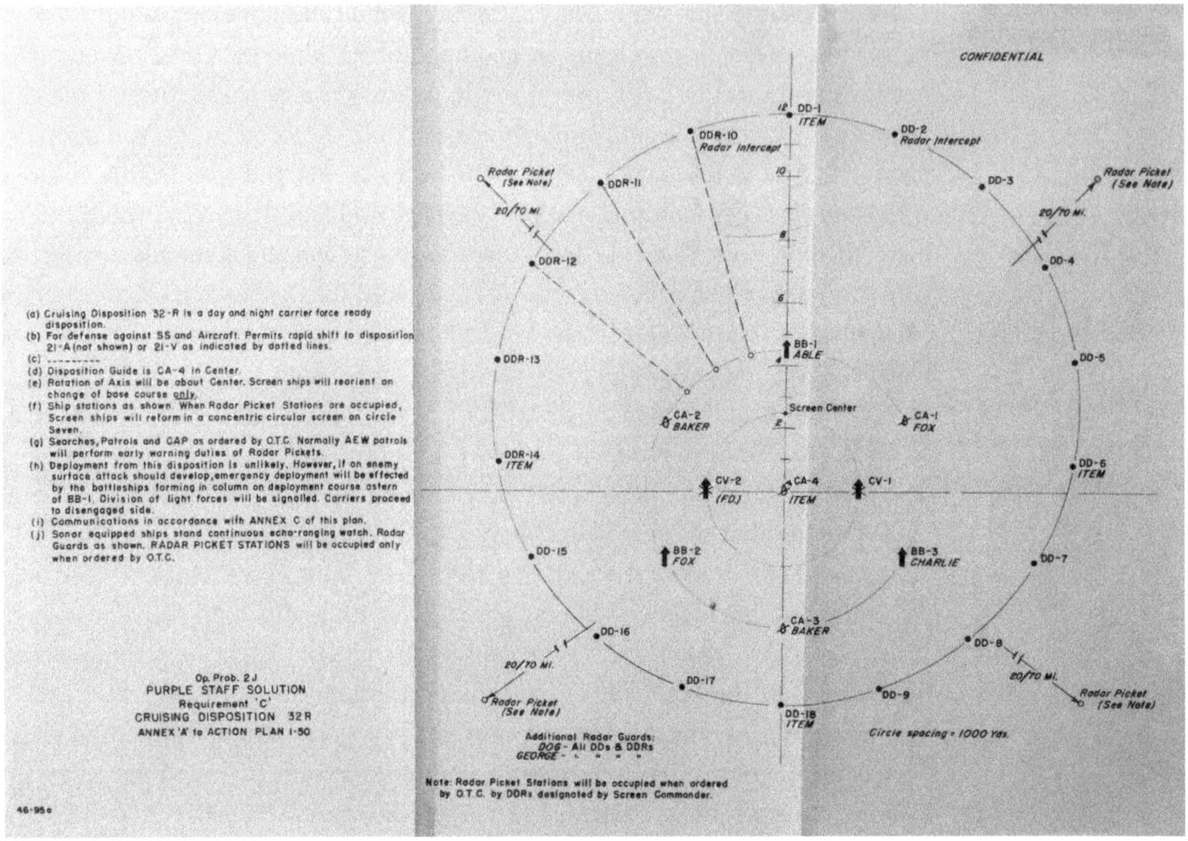

Fig. 146
Cruising Disposition 32-R, Purple Staff Solution, Operations Problem 2J

from DESDIV 42, and one radar-picket destroyer from DESDIV 41. Captain C-3's (not played by a student officer) Right Flank Group consisted of heavy cruiser CA-3 and five destroyers from DESDIV 21, while the Left Flank Group, under Captain C-4 (no designated student officer), comprised heavy cruiser CA-4 and four destroyers from DESDIV 22. SUBRONs 2 and 3 were to operate in support. Since the plan assumed that the carrier air strength on each side would have by now been neutralized and the Blue surface force would be seeking a surface engagement, the Purple Striking Force would engage in a daylight Normal Action. The Purple battle line would engage at extreme gun range and then at shorter ranges after the Purple light and air forces had reduced the Blue battle line's speed by coordinated attacks. (Page 44.)

The remaining planes from the Air Group would defend the Purple Striking Force from Blue air attack and support the light-force attacks on the Blue battle line. Additionally, they would continue to scout tactically. While the Purple battle line engaged the Blue battleships, the Center Group would defend the area ahead and between its own flanks from Blue attack and join the Purple Van upon deployment while the Right Flank Group held position. The Left Flank Group would defend the Purple battle line from Blue light-force torpedo attacks while it attacked

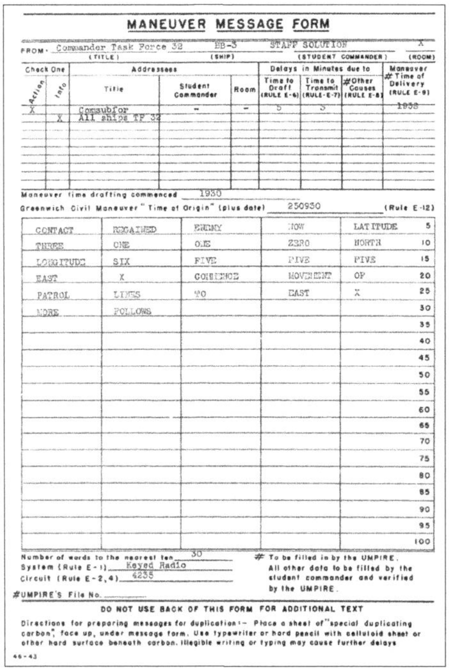

MANEUVER MESSAGE FORM

FROM: Commander Task Force 32 (TITLE) BB-3 (SHIP) STAFF SOLUTION (STUDENT COMMANDER) X (ROOM)

Action	Info	Title	Student Commander	Room	Time to Draft (RULE E-6)	Time to Transmit (RULE E-7)	#Other Causes (RULE E-8)	Maneuver #Time of Delivery (RULE E-9)
X		Comsubfor	-	-	5	3		1958
	X	All ships TF 32						

Maneuver time drafting commenced 1930

Greenwich Civil Maneuver "Time of Origin" (plus date) 250930 (Rule E-12)

CONTACT	REGAINED	ENEMY	NOW	LATITUDE	5
THREE	ONE	ONE	ZERO	NORTH	10
LONGITUDE	SIX	FIVE	FIVE	FIVE	15
EAST	X	COMMENCE	MOVEMENT	OF	20
PATROL	LINES	TO	EAST	X	25
MORE	FOLLOWS				30

Number of words to the nearest ten 30
System (Rule E-1) Keyed Radio
Circuit (Rule E-2,4) 4235
#UMPIRE'S File No.

To be filled in by the UMPIRE. All other data to be filled by the student commander and verified by the UMPIRE.

DO NOT USE BACK OF THIS FORM FOR ADDITIONAL TEXT

Directions for preparing messages for duplication:— Place a sheet of "special duplicating carbon", face up, under message form. Use typewriter or hard pencil with celluloid sheet or other hard surface beneath carbon. Illegible writing or typing may cause further delays

46-43

Fig. 147
Maneuver Message Form
1, Purple, Operations
Problem 2J

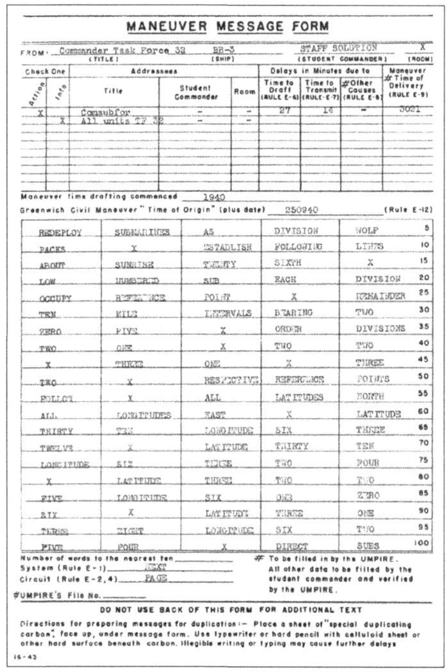

MANEUVER MESSAGE FORM

FROM: Commander Task Force 32 (TITLE) BB-3 (SHIP) STAFF SOLUTION (STUDENT COMMANDER) X (ROOM)

Action	Info	Title	Student Commander	Room	Time to Draft (RULE E-6)	Time to Transmit (RULE E-7)	#Other Causes (RULE E-8)	Maneuver #Time of Delivery (RULE E-9)
X		Comsubfor	-	-	27	14	-	3021
	X	All units TF 32	-	-				

Maneuver time drafting commenced 1940

Greenwich Civil Maneuver "Time of Origin" (plus date) 250940 (Rule E-12)

REDEPLOY	SUBMARINES	AS	DIVISION	WOLF	5
PACKS	X	ESTABLISH	FOLLOWING	LINES	10
ABOUT	SUNRISE	TWENTY	SIXTH	X	15
LOW	NUMBERED	SUB	EACH	DIVISION	20
OCCUPY	REFERENCE	POINT	X	REMAINDER	25
TEN	MILE	INTERVALS	BEARING	TWO	30
ZERO	FIVE	X	ORDER	DIVISIONS	35
TWO	ONE	X	TWO	TWO	40
X	THREE	ONE	X	THREE	45
TWO	X	RESPECTIVE	REFERENCE	POINTS	50
FOLLOW	X	ALL	LATITUDES	NORTH	55
ALL	LONGITUDES	EAST	X	LATITUDE	60
THIRTY	TEN	LONGITUDE	SIX	THREE	65
TWELVE	X	LATITUDE	THIRTY	TEN	70
LONGITUDE	SIX	THREE	TWO	FOUR	75
X	LATITUDE	THREE	TWO	TWO	80
FIVE	LONGITUDE	SIX	ONE	ZERO	85
SIX	X	LATITUDE	THREE	ONE	90
THREE	EIGHT	LONGITUDE	SIX	TWO	95
FIVE	FOUR	X	DIRECT	SUBS	100

Number of words to the nearest ten
System (Rule E-1) NEXT
Circuit (Rule E-2,4) PAGE
#UMPIRE'S File No.

To be filled in by the UMPIRE. All other data to be filled by the student commander and verified by the UMPIRE.

DO NOT USE BACK OF THIS FORM FOR ADDITIONAL TEXT

Directions for preparing messages for duplication:— Place a sheet of "special duplicating carbon", face up, under message form. Use typewriter or hard pencil with celluloid sheet or other hard surface beneath carbon. Illegible writing or typing may cause further delays

16-43

Fig. 148
Maneuver Message Form
2, Purple, Operations
Problem 2J

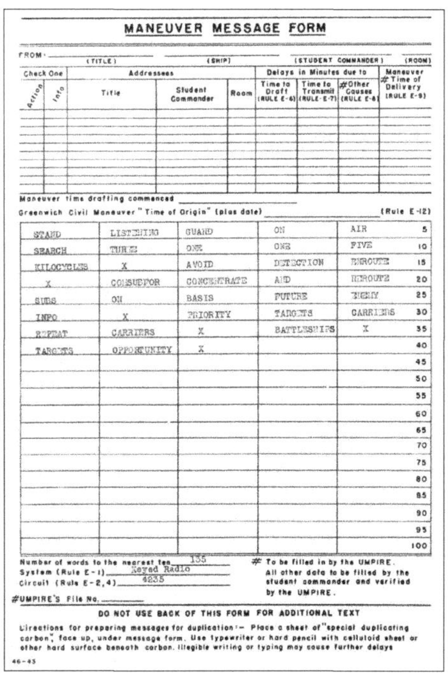

MANEUVER MESSAGE FORM

FROM: (TITLE) (SHIP) (STUDENT COMMANDER) (ROOM)

Action	Info	Title	Student Commander	Room	Time to Draft (RULE E-6)	Time to Transmit (RULE E-7)	#Other Causes (RULE E-8)	Maneuver #Time of Delivery (RULE E-9)

Maneuver time drafting commenced

Greenwich Civil Maneuver "Time of Origin" (plus date) (Rule E-12)

STAND	LISTENING	GUARD	ON	AIR	5
SEARCH	THREE	ONE	ONE	FIVE	10
KILOCYCLES	X	AVOID	DETECTION	ENROUTE	15
X	COMSUBFOR	CONCENTRATE	AND	REROUTE	20
SUBS	ON	BASIS	FUTURE	ENEMY	25
INFO	X	PRIORITY	TARGETS	CARRIERS	30
REPEAT	CARRIERS	X	BATTLESHIPS	X	35
TARGETS	OPPORTUNITY	X			40

Number of words to the nearest ten 135
System (Rule E-1) Keyed Radio
Circuit (Rule E-2,4) 4235
#UMPIRE'S File No.

To be filled in by the UMPIRE. All other data to be filled by the student commander and verified by the UMPIRE.

DO NOT USE BACK OF THIS FORM FOR ADDITIONAL TEXT

Directions for preparing messages for duplication:— Place a sheet of "special duplicating carbon", face up, under message form. Use typewriter or hard pencil with celluloid sheet or other hard surface beneath carbon. Illegible writing or typing may cause further delays

46-43

Fig. 149
Maneuver Message Form
3, Purple, Operations
Problem 2J

the Blue battleships with torpedoes, when ordered to do so by the OTC. It would also provide its own Air Spot and Anti-Submarine Patrols. The plan was to become effective upon signal; the Officer in Tactical Command would coordinate air support for the light-force attacks; and all units were to employ Approach Disposition 32B and Surface Battle Disposition 32C. (Pages 44–45.)

Student officers were assigned to play Commander, Task Force 32 (Vice Admiral PB); his Chief of Staff (again Lieutenant Colonel Street); and Commander, Battleline, who was also Commander, Battleship Division 1 (Rear Admiral PE). Other student officers played Commander, Center Forces, who was also Commander, Cruiser Division 3 (Rear Admiral PF); Commander, Detached Cruisers (Captain P-5—Commander Church); and Commanders, Destroyer Squadrons 2 and 4 (Captain P-2 [now Commander Lundgren] and Captain P-4 [now Lieutenant Commander Ramage]). Finally, student officers took the roles of Commander, Observation Aircraft (again Commander Owers); Commander,

Task Force 34 (also Commander, Submarine Force) (now Rear Admiral PS—Commander O'Connell); and Commanders, SUBDIVs 31 and 32 (Captain P-8 [again Commander O'Connell] and Captain P-9 [Commander Marvin-Smith]).[2]

The Critique

Operations Problem 2J ended here. As is fairly typical in war-game-based pedagogy, as soon as the Director of the Maneuver determined that no further educational value was to be derived from continuation, the exercise or maneuver was brought to an end. Since there were no written records kept for critiques at this point of the College's history, it is difficult to determine how even this scenario (much less a real-world one) would have played out, but these types of conclusions were not the purpose of the exercises and maneuvers in any case.

The Critique for Operations Problem 2J took place on 26–27 November 1946, in the Maneuver Room of Luce Hall. Again, while there is no written record of the Critique itself, a delineation of roles played by the College's staff and students gives some indication of what the critiques themselves attempted to accomplish. After an introduction by Captain Evans, the Director of the Maneuver, the student officers designated for Operations Problem 2J gave five-minute summaries of their own solutions as well as the Staff Solutions, paying particular attention to differences in methods of development and in what actually happened. Student officers also gave critiques of the Estimates of the Situation for both Blue and Purple. There was an examination of the mission itself, then a critique of the Situation and Opposing Courses of Action by Lieutenant Colonel Leary for Blue and Commander Foley for Purple. Colonel Edson for Blue and Lieutenant Colonel Thompson for Purple then critiqued Enemy Capabilities, while Lieutenant Commanders Taeusch and Aymond, for Blue and Purple, respectively, critiqued Own Courses of Action. Lieutenant Colonel Downey for Blue and Lieutenant Commander Bonner for Purple followed with an analysis of Opposing Courses of Action. Commander Coleman for Blue and Commander Rynd then presented a comparison of Own Courses of Action and the Decisions; Commander White for Blue and Colonel Mason for Purple critiqued the Complementary Problems. Commanders Kirkpatrick and Richardson for Blue and Purple, respectively, critiqued the Action Plans. Commander Dalton for Blue and Lieutenant Commander Howell for Purple critiqued the Cruising Dispositions.[3]

Moving to the Chart Maneuver itself, there were Blue and Purple analyses of the Officer in Tactical Command's Plans and Dispositions, at 1930 (it is unclear whether such evening sessions were the norm for the Naval War College at the time) by Commander Baldridge and Lieutenant Commander Wallace for Blue and Commander Loveland and Lieutenant Colonel Street for Purple. There were also Blue and Purple analyses of the Initial Searches and Contact Scouting, including a brief by Commander Miller; analyses of Air Operations by Commander Simonsen for Blue and Commander Owers for Purple; analyses of Submarine Operations by Commanders O'Connell and Gillette and Lieutenant Commander Ramage for Purple; and analyses of Surface Ship Operations by Commanders Baldridge and Morton for Blue and Commander Loveland for Purple. The officers who carried out the critiques of the Brief, the Air Operations, the Submarine Operations, and the Surface Ship Operations then critiqued the Blue Searches, the Purple Air Strike, Blue's attempted counterattack, and the Special Blue searches conducted in the early morning hours of the next day of the maneuver. These latter critiques included analyses of the actions of the Blue and Purple Air Group Commanders.[4]

Further, there would be analyses of intentions of the Blue and Purple Officers in Tactical Command by Commanders Baldridge and Loveland for Blue and Purple, respectively. Additional Comments on Air Operations, including Night Snooper Tactics, Intercepts, and Air Attacks, would be given by the Air Umpire, junior-class instructor Commander Kirkpatrick and his colleague and Assistant Air Umpire Lieutenant Commander Holzapfel, as well as their assistants Lieutenant Colonel Leary, Commander Richardson, Lieutenant Commander Bonner, Commander White, Colonel Mason, and Commanders Rynd, Coleman, and Foley. Comments on Submarine Operations, including Capabilities and Tactical Techniques of New Submarines, were given by the Submarine Umpire, Commander Reed, and his assistants, Lieutenant Commanders Taeusch and Aymond. Comments on Communications and Radio Countermeasures were provided by the Communication Umpire, Lieutenant Commander Curran. The conclusion was then conducted by Captain Evans.[5]

For the Board Maneuver, there would again be Blue and Purple analyses of each OTC's plans, this time by Commander Baldridge for Blue and Commander Loveland for Purple. Center Force operations during the approach would be analyzed by Commander Morton for Blue and Commander Gillette for Purple, while Light Force Consolidations for Blue were presented by Commanders Morton, Rogers, White, and Bridewell as well as Lieutenant Colonel Leary. Light Force Consolidations for Purple were analyzed by Commanders Gillette, Church, and Lundgren, as well as by Lieutenant Commander Ramage. Blue and Purple would then both comment on battle-line action, with Colonels Manzo and Hawthorne providing these comments. Concluding Light Force Action was analyzed by these same two officers, Submarine Operations for Purple by Commanders O'Connell and Marvin-Smith, and Convoy Operations for Blue by Commander Webber. Commander Baldridge for Blue and Commander Loveland for Purple then offered analyses of the Future Intentions of the Officers in Tactical Command. Following these presentations would be summaries on Fire Distribution Damage by the Chief Damage

Computer, Commander Mayberry; on Ship Handling, with Commander Dalton analyzing for Blue and Lieutenant Commander Howell doing the same for Purple; and Communications by Lieutenant Commander Curran, the Communications Umpire. Captain Evans then conducted the conclusion of the Critique.[6]

NOTES

1 Junior Class of June 1947, "Operations Problem 2J: The Purple Staff Solution," 21 November 1946, pp. 18–20, folder 2604-P, box 139, RG 4, NHC. Subsequent in-text page references until the next endnote are to this source.

2 Junior Class of June 1947, "Operations Problem 2J: Purple—Section B, Board Maneuver," 22 November 1946, p. 1, folder 2604-M, box 139, RG 4, NHC.

3 Junior Class of June 1947, "Operations Problem 2J: The Critique," 25 November 1946, p. 1, folder 2604-N, box 139, RG 4, NHC.

4 Ibid., p. 2.

5 Ibid.

6 Ibid., p. 3.

U.S.NAVAL WAR COLLEGE, NEWPORT, R.I. 4 April 1934.
Contract NOy-1700 - View looking west showing Central Des
in Maneuver Room. No. NTS."C" 34-112.

XXI *Operations Problem 3*

Statements, Estimates, Staff Solutions, and the History of the Maneuver, October–November 1946

In late October 1946, the senior class of June 1947 was issued materials for Operations Problem 3. Operations Problem 3 was the senior-class version of Operations Problem 1J. A Chart Maneuver, it took place between 25 October and 9 November. Introduced as usual by Rear Admiral Smith, Operations Problem 3 entailed the same mix of Staff Presentations, preparations for the maneuver and its conduct, and the Critique.

The Blue Statement of the Problem, Estimate of the Situation, and Staff Solution
The General and Special Situations were the same as Operations Problem 1J—Purple was going to attempt to invade the western Aleutians with an Amphibious Force escorted by a Covering Force. The Blue Alaskan Defense Force was again detailed to intercept and destroy these forces. The Blue Alaskan Defense Force comprised almost the exact same number of surface ships as in Operations Problem 1J: four fast battleships, two battle cruisers, four light cruisers, and thirty-six destroyers (instead of the twenty-seven assigned to the junior-class problem). The major difference in force composition, however, was in aircraft carriers. Instead of two fleet carriers, as in Operations Problem 1J, Commander, Heavy Surface Striking Force and Alaskan Defense Force now had three fleet carriers and two light carriers at his disposal. Intelligence information, tactical dispositions, purposes of the maneuver, and even Student Requirements mirrored those of Operations Problem 1J.[1]

In addition, the Blue Staff Solution was largely identical. The Blue commander, Admiral BA (Captain Madeira—see chapter 10), again reasoned that his best bet for preventing the Purple invasion of the Aleutians was to find, intercept, and destroy the Purple Covering Force so as to make the Purple Amphibious Force's position untenable. Blue submarines were not deployed very effectively, both Blue and Purple land-based air were neutralized by recent action, Admiral BA's force was on its own, and, he reasoned, Purple was in the same situation. Weather conditions told him that air activity might be severely curtailed while Blue and Purple surface-ship strengths and weaknesses were at a premium. The estimate of Purple surface ship strength in terms of numbers and characteristics was the same as in Operations Problem 1J aside from an assumption that Purple aircraft carrier strength

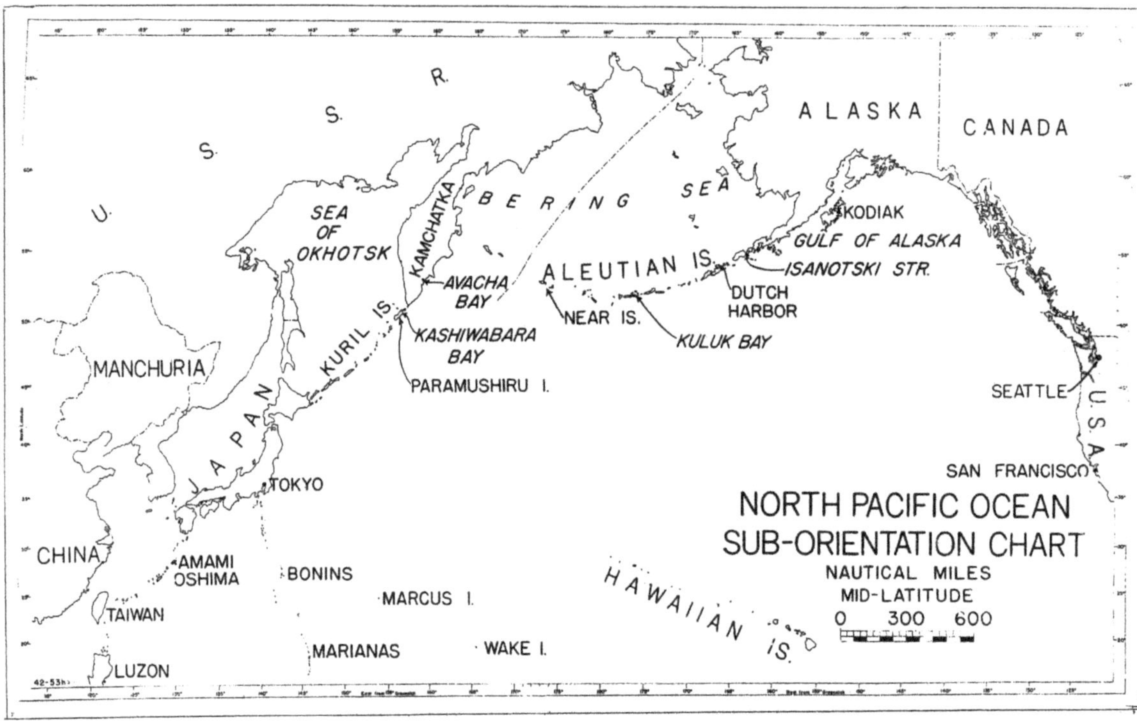

Map 14
North Pacific Ocean Sub-Orientation Chart

was four fleet carriers and two light carriers instead of the two fleet carriers of the earlier problem. There was more consideration of a night surface engagement this time but the primary emphasis was again on a day surface action with the Purple Covering Force.[2]

This time, however, there was a slightly longer analysis of the capabilities of Admiral BA's carrier aircraft and their role in potentially knocking out Purple battleship strength after destroying the Purple carriers. In general, there were more tasks assigned to his Carrier Group in the planned-for engagement. For example, the carriers this time were to launch an initial fighter strike to destroy Purple combat air patrols and reduce Purple antiaircraft defenses. The Blue fighters were then to take station as high cover for an initial glide-bombing strike by Blue attack planes. Returning with that strike, the fighters were to continue scouting tactically for Purple fleet units. Admiral BA was hoping to destroy the Purple flight decks with his initial glide-bombing strike—as opposed to the dive-bombing that occurred in the other Operations Problems—though he did not expect to destroy the carriers themselves. His attack planes, however, were to continue glide bombing until all the Purple carriers had been made nonoperational. His fighters were to maintain combat air patrols and his attack planes were to continue scouting, airborne-early-warning, and electronic-warfare missions. The attack planes were then to be prepared for general-purpose bomb strikes as well as aerial torpedo strikes on the Purple battle line. The remaining Battle Plan was again highly similar to Operations Problem 1J: Admiral BA would engage with his Battleline, Center Group, Right and Left Flank Groups, and Detached Wing. The Battle Plan again included a potential Reverse Action and had the Carrier Group on the disengaged side of the formation with a light destroyer screen as it assisted in the battle.[3]

The Purple Statement of the Problem, Estimate of the Situation, and Staff Solution

The Purple Schedule of Events was a mirror of Blue's, as was the Purple Statement. Blue surface-ship strength estimates by Purple Vice Admiral PS (Captain Taylor—see chapter 10) were also identical to those of Operations Problem 1J, the exception again being carrier strength. Whereas in Operations Problem 1J Admiral PS had thought the Blue Alaskan Defense Force would have two to three fleet carriers, he now assumed three to four fleet carriers, backed up by one to two light carriers. Mutual neutralization of Blue and Purple land-based airpower in recent Purple air offensives was again assumed, and it was again supposed that neither Blue nor Purple submarines would be in a position to give significant assistance. PS's estimates of Blue garrison forces in the Aleutians were also the same. The Purple Staff Solution was also nearly identical, Vice Admiral PS analyzing his mission of protecting the Purple Amphibious Force by the interception and destruction—or at least the driving off—of the Blue Alaskan Defense Force. Interestingly enough, although the comparisons of Blue and Purple ship characteristics were highly similar to those of Operations Problem 1J, Vice Admiral PS now speculated from intelligence estimates that a Blue BB-61-class fast battleship and CB-1-class battle cruiser were being armed with main batteries of rocket projectors in place of their guns, though he did not think that these ships were present in the Blue Alaskan Defense Force.

There was again an analysis of competing carrier air strength, weather, and logistics and, especially, the characteristics of the two sides' surface combatants; Admiral PS too saw the likelihood of a day surface engagement. Also not surprisingly, he saw this day action as very similar to that proposed in Operations Problem 1J; namely, that Purple—with a superiority in numbers of carrier planes—would be able to take out the Blue carriers' flight decks and then battleships, battle cruisers, light cruisers, and radar capability. There were also provisions for surface gun and torpedo action by the Purple battle line, Center Group, Right and Left Flank Groups, and Detached Wing as well as continued supporting action by the Carrier Group in renewed strikes, air spotting, and antisubmarine and combat air protection.[4]

A bit more involved, however, was the Detail of Maneuver Staff for the different phases of the Chart Maneuver itself since this exercise did not stop with the Solutions to the Problem but went on to a Chart Maneuver. The Maneuver Staff comprised the Director of the Maneuver, Capt. John Sweeney, head of the Naval War College's Department of Tactics (see chapter 1), along with Liaison Officers for both Blue and Purple: Capt. Cornell Sullivan, head of the Department of Tactics'

Liaison Officer for Purple, Operations Problem 3 (Cdr. Joseph Staley, Instructor, Department of Tactics)

War Gaming Section (see chapter 1); and Cdr. Joseph Staley, an instructor from the Department of Tactics.

The Air Umpire, Commander Johnson, also a Department of Tactics instructor (see chapter 1), had four Assistant Air Umpires—Captains Bruner (see chapter 12) and Tucker (see chapter 13) and Commanders Roudebush (see chapter 10) and Castree (see chapter 13), all senior-class students. The Communication Umpire, Commander Haskell of the Naval Reserve, a Department of Intelligence instructor (see chapter 10), only had one Assistant, his colleague from the Department of Intelligence, Lieutenant Commander Curran (see chapter 1). There was a Chief Damage Computer—Commander Pelling, from the Department of Tactics (see chapter 1)—along with Blue and Purple Move Umpires, Captains Smith (see chapter 10) and Stryker (see chapter 13) from the senior class, as well as Blue and Purple Force Damage Recorders, Captains May (see chapter 10) and Price (see chapter 13), respectively, of the senior class.[5]

The Chief Damage Recorder was assisted by a number of Blue and Purple Damage Computers from the senior class of June 1947. The Blue and Purple Damage Computers for the battle-line battleships and destroyers, for instance, were Captains Bruner and Greenacre (see chapter 13). The Blue and Purple Damage Computers for the Right Flank heavy cruisers were Colonel Vaughan (see chapter 1) and Lieutenant Colonel Negri (see chapter 13); Lieutenant Colonel Sawicki (see chapter 12) and Commander Castree were the Blue and Purple Damage Computers for the Right Flank destroyers. Commander Woodaman (see chapter 12) and Captain John Sweeney, a senior-class student (not the Director, also a John Sweeney—see chapter 13), were the Blue and Purple Damage Computers for the Center Force heavy cruisers. Commander Antrim (see chapter 1) and Captain Sweeney were the Blue and Purple Damage Computers for the Detached Wing battleships and destroyers; Commander Roudebush and Captain Tucker were the Blue and Purple Damage Computers for the Left Flank Force heavy cruisers and light cruisers. Completing this team were Commander Roudebush and Captain Purmort (see chapter 13) as the Blue and Purple Damage Computers for the Left Flank Force destroyers. Since this Chart Maneuver was to be conducted in two phases—Air and Board—the Damage Computers were to assist the Air Umpires during the Air Phase and these officers would be assigned specific positions by the Air Umpire. Additionally, there were four civilian draftsmen, three of them—Mr. Lawton (see chapter 9), Mr. Wilson (see chapter 10), and Mr. Wagner (see chapter 9)—serving as Plotters and one, Mr. Scannevin (see chapter 10), serving as the civilian Historian. All of these men were from the Department of Tactics' War Gaming Section and constituted the personnel of the Drafting Room.[6]

The Blue Staff for the Air Phase involved student officers playing Commander, Task Force 22 (Admiral BA—Captain Madeira), as well as his Chief of Staff (Colonel Shores—see chapter 10), Operations Officer (Captain Agnew—see chapter 12), and Communications Officer (Lieutenant Colonel Shell—see chapter 12). The role of Task Group 22.5 Air Group Commander was also filled (Captain Hoffheins—see chapter 10), along with those of his Chief of Staff (Lieutenant Colonel Lane—see chapter 10), Communications Officer (Captain Lane—see chapter 12), and Move Plotter (Colonel Grossetta—see chapter 10). Commander, Carrier Division 7 (Captain Snyder—see chapter 12) and Commander, Carrier Division 13 (Colonel Koonce—see chapter 10) had similar staffs. Commander Richards (see chapter 12) and Captain Hummer (see chapter 10) served as the respective Chiefs of Staff, while Colonel Erlenkotter (see chapter 10) and Captain Hardin (see chapter 10) served as the respective Operations Officers. Commander Kaplan (see chapter 12) and Colonel Privett (see chapter 10) filled the roles of the respective Communications Officers, while Commander Nisewaner (see chapter 10) and Commander Van Every (see chapter 12) performed the same roles as the respective Move Plotters. Commanders, Pickets Right (Captain Heberton—see chapter 10) and Pickets Left (Captain Outerbridge—see chapter 10) each had a Communications Officer (Commander Munholland [see chapter 10] and Colonel Wood [see chapter 1]).

There was additionally assigned a Move Umpire (again Captain Smith), a Force Damage Recorder (again Captain May), and six Damage Computers (again Captain Bruner, Colonel Vaughan, Commanders Woodaman, Antrim, and Roudebush, and Lieutenant Colonel Sawicki). For the Blue Board Phase, students played the Officer in Tactical Command, Alaskan Defense Force (again Captain Madeira), along with his Chief of Staff and Operations and Communications Officers (again Colonel Shores, Captain Agnew, and Lieutenant Colonel Shell, respectively). The role of Commander, Battleline was also filled by a student officer (Captain Sullivan—see chapter 10), as were those of his Communications Officer, Move Plotter, and Damage Computer (again Captain Hardin, Colonel Erlenkotter, and Captain Bruner, respectively). Commander, Center Group was played by a student without a staff (Captain Van Deurs—see chapter 10) but Commander, Heavy Cruisers, Center (also Captain Van Deurs) had Operations (again Colonel Privett) and Communications Officers (again Commander Richards), a Move Plotter (again Captain Hummer), and a Damage Computer (again Colonel Vaughan). Commander, Light Cruisers, Center (Commander Andrews—see chapter 10) also had a Communications Officer (again Commander Van Every), a Move Plotter (again Captain Snyder), and a Damage Computer (again Commander Antrim) as did Commander, Destroyers, Center (Commander Briggs [see chapter 12] with Lieutenant Colonel Lane, Commander Nisewaner, and Commander Woodaman as his staff officers); Commander, Right Flank (again Captain Heberton with Commander Munholland,

Colonel Koonce, and Commander Roudebush as the staff officers); Commander, Left Flank (again Captain Outerbridge with Colonel Wood, Commander Kaplan, and Lieutenant Colonel Sawicki as the staff officers); and Commander, Carrier Air Group (again Captain Hoffheins with Captain Lane and Colonel Grossetta as the staff officers).[7]

The Purple Staff for the Air Phase included Commander, Task Force (Vice Admiral PS, Captain Taylor), the Chief of Staff (Colonel Cooley—see chapter 13), and the Operations (Commander Woods—see chapter 13) and Communications Officers.

There were also two Task Group Air Coordinators (Captain Ricketts [see chapter 13] and Commander Hydeman [see chapter 1]); commanding officers of the three fleet and two light carriers (Captain Higgins [see chapter 15], Colonel Luckey [see chapter 13], and Colonel Lyle [see chapter 1] commanding the fleet carriers and Commander Lyons [see chapter 13] and Captain Ricketts commanding the light carriers); and Commander, Purple Air Forces (Colonel Bidwell—see chapter 13), his Air Officer (Commander Matter—see chapter 13), his Communications Officer (Commander Hay), and his Fire Director (Colonel Simonsen).

The Purple Board Phase roles consisted of the OTC, his Chief of Staff, and his Operations and Communications Officers (again Captain Taylor, Colonel Cooley, Commander Woods, and Colonel Johnson). Commander, Battleline (Rear Admiral PL—Captain Ashford [see chapter 13]) also had Operations and Communications Officers as well as a Move Plotter (all played by Commander Stevens).

These three staff billets were additionally filled for Commander, Center and Detached Wing (Rear Admiral PR—Captain Goulett [see chapter 13]) by Commander Hay; for Commander, Cruisers, Center (Captain P-1—Captain Greer [see chapter 13]) and for Commander, Destroyers, Detached Wing (Captain Heming) by Robert Strong; Commander, Right Flank (Rear Admiral PZ—Colonel Luckey) only had Operations and Communications Officers (both played by Captain Higgins), but Commander, Cruisers, Right Flank (also Colonel Luckey) had not only these slots filled but also a Move Plotter (all played by Captain Higgins). So did Commander, Destroyers, Right Flank (again Captain Ricketts, with Lieutenant Colonel Bowen [see chapter 13] as his various staff officers); Commander, Cruisers, Left Flank (Captain Fitzsimmons [see chapter 15] with Commander Lyons as his staff officers); and Commander, Destroyers, Left Flank (Commander Hydeman, with Colonel Lyle as his staff officers). Commander, Left Flank (Rear Admiral PT—again Captain Fitzsimmons) only had Operations and Communications Officers (again Commander Lyons in both billets) assisting him, but Commander, Carrier Air Group (Colonel Bidwell) had a Staff Operations Officer (Commander Matter), a Communications Officer (also Commander Matter), and a Move Plotter (Commander Matter again). Commander, Destroyer Squadron 17 (Captain

Communications Officer, Commander, Purple Task Force 82 (Col. Wilhelm Johnson, USA)

Communications Officer, Commander, Purple Air Forces (Cdr. Richard Hay, USN)

Fire Director, Commander, Purple Air Forces (Col. Edwin Simonsen, USA)

Operations Officer, Communications Officer, and Move Plotter, Commander, Purple Battleline (Cdr. Clyde Stevens, USN)

Operations Officer, Communications Officer, and Move Plotter, Commander, Purple Cruisers, Center (Robert Strong, U.S. State Department)

Commander, Destroyers, Purple Detached Wing (Capt. Harold Heming, USN)

Tackney—see chapter 13) had a Move Plotter (again Captain Tackney) and a Fire Director (again Colonel Simonsen).[8]

The History of the Maneuver

Unlike for Operations Problem 1J, a History of the Maneuver was recorded for Operations Problem 3. It began with a Preliminary History, which in turn began with weather conditions at the start of the Chart Maneuver. After giving wind force, sea conditions, visibility, ceiling, cloud cover, sunrise, and sunset, the Preliminary History noted the Blue force's task of bringing the Purple surface combatant forces in the northwest Pacific to decisive action in order to deny Purple the ability to establish forces ashore in the Aleutians. At 0600 on 28 July (Maneuver Time and Date), the Blue Alaskan Defense Force effected a rendezvous with Task Unit 11.2.2 and refueled on a southwesterly course. Just prior to the completion of refueling, Commander, Blue Alaskan Defense Force received a message from the Blue Commander-in-Chief, Pacific Fleet that Purple assault troops were landing on Attu. Upon completion of fueling, Commander, Blue Alaskan Defense Force ordered

Commander, TU 11.2.2 to operate to the eastward of 180 degrees east and south of 42 degrees north to await orders. Commander, Blue Alaskan Defense Force then proceeded at twenty knots to Point X-RAY, 47°31′ N, 173°40′ E. At 0300 on 29 July at Point X-RAY, Commander, Blue Alaskan Defense Force launched a search with bearing limits of 340 degrees and 010 degrees true out to 355 miles for the purpose of locating the Purple Covering Force. At 0430, he launched a day combat air patrol of twelve planes and landed eight night fighters. At 0500, a destroyer in TG 21.1's screen made a sound contact and conducted an unsuccessful depth-charge attack. Fifteen minutes later, Commander, Blue Alaskan Defense Force received a contact report from one of his search planes that a Purple task group of three to four fleet carriers, two to three fast battleships, three to four heavy or light cruisers, and sixteen to twenty destroyers had been spotted at 52°36′ N, 171°30′ E. At 0520, he received another contact report from the same search plane that a second group of Purple ships of similar composition had been sighted bearing 240 degrees, eight miles from the previously reported Purple group.[9]

Purple's Preliminary History recorded its task to seize, occupy, and defend Attu in order to develop it into a base for future long-range air operations against Blue bases in the eastern Aleutians and Alaska. The Purple Covering Force was more exactly ordered to cover the landings by the Purple Amphibious Force and exploit all opportunities to destroy Blue surface and air forces encountered north of 40 degrees north latitude and west of 170 degrees west longitude. At 2000 Zone M time, the Purple 8th Fleet—which included Task Force 81, the Covering Force—sortied from the Gulf of Kamchatka and Avacha Bay and proceeded to Attu. D-Day remained 28 July and H-Hour was set at 0500. The operation proceeded as planned and the first wave of Purple assault troops landed on Attu at 0510 on 28 July. At 0130 on 28 July, Commander, Purple Covering Force received a contact report from a submarine informing him that a large group of Blue ships—including several large vessels and destroyers—was sighted at 0100 on 28 July at 46°53′ N, 173°57′ E proceeding on a northerly course at about twenty knots. At about 0300, Commander, Purple Covering Force launched a search between 145 degrees and 180 degrees out to a distance of 340 miles to locate the Blue Task Force. At 0400, he also launched a day combat air patrol of twelve planes and landed his eight night fighters. At 0515, he received a report from one of his search planes that an enemy force composed of three fleet or light carriers, two to four battleships or battle cruisers, two light cruisers, and sixteen to twenty destroyers had been sighted at 48°10′ N, 173°25′ E. Five minutes later, the same pilot reported on a second Blue task group of the same composition about ten miles from the first group of Blue ships and bearing 270 degrees.[10]

As the maneuver began at 0515, game time, with Move 1, the Blue Heavy Surface Striking Force was in its Cruising Disposition on a course of 315 at a speed of sixteen knots. At 0600, it changed course to 347 and increased speed to twenty-seven knots. At the beginning of Move 1, there was one Blue destroyer division on picket lines out to thirty-five miles, thirty degrees each side of the Fleet Axis. Purple was on course 140 at twenty knots in its Cruising Disposition. At 0515, both forces turned into the wind and they both began launching planes at 0530. The Purple Covering Force was in several groups that were separated by ten miles: a section of destroyer pickets fourteen miles ahead of the Center, one section sixty degrees to the right of the right-hand group fifteen miles out, and one section sixty degrees to the left of the left-hand group, also fifteen miles out. At 0515, both Blue and Purple had two to three single bogies on their radar screens and by 0530 Purple had a combat air patrol of twenty-one planes aloft, while Blue had a CAP of thirty-nine planes in the air. Also at 0530, Blue Search Groups 4-VF and 3-VA reported the composition, course, and speed of the Purple Covering Force as they circled it. At the same time, the Purple night search plane noted above reported the composition, course, and speed of the Blue Alaskan Defense Force to his Air Group Commander. In addition, Blue night search plane 381-A-19, an airborne-early-warning aircraft, was shot down by Purple. Purple night search plane 471-A-16 was shot down by Blue at the same time. Blue AEW aircraft 301-A-10, on a bearing of 347 at thirty thousand feet and seventy-five miles ahead of the Blue Alaskan Defense Force, reported a "lot" of radar pips at approximately two hundred miles and appearing to orbit.[11]

Between 0530 and 0540, Blue began launching several strikes. Attack A consisted of thirty-six fighters, sixteen attack planes loaded with torpedoes, another sixteen attack planes each with a thousand-pound general-purpose bomb, two photographic reconnaissance fighters, and two AEW planes. Attack B consisted of twenty-four fighters each with a thousand-pound general-purpose bomb, twelve fighters each carrying eleven-inch rockets, eight attack planes with torpedoes, sixteen attack planes each with a thousand-pound general-purpose bomb, two photographic reconnaissance fighters, four tactical electronic-warfare aircraft, and two AEW aircraft. Attack C consisted of twenty-four fighters as escorts, twelve attack planes each with a thousand-pound general-purpose bomb, eight attack planes with torpedoes, two photographic reconnaissance fighters, four tactical electronic-warfare aircraft, and two AEW aircraft. Attack D consisted of twenty-four fighter escorts as well as eight attack planes carrying torpedoes, twelve more attack planes with a thousand-pound general-purpose bomb each, two photographic reconnaissance fighters, and two AEW aircraft. Finally, Attack E had twelve fighters with thousand-pound general-purpose bombs each, eight attack planes with the same ordnance load, and nine additional attack planes with torpedoes.

At 0540, Purple AEW aircraft 332-A-17—about thirty-five miles ahead of the Purple Covering Force—picked up many low-flying aircraft about two hundred miles away on a bearing of 167. At 0531, the Purple OTC had ordered his strike planes to attack in succession, those from TG 81.1 attacking first. At 0551, Purple launched thirty-seven fighters with orders to intercept Blue Raid 1. At 0600, Purple also launched 118 fighters and 111 attack planes, along with three radar countermeasures aircraft. All the attack planes had maximum loads of five-hundred-pound general-purpose bombs. At this time, Blue had sixty-two planes available for combat air patrols, and Purple had eighty-six. (Page 2.)

At 0635, the Purple OTC picked up Blue aircraft coming in at a distance of a hundred miles. At the same time, Blue vectored six fighters of its combat air patrol to a course of 270; at 0653, they began orbiting the Blue Alaskan Defense Force. Blue also diverted two of its AEW planes from Strike D to a position previously occupied by two AEW planes with which Blue had lost contact. Two of the same types of planes from Strike E were sent to a position previously occupied by another of the Blue AEW aircraft; the two relief planes were stationed at ten and thirty thousand feet, respectively. By 0635, the thirty-seven Purple planes vectored out at 0600 had made contact with Blue Attack Group A.

Blue lost fourteen fighters and seventeen attack planes in the engagement; Purple lost twenty-five fighters. Simultaneously, the Blue radar-picket line detected many planes flying high and the Purple AEW planes flying with their strike made contact with many planes on a northerly course. Weather conditions were also announced at 0635. Fifteen minutes later, Blue picked up Purple aircraft closing at a distance of a hundred miles. Two minutes later, Purple vectored out sixteen fighters to intercept Blue Strike B, which reportedly consisted of thirty-five fighters, twenty-four attack planes, two photographic reconnaissance planes, four radar countermeasures aircraft, and two weather planes. In the melee, Purple lost ten fighters while Blue lost four fighters and six attack planes. Also at 0652, the Blue fighter squadron commander gave "tally-ho" and reported fifty to two hundred "rats," "chicken," and "fish." Six minutes later, twenty Blue fighters intercepted the Purple raid. Blue lost twelve fighters and the remainder of its planes were out of ammunition. Purple lost fourteen fighters and five attack planes and had six additional fighters that were also out of ammunition. Simultaneously, Purple's sixteen-fighter combat air patrol intercepted Blue Strike C. It lost ten fighters and its six remaining planes were out of ammunition, while Blue lost four fighters, six attack planes, and one tactical electronic-warfare plane. In sum in Move 1, Blue had lost thirty-four fighters and thirty-one attack planes while Purple lost fifty-nine fighters and six attack planes. (Pages 3–4.)

In Move 2, Blue continued on a course of 347 at a speed of twenty-seven knots while Purple continued on a course of 140 at a speed of twenty knots. At 0705, thirteen Purple fighters intercepted Blue Raid D. In the engagement, Purple lost nine fighters and its remaining four fighters were out of ammunition. Blue lost four fighters, four attack planes, and one photographic reconnaissance fighter. At 0707, forty-two Blue fighters intercepted the Purple strike. In the fight, Blue lost twenty-eight fighters and Purple lost ten fighters and sixteen attack planes. At the same time, Blue Strikes A and B attacked Purple fleet carriers CV-40 and CV-47 as well as Purple light carrier CVL-29. Twelve Blue fighters, each armed with two Tiny Tim rockets, attacked light carrier CVL-29. Two of these planes were lost to five-inch gunfire and three were lost to automatic weapons. Five hits, however, were scored on the Purple ship to an effect of 95 percent Above Water Damage and light carrier CVL-29 was ruled sunk. Twelve Blue fighters strafed as well and eleven Blue attack planes with torpedoes attacked Purple fleet carrier CV-40. Four Blue aircraft were lost to five-inch gunfire and another three to automatic weapons but two hits were scored on CV-40, which was ruled to have 40 percent Above Water Damage, have its speed slowed to twenty knots, and have its two deck elevators out of commission for two hours. In addition, twelve Blue fighters strafed Purple CV-47 and twenty-three attack planes struck that ship with thousand-pound bombs. Six Blue aircraft were lost to five-inch gunfire and another seven to automatic weapons but the ten bombs dropped netted two hits for 24 percent damage on CV-47. Although Blue lost nineteen fighters and fifteen attack planes in total, the Purple OTC put CV-40 in the Center of the task group and slowed the formation to twenty knots to protect the damaged ship. At 0700, five Purple tactical electronic-warfare aircraft began dropping window as well. Seven minutes later, the Purple OTC also ordered four of his combat air patrol fighters to return to the fleet area to intercept and attack any Blue stragglers. (Pages 5–6.)

At 0717, the first Purple strike attacked in three sections from ahead and on each box. The first section of Purple planes attacked Blue CV-37 and light carrier CVL-48. The second section attacked CV-39 while the third attacked CV-38 and light carrier CVL-30. Fleet carriers CV-37, -38, and -39 were each damaged 35 percent. Purple began the attack with fifty fighters and forty-six attack planes. It lost sixteen planes to five-inch gunfire and another eighteen planes to "sub-caliber" guns. Twenty-five fighters and thirty-seven attack planes actually made the attack and they obtained fifteen hits on the Blue ships. After the attack, thirteen fighters and eighteen attack planes got away. Purple's second strike attacked Blue light carriers CVL-30 and CVL-48 with thirty-nine fighters and forty-six attack planes. It lost fourteen planes to five-inch gunfire and another nineteen to automatic weapons. Twenty-two fighters and thirty attack planes actually made the attack and achieved six hits on the Blue ships. Light carrier CVL-48 was 36 percent damaged and light carrier CVL-30 was 40 percent damaged. Eleven Purple fighters and fifteen attack planes got away, with a total of twenty-eight fighters and thirty-one attack planes lost. Also at 0717, four Purple fighters intercepted Blue Raid E. Purple lost all four of the fighters but Blue lost two fighters, an attack plane carrying a torpedo, and an attack plane armed with a bomb. At the same time, the Blue Task Force Commander directed the OTC to have his heavy carriers return to Dutch Harbor and effect repairs in the minimum possible time. He also requested that replacement aircraft be flown to Dutch Harbor at the earliest possible date and for the OTC to report to the Task Force Commander when the carriers completed this replacement. (Page 6.)

At 0717, Blue Strikes C and D attacked Purple CV-33 and light carrier CVL-49. Twelve fighters, eight attack planes with thousand-pound bombs, and six attack planes with torpedoes struck CV-33. Nine Blue planes were lost to five-inch gunfire and six more to automatic weapons. Eleven planes actually made the attack. One bomb hit and one torpedo hit on CV-33 resulted in 12 percent Above Water Damage and 20 percent Under Water Damage. Twelve more Blue fighters and six attack planes armed with torpedoes plus ten attack planes armed with thousand-pound bombs "started in on" light carrier CVL-49. Eight of these planes were lost to five-inch gunfire and eight more to automatic weapons. Twelve planes actually made the attack. One torpedo hit and one bomb hit caused 20 percent Above Water Damage and 23 percent Under Water Damage to light carrier CVL-49. Altogether, twelve of the Blue fighters and twenty of the attack planes were lost. One minute later, Commander, Blue Alaskan Defense Force ordered his Task Force to change formation positions when the carriers departed for Dutch Harbor. He was to become the OTC and the Fleet Guide was to shift to fast battleship BB-61, which would remain at its same point in the Center of the formation. Fast battleships BB-63 and BB-64 would close to Circle 3 on the same bearing, while light cruisers CL-86 and CL-91 would close to Circle 4 on the same bearing. In addition, light cruisers CL-89 and CL-101 would close to Circle 3 on the same bearing and Destroyer Squadron 21—less Destroyer Division 212—would form a screen ahead of the battleships in Circle 4 from bearing 314 to 045. (Pages 6–7.)

Ten minutes later, Blue Strike E—consisting of twelve fighters, eight attack planes with thousand-pound bombs, and nine attack planes with torpedoes—struck Purple CV-40. Eight of the Blue planes were lost to five-inch gunfire and six more to automatic weapons. Eleven planes, however, got in to achieve one torpedo hit, inflicting CV-40 with 20 percent more Under Water Damage for a total now of 60 percent Total Damage and reducing its speed to twelve knots. Blue lost three more aircraft on retirement for a total loss in this raid of six fighters and eleven attack planes. At 0730, Blue Strikes A and B were ordered to Shemya, as was Carrier Division 13, while Carrier Division 7 went to Dutch Harbor. Carrier Division 13 was on course 011 at a speed of twenty-four knots, while Carrier Division 7 was on course 072 at the same speed but zigzagging. The carriers were ordered away specifically to assist planes in getting ready for another strike and to influence the Purple OTC into thinking that Blue had formed its Detached Wing. Accordingly, at the same time as the Blue ships were ordered away, Commander, Purple Covering Force sent a message to one of his AEW planes to keep the enemy carriers proceeding toward the Aleutians under surveillance and verify the compositions of the formations. Damage was also assessed at this point. (Page 7.)

At the end of Move 2, Blue had lost seventy-two fighters and fifty-two attack planes while Purple had lost sixty fighters and forty-four attack planes. In addition, Blue fleet carriers CV-37, CV-38, and CV-39 had each sustained 35 percent damage, while light carrier CVL-48 had taken 36 percent damage and light carrier CVL-30 had taken 40 percent damage. Purple had seen light carrier CVL-29 sunk and CV-47 was 24 percent damaged. In addition, CV-33 had 12 percent Above Water and 20 percent Under Water Damage; light carrier CVL-49 had had 20 percent Above Water and 23 percent Under Water Damage; and CV-40 had taken 60 percent Total Damage, 20 percent Above Water and 40 percent Under Water Damage. (Page 7.)

In Move 3, Blue continued in its Cruising Disposition on course 347 at a speed of twenty-seven knots. At 0730, both Purple forces went into Cruising Disposition 5-R and changed course to 315. At 0750, Purple changed course to 160 and landed its combat air patrol. At 0830, Purple changed course back to 315 at a speed of twelve knots and landed its returning strike planes. Thirty minutes before, it had had two AEW planes shadowing Blue while Blue had five AEW planes shadowing Purple. Both forces received contact reports on their adversaries and the Purple AEW aircraft picked up the Blue carriers on a northerly course at twenty-four knots. Also at 0800, Blue sent fighters to reconnoiter the route from Attu to Shemya as Blue Carrier Division 13 changed course to 000 degrees at twenty-four knots and began zigzagging. Fifteen minutes later, Blue Carrier Division 7 changed course to 162 at twenty-four knots and was also zigzagging. At 0900, Blue changed to its Approach Disposition on course 016 at a speed of twenty-seven knots while Purple lost five of its attack planes on recovery twelve minutes later. At 0930, the beginning of Move 4, Blue was on course 000 at a speed of twenty-seven knots. Blue completed its shift to a modified Approach Disposition with its cruisers in the Van in a V formation and screened by eighteen destroyers. The Blue battle line was ten miles behind, also in a V formation that was screened by another eighteen destroyers. At the same time, Purple was on a course of 160 at a speed of twelve knots and one of its fighters shot down a Blue attack plane deployed as an AEW asset. At 1000, Purple changed course to 090 while maintaining its speed of twelve knots. (Pages 8–9.)

At 1009, the Blue Commander, Center Force received a message from the Blue OTC. The Blue leader was envisioning the worst possible situation he could encounter, one in which two Purple battleships were in the Van with the Purple heavy cruisers. In this event, the Blue OTC intended to advance one of his BB-61-class fast battleships toward the Purple battleships to a range of thirty-five thousand yards and take both under divide fire. He therefore "proposed" to Commander,

Center Force that the latter officer advance his battle cruisers up on the bow and concentrate on the leading Purple battleship with fire from both battle cruisers. He also wanted Commander, Center Force to make a torpedo attack on the Purple battleships at the earliest opportunity. With a combined assault from the Blue battleship, the battle cruisers, and the torpedo strikes, he felt confident that he could "finish" the Purple battleships. When these were out of action, he intended to have his BB-61-class fast battleship drop back on Purple's bow toward the remaining Purple battleships and he again proposed that Commander, Center Force have his battle cruisers drop back to thirty-five thousand yards from the Purple ships. At this point, the Blue OTC proposed to have his four BB-61-class fast battleships and his two CB-1-class battle cruisers concentrate on the remaining three Purple battleships. Commander, Center Force was then to attempt crossing the outside Purple range. (Page 9.)

At 1030, Purple changed course to 315, still at a speed of twelve knots, and the Purple fleet carriers and Destroyer Division 172 were ordered to withdraw from the Purple Disposition and act as the Carrier Group. The latter ships were, however, to conform to the Main Body's movement and other forces were to form the Approach Disposition in accordance with previous instructions. Destroyer Squadron 17—less Destroyer Division 172—was to join the Force Center and, upon deployment, join the Van Force destroyers in the outer area on approach axis 170 and course 090. At the same time, the Blue cruisers in their Van were in a circular position screened by their eighteen destroyers as was the Blue battle line, with the latter in a line of division and screened by the additional eighteen destroyers. At 1030, Blue Cruiser Division 17 launched eight observation planes to an altitude of eighteen thousand feet, each one carrying two hundred-pound general-purpose bombs and two 350-pound depth charges. These planes had orders to strike the deck of Purple CV-40. Simultaneously, Purple CV-40 began launching a strike on Blue light carriers CVL-30 and CVL-48. The strike actually departed at 1102 in two groups and arrived over the target at 1138. Group 1 consisted of six fighters with one Tiny Tim rocket each and eleven fighters with thousand-pound general-purpose bombs each. Two fighters were shot down going in but the strike made one hit on each light carrier. Three more Purple planes were shot down getting away. Group 2 attacked the Blue battle line. This strike consisted of eighteen fighters armed with Tiny Tims, thirty attack planes loaded with torpedoes, and another nine attack planes with one 1,600-pound armor-piercing bomb each. The Purple planes achieved five rocket and two torpedo hits on each of the two Blue column leaders. Ten fighters and twenty-four attack planes were lost from the Purple strike to antiaircraft fire and Purple fleet carriers lost four attack planes and five fighters upon launching its strike. (Pages 9–10.)

At 1030, one of the Purple AEW aircraft saw five to seven planes launched from the Blue cruisers. At 1100, Purple changed course to 090—still at a speed of twelve knots and in its Approach Disposition—with two battleships and two light cruisers in the Center Force screened by nine destroyers, and heavy cruiser CA-124 and light cruisers CL-56 and CL-57 in the Right Flank Force with Destroyer Divisions 71 and 72. Its Left Flank Force consisted of heavy cruiser CA-122, light cruisers CL-55 and CL-58, and Destroyer Divisions 91 and 92, while its battle line was composed of Battleship Division 3 and five screening destroyers. The four remaining Purple carriers were screened by four destroyers, with CV-40 in the Center of this circular formation. (Page 10.)

Also at 1100, the Purple OTC received a message from Commander, Detached Wing. The latter officer informed his superior that unless he was otherwise advised, he would use the following plan to initially deploy his Wing and Destroyer Division 112 during the approach: until the nearest Blue unit was within fifty miles, he would employ Destroyer Division 112 as a circular open screen out to a distance of four thousand yards from the Center of Battleship Division 4. When Blue had approached to within fifty miles, he would use Destroyer Division 112 in specific ways. He would put one section of destroyers on bearing 020 relative from his axis and the other on bearing 340 relative to his axis—both eight thousand yards from the Center of Battleship Division 4—so that under any circumstances forty torpedoes could be fired on any Blue heavy ships in the Van. At 1105, Commander, Detached Wing sent the Purple OTC another part of his message in which he stated that at "appropriate" ranges, he would form his two sections of destroyers so that they could provide crossing fire and he could use Destroyer Division 112 in what he saw as his only offensive opportunity. Battleship Division 4 would engage the Blue battle cruiser but then retire to the Purple battle line if Blue BB-61-class fast battleships were encountered. (Pages 10–11.)

Toward the end of Move 4, specifically at 1109, five of the Blue observation planes were shot down by the Purple combat air patrol. A minute later, however, the three remaining observation planes from Blue Cruiser Division 17 got in to make their attack. All three of these planes were shot down, but so were two Purple fighters and the attack resulted in Purple CV-40 taking on a twenty-degree list. At 1130, Blue also advanced one battleship with two screening destroyers to a position four miles astern of the Blue Center Group. During Move 4, Blue had lost all eight of its observation planes from Cruiser Division 17 as well as one attack plane. Purple, however, had lost twenty-two fighters and twenty-eight attack planes. For all four Moves, Blue air losses now amounted to 106 fighters, eighty-four attack planes, and the eight observation planes. Purple air losses now amounted to a total of 141 fighters and eighty-three attack planes. (Pages 11–11a.)

At the beginning of Move 5, Blue was on course 005 at a speed of twenty-seven knots, still in its modified Approach Disposition and with fast battleship BB-63 closing the Van Force. Purple was on course 130 at a speed of twenty knots in an axis of 170 in its Approach Disposition. At 1138, Blue fast battleships BB-61 and BB-62 were hit by two aerial torpedoes each and suffered 30 percent Under Water Damage. Rockets caused another 6.5 percent Above Water Damage and the Director of the Maneuver ruled that the Above Water Damage would remain but that the Under Water Damage would be restored. At 1140, Purple CV-40 took on such a list that it was considered out of commission. At 1224, the Blue and Purple forces were placed on the Maneuver Board. At the beginning of Move 6, this placement entailed the Blue Main Body of three battleships screened by Destroyer Squadrons 12 and 21 on Circle 10. Van Forces, led by Cruiser Division 4, were on Circle 24, while Cruiser Division 17 was on Circle 20. Fast battleship BB-63, with two screening destroyers, was four miles astern of the two Blue battle cruisers. Destroyer Squadrons 2 and 4 were also with the battle cruisers, and Blue by now had reduced speed to twenty knots. The Blue cruisers were in a line of bearing 135 and the destroyers in the Van were in a column of twos. The battleships were in line of Division Guide, with Battleship Division 1 in column. (Pages 12–13.)

At the beginning of Move 6, Purple was in an Approach Disposition. Battleship Division 3, the Battleline, was screened by Destroyer Division 111. The Center Force consisted of fast battleships BB-55 and BB-56, screened by Destroyer Division 112 about eleven miles on the axis from the battle line. Heavy cruisers CA-123 and CA-132 were about four thousand yards astern of the Van battleships, with Destroyer Division 171 in two sections, one section on each side of the heavy cruisers. The Right Flank Group consisted of two light cruisers bearing fifty degrees and six miles from the battle line as well as one heavy cruiser about eleven miles and fifty-five degrees from the battle line. The Right Flank Group also consisted of Destroyer Squadron 72 midway between the heavy and light cruisers. The Left Flank Group consisted of two light cruisers, CL-55 and CL-58; heavy cruiser CA-122; and Destroyer Squadron 9 in a similar formation to that of the Right Flank Group but on a course of 130 and an axis of 170. At 1224, carriers and carrier aircraft were ruled out of the Problem. Thereafter, Blue began a change of course from the Van to 035 at a Task Force speed of twenty knots. Fast battleship BB-63 was still closing the Blue battle cruisers in the Van and the destroyers in the battle line screen began taking normal flank approach positions on Circle 10, sixty degrees to the right and left of the battle line. At 1230, Purple changed course to 165, still at a speed of twenty knots but with no other change in disposition. At the end of Move 6, all available observation planes were airborne. (Page 13.)

At the beginning of Move 7, Blue shifted its battle line to a line of bearing. By this time, fast battleship BB-63 had closed the Blue battle cruisers to about two thousand yards and the Blue Task Force continued its change of course that was started in Move 6. Purple continued on its course of 165 with a speed of twenty knots but Purple Destroyer Division 171 with the Center Force was now maneuvering to form its two sections into one division. At 1200, Purple also had launched two observation planes and then launched twenty-three more at 1233. At 1236, the Blue battle line received a message from Commander, Blue Battleline. He ordered that fire on the Purple Van be undertaken at thirty-eight thousand yards and that slow fire be employed until the Purple ships were within "hitting gun range." His ships were also to use air spot modified radar control and then shift to full radar control when the range permitted. At 1236 and 1239, the end of Move 7, the heavy ships in both the Blue and Purple Vans made radar contact, with each force about fifty thousand yards from the other. By the beginning of Move 8, Blue fast battleship BB-63 was on the line with the Blue battle cruisers. Destroyers that had been escorting the battleship were falling back astern of the Center Force destroyers and changing speed to thirty knots on a course of 035. Purple by this time had changed course to 120 at a speed of twenty knots. At the end of Move 8, the range between the Blue and Purple Center Forces had closed to forty-one thousand yards. (Page 14.)

At the beginning of Move 9, Blue changed course to 080 at a speed of thirty knots. Purple Commander, Detached Wing requested permission at the beginning of the move for his destroyers to fire long-range torpedoes at the Blue force but permission was denied by the Purple OTC. Purple Commander, Detached Wing then retired his destroyer screen to a position two thousand yards on the disengaged side. He subsequently changed course to 045 to attempt to bring the Blue Van bearing to 180. The question was asked, however, why the Purple Detached Wing did not simply continue to close. At the beginning of Move 10, Blue changed course to 035, still at a speed of thirty knots. The Purple Detached Wing continued on course 045 and the Purple battle line changed course to 080 at a speed of twenty-two knots. Move 11 began with Blue still on course 080 at a speed of thirty knots and the Purple battle line still on a course of 080, but with a speed of fourteen knots. The Purple Detached Wing was now falling back on the battle line and continuing on course 045. By the end of Move 11, the range between the two Van forces was thirty-eight thousand yards, at which range Blue fast battleship BB-63 opened fire on Purple fast battleship BB-55. Halfway through Move 12, the Blue battle cruisers and fast battleship BB-63 changed course to 090 at a speed of thirty knots. The Purple Detached Wing also changed course to 090 at the same speed and the Purple battle line changed course to 160 while increasing speed to twenty-four knots. The Purple Detached Wing opened fire on Blue fast battleship BB-63 and the Blue battle cruisers at a range of thirty-five thousand yards, while the Blue ships did the same to the Purple Detached Wing at the same range. At 1302, the Blue battle line received a message from Commander, Battleline. He ordered his battleships to open fire on the Purple Center Force battleships at forty thousand yards with slow fire and with fire distribution from the Van fast battleship BB-61, then BB-62 and BB-64. The Blue battle line was to maneuver to remain outside of thirty-five thousand yards from the Purple battleships in the Center Force. (Pages 15–16.)

Move 13 saw Blue continuing on course 090 at thirty knots and the Purple battle line continuing on a course of 160 at a speed of twenty-seven knots. The Purple Right Flank heavy cruiser CA-124 was off about six or seven thousand yards from the nearest ship and the Purple Detached Wing was on course 090, while the Purple Center Force was on course 120. Move 14 began with the Blue battle line turning to the north, and its Left and Right Flank Forces maintaining position on the battle line. The Blue Van Forces were on an easterly course at a speed of thirty knots. The Purple battle line changed course to 120 at a speed of twenty-seven knots, while the Purple Center Force continued on course 120, the Detached Wing changed course to 103, and the Flank Forces kept station on the battle line. Move 15 saw Blue continuing on a northerly course at a speed of twenty-eight knots, its battle cruisers making thirty-three knots, and the attached fast battleship making thirty knots. The Purple battle line continued on course 120 at twenty-seven knots, though it was ruled that these units were separated because of "excessive" speed. The Purple Center Force continued on course 120 and the Detached Wing remained on course 103. Also at this time, Purple fast battleships BB-56 and BB-55 reopened fire on Blue fast battleship BB-63. The question was also asked of the student officers what would have happened if Purple had changed course to the west. Simultaneously, Blue battle cruisers CB-1 and CB-2 opened fire on Purple fast battleship BB-56, which was now under triple concentration fire from the Blue ships. At the end of Move 15, as a result of this gunfire, Purple fast battleship BB-56 had Above Water Damage of 10 percent. (Pages 17–18.)

In Move 16, Blue changed course to the northeast at a speed of twenty-eight knots. The Purple battle line continued on course 120 at a speed of twenty-four knots and the Purple Center Force changed course to 085 at a speed of thirty-two knots, while the Detached Wing continued on its course of 103 with a speed of twenty-seven knots. Purple fast battleships BB-55 and BB-56 were still concentrating their fire on Blue fast battleship BB-63. According to the Maneuver Rules, the Chance Factor would have sunk the Blue battleship but the Director of the Maneuver ruled that the sinking would not apply at this time. Instead, he awarded the Blue battleship 10 percent damage and reduced its speed to fifteen knots for one move. In Move 17, Blue continued on its northeasterly course at twenty-eight knots and the Blue Van Force took the Blue Fleet Course at a speed of thirty knots. The Purple battle line continued on course 120 at twenty-four knots, the Center Force continued on a course of 085 at thirty-two knots, and the Detached Wing continued on course 103 at twenty-seven knots. Blue fast battleship BB-63 now shifted its fire to Purple fast battleship BB-55 and Purple fast battleship BB-56 shifted its fire to Blue battle cruiser CB-1. By this time, Blue fast battleship BB-63 had 20 percent Total Damage. (Page 18.)

At the beginning of Move 18, the Blue battle line changed course to 050 at a speed of twenty-eight knots. The Purple battle line continued on course 120 at twenty-four knots but the Purple Center Force retired on the battle line to the northeastward at a speed of twenty-seven knots. The Purple Detached Wing also retired on the battle line to the northeastward and the Purple Left Flank Force was now due east of its battle line and retiring to the north. The Purple Right Flank Force was still maintaining its position relative to the Purple battle line. Blue fast battleship BB-63 was still under fire from Purple fast battleship BB-55 and was sunk by the Chance Factor but the Director of the Maneuver instead ruled an additional 10 percent damage and an arbitrary speed reduction to twenty-two knots instead. He also ruled that this damage could be restored later. At 1321, Commander, Blue Battleline sent his ships a message in which he ordered them to be prepared to take the Purple cruisers or battleships under fire at a maximum range of forty thousand yards. The fire distribution was to be Normal from the Purple Van, with slow fire until hitting gun range was established. (Page 19.)

At the end of the move, Blue fast battleship BB-63 had 30 percent Above Water Damage and had expended 24 percent of its ammunition. At the beginning of Move 19, Blue changed course to 010 and increased its speed to thirty-one knots. Blue light cruiser CL-101 left the division formation to take station 070 on Circle 6 from the battle line. The Purple battle line continued on course 120 at a speed of twenty-seven knots, while the Purple Center Force was still on its northeast course at thirty-two knots and the Purple Detached Wing was on the same northeasterly course at twenty-seven knots. Purple fast battleship BB-56 was now dividing its fire between Blue battle cruisers CB-1 and CB-2 but the Purple battleship was also 20 percent damaged by end of Move 19. Blue fast battleship BB-63 took another 10 percent damage during this move for Total Damage now of 40 percent. The Director of the Maneuver additionally ruled now that all ships that had lost radar because of 30 percent Above Water Damage would be assessed a two-tenths reduction for Fire Effect when outside the range of visibility, in this case twenty thousand yards. (Pages 19–20.)

In Move 20, Blue continued on course 010 at a speed of thirty-one knots, while the Purple battle line continued on course 120 at a speed of twenty-seven knots. The Purple Center Force changed course to 100 at a speed of thirty-two knots and the Purple Detached Wing changed course to 090 at a speed of twenty-three knots.

The Blue battle line then fired on Purple heavy cruiser CA-124 with a triple concentration. At the beginning of Move 21, Blue changed course to 050 at a speed of twenty-eight knots. The Purple battle line also changed course, this time to 215 at a speed of twenty-seven knots while the Purple Center Force changed course to 060 at twenty-seven knots and the Detached Wing changed course to 225 at twenty-four knots. Purple light forces in the Van then appeared to swing to the left out of fear of Blue Van heavy units that were closing in the direction of Purple. It also appeared at this point that the Purple OTC would welcome any maneuver by the Blue Van Commander that would place these three high-value Blue heavy units in a position where Purple could expedite a torpedo attack. It appeared just as important to the Purple Liaison Officer that those three Blue heavy units be eliminated as might any three units from the Blue battle line. When the Purple forces turned to course 215, Blue battle cruisers CB-1 and CB-2 were each 10 percent damaged and Purple heavy cruiser CA-124—in the Right Flank Force—opened fire on Blue Destroyer Division 122. At the end of the move, Purple heavy cruisers CA-122 and CA-124 were also each damaged 10 percent. (Pages 20–21.)

Move 22 saw Blue continuing on course 050 at a speed of twenty-eight knots, the Purple battle line change course to 205 at twenty-seven knots, the Purple Center Force continue on course 060 at twenty-seven knots, and the Purple Detached Wing change course to 245 at twenty-four knots. The Blue battle line also shifted its fire to the Purple Detached Wing while the Purple battle line fired on Blue fast battleship BB-63 and the Purple Right Flank Force heavy cruisers opened fire on Blue Destroyer Squadron 12. In addition, Blue battle cruiser CB-1 shifted fire to Purple heavy cruiser CA-123 and Blue battle cruiser CB-2 shifted fire to Purple heavy cruiser CA-122. The latter ship was left 10 percent damaged. At 1331, the Blue destroyer squadron on the left flank fired forty-five torpedoes, five from each ship on a base torpedo course of 314, with a one-degree spread, and on a twenty-foot depth setting. The target was the Purple battle line and the torpedoes' speed was thirty-two knots. Thirty-two of the torpedoes ran hot, straight, and normal but the Blue destroyers were under fire by the time they fired their torpedoes. By the end of this move, Blue fast battleship BB-63 was 56 percent damaged. Its guns could only fire from Local Control and it had no Plane Spot, searchlights, secondary batteries, or antiaircraft batteries. In addition, it had lost 23 percent of its speed. (Page 21.)

Move 23 began at 1333 and saw the Blue battle line continuing on course 050 at a speed of twenty-eight knots, with the exception of Blue fast battleship BB-63, which had changed course to 090 and had a speed of twenty-five knots. The Purple battle line continued on course 205 at a speed of twenty-seven knots, the Purple Center Force changed course to 160 with a speed of twenty-seven knots, and the Purple Detached Wing changed course to 205 at a speed of twenty-four knots. The

Purple battle line shifted fire to the Blue battle line as Blue Destroyer Divisions 211 and 212 fired thirty-five torpedoes, including five each at the two fast battleships in the Purple Rear. These torpedoes were fired on a base torpedo course of 303 with a one-degree spread, a twenty-two-foot depth setting, and a speed of thirty-two knots. The Blue destroyers at this time were not under fire and thirty-two of the torpedoes ran hot, straight, and normal. By this time, however, Blue fast battleship BB-63 was 94 percent damaged, and Blue Destroyer Divisions 121 and 122 were each 10 percent damaged. Purple heavy cruiser CA-122 was also 10 percent damaged. (Page 22.)

In Move 24, Blue changed course to 080 at a speed of twenty-eight knots and the Purple battle line changed course to 195 at a speed of twenty-seven knots. The Purple Center Force continued on course 160 but increased its speed to thirty-two knots while the Purple Detached Wing continued on course 205 at a speed of twenty-four knots. At 1337, Blue Destroyer Squadron 12 fired forty-five torpedoes at the Purple battle line on a base torpedo course of 325 with a one-degree spread and a twenty-foot depth setting. The torpedoes had a speed of thirty-two knots and thirty-two of them ran hot, straight, and normal. The Blue destroyers, however, were under fire by this time and each division had sustained 10 percent damage. Also at 1337, Blue fast battleship BB-63 sank and Purple heavy cruisers CA-122, -132, and -123 took three Blue light cruisers from Cruiser Division 17 under fire. The Purple Van battleships also took the leading Blue battleships under fire and during Moves 23 and 24, Blue battle cruisers CB-1 and CB-2 had been firing on Purple heavy cruisers CA-122, -123, and -132. However, the Blue ships were not considered to be in as advantageous a position for support as they could have been. In addition, Blue destroyer DD-748 from Destroyer Division 211 was sunk by its own torpedo when the division ran through its own torpedo water. At the end of the move, Purple fast battleships BB-55, -56, and -60 were firing on Blue fast battleship BB-61, which was assessed with 13 percent damage. Blue Destroyer Divisions 121 and 122 were also each assessed with 20 percent damage. (Pages 22–23.)

At the beginning of Move 25, Blue continued on course 080 at a speed of twenty-seven knots. By this time, the Purple battle line changed course to 095, also at a speed of twenty-seven knots. The Purple Center Force had changed course to 125 at a speed of thirty-two knots and the Purple Detached Wing changed course to 095 at a speed of twenty-four knots. Purple fast battleship BB-58 fired on Blue fast battleship BB-62, which was 10 percent damaged, while Purple fast battleship BB-59 fired on Blue fast battleship BB-64, which was also 10 percent damaged. Additionally, Purple heavy cruisers CA-122, -123, and -132 fired on Blue light cruisers CL-86, -89, and -91. Blue light cruiser CL-91 was damaged 10 percent as well. Simultaneously, Purple light cruisers CL-55 and CL-58 fired on Blue Destroyer Squadron 12 and Blue fast battleship BB-61 fired on Purple fast battleships BB-55 and BB-56.

Blue fast battleship BB-64 was also firing on Purple fast battleships BB-58 and BB-59, while Blue fast battleship BB-62 fired on Purple fast battleship BB-60. The result of all of this action was that among the Purple battleships, fast battleship BB-55 was 22 percent damaged, fast battleship BB-56 was 29 percent damaged, fast battleship BB-58 was 34 percent damaged, BB-59 was 4 percent damaged, and BB-60 was 6 percent damaged. In the same action, Blue battle cruiser CB-1 fired on Purple CA-123 and inflicted 30 percent damage on the Purple cruiser, while Blue battle cruiser CB-2 fired on Purple CA-122 and inflicted 20 percent damage on that ship. Finally, Blue Destroyer Divisions 211 and 212 were 20 and 30 percent damaged, respectively, while Blue fast battleship BB-61 took 30 percent damage. (Page 23.)

Move 26 saw the Blue battle line change course to 090 at a speed of thirty-one knots, while the Purple battle line continued on course 095 at a speed of twenty-seven knots. The Purple Center Force continued on course 125 at a speed of thirty-two knots and the Purple Detached Wing also continued on course 095 but at a speed of twenty-four knots. At 1343, Blue Destroyer Divisions 211 and 212 fired thirty-five torpedoes at the Purple Van battleships on a base torpedo course of 028 with a one-degree spread, a depth setting of twenty-two feet, and a speed of thirty-two knots. Thirty of the torpedoes ran hot, straight, and normal and the Blue destroyers were not under fire. In addition, Blue Destroyer Squadron 2 was firing on Purple Destroyer Division 171, Blue light cruiser CL-101 was firing on Purple Destroyer Division 111, and Blue light cruiser CL-91 was firing on Purple Destroyer Division 91. Purple Destroyer Division 92 was being fired on by Blue light cruisers CL-86 and CL-89 and Purple heavy cruisers CA-122 and CA-132 were being fired on by Blue battle cruisers CB-1 and CB-2. At the end of this move, Purple heavy cruiser CA-123 and Blue Destroyer Squadron 21 had both been 30 percent damaged, Blue fast battleship BB-61 had been 40 percent damaged, and Blue light cruiser CL-91 had been 10 percent damaged. (Pages 23–24.)

Move 27 began with Blue continuing on course 090 at a speed of thirty-one knots and the Purple battle line continuing on course 095 but with a reduced speed of twenty-four knots. The Purple Center Force changed course to 145 at a speed of twenty-seven knots and the Purple Detached Wing continued on course 095, also at a speed of twenty-four knots. Blue gunfire during this move entailed Blue Destroyer Division 22 and Blue light cruiser CL-91 firing on Purple Destroyer Division 91, Blue Destroyer Division 21 and Blue light cruisers CL-86 and CL-80 firing on Purple Destroyer Division 92, and Blue light cruiser CL-108 firing on Purple Destroyer Division 111. Move 27 Blue gun action also entailed Blue Destroyer Divisions 41 and 122 firing on Purple Destroyer Division 112, Blue Destroyer Divisions 42 and 121 firing on Purple Destroyer Division 171, and Blue battle cruiser CB-2 firing on Purple heavy cruisers CA-122 and CA-123. In addition, Blue battle cruiser CB-1 fired on Purple heavy cruisers CA-123 and CA-132 while Blue fast battleship BB-61 fired on Purple fast battleships BB-55 and BB-56. Finally, Blue fast battleship BB-62 fired on Purple fast battleship BB-60 while Blue fast battleship BB-64 fired on Purple fast battleships BB-58 and BB-59. (Page 24.)

Move 27 Purple gunfire saw Purple heavy cruiser CA-124 and light cruiser CL-56 firing on Blue Destroyer Division 212. In addition, Purple light cruiser CL-57 and Purple Destroyer Division 112 were firing on Blue Destroyer Division 211 while Purple light cruiser CL-58 and Purple Destroyer Division 92 were firing on Blue Destroyer Division 41. Move 27 Purple gunfire also entailed Purple Destroyer Divisions 91 and 111 as well as Purple light cruiser CL-55 firing on Blue Destroyer Division 42; Purple heavy cruiser CA-122 firing on Blue light cruiser CL-86; and Purple heavy cruiser CA-123 firing on Blue light cruiser CL-89. Additionally, Purple heavy cruiser CA-132 fired on Blue light cruiser CL-91; Purple fast battleship BB-58 fired on Blue fast battleship BB-62; Purple fast battleship BB-59 fired on Blue fast battleship BB-64; and Purple fast battleships BB-55, -56, and -60 were all firing on Blue fast battleship BB-61. (Page 25.)

The result of all of this gunfire was that at the end of Move 27, Blue fast battleship BB-61 was 46 percent damaged, Blue fast battleship BB-62 was 16 percent damaged, and Blue fast battleship BB-64 was 20 percent damaged. The Blue battle cruisers CB-1 and CB-2 now were each 11 percent damaged and Blue light cruiser CL-86 was 18 percent damaged. Blue light cruiser CL-89 was 9 percent damaged and Blue light cruiser CL-91 was 22 percent damaged but Blue light cruiser CL-101 was undamaged. Blue Destroyer Divisions 21 and 22 were also undamaged but Blue Destroyer Division 41 was 15 percent damaged, Blue Destroyer Division 42 was 24 percent damaged, and Blue Destroyer Division 211 was 40 percent damaged. Finally, Blue Destroyer Division 212 was 27 percent damaged, Blue Destroyer Division 121 was 33 percent damaged, and Blue Destroyer Division 122 was 43 percent damaged. As for Purple, Purple fast battleship BB-55 was 25 percent damaged, Purple fast battleship BB-56 was 36 percent damaged, and Purple fast battleship BB-58 was 42 percent damaged. Purple fast battleship BB-59 was only 8 percent damaged but Purple fast battleship BB-60 was 12 percent damaged. As for its heavy cruisers, Purple heavy cruiser CA-122 was 20 percent damaged, Purple heavy cruiser CA-123 had been sunk, Purple heavy cruiser CA-124 was 4 percent damaged, and Purple heavy cruiser CA-132 was undamaged. Purple light cruisers CL-55, -56, -57, and -58 were all undamaged, as were Purple Destroyer Divisions 71, 72, and 172. Purple Destroyer Division 91, however, was 54 percent damaged; Purple Destroyer Division 92 was 71 percent damaged; and Purple Destroyer Division 111 was 6 percent damaged. Finally, Purple Destroyer Division 112 was 46 percent damaged and Purple Destroyer Division 171 was 54 percent damaged. (Page 25.)

In Move 28, Blue continued on course 090 at a speed of twenty-six knots while the Purple battle line continued on course 095 at a speed of twenty-four knots. The

Purple Center Force saw Purple heavy cruiser CA-122 on course 120 and Purple heavy cruiser CA-132 on course 115, both at speeds of twenty-nine knots. The Purple Detached Wing continued on course 095 at a speed of twenty-four knots as well. At 1349, Blue Destroyer Division 41 fired twenty-five torpedoes at Purple fast battleship BB-55 on a base torpedo course of 320 with a one-degree spread and a twenty-two-foot depth setting. Eighteen of these torpedoes ran hot, straight, and normal while Blue Destroyer Division 42 fired another thirty torpedoes at the same target on a base torpedo course of 325 with the same spread and depth setting. Twenty-one of these torpedoes ran hot, straight, and normal. One minute later, Blue Destroyer Division 21 fired fifty torpedoes at the rear Purple battleship of the Purple Van Force on a base torpedo course of 310 and with the same spread and depth setting of the torpedoes from the other Blue destroyer divisions. Blue Destroyer Division 22 fired another fifty torpedoes at the Purple Van Force on the same course and with the same spread and depth setting of those from Blue Destroyer Division 21. Forty-five torpedoes in each salvo ran hot, straight, and normal. Purple heavy cruiser CA-124 passed through torpedo water but as there was only one chance in four for a hit, all of the torpedoes missed this ship. Purple light cruiser CL-57 also passed through torpedo waters and similarly escaped any damage. (Page 26.)

Gunfire in this move consisted of Blue light cruiser CL-89 firing with its main and secondary batteries on Purple Destroyer Division 91 while Blue light cruisers CL-86 and CL-91 also fired with their main and secondary batteries on Purple Destroyer Division 92. Blue Destroyer Squadron 2 fired on Purple Destroyer Division 111 while Blue Destroyer Divisions 121 and 42 as well as Blue Destroyer Squadron 21 fired on Purple Destroyer Division 171. Blue Destroyer Division 41 also fired on Purple Destroyer Division 112 while Blue Destroyer Division 122 fired on Purple light cruisers CL-55 and CL-58. Blue light cruiser CL-101 also fired on Purple light cruiser CL-58, Blue battle cruiser CB-1 fired on Purple heavy cruiser CA-132, and Blue battle cruiser CB-2 fired on Purple heavy cruiser CA-122. Finally, Blue fast battleships BB-61 and BB-62 fired on Purple fast battleship BB-56 while Blue fast battleship BB-64 fired on Purple fast battleship BB-55. Purple gunfire on Blue entailed Purple heavy cruiser CA-124, Purple light cruiser CL-56, and Purple Destroyer Division 71 firing on Blue Destroyer Division 212. Purple light cruiser CL-57 as well as Purple Destroyer Divisions 72 and 112 fired on Blue Destroyer Division 211 while Purple heavy cruisers CA-122 and CA-132 were firing on Blue Destroyer Division 21. Purple light cruiser CL-55 also fired on Blue Destroyer Division 42 and Purple light cruiser CL-58 fired on Blue Destroyer Division 41 while Purple heavy cruiser CA-122 fired on Blue light cruiser CL-86 and Purple heavy cruiser CA-132 fired on Blue light cruiser CL-91. As for the battleships, Purple fast battleships BB-55, -56, and -60 fired on Blue fast battleship

BB-61 while Blue fast battleship BB-62 was fired on by Purple fast battleships BB-58 and BB-59. (Page 26.)

Damage from Move 28 saw Blue fast battleship BB-61 now having a total of 58 percent damage, while Blue fast battleship BB-62 had a total of 18 percent damage. Blue fast battleship BB-64 now had 40 percent Total Damage, while Blue light cruiser CL-86 had 38 percent damage. Blue light cruiser CL-91 had 42 percent damage and Blue Destroyer Division 21 had taken 25 percent damage. Blue Destroyer Division 22 was undamaged but Blue Destroyer Divisions 41 and 42 had taken totals of 42 percent and 51 percent damage, respectively. Blue Destroyer Division 212 had a total of 57 percent damage, while Blue Destroyer Divisions 122 and 211 were entirely sunk. As for Purple, Purple fast battleship BB-55 was 25 percent damaged, Purple fast battleship BB-56 was 46 percent damaged, and Purple fast battleship BB-58 was 42 percent damaged. In addition, Purple heavy cruiser CA-122 was now 40 percent damaged, Purple heavy cruiser CA-132 was only 10 percent damaged, and Purple light cruiser CL-55 was only 5 percent damaged. However, Purple Destroyer Divisions 91, 92, 111, and 112 had been entirely sunk. (Page 27.)

At the beginning of Move 29, the Blue battle line changed course to 105 with a speed of twenty-five knots while the Purple battle line continued on course 095 at a speed of twenty-four knots. Purple heavy cruiser CA-122 changed course to 220 with a speed of twenty-seven knots and Purple heavy cruiser CA-132 changed course to 205 at the same speed. The Purple Detached Wing changed course to 135 with a speed of twenty-four knots. At 1352, the two remaining destroyers of Purple Destroyer Division 91—before they were sunk—fired ten torpedoes each on a base torpedo course of 129 with a one-degree spread, a depth setting of eight feet, and a speed of thirty-two knots. Fourteen of these torpedoes ran hot, straight, and normal but the Purple destroyers were already under fire. Blue ships firing on Purple ships now entailed Blue fast battleships BB-61 and BB-62 firing on Purple fast battleship BB-56; Blue fast battleship BB-64 firing on Purple fast battleship BB-55; Blue battle cruiser CB-1 firing on Purple heavy cruiser CA-132; and Blue battle cruiser CB-2 firing on CA-122. In addition, Blue light cruiser CL-86 was firing on Purple heavy cruiser CA-122 with its main and secondary batteries, while Blue light cruiser CL-89 fired with its main and secondary batteries on Purple heavy cruiser CA-132. Blue light cruiser CL-89 was also firing with its main battery on Purple light cruiser CL-58. Blue light cruiser CL-101 employed its main and secondary batteries on Purple light cruiser CL-55, which was also being fired on by Blue Destroyer Division 121. Blue Destroyer Division 21 also fired on Purple heavy cruiser CA-122 and Purple heavy cruiser CA-132 was fired on by Blue Destroyer Division 22. Finally, as concerned the Blue ships in this move, Blue Destroyer Division 41 was firing on Purple light cruiser CL-58, Blue Destroyer Division 42 was firing on Purple light cruiser CL-55, and Blue Destroyer Squadron 21 fired on Purple Destroyer Division 72. (Page 27.)

Purple ships firing at Blue during Move 29 entailed Purple fast battleship BB-58 firing with its main battery on Blue fast battleship BB-62. In addition, Purple fast battleships BB-59 and BB-60 also employed their main batteries on Blue fast battleships BB-64 and BB-61, respectively. All three of the Purple battleships were also employing their secondary batteries on Blue Destroyer Division 212, while Purple fast battleships BB-55 and BB-56 employed their main batteries on Blue fast battleship BB-61. Purple fast battleships BB-55 and BB-56 also deployed their secondary batteries against Blue Destroyer Divisions 42 and 41, respectively. As concerned the Purple cruisers, Purple heavy cruiser CA-132 fired with its main battery on Blue light cruisers CL-86 and CL-91 and with its secondary battery on Blue destroyers DD-757 and DD-775. Purple light cruisers CL-55 and CL-58 opened up with their main batteries on Blue light cruiser CL-101 and the two Purple light cruisers used their secondary batteries on Blue Destroyer Divisions 42 and 41, respectively. (Pages 27–28.)

Purple heavy cruiser CA-124 and light cruisers CL-56 and CL-57 also used their main batteries on Blue Destroyer Division 121, while Purple Destroyer Division 71 fired on Blue Destroyer Division 212. Blue Destroyer Division 121 was also fired on by Purple Destroyer Division 112 and Blue Destroyer Division 41 was fired on by Purple Destroyer Division 91. Damage from Move 29 saw Purple heavy cruisers CA-122 and CA-132 sunk, as were Purple light cruisers CL-55 and CL-58. In addition, Purple Destroyer Division 172 was 36 percent damaged. Blue ships saw Blue fast battleship BB-61 having 64 percent damage and Blue fast battleship BB-62 having 21 percent damage. In addition, Blue light cruiser CL-86 was sunk, Blue light cruiser CL-91 was 72 percent damaged, and Blue Destroyer Division 21 was 79 percent damaged. Blue Destroyer Division 41 was sunk, as were Blue Destroyer Divisions 42, 121, and 212. (Page 28.)[12]

Move 30, the last one of the maneuver, saw the Blue battle line changing course to 040 at a speed of twenty-five knots while the Purple battle line continued on course 095 at a speed of twenty-seven knots and the Purple Detached Wing continued on course 135 at a speed of twenty-four knots. In this move, Blue fast battleships BB-61, -62, and -64 fired on Purple fast battleships BB-60, -58, and -59, respectively, while Blue battle cruisers CB-2 and CB-1 fired on Purple fast battleships BB-56 and BB-55, respectively. Blue light cruiser CL-89 used its main and secondary batteries on Purple destroyer DD-730 while Blue light cruiser CL-91 did the same to Purple destroyer DD-731. Blue light cruiser CL-101 employed its secondary battery to engage Purple destroyer DD-731 and its main battery on Purple fast battleship BB-56 while Blue Destroyer Division 21 engaged Purple Destroyer

Division 91 with gunfire and Blue Destroyer Division 22 did the same to Purple fast battleship BB-56. Purple ships firing on Blue included Purple fast battleships BB-58, -59, and -60 engaging Blue fast battleships BB-62, -64, and -61, respectively. At the same time, the Purple battleships engaged Blue Destroyer Division 42 with their secondary batteries and Purple fast battleships BB-55 and BB-56 employed their main batteries on Blue fast battleship BB-61. The latter two Purple battleships also engaged Blue Destroyer Divisions 41 and 42, respectively, with their secondary batteries, while Purple heavy cruiser CA-124 fired on Blue fast battleship BB-64. In addition, Purple light cruiser CL-55 engaged Blue light cruiser CL-101 with both its main and secondary batteries and Blue Destroyer Division 42 was taken on by Purple Destroyer Divisions 71 and 72. Finally, Purple Destroyer Division 91 engaged Blue Destroyer Division 22. (Pages 28–30.)

In terms of damage in this last move, Blue fast battleship BB-61 was 64 percent damaged at the end of the maneuver, while Blue fast battleship BB-62 was 23 percent damaged and Blue fast battleship BB-64 was 40 percent damaged. Blue battle cruisers CB-1 and CB-2 were each 11 percent damaged, while Blue light cruiser CL-89 was 9 percent damaged, Blue light cruiser CL-91 was 72 percent damaged, and Blue light cruiser CL-101 was undamaged. Blue Destroyer Division 21, however, was 79 percent damaged and Blue Destroyer Division 22 was 38 percent damaged. As for the Purple ships, Purple fast battleship BB-55 was 43 percent damaged, Purple fast battleship BB-56 was 50 percent damaged, and Purple fast battleship BB-58 was 47 percent damaged. However, Purple fast battleships BB-59 and BB-60 were only 11 percent and 13 percent damaged, respectively, and Purple heavy cruiser CA-124 was only 4 percent damaged. While Purple light cruiser CL-55 was 83 percent damaged, Purple light cruisers CL-56 and CL-57 were undamaged, as was Purple Destroyer Division 71. Purple Destroyer Division 72, however, was 36 percent damaged and Purple Destroyer Division 91 had been sunk. (Page 30.)

Though this Critique was also not recorded, its organization again indicates what was discussed after the Conduct of the Maneuver. There were first to be discussions of both the Blue and Purple Staff Solutions by Captains Sweeney and Sullivan, respectively. Following these would be discussions of both the Blue and Purple Officer in Tactical Command Solutions by Captain Madeira and then Captain Taylor. These would then lead to discussions of the Staff and Student Cruising Dispositions as well as discussion of the Air Phases that entailed Moves 1 to 5. The latter points would be debated by the Department of Tactics as well as Colonel Cooley and Captain Hoffheins. Additionally, there would be Analysis and Summation of the Air Damage by Commander Pelling from the Department of Tactics, a discussion of the Surface Phases in Moves 6 to 30 by the Tactics Section and "interested" commanders, and an Analysis and Summation of Battle Damage, again by Commander Pelling. The Critique would conclude with a talk on Lessons Learned and Conclusions, all of this being organized by Captain Sweeney.[13]

NOTES

1 Senior Class of June 1947, "Operations Problem 3: Blue Statement Schedule of Events," 18 October 1946, pp. 1–2; idem, "Operations Problem 3: Blue Statement," 18 October 1946, pp. 1–6; both in folder 2605, box 139, RG 4, NHC.

2 Senior Class of June 1947, "Operations Problem 3: Blue Staff Solution," 18 October 1946, pp. 1–17, folder 2605-B, box 139, RG 4, NHC.

3 Ibid., pp. 17–25. See also Senior Class of June 1947, "Operations Problem 3: Section A—Blue," 31 October 1946, pp. 1–5, folder 2605-I, box 139, RG 4, NHC.

4 Senior Class of June 1947, "Operations Problem 3: Purple Statement Schedule of Events," 18 October 1946, pp. 1–2; idem, "Operations Problem 3: Purple Statement," 18 October 1946, pp. 1–7; both folder 2605-H, box 139, RG 4, NHC. Senior Class of June 1947, "Operations Problem 3: Purple Staff Solution," 18 October 1946, pp. 1–39, folder 2605-C, box 139, RG 4, NHC. See also Senior Class of June 1947, "Operations Problem 3: Section B—Purple," 1 November 1946, pp. 1–3, folder 2605-M, box 139, RG 4, NHC.

5 Senior Class of June 1947, "Operations Problem 3: Detail of Maneuver Staff," 1 November 1946, p. 1, folder 2605-D, box 139, RG 4, NHC.

6 Ibid.

7 Ibid., pp. 2–3.

8 Ibid., pp. 4–6.

9 Senior Class of June 1947, "Operations Problem 3: History of the Maneuver, Preliminary History," n.d., pp. I–II, folder 2605-N, box 139, RG 4, NHC.

10 Ibid., p. III.

11 Senior Class of June 1947, "Operations Problem 3: History of the Maneuver, Narrative," n.d., p. 1, folder 2605-N, box 139, RG 4, NHC. Subsequent in-text page references until the final endnote of the chapter are to this source.

12 Many of these ships had been sunk by the end of Move 28, especially on the Purple side. Though the History of the Maneuver does not indicate the exact circumstances, these ships had either not yet been sunk at this point or had been resurrected by the Director of the Maneuver.

13 Senior Class of June 1947, "Operations Problem 3: Critique of Operations Problem 3," 8–9 November 1946, p. 1, folder 2605, box 139, RG 4, NHC.

Blue versus Purple
Some Conclusions

In the late summer and fall of 1946, the U.S. Naval War College began a more complete transition from seeing Japan as its primary hypothetical enemy in the Pacific to the Soviet Union. While this transition had begun in the late spring of 1946 and Orange would appear again—in the summer of 1946 and in at least one exercise in the late fall—the 1946–47 academic year would see Purple largely become the number-one enemy of the U.S. Navy when it came to war plans and officer education.

What is highly noticeable in the war games described in this book—as in the games of the previous academic year at the Naval War College—is that major aspects of war gaming reflected the theoretical underpinnings of the activity itself, as well as more-practical applications of interwar and wartime doctrine. The theoretical and pedagogical bases of war gaming indicated that the strategic and tactical scenarios of the exercises and maneuvers themselves were less important than the simulation of real-world conditions under which naval commanders would have to make decisions. Therefore, exercises and maneuvers could be conducted under any strategic and tactical premises that the Naval War College staff and the director of a given maneuver thought would provide learning opportunities for and experiences with naval operational decision making. (These purposes would have justified, for instance, such striking artificialities as equipping Purple with *Essex*-like aircraft carriers, which the Soviet Union in fact gave no sign of building.) In fact, exercises and maneuvers did not even have to be played out to a conclusion, with one side winning or losing; such a conclusion might even undercut the learning experience to which the staff was attempting to expose the student officers.[1]

Nevertheless, the strategic and geographic foci of the various exercises and maneuvers—essentially operations in the North Pacific that were aimed at Purple striking Blue in the Aleutians, or even invading Japan, or Blue striking and intending to invade Purple on the Kamchatka Peninsula—reflected what the Naval War College staff thought a war in the Pacific against the Soviet Union would entail. Other aspects of the operations problems were very realistic as well in terms of those expectations—such as the Purple focus on using submarines and land-based naval aviation to compensate for a lack of aircraft carriers, or the use of atomic bombs by Blue against Purple.[2]

What seems less plausible upon initial examination was the operational and tactical focus on surface warfare in so many of the exercises and maneuvers, as had also been the case in the previous academic year. Since the U.S. Navy had transitioned from a prewar gunship-oriented fleet to one focused on carrier aviation and submarines, it seems anachronistic—pedagogy aside—that the Naval War College now created and conducted a series of war games in which carrier forces were neutralized in many cases. As in the previous academic year, in most cases, games simulated situations in which operational and tactical engagements were fought seemingly with interwar doctrine, with carriers restricted to fighting other carriers, providing reconnaissance for the surface fleets, and employing their airpower to aid surface ships in fighting decisive battles. Similarly, while submarines were more conspicuous and more often used as strike platforms than they had once been, they were still primarily employed as scouting platforms.[3]

Upon reexamination, however, the maneuvers again take on an aspect of "lessons learned" in wartime. The postwar Naval War College maneuvers harken back to Adm. Raymond Spruance's dispositions at the battle of the Philippine Sea on 17 June 1944, when he directed that the enemy carriers be knocked out first as a prelude to air strikes on heavy surface forces, which in turn were part of his intended preparations for a surface engagement. The war games designed by Vice Adm. William S. Pye and his staff were also highly similar to Spruance's operations off the Marianas as well as to Adm. William Halsey's initial preparations for the battle of Cape Engaño in October 1944.[4] Far from throwbacks to the interwar period, Naval War College exercises and operations problems in the summer and fall of 1946 combined interwar and wartime doctrine with some of the most recent experiences of the U.S. Pacific Fleet.[5]

Some of these operational and tactical scenarios need to be additionally viewed in postwar contexts. The U.S. Navy was going through a rapid demobilization in 1946 and the huge postwar fleet that had been planned for during the war was in fact being reduced in size significantly, primarily for fiscal reasons.[6] It can easily be seen that Admiral Spruance and his officers in the summer and fall of 1946 felt no more confident about having a large postwar fleet than had Admiral Pye and the Naval War College staff a year before. Given the postwar politics of defense unification and interservice rivalry over roles and missions, the Navy was by this time finding itself more under siege with respect to the future size of the U.S. Fleet.[7]

Against the context of a shrinking fleet beginning nonetheless to take on global responsibilities, the maneuvers seem even more realistic when certain periods of the Pacific War are recalled. As noted in my previous book in this series, at the end of November 1942, the U.S. Navy had been left with only two fleet carriers and two fast battleships operational because of the number of carriers sunk in 1942 and the small number of new battleships commissioned. In addition, there was a shortage of cruisers because of losses in the battles around Guadalcanal in 1942–43; by the summer of 1943, the Navy was being forced to rely heavily on destroyers as its primary combatant platform.[8] Moreover, the game scenarios in which Blue and Purple neutralized each other's carrier airpower were remarkably reminiscent of the mutual exhaustion of carrier forces in the battles of the Coral Sea, eastern Solomons, and Santa Cruz Islands in 1942. In fact, after Santa Cruz, the United States and Japan had such slender carrier forces available that the naval battles of Guadalcanal were entirely surface engagements.[9] In addition, in 1950, at the beginning of the Korean War, the Navy was to find itself with just two carriers deployed in the Pacific basin (though additional ships were provided as soon as President Harry Truman made the decision to use American forces to defend South Korea).[10]

The difference of the situation in 1950 from that of 1942, however, cannot be stressed enough. In 1950, there were additional carriers to deploy; most were deployed in the Atlantic and Mediterranean, because of the central focus on Europe in the context of the doctrine of "containment." In late 1942, however, there simply were no more fleet carriers available before the arrival of the *Essex* class in 1943.[11] Given the newfound global military responsibilities of the United States after 1945, the small number of carriers that had been available in 1942 may have been on the minds of Spruance and the Naval War College staff, much as it had influenced the thinking of Admiral Pye and his officers a year before. In other words, the faculty at the Naval War College in both academic years had to war-game a potential future in which the nation's foreign-policy commitments outstripped its military assets, as had occurred in the late interwar period. What if the U.S. Navy found itself ordered in the future to fight with such slim resources? To model this politico-military-fiscal situation, the Naval War College replayed the Navy's worst year in the war, 1942.

In addition, while the Pacific was now seen by the Navy as a secondary or even tertiary theater compared with the Atlantic, the Mediterranean, and the Persian Gulf, the Naval War College by this time thought that future naval engagements that did take place in the Pacific would occur in the far-northern part of the basin. There, more reliance would have to be placed on surface ships because of poor weather and the degraded performance of aircraft and other vital equipment, such as radar, in northern latitudes. These scenarios were a major theme during Admiral Spruance's tenure as President of the Naval War College since actual experience in the spring of 1946 was demonstrating that carrier operations were less than viable in Arctic waters.[12]

Uncertainty about the future also drove war gaming in a number of other respects. Looking back from our own time, it seems ridiculous that Naval War College staff and students were still in 1946 replaying what appear to be surface-warfare scenarios from the Pacific War as they had in the previous academic year. If, however, we realize that these were really combined-arms naval warfare scenarios that reflected the most recent experience, that of the late Pacific War, and that no one really knew what the next war would entail, the war games lose some of their anachronistic feel. No one in the American national-security establishment at this time had a crystal ball and no one could predict what the next war would be like any more than they could have a year before. Obviously, a majority understood the Soviet Union to be the next probable enemy but the Joint Chiefs of Staff did not completely dismiss a resurgent Japan as a potential enemy until well into 1946, after the College had planned its curriculum for that academic year.[13]

In addition, even assuming the Soviet Union as the probable next enemy, there was no guarantee by 1946–47 that the next war would be a limited, "brushfire" war as the United States was actually to encounter in Korea and Vietnam. Again, the Joint Chiefs of Staff between 1945 and 1950 assumed that the next war would most likely be a large conflict with the Soviet Union, fought by heavy ground, air, and naval forces in the Atlantic, the Mediterranean, mainland Europe, the Middle East, and the Persian Gulf—a conflict that almost exactly mirrored World War II in those theaters of operation.[14] If no one in the central American national-security establishment was accurately guessing what the next wars would actually look like, why would the Naval War College not assume, as its civilian overseers did, that the next conflict might look very much like World War II? The Soviet Union seemed to be building an oceangoing fleet of its own; why would the U.S. Navy assume it would never again have to fight another blue-water navy? If, therefore, the Navy might be forced to fight in areas like the Pacific, to fight an enemy like the Soviet Union in northern latitudes, or to fight some other blue-water fleet, combat such as that of the early Pacific War might again become a reality—in particular, one that needed to be rethought and refought.[15] We can comfortably say now that it did not happen that way or the Naval War College staff should have known better but this would be reading back into history rather than analyzing that history in its actual context.

Fear of a repetition of the early wartime strategic situation, fiscal uncertainty about the future force structure, and the inculcation by Spruance and the College's military faculty—many of whose members had just returned from the Pacific—of wartime American naval doctrine provided the Naval War College the basis for many of its scenarios. The result was a series of war games primarily oriented toward combined-arms surface engagements between opposing fleets against the background of a realization that although carriers had become the new focal point of American naval power, fast carriers could not perform all missions in the North Pacific and still might have to operate in conjunction with heavy surface forces. There was a realization—as there had been a year before—that while fast battleships were no longer at the center of naval warfare, they were not obsolete or useless.[16]

The Naval War College had a most difficult charge in this time. In an era of rapid demobilization, of domestic reconversion, and of a foreign policy that was changing in a revolutionary way at breakneck speed, staff and students at the Naval War College struggled to translate the lessons of the war into new strategy, tactics, and procedures. All of this had to be done on a slim budget and, in a new age of possible atomic warfare, in a way that shaped American naval forces as a deterrent. In the context of all of these transformations as the Naval War College was beginning to envision conflict against a new enemy after a forty-year focus on Orange, it does not seem incredible that it was still grappling with them in 1946. Nor does it seem out of the ordinary that the war games looking toward the future were so similar to the Navy's worst months of its last war.

NOTES

1 See Hal M. Friedman, *Blue versus Orange: The U.S. Naval War College, Japan, and the Old Enemy in the Pacific, 1945–1946* (Newport, R.I.: Naval War College Press, 2013), chap. 1.

2 For Soviet naval doctrine, see Milan Vego, *Soviet Naval Tactics* (Annapolis, Md.: Naval Institute Press, 1992), pp. 169–232. For the increased focus in the 1946–47 Naval War College curriculum on the use of atomic weapons in naval warfare, see Hal M. Friedman, *Digesting History: The U.S. Naval War College, the Lessons of World War Two, and Future Naval Warfare, 1945–1947* (Newport, R.I.: Naval War College Press, 2010), pp. 285–318.

3 These post-1945 war games were in sync with interwar American naval doctrine, which had evolved so that by 1941 the battleships were the strategic reserve, to be used only after auxiliary forces, such as aircraft carriers, had provided for fleet air defense, knocked out enemy carriers, and neutralized enemy land-based air forces. Subsequently, the carriers—along with cruisers, destroyers, and submarines—were to whittle down the enemy heavy and light surface forces so that the American battle line could finish the job in a surface gun fight; see John Kuehn, *Agents of Innovation: The General Board and the Design of the Fleet That Defeated the Japanese Navy* (Annapolis, Md.: Naval Institute Press, 2008), pp. 162–79. For the ascendance of the naval aviators in the wartime and immediate postwar Navy and the subsequent transformation of American naval doctrine to a carrier focus, see Vincent Davis, *Postwar Defense Policy and the U.S. Navy, 1943–1946* (Chapel Hill: Univ. of North Carolina Press, 1966), pp. 157–270.

4 Ivan Musicant, *Battleship at War: The Epic Story of the USS* Washington (New York: Harcourt Brace Jovanovich, 1986), pp. 257–59, 295–303. See also Friedman, *Blue versus Orange,* chap. 2, note 44, and Trent Hone, "U.S. Navy Surface Battle Doctrine and Victory in the Pacific," *NWCR* 62, no. 1 (Winter 2009), pp. 67–105. In addition, see Craig Symonds, *The Battle of Midway* (New York: Oxford Univ. Press, 2011), p. 104, for the Japanese plan in which Vice Admiral Nagumo's 1st Carrier Strike Force would neutralize American airpower on Midway, thus drawing out the remnants of the American Pacific Fleet, which in turn would be destroyed by a combination of carrier airpower under Nagumo and heavy surface forces under Adm. Isoroku Yamamoto, Commander in Chief, Combined Fleet.

5 Trent Hone, "The Evolution of Fleet Tactical Doctrine in the U.S. Navy, 1922–1941," *Journal of Military History* 67 (October 2003), pp. 1107–148; Hone, "U.S. Navy Surface Battle Doctrine and Victory in the Pacific," pp. 67–105; author e-mails with Hone, 28 February, 2 March, 3 March, and 14 March 2011.

6 Cdr. Dennis Ringle, USN (Ret.), conversation with author, 30 May 2010; Joel Davidson, *The Unsinkable Fleet: The Politics of U.S. Navy Expansion in World War II* (Annapolis, Md.: Naval Institute Press, 1996), pp. 141–81.

7 Davis, *Postwar Defense Policy,* pp. 207–70.

8 Vincent O'Hara, *The U.S. Navy against the Axis: Surface Combat, 1941–1945* (Annapolis, Md.: Naval Institute Press, 2007), pp. 67–137, 162–92. For a slight contradiction and argument that the cruiser situation was not quite as critical but nevertheless serious, see John Prados, *Islands of Destiny: The Solomons Campaign and the Eclipse of the Rising Sun* (New York: Penguin Books, 2012), p. 223.

9 For the battle of the Coral Sea, see Dallas Isom, *Midway Inquest: Why the Japanese Lost the Battle of Midway* (Bloomington: Indiana Univ. Press, 2007), pp. 79–91. For the battle of Santa Cruz and the naval battle of Guadalcanal, see James Hornfischer, *Neptune's Inferno: The U.S. Navy at Guadalcanal* (New York: Bantam Books, 2011), pp. 223–36 and 263–377. For the carrier situation being so critical that the U.S. Navy actually requested and received the loan of a British fleet carrier from the Royal Navy in mid-1943, see Prados, *Islands of Destiny,* pp. 158, 167–68, 217–59.

10 Thomas Cutler, "Sea Power and Defense of the Pusan Pocket, June–September 1950," in *The U.S. Navy in the Korean War,* ed. Edward Marolda (Annapolis, Md.: Naval Institute Press, 2007), pp. 1–51.

11 Dr. Jeffrey Barlow, History and Archives Division, Naval History and Heritage Command, Washington, D.C., e-mail to author, 22 September 2010.

12 For the Navy's reorientation of its global strategy toward the Atlantic, the Mediterranean, and the Persian Gulf in 1946, see Michael Palmer, *Origins of the Maritime Strategy: The Development of American Naval Strategy, 1945–1955* (Annapolis, Md.: Naval Institute Press, 1990), pp. 7–39. For Spruance's reasons for focusing on the northern Pacific in any Pacific basin battle, see Friedman, *Digesting History,* pp. 143–49, 156–61, 183–96, 199–207. For the Navy's largely failed attempts at carrier operations in the Arctic in March 1946, see Davis, *Postwar Defense Policy,* pp. 222–23.

13 Hal M. Friedman, *Creating an American Lake: United States Imperialism and Strategic Security in the Pacific Basin, 1945–1947* (Westport, Conn.: Greenwood, 2001), pp. 1–36.

14 Steven Ross, *American War Plans, 1945–1950* (London: Frank Cass, 1996), pp. 25–136.

15 See Milan L. Hauner, "Stalin's Big-Fleet Program," *NWCR* 57, no. 2 (Spring 2004), pp. 87–120; Hal M. Friedman, *Governing the American Lake: The US Defense and Administration of the Pacific, 1945–1947* (East Lansing: Michigan State Univ. Press, 2007), pp. 83–115; Friedman, *Creating an American Lake,* pp. 1–62; Eric Morris, *The Russian Navy: Myth and Reality* (New York: Stein and Day, 1977), p. 25; and Jürgen Rohwer and Mikhail Monakov, *Stalin's Ocean-Going Fleet: Soviet Naval Strategy and Shipbuilding Programmes, 1935–1953* (London: Frank Cass, 2001), pp. 178–220.

16 Hone, author e-mails, 28 February, 2 March, 3 March, and 14 March 2011.

BIBLIOGRAPHY

Archival and Manuscript Records

Naval Historical Collection, U.S. Naval War College, Newport, R.I.

Record Group 4: Publications

Record Group 16: Addresses

Naval History and Heritage Command, Navy Operational Archives, Washington Navy Yard, Washington, D.C.

Officer Biographical Files

Secondary Sources

Barber, James. "Mahan and Naval Strategy in the Nuclear Age: A Lecture Delivered at the Naval War College." *Naval War College Review* 24, no. 7 (March 1972), pp. 78–88.

———. "The School of Naval Warfare." *Naval War College Review* 21, no. 8 (April 1969), pp. 89–96.

Barlow, Jeffrey. *From Hot War to Cold: The U.S. Navy and National Security Affairs, 1945–1955*. Stanford, Calif.: Stanford Univ. Press, 2009.

Bredestege, Joseph. "Limited Nuclear War at Sea." *Naval War College Review* 19, no. 6 (February 1967), pp. 4–31.

Carney, Robert. "Role of the Navy in a Future War." *Naval War College Review* 6, no. 10 (June 1954), pp. 1–12.

Crowl, Philip. "Education versus Training at the Naval War College: 1884–1972." *Naval War College Review* 26, no. 3 (November–December 1973), pp. 2–10.

Cutler, Thomas. "Sea Power and Defense of the Pusan Pocket, June–September 1950." In *The U.S. Navy in the Korean War*, edited by Edward Marolda, pp. 1–51. Annapolis, Md.: Naval Institute Press, 2007.

Davidson, Joel. *The Unsinkable Fleet: The Politics of U.S. Navy Expansion in World War II*. Annapolis, Md.: Naval Institute Press, 1996.

Davis, Vincent. *Postwar Defense Policy and the U.S. Navy, 1943–1946*. Chapel Hill: Univ. of North Carolina Press, 1966.

Felker, Craig. *Testing American Sea Power: U.S. Navy Strategic Exercises, 1923–1940*. College Station: Texas A&M Univ. Press, 2007.

Field, James. "The Influence of Sea Power on Modern Strategy." *Naval War College Review* 10, no. 1 (September 1957), pp. 31–52.

———. "Origins of Maritime Strategy and the Development of Sea Power." *Naval War College Review* 7, no. 7 (March 1955), pp. 1–23.

———. "Seapower and Military Strategy Today." *Naval War College Review* 8, no. 8 (April 1956), pp. 21–39.

Fisher, Robert. "The U.S. Navy's Search for a Strategy, 1945–1947." *Naval War College Review* 48, no. 3 (Summer 1995), pp. 73–86.

Ford, Douglas. *The Elusive Enemy: U.S. Naval Intelligence and the Imperial Japanese Fleet*. Annapolis, Md.: Naval Institute Press, 2011.

Friedman, Hal M. *Blue versus Orange: The U.S. Naval War College, Japan, and the Old Enemy in the Pacific, 1945–1946*. Newport, R.I.: Naval War College Press, 2013.

———. *Creating an American Lake: United States Imperialism and Strategic Security in the Pacific Basin, 1945–1947*. Westport, Conn.: Greenwood, 2001.

———. *Digesting History: The U.S. Naval War College, the Lessons of World War Two, and Future Naval Warfare, 1945–1947*. Newport, R.I.: Naval War College Press, 2010.

———. *Governing the American Lake: The US Defense and Administration of the Pacific, 1945–1947*. East Lansing: Michigan State Univ. Press, 2007.

Gile, Robert H. *Global War Game: Second Series, 1984–1988*. Newport Paper 20. Newport, R.I.: Naval War College Press, 2004.

Hanley, John. "Creating the 1980s Maritime Strategy and Implications for Today." *Naval War College Review* 67, no. 2 (Spring 2014), pp. 11–29.

Hattendorf, John B. *The Evolution of the U.S. Navy's Maritime Strategy, 1977–1986*. Newport Paper 19. Newport, R.I.: Naval War College Press, 2004.

———, ed. *U.S. Naval Strategy in the 1970s: Selected Documents*. Newport Paper 30. Newport, R.I.: Naval War College Press, 2007.

Hattendorf, John B., B. Mitchell Simpson III, and John R. Wadleigh. *Sailors and Scholars: The Centennial History of the U.S. Naval War College*. Newport, R.I.: Naval War College Press, 1984.

Hauner, Milan L. "Stalin's Big-Fleet Program." *Naval War College Review* 57, no. 2 (Spring 2004), pp. 87–120.

Hay, Bud, and Bob Gile. *Global War Game: The First Five Years*. Newport Paper 4. Newport, R.I.: Naval War College Press, 1993.

Hegmann, Richard. "Reconsidering the Evolution of the US Maritime Strategy, 1955–1965." *Journal of Strategic Studies* 14 (September 1991), pp. 299–336.

Holwitt, Joel. *"Execute against Japan": The U.S. Decision to Conduct Unrestricted Submarine Warfare.* College Station: Texas A&M Univ. Press, 2009.

Hone, Thomas, and Trent Hone. *Battle Line: The United States Navy, 1919–1939.* Annapolis, Md.: Naval Institute Press, 2006.

Hone, Trent. "The Evolution of Fleet Tactical Doctrine in the U.S. Navy, 1922–1941." *Journal of Military History* 67 (October 2003), pp. 1107–48.

———. "U.S. Navy Surface Battle Doctrine and Victory in the Pacific." *Naval War College Review* 62, no. 1 (Winter 2009), pp. 67–105.

Hornfischer, James. *Neptune's Inferno: The U.S. Navy at Guadalcanal.* New York: Bantam Books, 2011.

Hoyt, Edwin. *Blue Skies and Blood: The Battle of the Coral Sea.* New York: Paul S. Eriksson, 1975.

Hurlburt, J. S. "War Gaming at the Naval War College, 1969–1989." *Naval War College Review* 42, no. 3 (Summer 1989), pp. 46–51.

Isom, Dallas. *Midway Inquest: Why the Japanese Lost the Battle of Midway.* Bloomington: Indiana Univ. Press, 2007.

James, Harry. "An Analysis of Limited Maritime War." *Naval War College Review* 19, no. 6 (February 1967), pp. 33–74.

Kraska, James. "How the United States Lost the Naval War of 2015." *Orbis* (Winter 2010), pp. 35–45.

Kuehn, John. *Agents of Innovation: The General Board and the Design of the Fleet That Defeated the Japanese Navy.* Annapolis, Md.: Naval Institute Press, 2008.

McBride, William. *Technological Change and the United States Navy, 1865–1945.* Baltimore: Johns Hopkins Univ. Press, 2000.

McInteer, J. F. "The Significance of Seapower to the United States." *Naval War College Review* 12, no. 1 (September 1959), pp. 1–32.

Miller, Edward S. *War Plan Orange: The U.S. Strategy to Defeat Japan, 1897–1945.* Annapolis, Md.: Naval Institute Press, 1991.

Morris, Eric. *The Russian Navy: Myth and Reality.* New York: Stein and Day, 1977.

Musicant, Ivan. *Battleship at War: The Epic Story of the USS* Washington. New York: Harcourt Brace Jovanovich, 1986.

Nofi, Albert. *To Train the Fleet for War: The U.S. Navy Fleet Problems, 1923–1940.* Newport, R.I.: Naval War College Press, 2010.

O'Hara, Vincent. *The U.S. Navy against the Axis: Surface Combat, 1941–1945.* Annapolis, Md.: Naval Institute Press, 2007.

Palmer, Michael. *Origins of the Maritime Strategy: The Development of American Naval Strategy, 1945–1955.* Annapolis, Md.: Naval Institute Press, 1990.

Phelan, George. "Sea Power and Strategies for the Control of the Seas." *Naval War College Review* 6, no. 10 (June 1954), pp. 15–36.

Prados, John. *Islands of Destiny: The Solomons Campaign and the Eclipse of the Rising Sun*. New York: Penguin Books, 2012.

Pradun, Vitaliy. "From Bottle Rockets to Lightning Bolts: China's Military Revolution and PLA Strategy against U.S. Military Intervention." *Naval War College Review* 64, no. 2 (Spring 2011), pp. 7–38.

Ringle, Cdr. Dennis, USN (Ret.). Author conversation, 30 May 2010.

Rohwer, Jürgen, and Mikhail Monakov. *Stalin's Ocean-Going Fleet: Soviet Naval Strategy and Shipbuilding Programmes, 1935–1953*. London: Frank Cass, 2001.

Rosenberg, David. "Being 'Red': The Challenge of Taking the Soviet Side in War Games at the Naval War College." *Naval War College Review* 41, no. 1 (Winter 1988), pp. 81–93.

Ross, Steven. *American War Plans, 1945–1950*. London: Frank Cass, 1996.

Smith, Douglas. *Carrier Battles: Command Decision in Harm's Way*. Annapolis, Md.: Naval Institute Press, 2006.

Smith, Nepier. "Historical Analysis of the Organizational Success of the Naval War College during the Twenty-Five Years following the Second World War." Naval War College Advanced Research Project. Naval Historical Collection, Naval War College, Newport, R.I., 1974.

Snyder, Frank. Introduction to *Sound Military Decision: U.S. Naval War College*. Annapolis, Md.: Naval Institute Press, 1992.

Solomon, Jonathan. "Maritime Deception and Concealment: Concepts for Defeating Wide-Area Oceanic Surveillance-Reconnaissance-Strike Networks." *Naval War College Review* 66, no. 4 (Autumn 2013), pp. 87–116.

Spector, Ronald. *Professors of War: The Naval War College and the Development of the Naval Profession*. Newport, R.I.: Naval War College Press, 1977.

Symonds, Craig. *The Battle of Midway*. New York: Oxford Univ. Press, 2011.

van Tol, Jan, with Mark Gunzinger, Andrew Krepinevich, and Jim Thomas. *AirSea Battle: A Point-of-Departure Operational Concept*. Washington, D.C.: Center for Strategic and Budgetary Assessments, 2010.

Vego, Milan. *Soviet Naval Tactics*. Annapolis, Md.: Naval Institute Press, 1992.

Vlahos, Michael. *The Blue Sword: The Naval War College and the American Mission, 1919–1941*. Newport, R.I.: Naval War College Press, 1980.

Watman, Kenneth. "Global 2000." *Naval War College Review* 54, no. 2 (Spring 2001), pp. 75–88.

White, Donald. "Admiral Richard L. Conolly: A Perspective on His Notions of Strategy." *Naval War College Review* 24, no. 3 (November 1971), pp. 73–79.

Wildenberg, Thomas. *Destined for Glory: Dive Bombing, Midway, and the Evolution of Carrier Airpower*. Annapolis, Md.: Naval Institute Press, 1998.

Wolters, Timothy. *Information at Sea: Shipboard Command and Control in the U.S. Navy, from Mobile Bay to Okinawa*. Baltimore: Johns Hopkins Univ. Press, 2013.

Wood, Robert. "The Conceptual Framework for Strategic Development at the Naval War College." *Naval War College Review* 40, no. 2 (Spring 1987), pp. 4–16.

ABOUT THE AUTHOR

Hal M. Friedman is Professor of Modern History, Associate Chair of History, and Social Science Academic Coordinator in the History & Political Science Department of the Social Science, Fine Arts & Fitness Division at Henry Ford College in Dearborn, Michigan. Educated at Eastern Michigan University with a bachelor of science in political science and history (magna cum laude) in 1987, Professor Friedman took his master of arts and doctor of philosophy degrees in the history of international relations at Michigan State University in 1991 and 1995, respectively. Additionally, he annually lectures as an adjunct lecturer of strategy and policy in the U.S. Naval War College's Non-Resident Strategy and Policy Seminar Program and a graduate senior instructor in Norwich University's Master of Arts in History Program. Professor Friedman formerly taught for Central Michigan University–Metro Detroit, Eastern Michigan University, Michigan State University, the University of Detroit Mercy, the University of Michigan–Dearborn, the U.S. Marine Corps University, and Washtenaw Community College.

Professor Friedman is the author of *Creating an American Lake: United States Imperialism and Strategic Security in the Pacific Basin, 1945–1947* (Westport, Conn.: Greenwood, 2001), *Governing the American Lake: The US Defense and Administration of the Pacific, 1945–1947* (East Lansing: Michigan State Univ. Press, 2007), *Arguing over the American Lake: Bureaucracy and Rivalry in the U.S. Pacific, 1945–1947* (College Station: Texas A&M Univ. Press, 2009), *Digesting History: The U.S. Naval War College, the Lessons of World War Two, and Future Naval Warfare, 1945–1947* (Newport, R.I.: Naval War College Press, 2010), and *Blue versus Orange: The U.S. Naval War College, Japan, and the Old Enemy in the Pacific, 1945–1946* (Newport, R.I.: Naval War College Press, 2013). Professor Friedman is also editing a collection of essays about the United States military in the Pacific and East Asia in the 1940s entitled *Establishing Hegemony: The American Military Presence in the Pacific Basin, 1940–1950*. He has published numerous additional articles in the *Hawaiian Journal of History, Intelligence and National Security, International Journal of Naval History, ISLA: A Journal of Micronesian Studies, Journal of Military History, Journal of Pacific History, Pacific Historical Review, Pacific Studies,* and *US Military History Review.* Professor Friedman has also written book reviews for *The Historian, The History Teacher,* H-Net, *International Journal of Naval History, Journal of American History, Journal of Military History, Journal of World History, Michigan Historical Review, Michigan War Studies Review, National Identities, Pacific Historical Review, World History Bulletin,* and *World War II.* Professor Friedman was an article abstractor for ABC-CLIO; an article referee for *Michigan Academician, Michigan Historical Review,* and *Pacific Historical Review;* a monograph referee for Texas A&M University Press; and a textbook reviewer for Houghton Mifflin, Longman Publishers, McGraw-Hill, and Prentice Hall.

Professor Friedman has been a conference program committee member for the Organization of American Historians, the Liberal Arts Network for Development, and the Ohio Academy of History; both a consulting historian and summer scholar with the National Endowment for the Humanities' Landmarks of American History and Culture Program; and a consulting historian for the Department of Education's Teaching American History Program. He is also a cofounder of the Metro Detroit Historians Collegium, a member of the *Michigan War Studies Review* Editorial Board, a former member of the Ohio Academy of History's Professional Development Committee, and the former two-year-college representative of that organization's Executive Council, and the Midwest regional coordinator for the Society for Military History.

Professor Friedman has also been honored a number of times, including with the 2011 John Lyman Book Award for Primary Source Materials, Reference Works, and Guide Books, North American Society for Oceanic History; with the 2008 Major General Thomas S. Jones USMC Distance Learning Instructor of the Year Award, Fort Worth Region, Command and Staff College Distance Education Program, Marine Corps College of Continuing Education; with honorable mentions in the 2007–2008 and 2010–11 Cold War Essay Contests from the John A. Adams '71 Center for Military History and Strategic Analysis at Virginia Military Institute; with a 2003 Edward S. Miller Fellowship in Naval History to the U.S. Naval War College's Naval Historical Collection; with the 2002 Alumni Award for Outstanding Academic and Professional Accomplishments from Eastern Michigan University's Department of History and Philosophy; and as an honorary member of Phi Theta Kappa, the Two-Year-College International Honor Society, since 1998.

Heavily involved with Henry Ford College's shared governance system, Professor Friedman has served on a variety of committees for HFC and his union local, the last as the secretary of the Henry Ford Community College Federation of Teachers, American Federation of Teachers Local 1650.

Professor Friedman served as a personnelman (PN) in the U.S. Naval Reserve, primarily as a pay and personnel clerk in several Naval Reserve Intelligence Command units. His service awards include the Naval Reserve Meritorious Service Medal, the National Defense Service Medal, and the Joint Meritorious Unit Award. He was also chosen as the Fleet Intelligence Rapid Support Team 1773 Junior Sailor of the Year for 1991.

INDEX

Note: Page numbers in *italics* indicate photographs, illustrations, and tables.

NAVAL WAR COLLEGE HISTORICAL MONOGRAPH SERIES

1. *The Writings of Stephen B. Luce,* edited by John D. Hayes and John B. Hattendorf (1975).
3. *Professors of War: The Naval War College and the Development of the Naval Profession,* Ronald Spector (1977).
4. *The Blue Sword: The Naval War College and the American Mission, 1919–1941,* Michael Vlahos (1980).
5. *On His Majesty's Service: Observations of the British Home Fleet from the Diary, Reports, and Letters of Joseph H. Wellings, Assistant U.S. Naval Attaché, London, 1940–1941,* edited by John B. Hattendorf (1983).
7. *A Bibliography of the Works of Alfred Thayer Mahan,* compiled by John B. Hattendorf and Lynn C. Hattendorf (1986).
8. *The Fraternity of the Blue Uniform: Admiral Richard G. Colbert, U.S. Navy and Allied Naval Cooperation,* Joel J. Sokolsky (1991).
9. *The Influence of History on Mahan: The Proceedings of a Conference Marking the Centenary of Alfred Thayer Mahan's* The Influence of Sea Power upon History, 1660–1783, edited by John B. Hattendorf (1991).
10. *Mahan Is Not Enough: The Proceedings of a Conference on the Works of Sir Julian Corbett and Admiral Sir Herbert Richmond,* edited by James Goldrick and John B. Hattendorf (1993).
11. *Ubi Sumus? The State of Naval and Maritime History,* edited by John B. Hattendorf (1994).
12. *The Queenstown Patrol, 1917: The Diary of Commander Joseph Knefler Taussig, U.S. Navy,* edited by William N. Still, Jr. (1996).
13. *Doing Naval History: Essays toward Improvement,* edited by John B. Hattendorf (1995).

14. *An Admiral's Yarn,* edited by Mark R. Shulman (1999).
15. *The Memoirs of Admiral H. Kent Hewitt,* edited by Evelyn Cherpak (2004).
16. *Three Splendid Little Wars: The Diary of Joseph K. Taussig, 1898–1901,* edited by Evelyn Cherpak (2009).
17. *Digesting History: The U.S. Naval War College, the Lessons of World War Two, and Future Naval Warfare, 1945–1947,* Hal. M. Friedman (2010).
18. *To Train the Fleet for War: The U.S. Navy Fleet Problems, 1923–1940,* Albert A. Nofi (2010).
19. *Talking about Naval History: A Collection of Essays,* John B. Hattendorf (2011).
20. *New Interpretations in Naval History: Selected Papers from the Sixteenth Naval History Symposium Held at the United States Naval Academy, 10–11 September 2009,* edited by Craig C. Felker and Marcus O. Jones (2012).
21. *Blue versus Orange: The U.S. Naval War College, Japan, and the Old Enemy in the Pacific, 1945–1946,* Hal M. Friedman (2013).
22. *Major Fleet-versus-Fleet Operations in the Pacific War, 1941–1945,* Milan Vego (2014).
23. *New Interpretations in Naval History: Selected Papers from the Seventeenth McMullen Naval History Symposium Held at the United States Naval Academy, 15–16 September 2011,* edited by Marcus O. Jones (2016).